REFLECTIONS & CONNECTIONS

Personal Journeys Through the Life Sciences
Volume I
Agricultural Economic & Plant Scientists

Editors

Otto J. Crocomo

Julius P. Kreier

William R. Sharp

ISBN: 1500459143
ISBN 13: 9781500459147

ACKNOWLEDGEMENTS

The editors are grateful to the following colleagues, family members and friends for their contributions and support during the two year book development journey.

Consuelho Baehr, Author and One of Four National
    Content Providers, Named by "Fast Company".
Dr. Roy S. Chaleff, Biotechnology Scientist and Author
Lauren Chomiuk, Associate Editor
Rebecca Butcher, Production Editor
Carla Maisa Crocomo, Educator
Ruth O'Heron, Publishing Assistant
Sally Sharp Holland, Educator and Author
Dr. Rachel Kreier, Healthcare Economist & Author
Walid Lofty, Engineer and Pilot
Rosa Shine Raskin, Microbiologist, Library
    Information Scientist and Author
Jeffrey W. Sharp, Entrepreneur and Film Producer
Edward Ramos Sousa, Attorney and Author
Dr. Douglas S. Steinbrech, Surgeon and Author

COVER DESIGN AND PHOTOGRAPHY

The editors provide special thanks to Yuan Hong Lin, Photographer, for providing the magnificent cover photograph entitled "Water Lily Reflections" and to Mauricio Diaz, Book Cover Designer, for his brilliant cover design.

Yuan Hong Lin resides in Taipei, Taiwan and Mauricio Diaz resides in New York City.

DEDICATION

The authors dedicate Reflections &
Connections to the legendary
author Maya Angelou, our mentors,
and our successors.

"When You Learn, Teach,
When You Get, Give"

Maya Angelou

PROLOGUE

R. S. Chaleff

We are taught from an early age about science; that it is an analytical process by which observation and experimentation increase human knowledge. It is a large and intimidating concept, but we get a sense of it early on because we are a science and technology-driven society. We are bombarded from birth about the importance and achievements of, and our dependency on, science. But who are the practitioners of that array of disciplines we call science? Yes, we learn the names of the most famous, those who made extraordinary discoveries or advances. But we don't learn who they were, or are, as individuals or how they became the monoliths we study and venerate. We are taught about the development of their accomplishments, but not about their personal development. We study science, but not scientists. Yet scientists are not like Athena who sprung fully formed and armed from the forehead of Zeus. What motivates people to become scientists? What is it like to labor in that profession? What are the challenges, frustrations, and the rewards? And how does one become a scientist? So while it is only natural for many young people in our science-dominated society to entertain the notion of becoming scientists, we provide little or no information on which to base such a life-forming decision.

This collection of autobiographical essays by scientists strives to address that void. Admittedly, it displays a bias toward the life, rather than the physical sciences, and within that domain the plant sciences were represented more than the medical sciences, but the the lessons are universal. In this book, we are offered insights into the lives of scientists and what influences motivated and shaped them and why they chose their particular fields. Moreover, we find how they were trained and about their failures and successes. The practice of science can be punishing: experimental designs may be flawed, long arduous hours are spent alone in the laboratory, and results can be inconclusive, misleading, or contradictory. Moreover, unconventional results and new theories that challenge established dogma can be met with brutal rejection. Yet the sense of personal fulfillment and achievement, and the awareness of making a meaningful contribution to humanity can be glorious. Open this book and you enter the lives of scientists, turn the pages and their lives and careers will unfold before you. Herein readers considering careers in science may find inspiration and acquire first-hand knowledge of what it is like not so much in the headlines, as in the trenches of science. Those already in the profession may find guidance for managing their careers, for knowing when to pursue and when to abandon an experimental strategy or interpretation, and perhaps most importantly, find courage and stamina to weather the inevitable storms and uncertainties of their ever challenging journey.

In American (and, as far as we know, elsewhere as well) graduate and postdoctoral programs train students and novitiates in experimental design and execution and in the analysis, interpretation, and presentation of results. However, it is only later, when they venture from beneath the protective wing of a mentor that these fledglings discover that additional personal skills are as important as skills in science for a successful career. Means and strategies for procuring funding, navigating departmental politics, promoting one's program, advancing one's career, and winning support for innovative ideas, are omitted from graduate school curricula. Many a promising career has been shipwrecked on these treacherous shoals. This collection of autobiographical essays hopes to rectify this lamentable deficiency in our system of graduate education. There is an art to the successful practice of science: an art that is essential, yet is not taught, that may come naturally to some, but that is acquired all too painfully by others.

It has been said that experience is something acquired only after it is needed. But it is also possible to learn from the experiences of others who have traveled the same path upon which one is embarking and to adopt methods employed by those predecessors to facilitate success, avoid pitfalls, and overcome obstacles. This book contains a trove of such experiences. It is not alone in that regard. Many valuable and informative books – some by individual authors and others comprising collections of personal narratives - provide lessons and advice for managing one's career in academia (e.g., *Rhythms of Academic Life: Personal accounts of careers in academia*, by Peter J. frost and M. Susan Taylor, 1996; *An Academic Life: a handbook for new academics*, by Robert H. Caldwell and Jill J. Scevak, 2010; *Academic Entrepreneurship: university spinoffs and wealth creation*, by Scott Andrew Shane, 2004). *Reflections and Connections* is distinct in that it chronicles career journeys in the natural sciences and is not a handbook, but a collection of intimate histories of the challenges, frustrations, and triumphs encountered on those journeys. It relates personal ordeals that inevitably obtrude on one's career, including bullying, competition, political infighting, racism and sexism as well as the human misfortunes of death, divorce, a family conflict, the Holocaust, medical disabilities, sickness and war. It also recounts the encouragement, inspiration, support, and counseling of colleagues, family, friends, and mentors that proved crucial in prevailing over these challenges. But above all, *Reflections and Connections* strives to capture and communicate the passion that drives scientists to succeed in the bittersweet, lifelong struggle of scientific research.

INTRODUCTION

Otto J. Crocomo, Julius P. Kreier & William R. Sharp

The book is organized in two volumes: Volume One - Agricultural Economic & Plant Scientists, Volume Two - Health Care Economic, Environmental & Medical Scientists and There is a back section which includes tributes to our influencers and mentors, biographical sketches about the authors and editors and an appendix with information about the long-term collaboration between The Ohio State University and the University of Sao Paulo.

The book chapters were written by various people in various stages of their lives. They were asked by the editors to reflect on how they got into their careers. What they were asked to write was loosely described. It was requested that they describe their family background and how they felt their families affected the careers they chose. They were asked to describe not only the effects their families had on their choices but also how encounters and connections with other people influenced what they chose to do. They further were asked to describe the results of their choices and how they handled these results. And to reflect on any other aspects of their personal and professional lives that they considered significant to their development.

The resulting chapters are quite heterogeneous as they are strongly autobiographical. There are also quite similar in many respects. This is because not only are job searches inherently similar but much of human development also is quite similar.

The editors of this book are all biologists but we are to some degree from different fields. Julius Kreier is medically oriented. His field was parasitology and he has a degree in veterinary medicine. He worked in animal disease control for some years before he returned to school to obtain a PhD in microbiology. After he obtained the PhD degree, he was hired as assistant professor of microbiology at the Ohio State University from which he retired in 1989 as a full professor. His colleague and friend Rod Sharp was trained as a plant cell biologist. They met in 1969 when Rod joined the department of microbiology at OSU because of his interest in cell biology and plant cell and tissue culture, a field not originally much thought of in botany. The third editor, Otto Crocomo, an agronomist and biochemist was also like Dr. Sharp a plant cell biologist but with a strong interest in agriculture and plant biology. Otto Crocomo and Rod Sharp have collaborated in research and teaching for almost 42 years since meeting in 1971. Dr. Crocomo recently retired as a full professor from the department of chemistry and from the position, Director of CEBTEC, the Biotechnology Center of the University of Sao Paulo. He was the founder of CEBTEC. Dr. Sharp has served as a full professor at Ohio State University in the microbiology department and at Rutgers University in the department of plant science along with stints in the food and biotechnology industries.

The result of the differences in the fields of the editors is that Sharp and Crocomo primarily picked the authors with an agricultural and plant biology slant while Julius Kreier picked primarily the authors who wrote chapters with a medical slant. A few of the chapters were authored by individuals with degrees in plant biology who later pursued medical research. The people we chose were people who had been our colleagues, collaborators, graduate students, and others whom we met at scientific meetings.

The author Rachel Kreier is the daughter of Julius Kreier and the chapters describing growing up with a life scientist are authored by the editor's children.

The leaders of various fields of science usually consist of a fairly small group of people who get to know each other as a result of reading journal papers and attending national and international scientific meetings in their respective fields. The subject matter of the chapters included in the book is largely determined by the choice of the contributors. What is really of interest to us is whether we created a book that will be useful to the reader. It is the hope of the editors that the chapters will be read and will aid some young people in their struggle to develop careers as well as secondary school and university career counselors in the provision of advice to students. It is always useful to learn about what others have experienced. It may help young people find that their unpleasant as well as pleasant experiences were not unique but similar to other's experiences. Some of us, already established or even retired, on reading these chapters may be comforted by finding that they were not unique in having gone through struggles and having encountered setbacks while trying to develop careers. Also the hope is that the reader will gain a better understanding of research and teaching institutions and the lives of career scientists.

There is something positive we would like to say about our experience while reading and editing the chapters in this book. Some of the authors were our students and they have made us happy when they said that our teaching and mentoring helped them in their careers and even in their personal lives.

CONTENTS

CHAPTER 1.

Developing a Career: Accepting the Challenges of Opportunities

Jerry R. Ladman

My story is about a fellow who grew up in rural Iowa with visions of becoming a farmer but who ended up spending a long academic career specializing in the economic development of Latin American countries and along the way serving in a number of internationally-related administrative and leadership posts. The last position was as the associate provost for International Affairs (the Senior International Officer) for The Ohio State University, a position that I held for the last seven years prior to my retirement in 2007. Throughout my life I have often reflected on how my professional career path developed and have concluded that it is not because of a careful planning nor sheer luck, but rather because I laid a good foundation, worked hard, was entrepreneurial in my tasks, developed people skills and then responded to opportunities that were presented to me by others. This begs the questions of "why were there opportunities; how did they come about?" I have given a lot of thought to this and conclude that my answers lie in the sage advice I received from my father during my youth. He said to me something like this: "Jerry, you do not know where the future will take you and you can't plan it out, but if you work hard and do a good job at whatever you are doing people will notice and new opportunities will be presented to you." In this retrospective essay I use this paradigm to explain my career. I begin with my formative careers as a youth and college student and then walk through the various positions and jobs that I have had.

My Youth: The Formative Years Where the Foundations Were Laid

These were the years where I developed my love for agriculture, learned how to work hard, gained good moral standards and was intrigued by foreign countries.

My father, Harry, grew up on a farm near Castana, a small northwestern Iowa town. He was the only member of his family to attend college, enrolling in agricultural education at Iowa State College, where he met my mother, Amy. He earned all his own way by getting his barber's license and then cutting hair and giving shaves while my mom also helped out by selling tickets at the local movie theatre. Upon graduation he took a position as a vocational agriculture teacher and the basketball and baseball coach in Blencoe, also in northwest Iowa. While there I was born in a Sioux City hospital and 15 months later my sister, Jan, joined us.

This was during the Great Depression and the federal government was developing a lot of new programs to help those in dire need. One such program was the Farm Security Administration (FSA) a supervised credit program designed to help small farmers, who had lots of promise but, because of a lack of guaranties were not considered by their local banks as eligible for loans for capital improvements and annual production. To give these farmers a chance to survive and advance the FSA provided credit at low subsidized interest rates but only after the development of a carefully crafted farm plan, which was done in conjunction with a loan officer. Furthermore the farmer had to agree to be subject to close supervision of the use of those funds by the loan officer. My father leapt at the opportunity to become a FSA loan officer. Little did he realize that someday his son would become a college professor who specialized in researching credit programs for peasant farmers in Latin America!

Dad would stay with the FSA (the name was later changed to FHA) until he retired. Our family moved often as he was transferred to different offices in the state. Two places stand out in my mind, and both importantly relate to my story. The first was Ames, the home of Iowa State College. I had my first four years of school there and as a young kid I vividly remember riding my bike to the campus and exploring the grounds and the buildings. From that time forward my sister and I knew that when we went to college, which our parents had always said that we must do, we would attend ISC. Although this was during World War II, with the rationing of gasoline, furnace fuel and food stuffs, our parents always provided for great family times. My father loved horses and passed this on to our family. He bought a Shetland pony, which we kept in an electric-fenced-in area back of our house and which I rode or hitched up to the cart and drove around the neighborhood collecting newspapers and tin cans for the war effort.

The second place was Clarion, a county seat town of about 3,000 inhabitants in north-central Iowa where I finished school. We lived on a small farm on the edge of town where we kept animals. Saddle horses were our favorites and my sister and I participated in 4H and showed dairy heifers at the country fair. I was a good student, but my real passion besides horses was athletics: football, basketball and track. I was fast, which served me well as a running back in football and a sprinter in track. I was elected captain of the track team in my senior year and our sprint medley and mile relay teams played well in the state meet. Of course there were girlfriends and I was active in the Methodist Youth Fellowship. I graduated from Clarion High School in 1954.

During my youth my father taught me how to work hard, and equally important how to love work. As a young boy he always had me helping him with chores. When I was in junior high school he built an addition on our house and he had me mixing cement, building rafters and putting up siding. He insisted that I get jobs. Early on I had a paper route and during junior high I was a bagger in a local grocery and really prided myself in being able to carry 100 lb. bags of potatoes to customers' cars. After I started high school I spent my summers and weekends working as a hired hand on farms. I loved it and couldn't work hard enough. I took pride in how many bales of hay I could toss on a wagon in a day or how fast I could load a manure spreader using a pitchfork.

Dad also taught me how to manage my money and save funds for college. His carrots were to give me a small allowance for weekly expenses, put the gas in his car, which he would let me use to go on dates, but with the condition that I put my earnings from my jobs into a savings account for college. In the years through junior high school he would match each dollar of my savings, after that I was on my own. The result was that I had enough money set aside to pay for my first year of college. The rest I earned by working both as a student on the campus and in summer jobs with the result that with my earnings I paid for my entire college education.

During these years I also formed my interest in foreign countries. I had the urge to travel, but would never have the chance to leave the United States until after I had graduated from college, when I made my first trip to Mexico. I developed this international interest in several ways. As I young boy I looked forward to visiting my maternal grandmother and reading her *National Geographic* magazines. The colored pictures of foreign lands were intoxicating and caused me to dream. In grade school geography was always my favorite class and in my church youth I gave a lot of thought to becoming a missionary. My aunt Lottie was a missionary in Panama and Costa Rica. She sent us letters and handcrafts, which intrigued me. I still have some of those pieces and credit this experience as an important influence on my passion for collecting Latin American folk art that I developed later in life.

The Formative Years of Developing Leadership Skills – College

I credit college as the time, when I built upon the foundations laid in my youth, to develop leadership, and entrepreneurial, people and management skills. I did this through my extensive participation in extra-curricular activities. These acquired skills contributed substantially to success and the opened doors to new opportunities, not only in college but also in my professional life. In this section I highlight some key moments in the process of developing these skills.

In 1954 in pursuit of my desire to be a farmer I enrolled at ISC in Farm Operations, which in essence was a general agriculture degree program designed to prepare one for farming. As an incoming freshman, despite holding two scholarships, my biggest fear was not being able to get good enough grades to obtain my degree. I had heard how tough college was so I studied hard and worked in the dormitory cafeteria to cover my board. My strategy in doing this was to use my mealtime to work so that I would not waste time in frivolous activities and have more time to hit the books. However, my commitment to studies did not impede my love for sports, so I tried out for the track team. Coach William Berry said that this meant that I would need to run cross-country in the fall and he sent me running with other aspiring tracksters on the College Golf Course. I quickly realized, despite my success in high school, that I was not going to be fast enough to compete well in college. After giving it a try for about a month I told Coach Berry that I would not continue. It is one of the few times in my life that I have quit something. This bothered me, but on the bright side I felt relieved of the time commitment and later

realized that this decision freed up time for me to actively participate in college leadership activities.

My hard studying paid off. When my report card came in the
mail during Christmas vacation I had all "A's" except for a "C"
in the one-credit-hour military science. In those days ROTC was
mandatory for all freshman and sophomores. Somehow I just
didn't get with it. But I made the commitment to improve and did
well in the subject and was recruited to enter the advanced ROTC
program. This was another opportunity. I hesitated and had an
internal debate within myself, but decided that I should go forward. I quickly realized that I had made the right decision. ROTC
was an opportunity to develop leadership skills and to eventually
serve my country as an officer, rather than to be subject to the
draft and serve in the enlisted ranks. When I graduated in 1958
I not only received my degree but also a commission as a second
lieutenant. In this program, along with my post-graduation experience on active duty and as a reservist, I took on responsibilities
and learned leadership military style.

My good grades that first quarter caught the attention of the
local chapter of Farm House Fraternity, a national social fraternity for agricultural students that places strong emphasis on
scholarship and leadership. I was invited to go to the Iowa State
Chapter house for dinner. I did not hesitate. Previously I had
gone through Greek Rush Week as a new freshman but had not
accepted any of the offers to pledge. I just did not feel like the
Greek System was right for me and I decided to remain an independent. However, when Farm House called it was different. By
that time I had heard about its reputation and how its fundamental values as expressed in its motto "Builders of Men" made it different from the typical social fraternity. When, after the dinner
visit, I was invited to pledge I couldn't pass it up. Accepting this
opportunity was one of the most critical decisions of my life. Farm
House would provide me with a supportive brotherhood that

would create opportunities and bring out the best in me. After all, these guys were among the best agriculture and veterinary medicine students on the campus, and, moreover, many were role models as prominent campus leaders. I credit Farm House with providing me the opportunities for building my confidence and skills. Once I joined, I never looked back.

One way this was manifested was through the fraternity's encouragement, or you might say expectations, that members not only be good students but also be involved in campus leadership. During my spring quarter of my freshman year I interviewed for a position on the VEISHEA Horse Show Committee. [VEISHEA was then and still is the major student run event in the nation. It is designed to showcase the campus and provide a venue for fun for students and visitors. Its name reflects the major areas of study that were in place when it was founded in 1922– *V*eterinary medicine, *I*ndustrial *S*cience, *H*ome *E*conomics and *A*griculture. For three days each spring it features open houses in all the colleges; programs to recruit new students; a major parade with a notable figure as parade marshal (such as Ronald Reagan), the ISC marching band, high school bands and numerous highly-decorated floats; a student performance of a Broadway musical; an elected queen; and a big all-campus dance. In my time the Horse Show, featuring entrants from around the Midwest, was also an important event.] I was asked to be chair of parking.

I must have done a good job because the next year as a sophomore I interviewed and was selected for a position on the VEISHEA Parade Committee as chair of Float Committee. Logan Van Sittert was the chair of the Parade Committee. He must have liked my work because he created a new opportunity for me when he suggested that I (as a member of the Greek System) join with Charles "Spike" Dodson (from the Men's Residence Association, i.e., dormitories) to apply as co-chairs of the Homecoming Celebration. Wow, a sophomore, and soon to be junior, to be chair of the second biggest student run event on campus was almost unthinkable. What a challenge and opportunity!

Spike and I developed a creative and entrepreneurial plan and interviewed before the selection committee. To our surprise we were selected! This was a daunting responsibility, there were many events to be held and as a first step we interviewed candidates for the Central Committee. Homecoming was a resounding success! Our confidence grew and we gained considerable experience in responding to the challenges presented and in learning-by-doing. One example stands out in my mind. Up to that time I was terrified to get up on my feet and make a speech, and one night that all changed. One public relations committee chair had organized a dinner in the Memorial Union for high-level college officials. President James Hilton and many deans were in attendance. Without any forewarning the committee chair asked me to stand up and speak. There was no way that I could turn him down. I sucked it up and delivered. This was both a challenge and an opportunity. Ever since then I have never felt that I could not get on my feet and speak. I may be a bit nervous but I can do it. This is something that has served me many times.

My participation in leadership roles continued. I was elected president of the Student Union Board, the student group that advised the director of the Memorial Union on matters relating to students use of and activities in the Union and was also responsible for organizing many student events held in the Union. We lobbied hard for and obtained some major remodeling in the Union. I was also a leader in the Agricultural Council, the student leadership organization for the College of Agriculture. As a junior I was selected as the incoming president of Cardinal Key, the Men's Leadership Honor Society, and was voted by my classmates as president of the Senior Class. There was also recognition in selected for membership in honor societies such as Alpha Zeta, Gamma Sigma Delta, Phi Eta Sigma and Scabbard and Blade.

As a graduating senior I believed that these new experiences and skills had given me management and leadership qualities that would serve me well in life. This would prove to be true. I also realized that farming was not going to be the career for me. There were two reasons. First and foremost, I did not have a family farm to go back to. This was a real disadvantage since starting as a renter did not seem promising. Second, with my leadership/administrative experiences I thought that I had some special skills that would serve me well in some organization or industry. The problem was that I did not have a job and, unfortunately, ISC had not provided its agricultural college students with good career services. Walter Falcon, president of Agricultural Council, and I decided to try to do something about this. We discussed our concerns with Associate Dean Roy Kottman and he organized a hearing among the department chairs for us to share our ideas. Little did I know that this may have been an important factor in changing my life?

Upon graduation I had a six-month active duty obligation, along with a commitment to serve in the Army Reserve for eight years. I was scheduled to depart in July for Fort Sam Houston in San Antonio for my Officers Basic Training.

In June the phone call came that changed my life. Dr. Louis Thompson, Head of the Farm Operations Program and my academic advisor, told me that Dr. Kottman had just accepted the deanship at West Virginia University and that he, Dr. Thompson, had been appointed to Dr. Kottman's former position at ISC, associate dean for resident instruction. He was forming his staff and he wanted me to be the new college placement officer with the charge to expand the career services for the College and also to serve as the secretary for two committees: scholarships and standards. He went on to say that I could enroll in any Ph.D. degree program and so my appointment in the office would be only 3/4 time, so that I could take classes. The salary was attractive.

Of course I was interested as I knew that this might well lead to a career in academia, something that I had not seriously considered previously. However, I told Dr. Thompson that I had this six-month military obligation. He said, "Don't let that stand in your way, I will hold the position for you and you can start in January." Of course I said "yes" and selected the field of economics for my graduate program. My life was now on a new trajectory that opened many doors for me. I have told Dr. Thompson many times how grateful I am to him for the opportunity that he presented to me.

Preparing for a Career in Academia: Graduate School and Working in the College of Agriculture

It took me nine years, until January of 1968, to finish my Ph.D. There were two reasons. First, because of my work load in the College I could not take more than two courses a quarter and second, because I spent two years (1965-67) on a Ford Foundation project in Mexico where, among my responsibilities, I gathered data in the field for my dissertation. Although this schedule slowed my entry in the profession I do not regret a minute of it as both my work at Iowa State and in Mexico were important factors in forming me for my future.

In my role as placement officer I was charged to improve the career services for graduating seniors and alums. I was proactive in bringing new prospective employers to the campus to interview our seniors, developed a database showing the employment profile of each senior class, and gained considerable experience in working with business and industry. This effort dovetailed nicely with the rapid growth of the relatively new agribusiness curriculum and agribusiness specializations in the traditional degree programs such that the employment figures for our graduates with business and industry rose sharply.

In my other assignments as secretary of the College's Scholarship and Standards Committees I worked closely with the faculty members of the committees, which helped me learn how to work effectively among academics. My jobs were to prepare the materials in advance of the meetings and then follow up with the record keeping. With over 200 scholarships awarded each year there was much work to be done in announcing them, reviewing the applications, selecting awardees and then following with recognition of the winners and publicity. This was a way to get to know the best students among both the incoming freshmen and the current student body. It was quite the opposite with the Standards Committee, where the work involved those students who were in academic trouble. Here the challenge was to recognize those who really had a chance of success and give them a chance to prove themselves. Many were successful. This taught me understanding and compassion.

In addition to these two assignments Dr. Thompson wanted me to serve as an academic advisor to some students in Farm Operations. In this role I learned the importance of role of advising in guiding students and helping them make good decisions. Some skeptics called it hand holding, but we saw it as a way to ensure success, a point of view that I carried forward throughout my academic career. Dr. Thompson also wanted me to get exposed to teaching, so he made it possible for me to teach beginning courses in economics and later money and banking

All of these experiences prepared me well for a career in academia. In my different roles I had learned more people skills through interfacing with business and industry, donors of scholarships, college administrators, faculty members and students. I was honored to be elected as the lecturer representative of the College of Agriculture to serve as a representative on the University's Faculty Advisory Council to the University President. As I finished my course work and passed my oral exams in 1965

I moved on to the next stage in my professional development with my two-year assignment in Mexico, where I enhanced my research skills, learned Spanish and how to live and operate in the Latin American culture. This experience and development of language skills set me on a path for professional work in this part of the world.

The Mexico project was part of a large joint effort by the Ford and Rockefeller Foundations to improve the infrastructure for the development of the Mexican agricultural sector, patterned after the U.S. land-grant university model, which is based on the three prongs of education, research and outreach through extension. The Rockefeller Foundation had already been active in corn and wheat research at CMMYT (The Center for Corn and Wheat Improvement) located near the city of Texcoco, which was close to Mexico City. Nearby was the National School of Agriculture, commonly known as Chapingo named after the ex-hacienda in which it is located, and its Graduate College, the Colegio, which was Chapingo's arm for research and the training of graduate students. Chapingo was a dependency of the Secretariat of Agriculture as was the Federal Extension Service. The cornerstone of the overall plan was to bring all of these – education, research and extension - together at Chapingo. New buildings and facilities were constructed. However, there were some important academic elements missing. Mexico had lagged in the development of graduate programs in economics and the Colegio did not have graduate programs in economics or statistics nor a modern computer center. The Ford Foundation took the responsibility for correcting this and approached Iowa State University (the name of Iowa State College had been changed to University in the early 1960's) to develop these programs. The pre-eminent agricultural economist Earl Heady was asked to lead it. Dr. Heady selected a team of faculty and graduate students to move to Mexico. I was one of the fortunate ones selected. This assignment fit well with the emphasis in my graduate program on economic development, an important field in economics that had expanded rapidly in the Post- World War II period in response to the needs of Third World countries for improving their economic growth and alleviating poverty. Now I had circled back to my international interests that were ignited in my youth.

My family and I moved to Mexico City and, after spending two months in an intensive Spanish program, I commuted to Chapingo each day. My job was to work with the some 10 Mexican students who had been carefully selected for the first class of the new economics graduate program, teach a course in macroeconomics (in Spanish) and begin my dissertation research. It was my good fortune that Ing. Ramón Fernández y Fernández, the head of the newly formed graduate program in economics, was a highly-recognized expert in Mexican agricultural credit. He was of great assistance in connecting me with the government banks that offered credit to farmers. On another front, the Ford Foundation considered us a part of their in-country team and we were regularly invited to meetings and social events hosted by the Foundation. It was in all these environments that I began to polish my professional and social skills for working in Latin America.

In the summer of 1967 my family and I returned to the United States and to my new job as an assistant professor of economics at Arizona State University. Among my several job offers I was attracted to ASU because of its proximity to Mexico and the opportunity to develop new undergraduate courses in economic development and Latin American economics. Also, the allure of the Phoenix area and the desert were appealing to a family with mid-western roots.

Arizona State University, Fulbright, Stanford and the Ford Foundation

I spent 23 years at Arizona State. During my first semester I finished my dissertation and defended it in January 1968. I advanced through the ranks, gaining the title of full professor in 1979. I had more than 50 publications of articles, three monographs, three edited books and numerous technical reports and book reviews. My research mostly focused on Latin America, at first on Mexican topics and later on Bolivia, the Dominican Republic and the Central American countries. Agricultural credit was a continuing topic but I also dealt with the Mexican internal migration, the economics of tourism as a spur for economic development, the economics of the U.S. – Mexican border region and the political economy of Mexico and Bolivia. I mentored numerous graduate students, many from Latin America.

The opportunities continued. The Ford Foundation apparently liked my work on the previous project and asked me to return to Mexico in 1971-1972 to work as a program assistant in project management. I accepted. In 1974 I was awarded a Fulbright Lecturer Fellowship in Ecuador, which provided me my first experience in South America. In 1975 I went to the Food Research Institute at Stanford University for my sabbatical. In 1976 I was asked by the selection committee to apply for the directorship of the ASU Center for Latin American Studies. Dean Charles Wolfe offered me the job and I continued in that role until 1990 when I left for Ohio State. While at ASU I did considerable consulting for USAID, the World Bank and other organizations. I received more than $7 million in grant funds from USAID, the National Institutes for Health and Resources for the Future, much of it to undertake research work on agricultural credit projects in Bolivia and Central America. I also co-taught a USDA course in Bolivia on agricultural credit. ASU, through my work in Bolivia, developed a special partnership with the Catholic University of Bolivia (UCB). We established a student exchange program that trained many persons who are now leaders in Bolivian business, finance, NGOs, government and international organizations. We received a small grant from the U.S. Embassy to help UCB start an MBA program, an experience that was very helpful to me when later I received a grant to develop a master's degree in agribusiness in Mexico.

I was active in national and regional organizations. I was president of the Pacific Coast Council for Latin American Studies and the U.S.-Mexico Borderlands Scholars Association. I served as a member of the boards of directors for the Rocky Mountain Council for Latin American Studies and the Consortium of U.S. Research Programs for Mexico (PROFMEX), a member of the Overseas Development Council Working Group on the U.S.-Mexico Border, and a member of the U.S. delegation for the United States/Soviet Union Conference on Latin America sponsored by the American Council for Learned Societies and the Soviet Academy of Sciences.

While on sabbatical at Stanford I met Dale Adams, a professor of agricultural economics at Ohio State and an agricultural credit specialist who was also on sabbatical. We hit it off well and our friendship opened doors for a very significant set of professional opportunities for me. He invited me to join the Ohio State Rural Finance Team that had recently obtained a large USAID Cooperative Agreement Grant to work in different USAID missions around the world. Under this arrangement I would remain at ASU except for a nine-month stint in 1979 as a visiting professor at Ohio State as well as carry out research in Bolivia and the Dominican Republic. This apparently worked out well because in 1990, I received a call from David Hansen, the director of the Office of International Programs in Agriculture at OSU, about an opportunity to join its regular faculty.

The Ohio State University

The acceptance of the position at OSU was another major career change. I was hired as a tenured professor in the Department of Agricultural Economics with a specific initial assignment to manage a five-year project in the Dominican Republic. This, however, was just the beginning. With time even more opportunities emerged for leadership and administrative positions at the University. Unfortunately my research and teaching suffered, but that was something I could accept since I was now heavily embroiled in the challenges of developing new programs and improving existing programs not to mention all of the regular tasks of managing budgets, personnel and operations in order to ensure that things moved forward as smoothly as possible.

My first assignment at OSU was to serve as Chief of Party for the USAID funded University Agribusiness Partnership Project at the Instituto Superior de Agricultura (ISA) located in Santiago, The Dominican Republic. The USAID grant was made to MUCIA (The Midwest Universities Consortium for International Assistance), which is comprised of Big Ten Universities and was headed by William Flinn at Ohio State. OSU was designated the lead institution for the ISA project. This wide-reaching, five-year endeavor was designed to help ISA modernize its research, teaching and outreach programs with special emphasis on its agribusiness components. There was an in-residence team of five professors from Big Ten institutions and we hired numerous consultants to carry out specific assignments. My position was to lead and manage all of this and to interface with both the ISA central administration and USAID.

Upon my return to the OSU campus in 1995 I was named the assistant director of the Ohio LEAD Program. I moved up to the position of director in 1967. This two-year leadership program was directed to a group of 30 highly-selected up-and-coming young adult professionals in the Ohio agriculture/agribusiness sectors. The program structure was for them to meet for a three-day training period each month where they concentrated on modules dealing with the development of leadership skills and knowledge of different aspects of Ohio agriculture. There was also a national tour to get to know agriculture as it was practiced in different parts of the United States and a two-week international tour to learn about agriculture and the culture of foreign countries. During my time we took the participants to the states of Washington and Louisiana on national tours and to Chile/Mexico and South Africa on the international tours. In this assignment I learned how to do programming for adults, which later came in handy in a project I managed in Mexico.

Simultaneously during this time period I was asked to put my knowledge of Mexico and Latin America to work in three other activities. First, to help further the College of Agriculture's goal to getting more students involved in study abroad I was asked to design and lead a six-week summer study abroad program in Mexico that focused on NAFTA and agriculture. As a partner we selected the Colegio de Postgraduados, the same institution with which I had worked as a graduate student. The program turned out very well and I led it, accompanied by my family, for three successive summers from 1998-2000. It was a wonderful experience for the students, we spent about half of the time on campus and the rest travelling to some 10 Mexican states to study NAFTA related activities enterprises, many of which were agricultural based.

Second, I received a grant from the Association Liaison Office, which was financed by USAID, for OSU to work with the Colegio to develop a new master's degree program in agribusiness. After a slow start this program took off like gangbusters. It has awarded many master's degrees, trained considerable numbers in non-degree programs and is the process of establishing a Ph.D. program.

Third, I was asked to head up the Center for Latin American Studies (CLAS). The expectation was that I would apply in the national competition for the Tinker Foundation and Title VI grants. At first I resisted, having spent 14 years leading the Center at ASU, but upon further reflection I decided to accept the challenge. I wrote the grant proposals and OSU was successful in both. The winning of the prestigious five-year Title VI grant to support the Latin America area studies program was a first for OSU after having made many previous attempts. It put us on the map among the elite Latin American Studies programs among U.S. universities. Since then OSU has repeated three times and is a current recipient.

The end result was that I was managing four things simultaneously – CLAS, ALO agribusiness project, Mexico Study Abroad Program and LEAD – as well as teaching a class each semester. Then came another life-changing opportunity. In 2000, while in Mexico leading the summer study abroad program I received a call from Vice-Provost Nancy Rogers asking me to apply for the job of associate provost for international affairs, which was the position of senior international officer of the University. I was dumbfounded. I had known of the search but had thought that I would not be competitive and so did not apply. But, after the phone call and careful consideration, decided to give it a try. Within a few short weeks after my return to campus I found myself in this position, a post that I served in for seven years until my retirement in 2007.

Office of International Affairs

My assignment as associate provost for international affairs was exciting and rewarding. It expanded my horizons; my international scope was no longer confined to Latin America but rather the whole world. As associate provost and the leader of the Office of International Affairs (OIA) I not only had direct responsibility for the five area studies centers (Africa, East Asia, Eastern Europe and the Soviet Union, Latin America, Middle East, and Western Europe) the Institute for Chinese Studies, the Institute for Japanese Studies, study abroad, international student services and the Mershon Center for International Securities Studies, but also was now in the driver's seat to work with the several colleges and university central administration in helping establish a new vision for OSU's multiple roles in the rapidly increasing globalized world. Examples of these roles are (1) preparing students as global citizens, both through programs of study, area studies and coursework taken on the campus as well as through study abroad, (2) providing excellent services for international students, (3) helping faculty to get more intensively involved in international/ global matters, including research, (4) enhancing the foreign presence of OSU overseas, including building partnerships with foreign institutions, (5) reaching out to Ohio entities – such as schools, organizations, communities and business -- with globally oriented information and training programs and (6) providing information on campus and to the Ohio and national communities about OSU's extensive international involvement. It included organizing trips for the president and other administrators to Brazil, China and India. The position required that I use my talents that I had learned and developed over the many years of my professional life. I had to be at once creative, entrepreneurial, planner, organizer, administrator and leader. I had a lot to learn and made some mistakes, but it was exciting, fun and challenging. There were many successes. It was a great way to cap off my professional career.

Epilogue

I was fortunate after my retirement to secure a two-year grant from the Office of Higher Education for Development (HED) and funded by USAID/Mexico to foster the development of microfinance services in the rural areas of Mexico. To do this I circled back once again to where I began as a graduate student by both working on rural finance and partnering again with the Colegio de Postgraduados. The project did very well and, based on upon its success, was eventually extended to a total of five years. This project took me back to my roots in agricultural credit. It was fulfilling, rewarding and an excellent way to slip into retirement. Moreover, it gave me the opportunity to be immersed once again in the Latin American culture, something that I am passionate about.

Now, I am moving in the direction of delving into another passion of mine, art, and have become a docent at the Columbus Museum of Art. I did not mention this previously, but my interest in art was strengthened by my travel and residing in Latin America and by those artifacts that my Aunt Lottie sent me from Central America during World War II. The profound and creative use of bright colors and the linking of art to the representation of culture and history make both the formal art and folk art of this region so appealing. I was "hooked" early on and became a collector. One could say that my experiences in the region opened the door to another opportunity, which I seized.

Concluding Remarks

I have had a wonderful personal and professional life. It was not like I had imagined as a youth, it took quite a different path. Nonetheless my formation as a youth has played a big role in my international interests and whatever successes I have had. My parents, by their examples, set the moral and behavioral standards and gave me my deep belief in and appreciation for family life and the need for a college education. It was college where I first honed my leadership skills. However, as I look back on it my father was right on with his advice of that whatever you do, work hard and do it well. People will recognize that and provide you with opportunities for better things. My self-analysis tells me that by hard work, diligence, entrepreneurship, a modicum of people skills, and some degree of intelligence combined with luck tells me that his advice has rung true as I have had a number of opportunities that I seized to move forward to even more exciting challenges and assignments. I am very grateful to those persons who have made those opportunities possible. Finally, in closing I want to say that the support on my home front has been important. I have been married twice. Both Mary and Carmen have been supportive of my career as have my three sons – Jeff, Jamie and Mike -- by first marriage and my daughter, Stephanie, in the second. My career caused a number of moves and often the need for travel, but the support was always there.

CHAPTER 2.

Participating in the Expanding Ethanol Industry

Henrique Vianna de Amorim

Henrique Vianna de Amorim considers himself a fulfilled man. In the course of his existence (74 years old in 2013), he had everything he has ever dreamed for his life. He also overcame his own expectations. Besides the personal and professional accomplishments, he has always wanted to play his role as a citizen, contributing to the development of his country, creating jobs and prosperity for the agribusiness sector, where he still acts. Thus, Henrique has fully attained these objectives.

At the beginning of his professional career, Henrique Amorim stood out in the scientific investigation about the coffee quality improvement (Coffea arabica), a subject which he deeply studied until the mid-1970s. Most of his researches came out of the scientific range and were incorporated into the grower's routine, turning them into good practices for the farming techniques, and also during the processing and storage, enhancing the quality of that beverage.

When the National Alcohol Program was launched in Brazil in the late 1970s, Henrique Amorim shifted the focus of his studies to the alcoholic fermentation, standing, overtime, as one of the greatest Brazilian leading specialists in this subject, with acceptance in Brazil and abroad.

In the fermentation area, a vital step into the industrial process, a team led by Henrique Amorim, who has provided technical assistance for forty years to a great number of alcohol and sugar Brazilian plants, performed groundwork in order to improve the process yield and efficiency, identifying and correcting the production bottlenecks, struggling against the contamination and the losses.

In the book Alcoholic Fermentation: Science and Technology (Fermentação Alcoólica: Ciência e Tecnologia), published in 2005, it's reported how the work performed by Henrique Amorim and his team was important to the fermentation development in the national sugarcane agroindustry, and how it's certainly contributed to the Brazilian leadership in the production of sugar and ethanol, with spectacular yield gains: sugarcane (Saccharum officinarum), which yielded 3,000 liters of alcohol per ha per year in 1975, currently yields from 7,000 to 10,000 liters per year in average, as progress occurred in the crops and industry.

As a respected professor, Henrique Amorim taught biochemistry at Luiz de Queiroz College of Agriculture (Escola Superior de Agricultura "Luiz de Queiroz") from 1970 to 2001, in Piracicaba (a city located in the State of São Paulo), when he had to cease his academic activities to completely devote time to his consultant, research and training company, Fermentec Ltda., which currently has more than fifty employees and more than seventy costumers in many regions of Brazil and abroad.

Nonetheless, Professor Henrique Amorim has always been enthusiastic about knowledge sharing production through publication (more than 80 scientific papers either authored by him or in collaboration with other scientists), including presentations at scientific meetings, invited lectures in Brazil and abroad (11 during 2012), and organizing courses at the company or at the client's production facilities.

At the beginning of 2013, Professor Henrique Amorim seems not to be interested in slowing down his working pace and does not intend to be retire either. Besides leading his team at Fermentec, in January 2013 he started chairing the Ethanol Cluster - APLA (Arranjo Produtivo Local do Álcool), which purpose is to stimulate the ethanol industry, services and marketing.

Professor Amorim still conducts cutting-edge research which has led to allowance of industrial patents in Brazil and abroad together with foreign researchers in vanguard areas such as customized yeast strains development for different fermentation conditions existing in the plants and for high alcohol content fermentation. In the research field, his company, Fermentec, operates in partnership with several Universities and Research Centers, although the experiments are usually performed within the Company's laboratories.

Throughout his professional career, Henrique Amorim has accumulated several awards in recognition of his outstanding work; however, one of the most desired opportunities is the sharing of knowledge with the younger generation about one's experiences and lessons learned all over his life, and to encourage young people to fight for their goals and to turn their dreams into realities.

"A father needs to pass moral values as honesty and working hard to their children, in order to prevent them from failing due to naivety, which is part of growing up, always respecting their inclinations and preferences".

Childhood

Henrique Vianna de Amorim was born in 1939, in the city of São Paulo. When he was about to complete one year, his father, the physician Henrique Berbert de Amorim, decided to move to the countryside of the State of São Paulo, since he was worried about the frequents asthma attacks of his child, which required extreme procedures, such as a tracheotomy, to allow him to better breath.

By listening to his colleagues' prediction he decided against raising his child in such a cold and humid climate, the physician moved his family to the city of Ribeirão Preto in 1941, in the northwestern region of the State, 300 kilometers from São Paulo. In Ribeirão Preto, where his father-in-law, the coffee farm agricultural manager Bruno Carneiro Vianna, already lived, the physician raised a large family with his wife Alda Luiza, they had five children: Álvaro, Ana Luiza, Vera Lucia and João Carlos, besides the firstborn, Henrique. He could, overtime, conquer good and loyal clients who came not only from the city but from the farming community in the neighboring States.

As a child, Henrique Amorim learned to like animals and plants. In the wide backyard of his home, he formed a kind of private zoo where he raised calm domestic chickens and even wild animals such as monkeys, alligators and deer. Many clients of his father, who knew his love for nature, used to give him wild baby animals, which is unacceptable nowadays.

When he was a boy he didn't like studying. As a frail and weak child he often missed classes and repeated the first year, which he had attended at Santa Úrsula School. The Ribeirão Preto environment provided a favorable climate which gradually led to his recovery. He transferred to another school, the Marista School, where he concluded the elementary level. Thereafter, he continued his education at Otoniel Mota High School.

During school vacation, Henrique constantly used to go to his paternal grandfather farm, in Ilhéus, State of Bahia, where his grandfather Virgílio Calazans de Amorim used to grow cacao tree and raise cattle. His grandfather was a strict man, as most of the big landowners of the Brazil northeast region, who were known as coronels. Despite having a bad reputation for being authoritarian and austere demeanor, the coronels were responsible for the progress and development of the region.

It was the case of Henrique's grandfather, who invested all of the monetary gains from the farm back into business opportunities in the region where he lived. Together with other local farmers he created job opportunities and provided important infrastructure improvements such as the first road to link the city of Ilhéus to the city of Itabuna, with the purpose of conveying the cacao beans which were transported previously in boats through the rivers.

The young Henrique developed an appreciation for the countryside while living on his grandfathers' lands. He used to ride in the trucks transporting crops with his grandfather and, when he was twelve, he owned a cow, the same as the other grandchildren, to start his own cattle herd.

The rural environment already attracted Henrique and became very familiar to him. In his home, in Ribeirão Preto, he also liked to ride in trucks with the farmer Eugênio do Val's grandchildren, the Santa Teresa Farm owner, and with the physician Oscar Figueiredo's children, Oscar and Luiz Otavio, whose father-in-law was the Floresta Farm owner.

"The circumstances change all over the time. No one can predict the future."

Choosing a Career

When it was time to choose a career, Henrique was on the horns of a dilemma: to become a physician, as his father, and to inherit a medical clinic, which was the easiest choice, or to study Agronomy, which he desired however if he decided to choose the second option he would face his father's severe opposition, who argued with him,

"An agronomist engineer doesn't earn enough money to raise his children. The Brazilian labor market is insufficient; you will run your own farm or will work for governmental agricultural development and inspection agencies."

His father shared another consideration with him that he needed someone to continue running his clinic, since his two other sons decided to become engineers.

In response to his father expectations, Henrique decided to apply for medical school, in Salvador, State of Bahia. There was a familiar plan of acquiring lands in the Una region, for rubber plantations, recently introduced in the State of Bahia, and Henrique was planning to run this farm while he studied to become a physician.

On the eve of his departure, his Uncle Silvio and his grandfather Virgílio wrote a letter to his father questioning if this was the best alternative for the young Henrique. His father, however, stuck steadfastly to his intent of having Henrique attend medical school and become a physician, but agreed with his departure to Rio de Janeiro. Henrique liked the idea. After all, the possibility of living in the "Marvelous City", with all its attractions, was not to be missed.

He lived and studied in the Brazilian old capital for a year, but he hadn't been accepted for admission to medical school. Then, he decided to face his father resistance and become an agronomist engineer. Nevertheless, his father didn't take the plunge: "If you want it, do it, but if you cannot get into the college of agronomy, I won't support you anymore!"

His paternal threat led him to intense study. He prepared himself for taking the college entrance examinations and, in 1962, he was accepted as a student at Luiz de Queiroz College of Agriculture (Escola Superior de Agricultura "Luiz de Queiroz", the ESALQ-USP), a University of São Paulo campus in Piracicaba, where he obtained the third position on the approved students list. This good news was communicated by his childhood friend

Luiz Otávio Figueiredo, who was already studying at that College, together with his brother Oscar.

Dr. Henrique Berbert died in 1992 and could follow his son's successful career, concluding that his concerns were unfounded. After all, the Brazilian economic scenario underwent change during those years. The agronomist engineer labor market increased dramatically in the 1960s, with the agricultural expansion to the unexplored cerrado lands, a Brazilian tropical savannah ecoregion, and with the applied technological developments in the field including the introduction of new fertilization and mechanization techniques, contributing to increase the productivity. Moreover, the coffee plantation (Coffea arabica), which was very important to the Brazilian trade balance maintenance, could hire half of the recently graduate agronomist engineers.

On the other hand, in those days, the Brazilian physicians started facing difficulties. Their career was not as appealing as before, since the public health care obligation and the competition reduced the income perspective. Thus, if Henrique's father wouldn't construct a sound financial basis, he would confront difficulties in the end of his life.

"No man is an island", as the English poet John Donne said (1572-1631).

Student's Life

As the most part of young people who came to study in Piracicaba, Henrique moved to a "república" (a kind of fraternity), named "Pito aceso" (Pipe lit). The "repúblicas" are part of ESALQ-USP student environment, but actually its origin is related to the old houses inhabited by students in Coimbra, Portugal, where universal values were applied, linking the past with the present time: the community life, sovereignty and democracy.

There, the decisions were taken unanimously and all the members were responsible for the house management.

The young Henrique, lively and sociable, didn't find any problems adapting himself. He soon became the graduation commission president and, together with other colleagues, he promoted several parties, balls and meetings, all of them well attended, which contributed to the student's social experience. Besides the busy social life, Henrique used to play sports, defending his College competing with running and high jump through the Luiz de Queiroz Academic Athletic Association – AAALQ (Associação Atlética Acadêmica "Luiz de Queiroz").

At that time, Henrique received a letter from Bahia, from his Uncle Sílvio, requesting that he scheduled a meeting with the Professor Eurípedes Malavolta (1926-2008), one of the biggest authorities in plant mineral nutrition and soil fertility in Brazil, and also the director of ESALQ-USP. One of the first books of this scientist (who published 45 books in many languages, besides 823 research papers), entitled "ABC of Fertilization", has had great success in Bahia, showing the mineral fertilization necessity for the cacao plantation, where this idea was accepted by many farmers who could significantly improve their productivity.

Professor Malavolta received Henrique and his Uncle Sílvio into the imposing boardroom at the ESALQ-USP main pavilion. During that conversation, the Professor invited Henrique to perform assays in regard to the cacao fertilization using a nutrient solution. This novelty of the approach had not yet been tested in Brazil.

Henrique didn't lose the valuable opportunity: he brought cacao beans from Bahia to perform the experiment, studied the subject exhaustedly and was assisted by Professor Henrique Paulo Haag, who was a member of the Professor Malavolta staff at the Chemistry Department to perform his first laboratory assay.

With Professor Malavolta, Henrique learned lessons which he carried to his own career as a professor. According to him, the professor never corrected the students' paper; he just used to mark the topics which had to be improved, asking the student to comeback after carrying out the alterations. During the elaboration of his first experiment, Henrique had to return almost ten times until the professor was convinced that his work was good; for his second experiment he returned three times with the requested alterations; then, he almost needn't to return anymore.

This first experiment called "Effect of different levels of nitrogen, phosphorus and potassium over young cacao trees (Theobroma cacao L.) In nutrient solution", was published in the Proceedings of the Luiz de Queiroz College of Agriculture, in 1964, and can be the first experiment performed in Brazil about cacao mineral nutrition in hydroponic solution. Besides Henrique Amorim, this work was signed by Luiz Carlos Scoton, Henrique Paulo Haag and Eurípedes Malavolta, and it was presented in a congress held in Bahia.

With the impact of his first experiment, Henrique Amorim's life would change forever. At that moment, he didn't think about studying agronomy to become a researcher or a professor, because he thought such activities would require complete devotion, and were incompatible with his extroverted personality and with his active social life. Over time, however, he realized that it wouldn't be so difficult to balance his different interests.

However, Henrique Amorim has gradually acquired the taste for scientific investigation. Maurílio Biagi (1914-1978), who was a great entrepreneur of the sugar and ethanol industry, was the person who encouraged him. Maurílio Biagi, besides being a great producer of sugar and ethanol industry, was an innovative industrialist, creating companies as Zanini and Sermatec, which provided equipments for the sugar and ethanol plants.

When he was a student, Henrique was a friend of Maurílio Biagi's sons. He was accustomed to arrive at the Biagi's home before the social commitments just for talking with the patriarch about agronomic subjects. Mr. Biagi wanted to study agronomy, but he needed to study accounting to assume management of the family business. Even so, he knew well the cane plantation and used to question Henrique about what he learned in practice.

One of his doubts was about fertilization. Mr. Biagi observed that the recently harvested sugarcane showed higher nitrogen quantities than the crops using chemical fertilization. He wanted to know why this had happened. Henrique suggested to him that the nitrogen content would most likely be incorporated from the rain. Henrique took this question to his Professor Eurípedes Malavolta, who was also in doubt, because the sum of nitrogen incorporation resulting from fertilization opposed to the average nitrogen from the rain was less than his laboratory found in studies of cane nutrition. Maybe the rain was richer in nitrogen oxides in the region of Ribeirão Preto?

Studies have proven that these factors could not be sufficient to explain this fact. The question was answered years later by a Czech researcher named Johanna Liesbeth Kubelka Döbereiner (1924-2000), an agronomist engineer, hired by the Ecology and Agricultural Research Institute, currently known as Embrapa Agrobiology, in the city of Seropédica, Rio de Janeiro. Her studies led to the discovery of nine species of nitrogen fixing bacteria associated with grasses, cereal and tuberous plants.

In 1988, this scientist described the association between the endophytic and nitrogen fixing bacterium Gluconacetobacter diazotrophicus and the sugarcane. The most spectacular results were observed for some sugarcane varieties, capable of presenting high productivity (over 160 tons per hectare), with up to 200 kilos of nitrogen, coming from its symbiotic association with the nitrogen-fixing bacteria. These researches were essential for the sugarcane production increase in Brazil and provided Dr. Dobereiner to be nominated for the Nobel Prize for Chemistry, in 1997, but, unfortunately, she didn't win the prize.

"It's necessary to understand quite well the nature processes, in order to intervene in them safely"

Working with Coffee

In order to answer the questions related to the agricultural practices which were proposed by the farmers who were part of his social circle, as Maurílio Biagi, Henrique Amorim became ever more motivated to proceed with the research work. When Professor Malavolta realized the potential of his pupil, he suggested the opportunity to study coffee, instead of cacao, because, at that time, the Brazilian Coffee Institute (IBC, which was founded by the former president Fernando Collor, in 1990), had responsibility for setting the Brazilian policy and providing financial resources for those interested in research of coffee.

The researches performed in Brazil, unlike other producing countries, were chiefly intended to obtain coffee record crops and not to improve its quality. With the guaranteed funding to work, Professor Malavolta suggested to Henrique about the opportunity to solve practical issues for the growers: many of them at the time thought the chemical fertilization impaired the beverage quality; therefore they preferred to nourish the plants only with organic fertilizers.

The only way to prove the role of chemical or organic fertilizers was to conduct comparative studies. Henrique started a careful and detailed experiment at the ESALQ-USP Chemistry Department. He contacted the lead of the Coffee Experimental Farm in Ribeirão Preto, Antonio Junqueira, to obtain the adequate material for the experiment and went personally to harvest and pulp the coffee intended for the tests.

When the degustation was necessary to verify the beverage quality, Henrique had joined forces with a coffee taster, the technician Aldir Alves Teixeira from the Secretary of Agriculture of the State of São Paulo, who was a post-graduate student at ESALQ-USP, advised by Professor Pimentel Gomes.

The scientific work resulting from these experiments, in 1965, was called "Studies on the mineral feed of coffee. Effect of N P K fertilization in the chemical composition of the soil, coffee beans and beverage quality". It was signed by the researchers H. V. Amorim, L. C. Scoton, A. Castilho, F.P. Gomes and E. Malavolta, and it was the first of a series of articles about this subject.

At that time, the lessons of one of his professors were very useful to the young Henrique, Professor Anivaldo Pedro Cobra, who taught mechanics. The professor emphasized to his students the importance of understanding systems and the processes functioning, relating to agricultural machinery, so it could be used to its fullest potential. The importance of understanding the fundamentals on how and why things work remained in the future researcher's mind.

"In the scientific investigation, you cannot be restricted to the main subject of a research, because you take chances of being stuck to that specific knowledge only. It's necessary to whet the curiosity, to look for links with other fields, other subjects, to open your mind to new ideas to evolve".

At that point, Henrique had graduated in Agronomic Engineering (1966) and, as a National Council of Scientific and Technological Development (CNPq) scholar; he had already read the literature and developed an extensive knowledge about coffee and cacao. Then, he started to examine the similarities and differences between both cultures. As tropical plants, they often presented the same behavior, but there were some differences: in the case of the seeds, for example, the cacao bean needed to be fermented to produce a good chocolate; the coffee beans couldn't be fermented before the specified time, or for a long time, since it could be impair the beverage quality.

Other cultures which also presented similarities between coffee and cacao were tobacco and tea. Comparing with coffee and cacao, all of them had phenolic compounds which influenced the quality of the final product. Both in the case of tea leaves and in the case of tobacco leaves, when they were subjected to drying, these compounds were modified.

Henrique studied the enzyme polyphenol oxidase that causes oxidation in fruits and vegetables that results in browning, and its relation to the phenolic compounds. He considered that the enzyme activity could influence coffee quality, making a hypothesis: the lower the polyphenol oxidase activity, better the coffee quality would be. He had come to this conclusion because, in the other cultures (tobacco and tea), if the enzyme activity was low, the phenol oxidation wouldn't occur.

However, the hypothesis veracity had to be proved. Then again, Henrique requested the advice of the agronomist engineer Aldir Alves Teixeira, who brought him five different coffees in regard to quality – strictly soft, soft, hard, riado and rio – to perform the experiments. On the day that he received the material, he locked himself into the ESALQ-USP Chemistry Department laboratory and, at 4 p.m., he started the tests with the enzymes, whose procedures were advised by Professor Darcy da Silva. At 10 p.m., the results arose, completely different from what he had expected: the better coffees had high polyphenol oxidase activity.

Henrique thought the tests were run improperly, so he phoned Professor Malavolta at his home who suggested reversing the analyses sequence, starting from the best to the worst coffee. At 2 a.m., he called the professor again to confirm the same results: the higher the enzyme activity, the better the coffee quality. Professor Malavolta said to him that he had made an important discovery.

Moreover, he said, "The results are so unexpected that the experiments deserve publication in a world renowned journal."

Just in case, in order to document the Brazilian discovery, Henrique wrote an article very quickly to the ESALQ-USP Bulletin; he also sent it to the Nature British journal with the title "Relationship between the polyphenol oxidase activity of coffee beans and the quality of the beverage". It was published in 1968, and the article was well received. Similar experiments were performed subsequently in Rio de Janeiro, by A. Iachan, from the National Institute of Technology and, also, by other researchers, in Colombia, who obtained similar results.

With the continuous development of this work, Henrique elaborated a methodology for coffee quality rating in several ranges. The chemical analysis, which lasted only half an hour, allowed the farmers to know exactly the product quality that they were placing on the market, in spite of not having, of course, the same sensitivity and degree of refinement, obtained from the coffee aroma and taste evaluation by a connoisseur. Up until now, however, this kind of analysis is used by many growers, especially in the State of Minas Gerais, and also, by several national food companies.

"Things Do Not Just Happen by Chance…"

Following that, Henrique's professional career received a great impulse. He got three invitations to stay at ESALQ-USP as a professor, while he was waiting for a scholarship to start a post-graduate course abroad: one from Professor Ernesto Paterniani (Genetics Department), another from Professor E. A. Graner (Agriculture Department) and the third from his advisor, Professor Eurípedes Malavolta, who offered him a position as Assistant Professor in the Chemistry Department.

There were other possibilities to Henrique: to work with coffee outside Brazil, in Ceylon, or in Cuba, for the Food and

Agriculture Organization (FAO), a body of the United Nations, or to produce natural rubber in Nigeria, hired by the multinational company Firestone. As he was in doubt, he asked a friend who was a FAO employee, and he advised him to go to the United States to take the post-graduate course, where his professional development would be greater.

Thus, the master degree in Plant Physiology was performed between 1968 and 1970, at Ohio State University, OSU, in the United States, under the guidance of Professor Donald Dougall, focusing on the tissue culture studies, which was a scientific novelty at that time. The title of his Master's dissertation was "The Effect of Nitrogen and Carbohydrates on Production of Phenols by Plant Cell Cultures".

The research sought to expand knowledge of the metabolism of phenolic compounds – which were important to the beverage quality – in different stress conditions.

This work represented another turning point in his career. Initially, it was agreed that the thesis research topic at Ohio State University would focus on Plant Nutrition, but as Henrique decided to change the topic of research to plant biochemistry as he already had a thorough knowledge of plant nutrition from his research studies in Brazil.

In order to achieve this change, it was essential the intervention of Professors Eurípedes Malavolta and the renowned Fred Deathridge, from Ohio State University, which spent some time in Brazil, doing research in food biochemistry, through an agreement between the Brazilian government and United States Agency for International Development, the USAID.

It was a moment that required a great deal of commitment and effort of the young researcher. Besides facing some difficulties with the language, which he had not yet mastered, he needed to strengthen his knowledge about Organic Chemistry and Biochemistry, key disciplines for the development of his research.

He was helped a lot by two people at that time: the Brazilian Ruy de Araújo Caldas, from ESALQ-USP, who was seeking a Ph.D. in that country, and the North American biologist Rod Sharp, a professor newly hired by Ohio State University, expert in tissue culture.

Besides providing his highly specialized knowledge on this field, Professor Sharp, who became a great friend of Henrique, lent his laboratory and equipment to perform experiments, what normally happened during overnight, when the place was not used by the local researchers.

When there were four months remaining to conclude his thesis, Henrique was surprised: the tissue culture he was working on suffered degeneration. When his advisor, Professor Dougall, knew it, he asked him to discard all of the expermentental data and said him not to rely on the obtained results.

The only one to give him a vote of confidence was Professor Sharp: he sent the material to be analyzed in two renowned laboratories in New York and California, and after a month both laboratories returned a letter declaring that the material was normal, validating the results. Thus, Henrique Amorim could finish his Master's dissertation, which was published in a reputable journal, the Physiologia Plantarum.

"This Life is a Chain of Relationships"

When he was still studying in the United States, Henrique wrote a letter to Professor Otto Crocomo, a specialist in biochemistry at ESALQ-USP, who was really interested in tissue culture, telling about his friendship with the specialist Rod Sharp, since he knew Professor Crocomo interests in establish courses about this subject in Brazil. Then, a synergy was built between the two scientists. The result was the first course of tissue culture at ESALQ-USP, an expertise that has expanded throughout the Latin America from this event.

Besides Professor Crocomo, Henrique introduced Professor Rod Sharp to Paulo Alvim, one of the greatest Brazilian plant physiologists, director at Executive Committee of the Cacao Plantation Plan (CEPLAC), in Bahia, who was interested in studying the tissue culture in order to apply it to cacao diseases. The North American specialist became a consultant in that institution for a while and was responsible for the advance of this methodology with the cacao plantation in Bahia.

"Knowing and not doing is the same as not knowing" advises the *Zen philosophy.*

Doctorate

Back to Brazil, in 1970, Henrique Amorim held the position of Assistant Professor in the old ESALQ-USP Chemistry Department, in the Biochemistry field, where he started executing academic and research activities at the same time.

In the area of education, Henrique decided to make a few changes in the didactic structure used until that time by professors. The programme contents, which the students needed to know by heart, was too theoretical and boring. Looking for a more attractive and realistic approach to the discipline and showing its practical application in the people's routine, the young professor engaged in thoroughly studying the subject in order to turn it more comprehensive. After all, his students were going to specialize themselves in different fields, requiring practical examples of biochemistry studies which should cover all the agronomy fields.

Due to the great acceptance of the students to the new way of learning biochemistry, other professors who succeed him in that function adopted the methodology. This efficiency was also awarded: by the end of the first year of teaching Professor

Amorim received a tribute from his students, on the occasion of their graduation ceremony.

Another initiative was to establish, in the Chemistry Department, an annual scientific meeting, where all the students could present their ongoing work, with the aim of stimulating the interests for research. The success of that kind of event made other students of different areas to request that other departments could be involved in the meeting.

One of the most enthusiastic students about these effective scientific meetings was Raul Machado Neto, who would graduate in Agronomic Engineering in 1973, with an outstanding academic career. Raul Machado Neto became a professor in the Department of Zoology (currently Animal Science); he was the Chairman of the Research Committee and ESALQ-USP Deputy Director; nowadays he is an advisor of the Pro-Rectory of Post-Graduate Studies, Pro-Rectory of Research, and Deputy Chairman of the University of São Paulo International Relations.

In the research area, Henrique's aim was to better understand the quality coffee deterioration mechanisms. Therefore, he involved his scholars in this work, creating a kind of community. Their interaction included frequent gatherings out of the academic boundaries. Thus, the distance between professor and students was decreased, creating a more relaxed atmosphere which favored the exchange of ideas.

As a result of this joint work, Henrique could conclude, in 1972, his Ph.D. thesis in Biological Sciences (Biochemistry): "Relationship between organic compounds of green coffee beans and the beverage quality", under the advisement of Professor Eurípedes Malavolta.

Aiming to make the application of some of his researches concerning to the coffee quality financially viable, Henrique received a support of a friend of his brother Álvaro, the economist

Carlos Viacava, who was a Marketing Director at Brazilian Coffee Institute (IBC). Through his own way, Henrique was received several times at that institution, in Rio de Janeiro, to discuss his projects personally and to show to the technicians its relevance in order to obtain a great knowledge about the subject.

"People come to our lives by chance, but it is not by chance that they remain" Lilian Tonet

Post-doctorate and Habilitation ("Livre-docência")

The research work, still investigating the coffee quality, continued in the following years, always together with the study group formed by the agronomic engineering students. One of his students was Vera Lúcia Natal, from the city of Oswaldo Cruz in the State of São Paulo, a scholar who asked him for an internship and became a very important person to his life. In the first moment, in order to know her abilities, Henrique offered her a very tiring job: she had to prepare a huge bibliographic review about the subject "caffeine in coffee".

The young Vera locked herself in the ESALQ-USP Library to search for articles during three months and, in face of the great demands of Henrique, she renounced to her own vacation to complete the job. Besides the dedication, the intern showed organization, discipline, persistence and many other qualities which led the professor to fall in love with her.

As the University of São Paulo regulations did not allow this kind of relationship, Vera Lucia had to work with a professor from another area, but she soon had to abandon her studies, since her dating with Henrique developed rapidly in a marriage, performed in 1974. At that time, he decided to go to the United States to evolve with his researches about coffee, obtaining a scholarship as a visiting professor at the Ohio State University, for a year.

After a fast honeymoon in Acapulco, Henrique started working in his friend's laboratory, Professor Rod Sharp, in Columbus, having as an immediate assistant his wife Vera Lúcia, who proved to be an excellent laboratory technician. Under his advisement, she started analyzing the coffee samples brought from Brazil, using electrophoresis, a technique used to separate proteins through electrical current.

At the same time, Professor Henrique also acted in another laboratory: the Food Science, at the same university, together with the specialist R.V. Josephson. The aim was to separate the different proteins and to relate the coffee beans chemical composition which provided different qualities to the beverage.

In the following six months in the United States, Henrique and Vera Lúcia worked in a research centre in New Orleans, ran by the United States Department of Agriculture (USDA) – the North American body responsible for agricultural research -, where there were specialists in phenolic compounds to support him with the Brazilian coffee sample analyses. An invitation came from Robert Ory, head of the food section of the American Chemical Society, who Henrique met in a congress which he co-ordinated in San Francisco. Bob Ory was a peanut crop specialist, which in some aspect was similar to coffee.

In the New Orleans laboratory, Henrique was searching for new techniques to determine the mechanisms linked to the coffee quality. In this work, the cooperation of his wife Vera was essential, since she performed the electrophoretic analyses with patience and care. On one occasion, there were many difficulties on making the gel for the experiment. After losing twenty valuable working days, she had the idea of putting the material in an ice bath; in spite of the researchers discredit, her initiative was successful and Henrique could continue with the analyses, which were essential for his future works.

His staying in the United States was a very rich period in terms of scientific production. Besides the chapter "Coffee Enzymes and

Coffee Quality" (1977), that he wrote together with his wife Vera Lúcia for the book "Enzymes in Food and Beverage Processing". He published also "Some Physical Aspects of Brazilian Green Coffee Beans and the Quality of Beverage" together with R. Smucker and R. Pfister (1976), about electron microscopy and, "Water Soluble Protein and Nonprotein Components of Brazilian Green Coffee Beans" (1975) together with R.V. Josephson.

Many years later, in 2004, Henrique Amorim would receive an acknowledgement for his work, a tribute from Ohio State University: the prize University Ambassador Medal, awarded to his pioneer contribution in biological sciences research and entrepreneurship, during the celebration of the cooperation agreement 40th anniversary between the Ohio State University and the ESALQ-USP.

New Challenges
When he still lived in New Orleans, Henrique received a visit that would cause a real turnaround in his career: his friend and brother-in-law Maurílio Biagi Filho (who married his sister Vera Lúcia), was running a family-owned sugar and ethanol plant.

During his visit, Maurílio suggested that Henrique began to work with ethanol, because the potential for growth in Brazil was great. Since the international oil crisis was taking place in that country, the challenge was to increase the ethanol production, to use it in a first stage, as an additive to gasoline, and afterwards, as an exclusive fuel. Therefore, the government set up the Ethanol National Program (Pró-Álcool), in 1975, which would be one of the successful initiatives to use renewable energy in the world.

At that moment, after twelve years of studies, Henrique became a respected researcher in coffee quality, and also internationally renowned. Because of this work, he provided technical advice to the North American multinational General Foods Corporation, which had been installed a chemistry laboratory to investigate this subject. However, he was disappointed with the Brazilian government policy at that time, since it placed emphasis on production increase rather than the beverage quality, to serve the market needs.

Back to Brazil, Professor Amorim started teaching again at ESALQ-USP and also concluded his full professorship thesis, based on the information collected and the experiments performed in the United States. He was trying to prove that poor quality beverages occur because of protein degradation, since such compounds are very important to the aroma and taste of the coffee. His thesis, published in 1978, had the title "Biochemical and Histochemical Aspects of the Green Coffee Beans Related to the Quality Deterioration".

In order to develop his research, Professor Amorim performed assays in the Agronomic Institute of Campinas (IAC), together with the researchers Alcides Carvalho and L.C. Mônaco. They examined some aspects in regard to the beans harvested in different regions and in different climates.

At that time, Henrique Amorim received an important collaboration from Murilo de Mello, his former student and intern, who became also a biochemistry professor at ESALQ-USP, after taking a post-graduation course in the United States. He helped Henrique to lead the experiments at IAC, in the city of Campinas; the experiments were related to the storage of coffee in different kinds of packaging. The performed tests showed that the more adequate material to wrap the product was not jute sacks–widely used at that time by the growers -, but plastic sacks, since it protected coffee from humidity variation avoiding damages to its quality.

Moreover, they studied the temperature influence on the coffee drying for the beverage quality and germination. In 1953, researchers from IAC, discovered that best beans would give a more tasteful beverage, but there was not a specific methodology to detect the beans germination potential yet. In the published literature, Professor Henrique Amorim found a simple and fast method for that purpose, developed for soy bean (Glycine max), showing that the beans germination ability was linked to its cellular membranes integrity.

Professor Amorim applied the same method to coffee beans and was able to prove his hypothesis successfully: the good harvesting practices as pulping, drying, storage and processing are essential to improve the quality of the final product, the beverage.

Thus, some simple and practical recommendations were provided to the growers, as for example, to storage the coffee beans in plastic sacks to better conserve it, protecting it from humidity and temperature variations, aiming to preserve its quality.

Another practice already employed by the growers was storage of the beans with the skin, avoiding the variations of the air temperature and humidity, such procedure showed to be effective and was validated by the scientific research.

At that time, Professor Amorim had already decided to change his focus, abandoning his studies about coffee and accepting the sugar and ethanol producers' invitation to optimize the fermentation process, whose yield was very low. He had the foresight that his biochemistry knowledge would be useful to improve this step in the industrial production, since biochemistry is a science that it was born from an attempt to explain why yeast produces alcohol.

"If you can't measure, you can't manage" (Peter Drucker, 1909- 2005)

Pioneer Studies in Fermentation

During the year of 1976, while he was finishing his scientific works with coffee, Professor Amorim was studying fermentation at the same time, looking for the theoretical basis of this subject. On that bibliographic review, he translated and analyzed more than 2,000 papers related to the field.

In 1977, he also became a technical consultant of three sugar and ethanol plants of the State of São Paulo: the "Santa Elisa" Plant, in the city of Sertãozinho; the "Vale do Rosário" Plant, in the city of Morro Agudo; and the "Pedra Plant", in the city of Serrana. The directors of those companies, Maurílio Biagi Filho, Cícero Junqueira Franco and Pedro Biagi Neto, respectively, had already thought to create a fund in order to finance the academic researches needed to improve the alcoholic fermentation process.

This initiative became viable after the hiring of Professor Henrique Amorim. At that time, he was impressed by what he had observed in the plants routine. During his visits, he verified that there was not any kind of process control and any adequate measurement for the obtained results, which made it impossible to intervene in order to improve it.

Moreover, it was a consensus in the sugar and ethanol industry, in the 1970s that Brazilians knew how to conduct the fermentation very well, because they perfectly mastered the production technology of cachaça, which was made in Brazil since the colonial period. However, when large volumes of alcohol started being produced, reaching up to 200,000 liters per day in some plants, the producers realized that the reality was different and the entire process needed to be adjusted to the new conditions.

The traditional fermentation performed to the cachaça manufacturing lasted more than 25 hours, because the yeasts were not recycled. Several attempts were made in order to improve this process, but they were not always effective, since the knowledge inside the books couldn't be practically employed. It was necessary to perform laboratory systematic tests in the industry to evaluate its response, before applying the technological innovations to the process.

In conclusion, Professor Amorim demonstrated to the industrialists that large scale ethanol production was quite distinct and requires alterations. Therefore, he started developing and adapting specific methodologies to measure the results of the alcoholic fermentation performed in the plants.

For this job, he invited Professor Humberto de Campos, from the ESALQ-USP Mathematics Department, to help him to develop specific statistical analyses (regression and correlation) in order to evaluate several parameters inside the industry. By that time, with the systematic monitoring, some modifications were introduced into the process and brought positive results, as the yield increased.

In 1978, the graphs that showed the fermentation results were handmade and it continued until the acquisition of the first computer, which made it automatic. His brother Álvaro Amorim was the one who convinced him to buy the computer to expedite the work; he already had one for his job on defining more economical mixtures for livestock feed.

In these first years, the yield measurements taken in the three plants were capable of analyzing ten parameters, approximately. Over time, they became more complex and nowadays more than 100 parameters are analyzed, since they can influence the industrial process.

Right after, Professor Amorim faced a problem: the available methods at that time were not appropriate and many of them, which were commonly used in biochemistry, were adapted to measure the sugar and ethanol industry data. It was the case of a methodology developed in 1940 by Somogy-Nelson, used nowadays to measure small quantities of sugar, mainly residuals.

The bacterial contamination, on the other hand, was evaluated through a chemical test called acidity. The counting of living bacteria were done through plating and lasted up to 3 days. For the purpose of making this work more efficient, Professor Amorim invited Professor Antonio Joaquim de Oliveira, who had concluded his Ph.D. at Ohio State University, studying the milk microbiology.

After six months of research, this specialist developed a fast methodology to determine the number of living bacterium in the fermentation that is still used nowadays. He adapted a test applied in the Lactobacillus measuring that enables the direct counting using an optical microscope in 15 minutes.

The entrepreneurs of the sugar and ethanol industry were astonished by the bacterial contamination level inside the fermenters and, as suggested by Professor Amorim, they implemented several changes to decrease the contamination. One of the alterations was the optimization of centrifuge use, high value equipment that is crucial for the process yield. Nowadays, controlling the contamination is a great challenge and the efforts to fight against it are essentials in order to increase the industrial efficiency.

With the increase of customers and subsequent volume, Professor Amorim realized that he had to enlarge his team. He went to the city of Jaboticabal to invite a former post-graduation student, Edvaldo Zago, who taught biochemistry in a university. This researcher performed the final adjustments to the colorimetric method to analyze sugars in the must (a sugar solution to be fermented) and the "wine" (the fermented mixture which goes to the distillers).

In 1977, Professor Amorim established his company, Fermentec Ltda., in a very simple way, which operated in a single room, but with a very comprehensive aim: "generating and transferring alcoholic fermentation and production control technologies through the continuous improvement of its quality management system, qualifying people and organisations". In that year, his wife Vera Lúcia abandoned her academic activities to take care of the children (Henrique Berbert de Amorim Neto and Flávia Amorim), due to the frequent absences of her husband, who was visiting the plants. Later, in 1990, she returned to school to become a lawyer.

In 1984, the accumulation of work forced Professor Amorim to reduce his time dedicated to the university (he would teach until 2001), because the challenge of processing the fermentation in a large scale, as it was done in Brazil in the Pró-Álcool period, an unprecedented event in the world. By that time, the techniques were rudimentary and inaccurate, since the national production was small and did not exceed 30,000 liters of ethanol per day. For that reason, the investments made to improve it were not rewarding to the industry.

New Fermentation Technologies

In 1978, advanced equipment was introduced to the international market with the capacity to measure, with accuracy, the alcohol content in "wine", treated yeast, vinasse, flegmass and alcohol: the density meter. It was developed in Austria by a biochemist at the University of Gratz and the equipment was digital and electronic. The problem was that to perform this analysis it was necessary to distil the sample, before taking it into the density meter, which lasted almost an hour; for that reason, all the equipment advantage was impaired by the slow distillation progress of the operation.

Helped by Professor Luiz Eduardo Gutierrez, from the ESALQ-USP Chemistry Department, Professor Amorim could design faster equipment from a model used to distil ammonia. On that task job, they were supported by a glass specialist – a German technician who lived in the city of Vinhedo (SP), near Campinas -, who performed the necessary changes.

Therefore, after the manufacturing of several prototypes, a device was developed allowed distilled "wine" in three minutes and had to be attached to the density meter. This set of equipments allowed the plants to realize the significant quantities of alcohol losses in the residues (vinasse and flegmass), which went unnoticed using the older technology.

In order to get a better view of the importance of this methodology, it's necessary to remember that the Santa Elisa Plant, at that time, had a 3% to 4% loss of the produced ethanol in the stillage (vinasse) or even more. The distillery producers' catalogues, on the other hand, considered that a 4%-loss was something normal. Here, the analytical step was modified and also the way of expressing the yield loss in the vinasse, in order to demonstrate the importance to the ethanol production yield.

Increasing the Efficiency

When the Brazilian ethanol production started growing significantly, the same happened to the fermentation losses and some of the entrepreneurs realized that it was necessary to take some actions to minimize them. Some changes were accomplished over time, in order to reduce the production process bottlenecks, which impair the efficiency and consequently the productivity.

Then, in terms of fermentation yield, the situation improved substantially in the 1970s and varied from 75% to 80%; nowadays, it turned from 90% to 92%. The bacterial contamination level also decreased considerably, because of investment in asepsis procedures within the factory. At the same time, the process became faster than before. At the beginning it lasted 25 hours; nowadays it takes from 6 to 12 hours, depending on the alcoholic content at the end of the fermentation.

For more than 35 years, during the sugarcane season (around 180 days), usually between April and November, Professor Amorim systematically visited his customers at the plants located in the State of São Paulo, State of Paraná, State of Mato Grosso do Sul, State of Minas Gerais and State of Goiás, personally verifying the fermentation process nonconformities and suggesting immediate changes.

Also in 1982, Professor Henrique Amorim started working with sugar, besides alcohol, estimating the industrial losses and trying to solve the problems in order to cut down costs and increase the company's profitability. When there were no computers yet, he used to write the technical reports by hand, illustrating with photos taken with a Polaroid camera. Subsequently, with the information technology advances, Professor Amorim started monitoring the process and solving them in real time.

Therefore, Fermentec, succeeded in gaining a prominent place on the national scene. Nowadays, it advises around 70 plants from the sugar and ethanol industry in various regions of Brazil, which produces approximately 30-40% of the Brazilian sugar and ethanol. These companies crushes approximately 180 million tons of sugarcane every year. As an example, around 600 million tons of sugarcane is processed in Brazil each season (2013 data). Fermentec also operates abroad in several countries, i.e., United States, Canada, Austria, Mexico, El Salvador, Argentina, Peru, Equator, Guatemala, among others.

The success of Fermentec required construction of expanded facilities in 2006. The new expanded 2,500 m² state headquarters were constructed in the city of Piracicaba, State of São Paulo, with five state-of-the-art laboratories, classrooms and library. The Company's technology/business team includes 54 employees with 15 employees assigned to the research operation (2013 data). Fermentec promotes two annual technical meetings, training and continuing education and continuing education programs at the Fermentec Corporate Facility along with remote training and continuing education programs at the facilities of our clients. The education programs focus on extending and updating the professional's knowledge of the fermentation area.

The technology transfer work performed by Fermentec allows our customers to achieve high efficiency to the industrial process the sugar, ethanol and beverage quality improvement, produced in Brazil and abroad along with quality improvement and reduction of the production costs for.

Exporting Technology

In order to disseminate knowhow and conquer new markets abroad for his company, Professor Henrique Amorim participated in several international events during the 1980s. In one of them, placed in Guadalajara, Mexico, promoted by the United States Distillers Association, which gathers maize ethanol producers, he gave a lecture describing the extraordinary evolution of the Brazilian alcoholic fermentation.

At that time, directors from two José Cuervo distilleries, who were attending to the lecture, challenged him about the possibility of incorporating new technologies to the tequila production to improve its quality. They invited him to visit one of José Cuervo plants, in Los Altos.

That was the beginning of a successful partnership between Fermentec and José Cuervo that lasted more than twenty years and resulted in a huge technological breakthrough. In that period, the company achieved more efficiency to the production process and more uniformity to the beverage quality.

According to Professor Amorim, the purpose was not to teach the Mexicans how they should make tequila, because they had already been making it very well since 1795, but the accumulated knowhow in alcoholic fermentation in Brazil could improve the productivity in whatever beverage industry, especially in the distillate industry.

This work caught the attention of foreign producers and many of them requested Fermentec Consultancy and research, as the Casa Bacardi, in Puerto Rico, and the Cazadores distillery, in Mexico, to solve problems relating to the fermentation and also to improve the quality and standardization of their products.

To observe the nature is to discover the true meaning of life. On trying to understand what goes on around us and understand our connection with the whole, we can get a better basis to innovate.

Yeast Genetic Improvement

Until the end of the 1970s, the Brazilian distilleries employed the baker's yeast to convert sugar into alcohol, with renowned trademarks, as Fleischmann or Itaiquara. ESALQ-USP, on the other hand, produced and marketed another kind of yeast imported from Germany, the IZ-904, also used by the industry. There is a story about how this yeast was introduced in Brazil: in the 1940s, Professor Jaime da Rocha Almeida, head of the Alcohol Technology Department, during his visit to a German distillery, threw a handkerchief inside the fermenter to collect a sample of that special yeast, which started being cultivated in Brazil later. This story was never confirmed, but the truth is that there was no methodology to measure its fermentation performance until that time.

When he was a student, Henrique Amorim admired the memorable zoology classes of Emeritus Professor Salvador de Toledo Piza Jr. (1898-1988). The professor, which had a soft and slow speech, but very convincing, always spoke to an attentive and crowded classroom, defended the hypothesis that the gene expression acted as a whole chromosome and not as separate parts, working with the observation of nature.

This statement caused an intensive debate that lasted several years including other professors from the ESALQ-USP Genetics Department, the brilliant scientists Friedrich Gustav Brieger and Ernesto Paterniani, who disagreed with Professor Piza. At that time, Professor Piza was not able to prove how the DNA acted; and his controversy with the other professors never came to an end. The scientific investigation, however, led to the discovery of the DNA structure, called double helix, by James Watson and Francis Crick, in 1953, in England, which clarified partially the subject.

At that time many scientists believed that the proteins, due to their complexity, were responsible for carrying the genetic code. The biochemists were the ones that realized the importance of the DNA to the protein synthesis, which are responsible for transferring, to the cells, the information about what they should be and do.

When Professor Amorim started studying the selection of more efficient yeasts to the process of alcoholic fermentation, he remembered that Professor Piza taught, many times, that the answer to the scientific enquiries could be solved by carefully observation of the natural process, followed by an adequate cataloguing and interpretation of the information achieved.

Therefore, Professor Henrique teamed up to the Professor Flávio César Almeida Tavares, from the ESALQ-USP Genetics Department, aiming to select a more appropriated yeast to the sugarcane juice and molasses. Thus, he provided to Professor Flávio all the information about the best qualities that should be incorporated on yeasts to make them resist to the fermentation process: it cannot produce foam, either flocculation, inside the fermenters, and should remain unchanged during various fermentation cycles. At the beginning, the yeast genetic improvement was obtained by the classical method of hybridization. In the laboratory, the selected yeasts revealed good yields.

The research project was performed with government agencies funding, as Funding for Studies and Projects (FINEP) and the São Paulo Research Foundation (FAPESP). Professor Flávio Tavares performed more than 400 crossbreeding and laboratory experiments, with molasses samples and alcohol contents from 7% to 8% in order to select the best yeast. After several tests, it was chosen the one that seemed to be more suitable, called "M-300A", but it was well known as "TA-79", since it carries the initials of both scientists: Tavares and Amorim.

Simultaneously, a Master's dissertation was done at ESALQ-USP, to follow the performance of that yeast during the fermentation process and to verify its survival after various fermentation cycles. In face of the positive results, the "TA-79" was provided for more than ten years to many plants and accounted for 30% of the national alcohol production, according to the "Revista Brasileira de Tecnologia" (1988, issue 19, p. 19-22).

For the period of 1985 to 1990, the researchers unsuccessfully searched for a suitable methodology to distinguish the different strains of Saccharomyces cerevisiae, which became possible with the nuclear DNA karyotyping technique.

It was discovered that both the baker's yeast and the "TA-79" did not remain during the fermentation in the industry. At that time, it was an important finding to researchers. Nowadays, the baker's yeast is still used by many distilleries to start up the process, but, generally, they are introduced together with selected yeasts, which are more efficient, stable and capable of dominating the fermentation environment.

Generally speaking, the work of genetic improvement research didn't evolve for the most part as expected, The yeast strains used by the sugarcane fermentation industry now days in the State of São Paulo didn't come from laboratory genetic modifications, but, actually, from a process driven selection process performed inside our client's plants.

We can learn from our mistakes and, for that reason, we persistently use research to best know the truth.

Karyotyping

The use of nuclear DNA karyotyping techniques on Brazilian industrial yeasts started in 1989. During a congress in France, Professor Amorim took notice of a work performed by Dr. Pierre Barre, who was searching for a way for identifying yeast from the genus Saccharomyces, in wine, through a DNA isolation technique. This methodology that was also still in development in the United States, had already aroused interests: the Brazilian researchers were monitoring its improvement because they noticed that it would be very useful to the fermentation process.

Dr. Pierre Barre invited the Professor Amorim team to better know his work performed with the alcoholic fermentation for wine production, in his Montpellier laboratory, in France. Professor Luiz Carlos Basso from ESALQ-USP, who worked as a Fermentec partner on through an agreement formed with the "Fundação de Estudos Agrários Luiz de Queiroz – FEALQ", went to France to learn the technique.

In France, Professor Basso was received by the researcher Françoise Vezinhet and, during fifteen days, he familiarized himself to the new methodology that analyzed the chromosomal DNA from the yeast Saccharomyces cerevisiae, through the pulsed-field electrophoresis technique.

Shortly after, the equipment necessary to perform karyotyping in Brazil was imported from the United States. Subsequently, it was possible to start selecting more efficient yeasts for the perfect development of fermentation process and discarding those which did not remain in the process.

The first evaluation tests of the yeast "M-300A", through karyotyping, were done for the Jardest Plant, located in the region of Ribeirão Preto (SP), verifying that this yeast disappeared in some weeks. The yeast which dominates the fermentation process and persisted during various cycles was one considered wild yeast. Professor Amorim's team concluded that this phenomenon also occurred in other plants and he decided to isolate these yeasts in the laboratory to use them in the next season.

Therefore, the yeast "M-300A" was definitely abandoned by industry at the beginning of 1990s and, in its place, more efficient yeasts appeared. During ten years, Professor Amorim's team and the researchers from ESALQ-USP analyzed approximately 400 yeast strains considered wild yeasts", but only two – the ones called "Pedra-2" (isolated from the Pedra Plant) and the "VR-1" (isolated from the Vale do Rosário Plant) – had proved to be efficient, demonstrating how difficult this work was. In 1998, another yeast was selected, the "CAT-1", from the Catanduva Plant.

Due to the great knowledge acquired about yeast behavior during fermentation, it was allowed to determine its permanence period inside the fermenter and its survival capacity, in the case of process interruption at the plant, optimizing the industrial process. These information about the performance were essentials in order to fit the fermentation with all the sugarcane crushing process.

According to Professor Amorim, the karyotyping methodology was useful to identify many yeast strains and to select the best ones, but the researchers want to move beyond. Over time, it was verified that even the selected yeasts were being replaced by wild yeasts during the sugarcane season.

This change was noticed after analyzing the karyotyping results and the chromosomal variation on the yeast population. Thus, it was necessary to deeply investigate the reason for these changes, his behavior during the process, the must composition and, for all these reasons, it was essential to adopt more accurate and modern techniques.

The researchers observed, from the nuclear DNA karyotyping analysis, that yeast chromosomal rearrangements were frequent, as they observed different profiles; sometimes there were changes on physical features; sometimes the chromosomal profiles were modified.

In order to determine the chromosomal variations with accuracy, the Ph.D. thesis developed by the Fermentec Scientific Director and agronomist engineer Mário Lúcio Lopes, entitled "Chromosomal Polymorphism Study in S. cerevisiae (PE-2 strain) used in the Industrial Process of Ethanol Production" (2000), was very useful. For this work he analyzed the changes occurred on the PE-2 yeast strain, concluding, theoretically, what Professor

Amorim team had observed in the industry: yeasts suffered from changes in some chromosomes, which could modify its features, impairing the strains identification by the karyotyping analysis.

Nothing is handed to you on a silver platter; the success is a result of hard work, persistency and learning.

Mitochondrial DNA Analysis

By the year 2005, Dr. Mário Lúcio Lopes and Professor Basso knew a technique developed by the researcher Amparo Querol Simón, from the Agrochemical and Food Technology Institute of Valencia (Spain), to identify the main features of several yeast strains in a simple way: instead of analyzing the nuclear DNA, her team worked with the mitochondrial DNA. This genetic material was found inside the mitochondrion, cytoplasmic organelle within all yeast cells,

Unlike chromosomes, the mitochondrial DNA does not follow the Mendelian inheritance laws as it's known in classical genetics, in other words it's not subject to the mechanisms of variability which occur in the chromosomes, as is the case of the genetic recombination and chromosome segregation. For this reason, the mitochondrial DNA remains more conserved than the nuclear DNA.

The mitochondrial DNA methodology was adapted to the Brazilian conditions of alcoholic fermentation and started to be used as a complement to the karyotyping, from 2005. Therefore, it was possible to identify the various yeast strains and to determine the degree of relatedness among them. With this new technique, discovery was made, for example, why the PE-2 strain was the most prominent seller in Brazil: the strain carries significant phenotypic plasticity when compared with other yeast strains, which suffers genetic instability during adoption to the fermentation conditions.

With the mitochondrial DNA analysis, it was possible to discover when selected yeasts undergo genetic modification as distinguished from the contaminant yeasts. If only the nuclear DNA technique was used, it wouldn't be possible to determine which yeasts originated from a selected strain and which ones originated from a contaminant strain.

Besides the industrial yeasts, this technology is normally used to investigate and trace the genetic origin of yeasts, plants, animals or even humans. Recently, this method was chosen, for example, to identify King Richard III's bones, who governed England from 1483 to 1485. His remains were missing since the XVI century, when the church where they was placed was demolished, and they were found in 2012 in a vehicle parking lot. Researchers from the University of Leicester performed the mitochondrial DNA tests and confirmed the similarity between the genetic materials from the bones with the king's living descendants, achieving the most important archaeological discovery for the last decades in England.

Genetically Modified Yeasts

The employment of genetic engineering to the alcoholic fermentation started in 1970. In this area, the great challenge to the researchers is to build more efficient yeasts, with new genetic features.

During the 1980s, Professor Amorim performed a work together with the biologist Ana Clara Schenberg, from the Biomedical Sciences Institute at the University of São Paulo, and with the chemist Elisabete Vicente, from the Microbiology Department at the University of São Paulo, aiming to promote genetic transformations to improve the PE-2 strain, concerning to the bacterial contamination. The modified yeas strain, however, was not introduced in the industry because of genetic instability and loss of some of its favorable fermentation features.

Another work in genetic engineering which resulted in an industrial patent was performed with the biologist Boris Juan Carlos Ugarte Stambuck's team, from the Yeast Molecular Biology and Biotechnology Laboratory at the Federal University of Santa Catarina. Professor Henrique Amorim encouraged him to go to Stanford University, in 2005, to master the DNA sequencing technique for yeasts. Stambuck discovered that yeasts which dominated and persisted in the process held two genes related to the vitamin production: biotin and pyridoxine.

The scientific articles published about this research were "Whole-genome sequencing of the efficient industrial fuel-ethanol fermentative Sacharomyces cerevisiae strain CAT-1" (2012), whose authors are F. Babrzadeh; R. Jalili; C. Wang; S. Shokralla; S. Pierce; A. Robinson-Mosher; P. Nyren; R. W. Shafer; L. C. Basso; H. V. Amorim; A. J. Oliveira; R. W. Davis; M. Ronaghi; B. Gharizadeh; B. U. Stambuk; and "Industrial fuel ethanol yeasts contain adaptive copy number changes in genes involved in vitamin B1 and B6 biosynthesis" (2009), signed by B.U. Stambuck; B. Dunn; S.L. Alves Junior; E.H. Duval; G. Sherlock.

Together with this research team, professor Amorim also developed and patented a Saccharomyces genetic modification process by using genetic engineering, in order to change specific yeast chromosomal genes, related to the production of an enzyme called invertase that catalyzes the hydrolysis of sucrose.

The aim of the research, which lasted two years, was to induce the CAT-1 strain to absorb the entire sucrose molecule, resulting in two advantages: it would favor the yeast energy balance and would avoid the sucrose hydrolysis, whose products (glucose and fructose) stimulate the bacterial propagation and the fermentation process contamination. Nowadays, the efficiency of this microorganism is being tested in the industry.

By the year 2005, Professor Amorim researches related to yeast selection in the fermentation process caught the attention of Professor Johan Thevelein, who works in the Institute of Botany and Microbiology at Katholieke Universiteit, in Leuven, Belgium. Studying the subject, he just switched his research focus to industrial selected yeast to his studies, because, until that time, he only used in his laboratory yeast for ethanol production.

His work in partnership with Professor Amorim's team started with the Brazilian yeasts, isolated from the industrial fermentation process, and resulted in genetically modified strains development. These microorganisms had already been patented in Europe (CAT-1CD and PE-2CD) and they will be subjected to laboratory tests at Fermentec to verify if they still carry the good fermentative traits.

When a license to operate with these genetically modified organisms is approved in Brazil, these yeasts, also will be patented in the country, will be produced in large scale in order to be marketed for plants and distilleries.

Technology Transfer

One of the achievements that Professor Amorim considers to be the most relevant is the significant contribution of his company, Fermentec, to the increment of industrial yields in the ethanol industry. This relevant work was scientifically proved through a research investigation developed by the student Marina Biagi Barros (final term paper), in the Department of Economy, Management and Sociology at ESALQ-USP, advised by Professor Lucílio R. A. Alves, in 2009.

This study sought to analyze the economic gains from the industrial efficiency over 32 years of Fermentec operation together with some of its customers. The results of the performed analyses were very expressive: for each R$ 1 applied, the return rate reached R$ 240 (2009 data). In other words: the increase in profit after introduction of technological innovations in the 33 analyzed plants was R$ 7.8 billion in 2009. If we consider these values corrected to the year 2013, they will represent R$ 9.7 billion or US$ 4.8 billion.

The study showed a positive correlation between the period which the companies were followed by the Fermentec consultant team and the distilleries overall return: the greater the time, the greater the yield obtained. During more than 30 years of activity, Fermentec transferred technology for more than 170 companies.

High Gravity Fermentation

In the past decade, an issue that puzzles Professor Amorim is the yeast viability in a high gravity fermentation. By the year 2005, he started the investigation about this subject using his own resources. A pilot project was set up at the Pedra Plant (Pedra Agroindustrial Group), in the city of Serrana, São Paulo, to evaluate its viability.

In 2010, Professor Amorim developed, together with Dedini S.A. Indústrias de Base, a traditional producer of equipment for sugar and ethanol industry based in Piracicaba, São Paulo, and an industrial scale system called Ecoferm. The aim was to produce ethanol in large volumes, with the same or higher yield, reducing significantly the amount of vinasse generated from the alcohol distillation.

Currently, the Brazilian plants operate with an alcohol content in the beer between 8% and 10%, generating from 10 to 12 liters of vinasse per liter of ethanol. Through the Ecoferm system, it's possible to perform the fermentation with 16% of alcohol content, without compromising the yield, generating only 5 liters of vinasse per liter of ethanol. It also provides a great reduction in the steam consumption during the distillation step.

The new system, created to operate with high alcohol content, is based on the fermentation process redefinition, performing it at a lower and constant temperature. Coupled with a new project of fermentation tanks and connections without dead ends, which avoid the biofilm formation. Moreover, it reduces the bacterial contamination, restricting the use of antibiotics, acids and anti-foams.

This technology, already available for industrial use, only became viable with the selection of suitable yeasts by Fermentec. Dedini Company developed a temperature reduction system that allows the energy optimization to take advantage of thermal currents.

On the other hand, Fermentec is responsible for the fermentation, defining the system operational parameters to reach defined goals, providing yeasts and also continuing the technological development. The use of the Ecoferm system allows the water consumption reduction by half in the distillery, but Professor Amorim believes that is possible to achieve further reduction. At the moment, his team is performing tests toward this goal.

For this investigation, Fermentec has ongoing research collaboration with the Federal University of Santa Catarina, from ESALQ-USP and from the Institute of Chemistry, University of São Paulo. Fermentec also received the support of the National Council of Scientific and Technological Development (CNPq) and of the São Paulo Research Foundation (FAPESP), through the Innovative Research in Small Companies (PIPE) program, which provides funding, as non-refundable grants, to companies.

"Search and You Will Find…"

Customized Yeasts

The use of modern techniques to identify dominant yeasts in the industrial process allows Fermentec to develop customized yeasts for its customers, which are adapted to the specific process of each plant. The work starts with the selection of the dominant yeasts during a sugarcane season; after its propagation in the laboratory, its efficiency is measured. The yeasts considered to be more efficient are reintroduced in the fermentation operation in the next seasons.

After performing several tests, Professor Amorim's team achieved six cases of success: two of its customers have been able to continue use of the same yeast strain for seven years; one customer for three years and another customers for two years.

At the beginning of 2013, Fermentec was successful in performing the patient work of identifying and measuring the efficiency to select suitable yeast strains in twenty plants. Besides the Brazilian specialists, researchers from Italy, France and Spain are trying to perform similar researches in this field for wine.

According to Professor Amorim, this work of prospecting more appropriate yeas strains to remain in the fermentation process is of great importance, considering the production of second generation ethanol, which uses cellulose and the sugarcane residues as raw materials. In his opinion, the possibilities of finding more efficient new yeast strains are very challenging.

Extending the Range of Action

By the year 2008, Professor Henrique thought about decreasing his accelerated pace of work, reducing his trips and sharing the management job at Fermentec with his son Henrique Berbert de Amorim Neto, who was preparing himself, since the student times, to take his place at the company leadership, together with his father. Henrique Neto, graduated in agronomic engineering at the University of Lavras (UFLA), in the State of Minas Gerais, in 2004, and he achieved the Master of Science degree at the University of Abertay, in Dundee, Scotland (2005). His work entitled "Evaluation of a Brazilian Fuel Alcohol yeast strains for scotch whisky fermentation" has tested the efficiency of yeasts selected by Fermentec for whisky production. Henrique Neto, also has trained in several plants and distilleries and at ESALQ-USP, acquiring knowledge and experience to become the Chief

Operating Officer of Fermentec, his present occupation at the company.

At the same time, Professor Henrique created a committee-like organization to provide orientation and assist him with the business management and with the decision-making processes, composed of some of the Fermentec partners: Claudemir Bernardino, Alexandre Godoy, Mário Lúcio Lopes, Luiz Francisco Ferreira da Silva and his son Henrique Neto.

Recently, Fermentec extended its range of services offered to its customers and, with more hiring and internal staff rearrangements, the company is capable of developing, from the industrial process idea to the elaboration of an engineering project intended to the enterprise implementation. It's a new work direction, which is requiring all the attention of Professor Henrique Amorim, who is already accustomed to overcome challenges. After all, his work is motivated by these challenges and they conserve his spreading enthusiasm.

Em 1965, ainda estudante, Henrique Amorim apresenta
trabalho científico sobre cacau em Itabuna (Bahia) no
XVI Congresso da Sociedade Botânica do Brasil.

2. Henrique Amorim next to his advisor and master,
Professor Eurípedes Malavolta (1926-2008), when both
were honored by the Agronomist Engineer Association
of the State of São Paulo, in 2000, in Campinas (SP).

Henrique Amorim Neto, ao lado do pai, também Henrique, que dividem o comando da empresa *Fermentec,* criada na década de 1970, em Piracicaba

3. Henrique Berbert de Amorim Neto next to his father Henrique, both in charge of Fermentec, created in the 1970s, in Piracicaba

No primeiro ano em que lecionou na Esalq-USP,
o prof. Henrique Amorim agradece, em nome de
todo o corpo docente, a homenagem recebida dos
alunos pelo trabalho didático desenvolvido

4. In his first year as a Professor at ESALQ-USP, Henrique
Amorim thanks the honor received, on behalf of the entire
faculty, from the students for his educational work developed.

O prof. Henrique Amorim e sua esposa Vera por ocasião
da apresentação de trabalho da autoria dos dois, sobre a
qualidade do café em Hamburgo, Alemanha, em 1975

5. Professor Henrique Amorim and his wife Vera Lúcia
on the occasion of a scientific paper presentation
about coffee quality, which work was performed
by the couple, in 1975, in Hamburg, Germany

Acknowledgement

The author thanks Mrs Regina Machado Leão for the excellent development of this biography and acknowledges the Portuguese to English translation services provided by Mrs Bruna Buch during the preparation of the manuscript.

CHAPTER 3.

Genomics and Molecular Biology at CEBTEC

Helaine Carrer

The decade of 70's is recognized by the huge advances in the area of molecular biology. With the discovery of a new class of molecules able to cut DNA in specific locations, the possibility to ligate these fragments by specific enzymes and the creation of a recombinant plasmid able to generate multiple copies when transferred to bacteria, a new Era of biology and biotechnology started. Also, these accomplishments were accompanied by the development of a technique able to read the chemical bases of DNA, adenine (A), thymine (T), guanine (G) and cytosine (C) in 1977 by Dr. Frederick Sanger and Walter Gilbert working independently. In this same year a bacteriophage (virus) was the first organism to have the complete genome sequenced. Since then the goal has been the complete genome sequence of the entire living organisms. It was in 1988 when the human genome sequencing was launched by a Consortium formed by recognized laboratories in the United States and Europe. At that time, only few laboratories in developed countries were able to access this expensive and high technology methodology.

While technological advances were happening in the world not only in biology but also in computer sciences and engineering areas, in Brazil only very few scientists had such skills. Luckily, with the goal to improve human resources in strategic areas of knowledge in science and technology in the country, a new program called PADCT supported by CNPq was created in middle 80's. Also some equivalent programs were established in state granting agencies financing research such as FAPESP in Sao Paulo. At that time I was among the selected students to receive a PADCT fellowship to study one year in the United States. It happened because of my visionary, enthusiastic and intelligent adviser, Prof. Otto Jesu Crocomo. I started working with Dr. Otto when I was an undergraduate student in 1982. It was during a plant physiology class when Prof. Paulo Roberto de Camargo e Castro presented the technique of plant tissue culture as a relevant tool for studies on plant physiology and biochemistry as well as a powerful tool to propagate selected plants as clones. During the class Prof. Paulo Castro showed wonderful slides (photos) of *in vitro* plants mostly ornamental that made my eyes sparkling and instigate me to know more about this technology. Dr. Paulo Castro suggested I should talk to Dr. Otto at the biochemistry Sector at the former Chemistry Department, which now is the biochemistry area of the Biological Sciences Dept. with Botany and Zoology.

I was very concerned to talk to Dr. Otto Crocomo who was the chair of the Department and the coordinator of the biochemistry and biotechnology Laboratory but I got courage to follow my dream and finally I went to his office. Although he was very busy organizing the International Genetic Engineering Symposium he spent some time talking to me and I found him to be the most enthusiastic and driven person that I had ever met at ESALQ. He accepted me to his laboratory and to attend the Genetic Engineering Symposium.

The Plant Tissue Culture Laboratory was located at The Center for Nuclear Energy in Agriculture (CENA), it was a pioneer lab in the field of Plant tissue culture in Brazil. It was established in 1971 by William Rod Sharp and Otto J. Crocomo, from The Ohio State University and ESALQ. Several students were trained in this lab and became recognized scientists in different universities and Institutes of research in Brazil and abroad. At the lab, I was introduced to Enio Tiago de Oliveira who taught me most of my lab skills. I learned how to prepare culture medium, proper washing of glassware and transfer of the *in vitro* plants to fresh medium. After a few months I was able to have my own project studying the effects of gamma radiation on soybean and bean embryo germination *in vitro*. The laboratory training was part of my practical requirements for the course entitled: "Introduction to Nuclear Energy in Agriculture Course (CIENA)". I was taking the course parallel to my undergraduate courses. Later I started another project on papaya plants regeneration in tissue culture, which became the subject of my M.Sc. thesis. The goal was to regenerate plants through *in vitro* somatic embryogenesis of papaya, a huge challenge for most of crops at that time, especially to papaya. Embryogenesis was dependent on development of customized culture media and growth regulators. Considering papaya is an important commercial fruit, a source of papain, an important secondary product for industry uses but prone to several diseases, we considered it was an interesting project for my M.Sc. research program. Dr. Otto made all his efforts to obtain my fellowship from the PADCT/CNPq Program to spend a year working with Dr. Phillip Ammirato at DNA Plant Technology Corporation (DNAP). Dr. Ammirato at the time had joint research appointments at DNAP and Barnard College, Columbia University, NY in the United States. Dr. Ammirato knew about our lab through his research collaboration with William Rod Sharp.

I have very nice memories of happy moments with colleagues such as Marina Murayama, Leonardo Alves Carneiro, Albenisio Silveira, Irenice Vieira, Jose Barbosa Cabral partner's graduate students at Crocomo's Laboratory. Also there were many good friends from the labs around ours, Maria Helena Goldman, Marli Fiore and Beatriz Mendes. Most of them became professors and we continue interactions.

Just after the defense of my master thesis and appointment to the position of assistant professor at ESALQ I spent one year with Dr. Ammirato. The training was not at Barnard College but at the DNA Plant Technology Corp. Company in New Jersey where Dr. Ammirato was spending a one year sabbatical and I was also kindly accepted to stay at the company where Dr. Rod Sharp was the Scientific Director. The experience of being in a private company was interesting and fruitful. There I met several outstanding researchers. The labs were very active on plant tissue culture research and there was a group working on molecular markers for genetic characterization of genotypes for food quality.

When I arrived at DNAP I met Dr. Maro Sondahl, DNAP scientific director of tropical plants, Clemencia Noriega, a scientist from Colombia, Claudia Bellato, a Brazilian researcher working at DNAP, Dr. Laudenir Prioli (Lau), a visitor researcher from UNICAMP and João Batista Teixeira from EMBRAPA-Cenargen. All were very helpful during my arrival and living experience in the New Jersey. Lau in special, kindly received me in her house until I could find a place to live but she invited me to stay with her and her daughter Juliana. Afterwards, I was admitted to the Ph.D. Program at Rutgers University, in New Brunswick, NJ in August 1989. Actually, Lau played an important role in encouraging me to apply for the Ph.D. Program at Rutgers. She introduced me to Prof. Pal Maliga who became my PhD adviser at the Waksman Institute.

I remember when Lau took me to the Waksman Institute to talk to Prof. Pal Maliga. I was very impressed and thought it would be very nice studying there. I went to Maliga's laboratory and he nicely explained his projects of research and my thoughts took a different direction. I felt I could never survive a PhD in this new subject of molecular biology because I had never learned about plasmids, transposons and the gene gun. When I said Lau how scared I was she made me understand the importance of the opportunity for enhancement of my scientific career. She emphasized that only a few students had the opportunity to study in one of the best Institutes of research in the United States with a very prestigious and intelligent mentor. At night, talking to my pillow I thought about a sentence I read once and always come to me "we only grow when we accept a challenge" and decided to follow up this opportunity, I decided to dedicate myself to learn the new and mysterious area of plant molecular biology. Also, talking to Dr. Otto, he reinforced mostly of Lau's words and requested the Chemistry Department to extend the fellowship to include a Ph.D. degree. I can say that I am very grateful to Otto Crocomo, Laudenir, Rod Sharp, Philip Ammirato and everyone who put me in this direction. Pal Maliga and his wife Zora Svab were helpful and they are the greatest teachers and supervisors I could have had at that moment. Also they quickly became special friends which continues today. They gave me the chance to learn a very exciting and challenging technology of genetic modification of chloroplasts in higher plants, in my opinion one of the most fascinating approaches in plant genetics and biotechnology.

During the time that I was in the United States, CEBTEC-Center of Biotechnology in Agriculture – now named for Prof. Otto Jesu Crocomo in recognition of his role as founder, was inaugurated in October, 1988. When I returned in March 1994 CEBTEC was a very active lab with several students and I joined Luiz Antonio Gallo and Murilo Melo already settled as faculty members. It was like a dream that the building idealized by Dr. Otto in 1981 was a reality. Although CEBTEC had a well-established infra-structure for plant tissue culture, it was not the case for molecular biology research. There was a need for acquire specialized equipment and supplies. After my arrival I submitted a grant proposal to FAPESP but it was rejected because I was considered a young scientist and the reviewer suggested I join a senior researcher leading the project to obtain experience. Nowadays, FAPESP has a contrary view, with incentives for young investigators, but at that time it was not possible. Feeling frustrated, I received an offer to a post-doc position with Prof. Michael Lawton at the AgBiotech Center, Cook College at Rutgers University. I spent a year studying the activity of a transmembrane protein involved in the response of plant pathogen infection. The project provided important research experience in protein expression and plant pathogen interaction that contributed to improve my knowledge in plant pathology, an important area of research at ESALQ. This year away was very important to my maturity as an independent investigator and provided me with the confidence to return and struggle for resources to start my own line of research at CEBTEC and by middle of 1997 I was back in Brazil.

The year of 1997 was remarkable for at least two essential scientific events that introduced molecular biology in CEBTEC and ESALQ. One was the announcement by FAPESP about the genome project to initiate genomics in selected labs and the other, was an event occurred at ESALQ organized by Prof. Raul Machado Neto, vice dean of ESALQ and William Rod Sharp, dean of research at Cook College, Rutgers University. They organized the first Workshop in Biotechnology with the presence of more than 30 research faculty members and deans from Rutgers and sister institutions at ESALQ. This workshop has been continued until now. In 2003, The Ohio State University joined this initiative and already hosted two of the workshops in Columbus, OH. These workshops happen every other year alternating the locations, USP, Rutgers and OSU. This initiative fostered important collaboration among researchers inside USP and internationally resulting in the creation of the First International Tripartite Graduate Program in Plant Cell and Molecular Biology in 2009.

An announcement was made by FAPESP in October 1997 about the Genome Project to sequence the whole genome of the first plant pathogen bacteria, *Luella fastidiosa*, responsible for Citrus Variegated Chlorosis (CVC) disease in sweet orange in Brazil made all scientific community willing to participate. I submitted a proposal with the collaboration of two ESALQ colleagues, Marcio Rodrigues Lambais from The Soil Sciences Dept. and Weber Antonio Neves do Amaral from the Forest Sciences Dept. It was one of the greatest moments of my life when I knew we were selected among more than a hundred applicants to be part of ONSA (Organization for Nucleotides Sequencing and Analysis).

At that time such a project was a huge challenge, especially because not only did the laboratories require infrastructure updates and specialized equipment however most of the participants did not have the skills in Automated DNA Sequencing and assembly of a whole genome using bioinformatics tools. Although, the bacterial genome is minuscule in comparison to the human genome that was under way at that time was a challenge because only large laboratories in The United States, Europe and Japan dominated genomics technology. All the chosen labs accepted to be part of ONSA Network and followed our goal lead by Andrew Simpson (Ludwig Institute), Fernando Reinach (former Chemistry Institute, USP) and Paulo Arruda (UNICAMP). FAPESP by the Scientific Director Jose Perez provided us with all the essential supplies and specialized equipment required in the short term. At the beginning of 1998 we started preparing the shotgun libraries and DNA sequencing. Shortly after the project started, Weber Amaral returned to Harvard to accomplish his PhD. Marcio Lambais was a great partner supporting our group learning tools of bioinformatics. He was in charge of preparing the sequences to be submitted to the central lab at UNICAMP. Also as a microbiologist Márcio contributed on the gene annotation and we worked as a team. My lab was in charge of preparing the DNA shotgun libraries and DNA sequencing. At this moment a very competent scientist, Dirce Maria Carraro, joined our group as postdoc and really made a difference. Dirce was very organized (maybe because part of her doctor degree was in Germany) and had deep experience in molecular biology. She was a very driven person (and still is as a senior scientist at the A.C. Camargo- Cancer Hospital) with great ways to motivate everybody to work really hard. Also our group counted with Valentina de Fatima D'Martin and Enio Tiago de Oliveira, both very dedicated and competent technicians. I am very proud to say that we were one of the most successful groups of ONSA.

The project was concluded in 1999 when the 2, 7 million bp was assembled and published in the respectful scientific journal – Nature - as the first plant pathogen completely sequenced. More than 200 participants received recognition from the scientific community in Sao Paulo, Brazil and outside. The Governor Mario Covas (*in memorian*) and Ronaldo Hardenberg, Minister of Science and Technology at that time, recognized the participants with a Medal and to the head of each group also with a sculpture prize "*The Tree of Life*" that at this moment I would like to dedicate this prize to Prof. Otto Jesu Crocomo who was the main responsible for me to be there with that prize on my hands. He gave me a lot of incentive to participate in this program and the confidence that we were going to succeed. Thank you!

The importance of the Genome project besides providing Brazil membership in the select fraternity of countries working on genomics, I consider that the project provided a model for transfer of knowledge among participants exploring a new field of science which proved to be both expedient and efficient in the improvement of the team member's scientific skills. The connections formed were unique. Most of the undergraduate and graduate students participating in the program at that time are now professors and scientists in different Institutions not only in Brazil but also internationally.

There was a significant escalation of research activity in my laboratory after Luella, we subsequently participated in the SUCEST to sequence ESTs of sugarcane, the human cancer genome project, the Agronomical and Environmental Genome Program when we sequenced complete genome of *Leifsonia xyli*, a bacteria that causes ratoon stunting disease in sugarcane, Leptospira, bovine ESTs and I was the main coordinator of the FORESTs Project to sequence ESTs of Eucalyptus. In this last project we had the participation of four paper companies, Duratex, Votorantin, Suzano and Ripasa. It was an interesting format and we succeeded in development of a collaborative approach in the accommodation of competitive companies with the one goal of producing the transcriptome of Eucalyptus. In this project prof. Carlos Alberto Labate, Luis Lehman Coutinho from ESALQ and Prof. Celso Marino from UNESP-Botucatu were skilled partners and key people to contributing to the success of the project. For almost 10 years my group has been involved in genome sequencing projects and some branched projects related to functional genomics in citrus pathogen interaction. These projects resulted in a large number of publications and the conclusion of 10 M.Sc. and 04 Ph.D. dissertations.

Nowadays, at CEBTEC we have one more member, Prof. Daniel Scherer Moura, who joined us in 2009 after an opening for a position in the biochemistry area to support the Biological Sciences Course established in 2005. Prof. Daniel works with peptides with hormones function in plants. It is nice having a competent and enthusiastic colleague with us introducing new expertise and preparing students to science.

Actually, my laboratory is working very actively as part of the BIOEN Program supported by FAPESP and in the INCT for Bioethanol supported by CNPq/FAPESP. These are challenging projects with hundreds of participants to committed to research for improvement of energy crops with special emphasis on sugarcane. My project is related to the functional genomics of photosynthetic genes in sugarcane. I have to say that the science of photosynthesis has always attracted my attention and interest. Not only because my Ph.D. research was on chloroplast genetics and molecular biology but also I remember back when I was a M.Sc. student I attended a seminar given by Carlos Alberto Labate about the origin of chloroplasts in higher plants just after he returned from Canada. It is a fascinating area to study. In my BIOEN Project, one of the objectives is the establishment of a protocol for genetic transformation of sugarcane and CEBTEC became a reference lab on this technology. How? Because of the great skills on plant tissue culture of sugarcane by Enio Tiago de Oliveira acquired by working with Prof. Otto Crocomo in Marina Murayama's thesis and together we trained graduate and undergraduate students on this technology. We successfully achieved high efficiency of sugarcane genetic transformation by both, biolistic and *Agrobacterium* approaches. The BIOEN and INCT Programs also created the opportunity to establish scientific collaboration with several colleagues from USP, Glaucia Mendes Souza from Chemistry Institute, Marcos Silveira Buckeridge, Marie Anne Van Sluys from Biosciences Institute, from UNICAP, Marcelo Menossi, Anete Pereira, from UNESP-Botucatu, Fabio Tebaldi Nogueira Silveira, and so many others.

All these highly successful programs for developing human resources during the last decade made possible the creation of the International Tripartite Graduate Program – University of Sao Paulo, Rutgers University and The Ohio State University under the oversight of Raul Machado Neto and William Rod Sharp (RU and OSU) two visionary men with devoted hearts to the improvement of science beyond the borders of countries. This program is a reality with ten registered students registered and with participation of several professors from different institutes at USP and similar participation at the North American universities.

About all of what I wrote above related to myself and to CEBTEC, were possible because of the connections of so many committed people and, in special by the network created by Otto Crocomo, William Rod Sharp and Raul Machado Neto, three leaders and visionary men of science.

CHAPTER 4.

Science without Borders

Roy Chaleff

What follows is an admittedly imperfect recollection of events of which I retained no records and that occurred in what now seems a distant place long ago. It was 1987. I had just left DuPont to direct the fledgling plant biotechnology program at American Cyanamid. The program was born of excitement over promising results obtained with herbicide tolerant corn mutants generated by tissue culture through an external project sponsored by the Company. It was initiated by visionaries despite strong opposition from marketing executives who didn't believe in the product, hadn't the patience for the long time frame required for its commercialization, and feared entanglement in the seed business, an alien industry of which they had no knowledge or understanding and in which they had no interest. Traditionalists were simply unable to transition from chemistry-based to biology-based products. This resistance was reinforced by scientists who contributed to the program initially, but resented recruitment of an outsider to a leadership position that they deemed rightfully their own. They formed a powerful fifth column that strove to destroy that which they could not control - to drown the infant taken from them at birth. And so the program, under attack both from within and without, was reluctantly given a small budget to explore the dubious potential of biotechnology for generating profits for the agrochemical industry with hostile onlookers in many quarters hoping to see it fail to vindicate their opposition. Thus, the plant biotechnology program itself and efforts to cooperate with allied programs within the Agricultural Division were rent by petty jealousies and rivalries, ignorance, enmity, and misunderstanding. In other words, it was a typical human endeavor. But I digress.

As well as being underfunded and understaffed (5 Ph.D.'s and perhaps 7 technicians), the Cyanamid plant biotechnology program was blocked from using prevailing genetic manipulation technologies that were patented by competitors with more extensive, better supported, and long established programs that dominated the field. Yet this puny orphan child was expected to prevail against those very Goliaths, such as Monsanto, DuPont, AstraZeneca, and Novartis.

The only way such a disadvantaged program could hope to succeed was through collaboration. Collaboration would not only amplify manpower and other resources, but provide a development, commercialization, and marketing partner with expertise, facilities, and technology otherwise not available. For example, by partnering with seed companies to introduce into crops a genetic trait such as herbicide tolerance, an agrochemical company would acquire access to proprietary commercially advanced germplasm, be able to perform more field trials under diverse environmental conditions, obtain genetic manipulation technologies (i.e., techniques for genetic transformation and plant regeneration for commercially important varieties), be able to combine the herbicide tolerance trait with other agronomically desirable traits, such as disease resistance or high yield, to increase market penetration, and have a means for producing and selling the modified seed. Probes were sent out and emissaries dispatched. But plant breeders in both the public (university) and private (company) sectors were skeptical. Our contention that herbicide tolerance would prove a desirable trait was greeted with suspicion. After all, who were we – an agrochemical company with transparent ulterior motives and no knowledge of the seed business – to advise them on how to improve and market their varieties? And why should they believe that our herbicides were truly superior to competitive products, as we claimed, and that they wouldn't in fact be disadvantaging themselves by conferring tolerance for a Cyanamid herbicide rather than for a competitive product? And how could we guarantee the security of their proprietary germplasm if it were released to us for experimentation? And so on. Added to these obstacles was Cyanamid's haughty stance that tolerance for its herbicides conferred such a singular commercial benefit that collaborators could not expect any other form of compensation for their efforts. Needless to say, negotiations were difficult. And interminable. There were innumerable meetings and presentations. Suspicions had to be allayed and performance of the herbicides convincingly demonstrated. But eventually partnerships were formed with several university and company breeding programs in North America, Europe, and Japan to introduce tolerance for Cyanamid herbicides into corn, rapeseed, cotton, rice, and sugarbeet. The joint programs varied in nature. In some cases, Cyanamid merely provided the resistance gene under the guise of its value as a genetic marker or research tool and hoped that the trait would find its way into a commercial variety. In other cases, a more comprehensive collaboration was developed with shared research responsibilities and clearly defined commercial goals and time frames.

As collaboration is the theme of this chapter, perhaps it would not be inappropriate at this point to interject a note on negotiating agreements. Clearly, negotiations must be approached with desired terms and objectives in mind. However, it is a mistake to propose terms that are one-sided and seek to take advantage of the other party. Regrettably, negotiation is sometimes viewed as an opportunity to demonstrate one's superiority and dominance rather than to develop a workable agreement. While that tactic may succeed on rare occasions where the other party is extremely desperate or stupid, such lopsided victories are only short term. Eventually the subordinate partner will realize and resent his position and, unless the agreement is renegotiated, will cease to cooperate. It is best to try to establish a long-term, amicable, and fruitful collaboration from the outset by negotiating with respect for the other party, displaying a willingness to strive for an outcome that benefits both, and exercising creativity and flexibility to achieve that outcome. For a partnership to succeed, both parties must feel that they are winning.

An extraordinary collaborative opportunity arose in a most unexpected place. It was the time of *Perestroika* in the Soviet Union. Wrenched by internal economic and political upheaval, the once impenetrable country was throwing itself open to West. I received a call from two young Soviet scientists, one from Moscow and the other from Kiev, who were traveling throughout the U.S. and wished to visit our laboratory. We obliged and, in discussions after their seminars, were told that they were seeking employment in the U.S. and/or support for their research programs in the USSR. As an internationally recognized plant biologist who was Director of the Institute for Cell Biology in Kiev and fluent in English, German, and Russian, the Ukrainian scientist, Andrei Rybin , had particularly impressive credentials. The Company decided to hire him (reporting to me), as it had been considering the possibility of expanding its business in the Soviet Union and saw value in Andrei's political connections and familiarity with that country's extensive and Byzantine administrative policies, regulations and procedures for foreign commercial enterprises. Thus, Andrei's position as Director of a Kiev research institute was considered an asset, rather than a conflict, and retention of that position (to his delight) was made one of the responsibilities of his Cyanamid employment.

The emphasis of our program at that time was making the world safe for Cyanamid herbicides by introducing tolerance for those products by whatever means possible (conventional breeding, seed mutation, or genetic engineering) into as many crops as possible. One crop that seemed a good target for this program was potato. You see, potato is a "secondary crop," in that it does not represent a large enough market to justify the expense of developing and registering a herbicide specifically for that crop alone. However, it does provide sufficient commercial enticement to extend to potato the range of application of an herbicide with a suitable weed control spectrum that was developed for use on a major crop, such as corn. But Cyanamid's program hadn't the knowledge, techniques, or experience to genetically engineer potato. Opportunely, Andrei came forward and offered his Institute in Kiev for the project. We presented a proposal to management that, like all innovative initiatives, such as tying one's shoelace, stirred intense debate within the Company. Some feared giving our proprietary resistance gene to "the Russians." Others were concerned about overpaying, as they heard reports of technology companies paying $10,000 per man-year for contract research in the Soviet Union. And so our proposal's already modest budget request was slashed to $35,000 per year and we were grudgingly given permission to proceed.

It was, after all, an inevitable development. A fundamental law of economics is that capital will flow to where it can realize the greatest return. For more than a century, this law had been brutally obeyed by manufacturing as industries relocated from Great Britain and the United States to Southeast Asia. Now research investment was abiding by the same principle.

Despite its many practical contributions, science, like the arts, is considered an extravagance affordable only by a prosperous society and is one of the first activities eliminated in periods of economic crisis. In the way that animals will eat their young, starving countries will dispense with anything not essential for survival. The Soviet Union was in the throes of such a crisis at this time and support for science was being sharply curtailed. Cyanamid's pathetically small grant was a lifeline for his Institute and Andrei seized it. And, because of the peculiar nature of the Soviet economy, the project was able to proceed on that meager budget. In a socialist country workers, including scientists, are never fired. No matter how desperate the situation and how devalued the currency, the state would continue to pay the scientists' salaries in rubles. Budget cuts were achieved by eliminating funding for supplies and utilities. The Cyanamid grant would be used to purchase equipment, chemical reagents, electricity, and coal.

And so, with Andrei as our guide and interpreter, the Agriculture Division's Research Vice President and I embarked on our first trip to Kiev. (A senior Company executive was required to impress our hosts with the depth of the Company's commitment and to avoid insulting the high level officials whom we hoped to meet.) Our purpose was to view the Institute's facilities, meet the scientists and congratulate them on being the recipients of Cyanamid's munificence, and obtain approval for the collaborative project from the regional authorities.

The Kiev airport was being renovated, modernized, and expanded for the anticipated increase in traffic from the West. Improvements were already evident. A sign in English directed handicapped travelers to the "disabled toilets."

We first went to see Andrei's former mentor Pavel Bakin , who, as Director of the Institute of Botany and Vice President of the Ukrainian Academy of Sciences, was very influential. Professor Bakin personally guided us through his Institute and proudly displayed the extraordinary herbarium, which housed a vast collection of pressed specimens dating from the 17th century.

He then led us on a brisk walking tour of the city. He was a short, stocky, and energetic man who had been an officer in the Soviet Army during the Second World War. When we were unable to keep pace with him and lagged behind, he would pause, raise his right clenched fist, and shout, "Forward!" in Russian (the closest I can come to phonetically is fperyod). Not much imagination was required to see him leading Red Army soldiers in an attack against German gun emplacements during WWII. That evening we were introduced to a charming Russian (and Ukrainian) custom of toasting, which took the form of an amicable competition in which contestants consumed copious quantities of vodka and vied to outdo one another to deliver the most florid and flattering toast. Dinners were paid for by our Ukrainian hosts with suitcases of the local currency. Shades of the Weimar Republic.

The next day, feeling the effects of the previous night's excesses, but able to walk, we found our way to the Ministry of Agriculture, that institution's blessing being essential for the conduct of our program. The Minister received us warmly and spoke excellent English, save for his unfamiliarity with English names for crops, which accordingly had to be communicated in Latin. So we explained our program to introduce herbicide tolerance into *Solanum tuberosum* (potato). He was very supportive, as Ukraine was experiencing a severe weed control problem in potato cultivation for which herbicides imported from the West at great cost were proving ineffective. He then inquired if we could also introduce tolerance into *Ipomoea batatas* (sweet potato), which, although similar in name, is genetically unrelated. Despite our disappointing response to this last request, we were assured of full support for our project.

(This use of Latin names to identify plants provided the basis for an amusing anecdote of a trip with Andrei to China several years later. Astounded by the sight of enormous pine trees in a central square in Beijing, Andrei exclaimed quite loudly, "Look at the giant *Pinus!*" However, he pronounced the "i" not as in *wine* or *mine*, as Americans do, but in the European manner as an "ea" as in *mean* or an "ee" as in *green*. Realizing what he had said as soon as the words left his mouth, Andrei turned bright red. But passersby registered no reaction, either not understanding or accustomed to profane outbursts from foreigners on crowded streets.)

Over the next few years, the potato genetic engineering project proceeded well and yielded herbicide tolerant lines of the important cultivar *Russet Burbank*, which is widely grown in the U.S. Pacific Northwest, Canada, Maine, and some regions of Europe. I traveled to Kiev again to evaluate the transgenic (containing a foreign gene inserted by genetic engineering) plants and to visit a communal farm that had offered to conduct field trials. The head of the commune was a warm, outgoing, avuncular man, perhaps 60 years of age. The commune was like an extended family over which he reigned as the benevolent father. Following a wholesome lunch at the commune refectory, we trudged out to the potato fields, where we were told of their need for better weed control and their enthusiasm for the project. This was a welcome development because of the hostility toward genetically engineered plants throughout Western Europe and, as a consequence, the severe restrictions that had been imposed on their production and cultivation. But any concerns that the Ukrainians may have had were overcome by their strong desire for the product and the knowledge that the potato cultivar was sterile, making escape or transmission of the introduced gene highly improbable.

The results of the field trials were positive; the transgenic potato lines were unaffected by field use rates of the herbicide that provided excellent weed control. However, the marketing group was unimpressed by this success. The colder temperatures of the northern potato growing regions of the U.S. and Europe slowed the degradation of herbicide in the soil, leaving residual levels that might damage crops grown in rotation. And whereas the utility for the warmer regions of southeastern Europe was not disputed, that market was considered too small to justify the expense of product registration, and moreover lacked hard (i.e., Western) currency to purchase the product. Hence, to the consternation of our Ukrainian collaborators, the project was deemed a technical success, but a commercial failure.

Regardless of the commercial decision, an invaluable and indelible outcome of our collaboration was the respect that we gained for Soviet scientists. We found them to be superbly trained (especially in classical disciplines such as taxonomy and botany, which in general are no longer taught in the U.S.), technically skilled, and highly dedicated. Restrictions on travel and communication had isolated them from the Western science community, but our collaboration had broken through that barrier. Many Kiev scientists were brought over by Cyanamid for periods of rotation in the Company's laboratories to improve their English, acquire familiarity with the Company, and exchange knowledge and techniques with our staff scientists. Eventually, some were hired permanently by Cyanamid. Others went on to positions elsewhere in U.S.

Now, it cannot be denied that, quite apart from the science, there are many pleasant aspects of foreign collaborations, such as tourism and meeting wonderful and interesting people. There also are less pleasant, but mundane and bearable, aspects, including prolonged and uncomfortable travel, primitive and unhygienic accommodations and amenities, sleep deprivation, poor food, infection by virulent exotic pathogens, and endless lectures, speeches, dinners, and the like. And then there are the occasional sacrifices that transcend the call of duty. An experience of this last type occurred on an early trip to Kiev. As Director of the International Institute of Cell Biology, Andrei was perennially advertising its accomplishments and capabilities. To that end, he organized an international congress at his Institute to which I, among other Western scientists, was invited as a speaker. Such international congresses are logistically complex affairs, especially in countries that lack the extensive infrastructure that we have grown to take for granted in the west, and involve not only organizing the scientific sessions, but also providing accommodations, meals, and diversions for the attendees. However, the tours of Kiev's beautiful churches and ancient monastery with catacombs containing mummified remains proved insufficient for an influential Indian scientist from a large American university, who requested that a visit to Chernobyl be arranged. As you surely know, Chernobyl, a cluster of nuclear reactors some 140 km and 2 hours north of Kiev, was the site of a terrible accident that spewed enormous quantities of radioactive contamination into the environment just the previous year (1986). For the rest of us, Kiev was not far enough from Chernobyl. As a tourist attraction, Babi Yar seemed far safer. But being a gracious host eager to accommodate all the wishes of his guests, no matter how deranged or suicidal, Andrei set out to organize the excursion. His problem seemed not to be hiring a bus and obtaining permission to visit Chernobyl, formidable challenges for ordinary humans in Kiev at that time, but filling the bus. But as Andrei explained, he couldn't justify the expense or the elaborate arrangements simply to transport one person. And so he made the rounds of the attendees, appealing to us individually until he assembled a group, of which I was one, whose sense of obligation to our host overcame our better judgment. The next day, this band of foolhardy (or intrepid depending on your viewpoint or generosity) scientists reluctantly boarded the chartered bus and headed north.

The Chernobyl accident, while a catastrophe in many respects, was a boon for radiation botanists. This small, neglected, diminishing, and somewhat outdated field was suddenly propelled to the forefront of international funding and attention. One cheerful beneficiary of this reversal of fortune climbed aboard our bus as a tour guide. His narrative – clearly intended to put us at ease – began by stating that the levels radioactivity at Chernobyl were lower than the level of natural background radiation in New York City and, therefore, harmless. Of course, we wanted to believe him, but his credibility vanished with the somewhat contradictory assurance that as a precautionary measure the buses at the site were washed down daily.

Our first stop was the regional Soviet Army command post, where we were greeted formally by a sternly xenophobic Soviet Army officer. One could see this man, conditioned as we were by Cold War hostility, yet also still shaken by the tragedy his country had recently suffered, grappling with his new orders to open his region to camera-toting western scientists in Izod shirts and Docker chinos. He gave a presentation on the history of the Chernobyl reactor and the recent accident, and handed us books about Chernobyl as souvenirs. His somber dignity and professionalism commanded admiration.

An exclusion zone had been demarcated around Chernobyl with guard stations placed on the approach roads at the boundary of the zone. Our bus stopped briefly at one such station before being waved on. Our guide explained (in perfect BBC English) that the purpose of these stations was not to prevent vehicles from entering the zone, but to provide the driver with 100 ml of vodka and instructions to gulp it down and drive as quickly as possible through the zone to the other side. To the many therapeutic attributes of vodka had now been added the flushing of radionuclides from the body. However, we were not offered the antidote. Either the guard's supply was inadequate for a busload, or, as foreigners, we were not entitled. But for whatever reason, we pressed on in distraught sobriety.

The next stop was Pripyat, a city of 50,000 that lay directly in the path of the radioactive cloud released by the damaged reactor. Signs of hasty abandonment were evident everywhere.

Laundry hanging on lines stretching from open apartment windows flapped gently in the breeze. Titles of current films were displayed on the marquee of the local cinema. Variegated weeds with sectors of albino tissue, a classic indicator of genetic damage, pushed through cracks in the pavement. Intact, yet devoid of life and eerily silent, the empty city stood as a monument to tific truth. But, as a professor of biochemistry, his constant concern was how to explain to his students the chemical equations that represent, on the paper and on the blackboard, the chemical reactions. Maybe it was something related to his incessant dwell on the accuracy of mathematical equations to solve algebra problems. He had had excellent math teachers. He had always been aware that although not perceptible to our senses, mathematical truths exist in human minds and are materialized in the great scientific achievements that have promoted the development of humanity. One example should suffice: the theories of Albert Einstein, fruits of his deep intuition, were brought to life by mathematical equations! The modern world is what it is today due to the application of mathematical principles. Computers and interplanetary travels are always headlines. How to "handle" mathematical formulas? His inquisitive mind asked the same about "handling" chemical formulas. He had been trained in solving mathematical problems, which facilitated the transmission of biochemical events to his students while manipulating the formulas on the blackboard of the chemical components of biochemical reactions. The chemical formulas and equations are models that serve the purpose of facilitating understanding of what happens in the microscopic world of cells which are not seed. He had always been aware of that!

CHAPTER 5.

Merging Chemistry, Cell Biology & the Understanding of Life

Otto J. Crocomo

I
In Search for His Scientific Identity

1956

The train moved very slowly along the narrow-gauge rails, tearing the bowels of the mountains that seemed, without any vegetation, naked to the 24 year-old man. He took the train at 6 a.m. on one day of August 1956 at Leopoldina Railway Station, in Rio de Janeiro, bound for Vicosa, Minas Gerais, in central western Brazil, where he would participate in the Ill Congress of Students of Agronomy at Federal University of Viçosa, while presenting two scientific papers that he had done under the supervision of Professors Eurípedes Malavolta and Jose Dall Pozzo Arzolla and in collaboration with his colleague Ary A. Salibe. The train was nicknamed "Smokey Mary" because of the dense cloud of steam and soot expelled by the engine and that soiled the clothes of the passengers in all cars. The railway station was named in honor of Leopoldina, the first wife of the first Emperor of Brazil,

D. Pedro I, who in 1822 proclaimed the independence of Brazil from Portugal. The arrival in Vicosa was scheduled for 10 p.m.

As the slow train proceeded on course, the heat and the constant smoke provided the young man with dreams and recurring visits to his past in search of reasons why he was there sitting on an uncomfortable wooden bench. He was in his fourth year at the Escola Superior de Agricultura "Luiz de Queiroz" (ESALQ), University of Sao Paulo (USP). The University was established in 1934, but his School, ESALQ, was created in 1900 and inaugurated in 1901, with classes beginning on June 3rd of that year. ESALQ was born from out of the desire of a visionary, Luiz Vicente de Souza Queiroz, with technical training in agriculture in France and Switzerland, who in 1892 donated to the State of Sao Paulo a farm on the outskirts of Piracicaba to be used as the base for an agricultural school. ESALQ was among the first Colleges of Higher Education in the State of Sao Paulo to integrate into the University of Sao Paulo System in 1934.

At a certain moment of reverie, it came to the memory of the young man, one dream he had as a child during sleep: he, who had lived in the center of the city of Piracicaba, State of Sao Paulo, since he was born, saw himself in a big house in the middle of trees lining the alleys that led to it, just on the side where now stood "his" School. In his childhood, while taking the streetcar that wound the streets of his city, he constantly visited the Agricultural School in the company of his mother or one of his seven siblings. It seemed to him that that dream was a foresight of the future: his dream had become reality since the day that taking the same streetcar; he went to School to submit his examination for admission, later be approved and penetrate into that sacred temple of agronomical knowledge.

The memories of the young man, interspersed with little sleep and stops for lunch and dinner in the towns that lined the railroad, was interrupted when the train finally arrived at the train station of Viçosa. But it did not end there. Owner of an inquiring mind, after the presentation of his scientific work, he was determined to get acquainted with the campus and the city. A grove of tall pines, their branches intertwined, casting a welcome shade, seduced him--that was the refuge he chose to meditate for hours about the meaning of his life and what his future would be like in the scientific world. Indeed, this was a feature of the young man: being able to isolate himself from the world, just like then, as well as when in a crowd.

1932 – 1956

The son of Italian immigrants, both from southern Italy-his mother Teresa, from the region of Calabria and his father Joao, a native of Ravello in the region of Naples--the young man was the last to be born in the family. He was introduced to the world in Piracicaba at 8 am on September 23, 1932, through the hands of a doctor at his parent's home. As the eighth child, his father gave him the name of Otto (eight, in Italian). His seven siblings were Teresa, Salvador, Maria, Leticia, Francisco, Lidia and Ada. A large family. Maria, one of his sisters, became fond of him while helping her mother to care for him, replacing her after his mother's death on August 2, 1944, when Otto was 12 years old. In turn Otto was very devoted to taking care of her until she passed away at 81 years of age. He considered her as his "second mother." His father João, an industrial worker, owner of an industry of manufactured copper appliances for production of alcohol and spirits distilled from cane sugar. The Piracicaba City Hall honored him naming a city street after him. His mother Teresa, a descendant of a wealthy Italian family, definitely influenced Otto's intellectual development: at age 11, he wrote a novel based on the tales that his mother recounted to him while sitting in the doorway of their home in the evenings. Even before primary school, he could read and write. In his adolescence, not only literature charmed him (in his first year in high school, mathematics, geography and history were his favorite subjects), but also, during those school times, the biological sciences, chemistry and physics: these were essential for his vision and understanding of the physical world, influencing the definition of his scientific career.

1950 -1952.

On the train, which swallowed the iron rails as it departed from Rio de Janeiro, where he arrived at midnight coming via Sao Paulo from Piracicaba, the young man had glimpses of his activities as a high school student and his interest in theoretical and practical chemistry taught superbly by Demosthenes Santos Correia at the High School "O Piracicabano" under the orientation of the Methodists, in the years 1950 to 1952. The five chemistry debates, performed during this period, with the State High School "Sud Mennucci," or the interclasses of the School where he studied, for which he was summoned by his teacher to participate and be a team leader, indelibly marked his future.

1956.

At that moment, on the train, more than at any other moment, like in the pine woods in which he, unaware at that time, would later meditate, he was sure that beyond the fact that his teams had always won the debates, the seeds planted in his teenager mind in those inevitable tense times for his age, had gradually increased his desire to know the reason for his existence and for the things around him: how they are made and what makes them what they are. And he wondered about the controls of the chemical reactions that take place within the plant cells which lead to the final product, many of them sensitive to our senses when extracted from the cells. "The scientific progress is due to the curiosity of the man in the pursuit of truth" - those were the words with which he began his first "Debate of Chemistry" in the Amphitheatre of the "O Piracicabano" on the morning of September 19, 1950. He was then 18 years old, and from his own mind he drew his deep concern about the meaning of the existence of the man and the living beings that surround them.

This desire, the result of his restless mind, was why he was there, he thought while sitting on a hard wooden bench in a train that spewed smoke, on the way to the conquest of new scientific evidence that the Congress of Vicosa could provide him. This would be the third and last Congress in which, as an undergraduate student, he would participate before graduating; the first Congress was carried out at ESALQ in 1954 and the second at the National School of Agriculture (ENA) in Rio de Janeiro in 1955, when he met for the first time with Paulo Fernando Cidade de Araujo, who years later would become his great partner in managing their scientific projects. All these conferences were attended by the Center for Debates and Agronomic Studies which he helped create at the Academic Center "Luiz de Queiroz" (CALQ), of the ESALQ's students in 1954.

The memory of recent events came to his mind. As a student, even in the third and also in the last year of the agronomical course, he found himself taking an active part of weekly meetings for the presentation and discussion of scientific papers at the "Institute Zimotécnico", under the leadership of its Director Jaime Rocha de Almeida. He recalled that he was the only graduate student accepted to attend these meetings. At that point as he wandered, he took a new breath of life to remember that, even as an undergraduate student, he had attended a course on Enzymology, at the same Institute, which was taught by Metry Bacila, a professor at the University of the State of Parana in southern Brazil, one of the most respected researcher in biochemistry of microorganisms at the time.

At that moment he was about to become an Agronomist. The work he presented in Viçosa would be his last as an undergraduate student. He felt assured of himself, with plans for the near future: he wanted to devote himself to the study of chemistry as soon as the course ended. And that, for him, ended on November 30, 1956. The graduation ceremony happened four months later, on March 16, 1957, when he received the "Manah Award" for his performance in the disciplines of chemistry during the undergraduate course. He would then be an Agronomist.

II
The Academic Experiences

1957 – 1962

October 1961. The plane, a Panair's Constellation, was ready to take off from Congonhas Airport in Sao Paulo heading to Venezuela. In it were Otto and his young wife Diva. Their families came to say goodbye to them: Otto's father and two of his sisters, Theresa and Maria. Also present were Diva's parents, Gelindo and Maria, both sons of immigrants from Northern Italy. The farewell was both sad and happy. Sad because they would leave their loved ones for a period of 14 months and happy because he had been hired as a Professor of Biochemistry and Agricultural Chemistry at the Faculty of Agronomy at the Universidad del Zulia in Maracaibo, on the shores of the Lake of the same name, in a rich region of Venezuela's oil industry.

Otto and Diva met on November 9, 1957 (Photo 1). The musical success at that time was "Unchained Melody" by Alex North and Hy Zaret; they elected as "their music" and have been repeatedly sung by Diva, owner of a single tone of voice with the sound of outstanding soprano that she is. About four years later, their civil marriage took place on March 25, 1961. Among the witnessed were the great collaborator of Otto in several of his scientific works, Andre Martin Louis Neptune, of Haitian origin and naturalized Brazilian, and his wife, Nair. The religious marriage, celebrated by D. Aniger Melillo, the then a Catholic Bishop of the Diocese of Piracicaba, took place on May 23 of that year. Eurípedes Malavolta, of whom Otto was an assistant, and his wife Leila, and Ben-Hur Carvalhaes de Paiva and his wife Maria Helena, were among their men and maids of honor. Maria Helena was, in the past, a colleague of Otto during his High School years at Colegio "O Piracicabano".

June 1958. On the morning of the 22, as he would usually do, upon reaching ESALQ Otto got off the streetcar that took him from the center of Piracicaba to his destination. He walked toward the Chemistry building, went up the steep stairway with rungs of red brick located on the back of the building, and entered the Laboratory of Agricultural Chemistry. Eurípedes Malavolta was sitting on a footstool while poring over the books and journals that were spread all over the counter. He was preparing for a contest for the position of a Full Professor that would take place in the coming months.

Otto, I have an invitation to make you a "Livre Docente"! Astounded, Otto nodded and drew closer to Malavolta.

- I'd like you to submit your work to the contest for the "Livre Docente Professorship," said Malavolta looking directly at the young man, who surprised, stammered: "But how? I'm a recent undergraduate, and Professor Arzolla is the one who is preparing for this contest",

- I want both of you take the examination at the same time, said Malavolta. And continued: "I really enjoyed your seminar on June 8th, at the "Instituto Zimotecnico" on the results of your work with 14C-urea. You impressed my colleagues."

But those are the results of my Dr. Thesis in Agronomy, Otto said, leaving a question in the air, is...?"

You can use them for the contest, said Malavolta. Returning to his books, he added: "I trust you and give you a few days to decide."

September, 1959. Having accepted Malavolta's invitation, he took for 4 days, from the 19 to the 22, the theoretical and practical examinations in Organic Chemistry and Biological Chemistry, before a Committee formed by five professors: Eurípedes Malavolta, Tufi Cury, both from ESALQ, Veronica Rapp de Eston, Henrique Tastaldi, a professor at USP at the Sao Paulo campus, and Metry Bacila, from the Federal University of Parana, in Curitiba. One of the five tests was the public defense of his thesis, the first in ESALQ to use radioisotopes in the biological field: "^{14}C-urea metabolism in coffee leaves", highlighting the presence of the urea cycle in higher plants. The metabolic transformations: the constant questions he always had on the operation of chemical reactions within living cells, and how they are controlled, were beginning to be answered to him! Understanding the chemical reactions within the plant cell would be the purpose of his scientific involvement in the future. At that moment, a happy and everlasting marriage between Otto and Biochemistry had been established.

The University of Sao Paulo hired him as Assistant Professor of the Chair of Organic Chemistry and Biological Chemistry of his school, the ESALQ, on April 16, 1960. In that same year, he was granted a Doctorate Degree in Agronomy by the University of Sao Paulo as a result of his having already earned the title of "Livre Docente" from the same University. For the same reason, he was not required to undergo the probationary period of five years, which was a regular practice to Assistant Professors hired for full time work. At last, he would give the ESALQ's students his Biochemistry lessons.

Throughout his life he often reflected on the reason for his deep interest in the chemical reactions that occur within the cells. That was the consequence of his incessant search for the scientific truth. But, as a professor of biochemistry, his constant concern was how to explain to his students the chemical equations that represent, on the paper and on the blackboard, the chemical reactions. Maybe it was something related to his incessant dwell on the accuracy of mathematical equations to solve algebra problems. He had had excellent math teachers. He had always been aware that although not perceptible to our senses, mathematical truths exist in human minds and are materialized in the great scientific achievements that have promoted the development of humanity. One example should suffice: the theories of Albert Einstein, fruits of his deep intuition, were brought to life by mathematical equations! The modern world is what it is today due to the application of mathematical principles. Computers and interplanetary travels are always headlines. How to "handle" mathematical formulas? His inquisitive mind asked the same about "handling" chemical formulas. He had been trained in solving mathematical problems, which facilitated the transmission of biochemical events to his students while manipulating the formulas on the blackboard of the chemical components of biochemical reactions. The chemical formulas and equations are models that serve the purpose of facilitating understanding of what happens in the microscopic world of cells which are not seed. He had always been aware of that!

On the plane that led them to Venezuela, he recalled and reflected on all this while Diva took notes for a letter that later would be sent to her parents. In the letter, Diva penned the wonders of flying over the Amazon jungle: the green canopy of flora that she observed from the airplane cabin window. The air route was being done at low altitude.

Maracaibo, Venezuela. The invitation to teach in Venezuela was made by the Dean of the Faculty of Agronomy, Jose Gonzalez Mateus, orchestrated by Andre M. Louis Neptune, who had previously visited there. At the International Airport of Maiquetia "Simon Bolivar" in Caracas, they were greeted by Felix Taborda, a former student of ESALQ. That evening they met with Gonzalez

Mateus who soon put at his disposal all the facilities of the Facultad in which he would install a Radioisotope Laboratory. He was responsible for the Departments of Biochemistry and Agricultural Chemistry.

Due to the use of texts and books in Spanish, in which he studied inorganic chemistry, and his interest in Caribbean rhythms such as boleros, cumbias and salsas, Otto felt comfortable in the new environment. This in turn greatly facilitated the adjustment to his and Diva's new lifestyle.

The teachers of the Faculty of Agronomy and the administrative staff gave him all their support. Festive gatherings and the very cheerful and affectionate spirit of Venezuelans were frequent. Perhaps because he was a foreign teacher, his students were enthusiastic about his classes. Two in particular were interested in being his monitors in practical classes. One, J. Villasmill, was the most dedicated, an excellent student who later became involved in the administration of the Faculty. The other, Tiberio Perozo Yori, who was not considered a good student, became interested in his teachings, including assisting him in the installation and monitoring of a field experiment on the mineral nutrition of corn (Zea mays) in a farm his father owned. His dedication led him to a successful performance, under the guidance of Otto, at a seminar he presented before his classmates and teachers of the Faculty and attended by the Dean.

On January and February 1964, Tiberio spent time as an apprentice at Otto's laboratories at CENA, in Piracicaba. In August of that year, Otto and Diva, in their way to the United States of America, visited Tiberio's family in Maracaibo, whose father regarded Otto as the "Tiberio's second father." On February 1972, Tiberio was developing his Master's program in the American Institute of Agricultural Sciences (llCA), in Turrialba, Costa Rica. On this that occasion, he met with Otto who was teaching a course on "Biochemistry of the Mineral Nutrition of Plants" at that Institute. Tiberio would pass away a few years later, still quite young.

Costa Rica, 1972. The car waited for Otto at the door of the guest house where he was staying during the course which he administered to the Institute, in Turrialba. They were heading to San Jose. . It was late February, the sun insisted on appearing in among the clouds that framed the curves of the road, the asphalt being swallowed up by the wheels of the car while he was delighted with the magnificent view of the valley below where, at some point, was Turrialba in the province of Cartage, known as "a piece of paradise in the Caribbean, the ports of Costa Rica to the Atlantic Ocean." He sighed at the sound of Guantanamera from the car radio. One of the most preferred by him among those who formed his collection of songs with lyrics in Spanish, authored by the Cuban professor and intellectual Jose Mautas Marti and music by Josito Fernandez. Yes, Marti, an important participant in the struggle for Cuban independence from Spain rule in the late nineteenth century! He couldn't imagine that 21 years later, in May 1993, accompanied by Diva, he would go on a scientific mission as United Nations' Expert, to Havana, to oversee the development of a research program on sugarcane at the Institute of Biotechnology and live with the generous Cuban people, who sings the verses of Jose Marti, a poet whom they worships.

The car continued to meander the mountain slopes until they arrived at the airport of San Jose, capital of Costa Rica, located more than one thousand meters of altitude.

In Brazil, his family and ESALQ Awaited

May, 1962. Marco Augusto, his first son, was born on the 31st at the Hospital of Our Lady of Coromoto. Being born in Maracaibo earned him the nickname "el Maracucho" by their Venezuelan friends, including Felix Taborda and Hiram Reyes-Zumeta, both former ESALQ students and Otto's colleagues in the Facultad, and Emiro Gomez also a former student at ESALQ. Emiro and his wife Maria Jose, from Piracicaba, were Marco Augusto's god-parents at baptism.

The experience of being a professor in a country that was not his, at the age of 29, being in charge of two disciplines and of the radioisotope laboratory, coordinating a Symposium on "The use of Radioisotopes in Agriculture ", installing a field experiment, and speaking a language that was not his, was essential for the development of his academic life, since he was alone to make decisions. There were no mentors or colleagues who could assist him in his self-criticism. He began to look at the world with new perspectives, with desire for new achievements. This experience was present in his mind and his heart when, in the company of Diva and Marco Augusto, he left Maracaibo heading to Caracas and then to Sao Paulo, in Brasil, where, at Congonhas Airport, he hugged his family. It was the morning of December 19, 1962.

He had already had such an experience, though not as great as that experienced in Venezuela. He had been invited by the cardiologist Ben-Hur Carvalhaes de Paiva to be his assistant and responsible for the Biochemistry Discipline at the then College of Pharmacy and Dentistry of Piracicaba, in the years of 1959-1960. This was his first job after he graduated from ESALQ. Ben-Hur was then Professor of Physiology. This School was established in 1955 as part of the Institutes of Higher Education of the State of Sao Paulo. In 1967 it was transformed into the Faculty of Odontology of Piracicaba, becoming part of the State University of Campinas (UNICAMP). At that time he had several very good students, such as Antonio Carlos Neder, his monitor in the laboratory practices who, according to him, devoted himself to pharmacology due to the influence of Crocomo's biochemistry classes. Neder would become in the future the Director of that Faculty and an internationally respected name in the field of dental pharmacology, which he introduced as a discipline in dentistry courses.

III
The Continued Pursuit of Scientific Knowledge

1964

His inquisitive nature was still present in all his personal and professional activities. Along with his philosophical questions, which Otto would often discuss with his friend and journalist Cecilio Elias Neto, a brilliant thinker, a desire to deepen more the study of plant biochemistry made him even more uneasy. In Venezuela, where he published his first book, "Bioquimica de la Respiration Cellular" edited by the Facultad de Agronomia in Maracaibo (1962), Otto had drawn plans that could lead him to study in the United States: this was one of the reasons that he had not accepted the renewal of his contract with the Faculad, adding to the fact that ESALQ wanted him back in Piracicaba.

August. Otto, holding hands with his 2 years old son Marco Augusto, and Diva carrying on her lap their son Adolfo Egidio, who was born on September 6, 1963, in Piracicaba, climbed the DC-8 PANAM high ladder at Congonhas Airport in Sao Paulo. The couple's second international trip would take them to the United States of America. As a Rockefeller Foundation Fellow Researcher, he would develop scientific works at the Department of Biophysics and Biochemistry, University of California, Davis campus.

On his way to the United States, he made a series of lectures at the Facultad de Agronomia of the Universidad Del Zulia, in Maracaibo. The lectures were published by the Facultad in the book "Absorcion de ions por las plantas", in collaboration with Andre Martin Louis Neptune and Hiram Reyes-Zumeta.

In a condominium apartment in Orchard Park, Davis, he and Diva lived among families of students from several European, African and Eastern countries. They often met with Brazilian friends from different regions of Brazil, with their small children who became Marco Augusto's playmates. Their second son, Adolfo Egidio, learned to walk a few months after arriving in Davis. Friendships were established and others narrowed with Brazilians who did their doctoral programs on campus, including Lourival Carmo Monaco and his wife Mercedes. Lourival later developed intense and important work in plant genetics at the Agronomic Institute of Campinas (IAC), in Brasil. He and Otto had been colleagues in times of high school and during the undergraduate course at ESALQ.

In Davis, besides developing research on microorganisms with C. C. Delwiche, Otto took theoretical and biochemistry laboratory courses, taught by E. E. Conn and P. K. Stumpf and their assistants, and also the course of Conceptual Physical Chemistry and the course on Principles of Mineral Nutrition of Plants, the latter given by E. Epstein. At that time, Otto learned the computer language FORTRAN, necessary for the elaboration of data obtained in his laboratory works. The time he spent at the University of California was essential to open new perspectives, preparing him to take over, immediately after his return to Brazil, the Discipline of "Biochemistry of Plants", in the Graduate Course in ESALQ, which he taught for 23 years, also guiding students for Master and Doctor Degrees. In collaboration with Euripedes Malavolta and Darcy Martins da Silva, he translated into Portuguese the book "Outlines of Biochemistry" by E. E. Conn and P. K. Stumpf (1972 edition, John Wiley & Sons Inc.), with several editions in Brazil.

The military revolution. The year 1964 has indelibly marked the history of Brazil. On March 31, the military, from several Brazilian states, mainly from Minas Gerais and Sao Paulo, rebelled against what they considered authoritarian excesses of the then President of the Republic of Brazil, João Goulart. The military had always been opposed to him since 1961 when, being the Vice President, he assumed the presidency at the resignation of Janio Quadros. He was allowed to govern Brazil only within the parliamentary system established by the forces opposing him. Goulart then proceeded to rule with the machinery of social security in his hands and making alliances with the left in control of the unions.

It was in this political scenario that, on 21 August 21, 1965, shortly after the return of Otto and family to Brazil from the United States, Maria Paula, their third child was born. On that same year, from October to December, he gave lectures and laboratory practices on the "Application of Radioisotopes in Soil/Plant Relationships" at the Institute of Nuclear Affairs in Bogota, Colombia.

As a consequence of the military rebellion, a military regime with broad dictatorial powers was established in Brazil, especially the press censorship, political persecution of students, union leaders, musicians and intellectuals. Among these, many were professors at the University of Sao Paulo who were exiled in foreign countries, such as Fernando Henrique Cardoso. The union leader Luis Inacio Lula da Silva was arrested for a few days. When the military regime ended and democracy was restored in Brazil, which occurred in 1985, Fernando Henrique Cardoso was the fourth President for two terms (1994-2002), elected directly and freely by the people and responsible for implementing the Real Plan, created by his predecessor, ltamar Franco. The new currency was appreciated while stabilizing the economy. This resulted from implementation of the Fiscal Responsibility Law, to which all rulers were obligated obey. Luiz Inacio Lula da Silva was elected the fifth president and ruled Brazil for two terms (2002-2010), with strong popular appeal. Lula supported the candidacy of Dilma Roussef, a former fighter against the military dictatorship, to the presidency, becoming the first woman President of Brazil in 2011.

IV
The Early Years at CENA

December, 1973. He left his office, and in short steps walked down the hall of the Section of Plant Biochemistry, of which he was responsible, until the door to the park. Here lies the site of the Center for Nuclear Energy in Agriculture (CENA). It was a Saturday morning in early December. Despite the summer, an intense mist covered the trees and shrubs framing the few buildings that could still be seen. Through the mist he slowly walked, taking in the moist air. He instantly recalled the recent train trip that he had made in the route between Geneva and Zurich in the Swiss Alps among trees covered by the snow. While strolling around Otto skirted the buildings recalling that his scientific travels around the world (which were just beginning) were due mainly to that place, now flooded by the fog on an unusual Piracicaba summer morning. And once again he was reminded of how his scientific work had begun at CENA.

In 1967, as Associate Professor since 1966, Otto was working in the laboratories of biochemistry at the Chemistry Building at ESALQ, with three of his undergraduate students, Luiz Carlos Basso, Oswaldo Galvão Brazil and Celso Rossi. Malavolta approached him and asked him and his students, and all the materials being used, to be moved to a new laboratory at CENA. Thus his scientific life began again elsewhere. Yes, once again, because he had previously used the laboratories of the "Institute Zimotécnico" in ESALQ during a long period so that he could carry out experiments for his thesis of "Livre Docencia" in 1959. And another change in his life happened: that same year of 1967 was born their second daughter, Carla Maisa, on September 25th, in Piracicaba.

As Otto walked through the alleys of the park around the buildings, while absorbed in his memories, the fog slowly dissipated and he suddenly found himself surrounded by a bright light: it was the sun that embraced him and heralded the new era of scientific adventures. He entered through the main door of his Section, went to his office and contemplated new ideas.

Now, however, his thoughts were scattered, so he revisited the initial scientific events that gave birth to CENA.

The pioneers. The steps taken for the creation of CENA had begun much earlier. In 1953, Eurípides Malavolta introduced the techniques for the use of radioisotopes at ESALQ. As a research associate at the University of California, Davis. He has used these techniques in his studies of mineral nutrition of plants in collaboration with E. R. Stout. Malavolta was among the first in Brazil to publish a scientific paper on the use of radioisotopes in plants: radioactive zinc absorption by the leaves of orchids. In 1956, Constant C. Delwiche, from the IBEC Research Institute, University of California, with whom Otto would work later (1964-1965), came to Piracicaba. This was crucial to implement the use of nuclear techniques in higher plants in ESALQ. In 1958, Malavolta's work on the absorption of radioactive superphosphate coffee, in the field, revealed the best location of the phosphate fertilizer in the soil for this crop. In September de1959, Otto defended his "Livre-Docencia" Thesis whose results were obtained with the use of radioisotopes.

The pioneering scientific works using radioisotope techniques in the fields of plant nutrition and biochemistry, and others in the field of physics made by Ademar Cervellini, Anivaldo Pedro Cobra, Eneas Salati and Jose Goldenberg, were conducted at the Isotope Laboratory of the Chair of Physics and Meteorology, and in the laboratories of the Chair of Organic Chemistry and Biological Chemistry and at the "Institute Zimotécnico," all at ESALQ in the 1950s. One of his collaborators was Otto's close friend Andre Martin Louis Neptune. Adequately protected against nuclear radiation, both Crocomo and Neptune and their students would leave their homes at dawn, to brush coffee leaves with a solution of radioactive sulfur (^{35}S), in order to demonstrate the foliar absorption of minerals and monitor the metabolism of sulfur compounds in higher plants. In January and February 1960, Otto participated as a student in the International Course on Nuclear Medicine, at the Nuclear Medicine Center of the University of São Paulo in the city of São Paulo. The following year, in January 1961, he taught theory and practice regarding the use of radioisotopes in the metabolism of plants and methods of measurement of radioactivity in plant extracts, in the same International Course.

Along with Malavolta, Otto was one of many Cervellini's collaborators in the creation and consolidation of CENA, among who were Almira Blumenschein, Akihiko Ando, Andre Martin Louis Neptune, Darcy Martins da Silva, Eneas Salati, Epaminondas Sansigolo Ferraz, Frederico Maximiliano Wiendl, Henrique Bergamin Filho, Klaus Reichardt, Renato Amilcare Catani and Valdomiro Correa Bittencourt.

Despite the advances of dictatorial military governments, whose Presidents succeeded one another during the period of about 20 years, Brazil experienced great advances in many areas including the scientific. It was then that, on September 22, 1966, the Center for Nuclear Energy in Agriculture (CENA) was established as an Institute attached to ESALQ. In 1968, the Agreement signed between the University of São Paulo and the National Commission of Nuclear Energy (CNEN), a strategic entity of the Federal Government in the area of nuclear energy, was essential for the consolidation of the institution. In 1985, CENA became part of the campus "Luiz de Queiroz" and has been an USP's "Instituto Especializado" since 1988.

V
The Scientific and Cultural Experiences

1968

October. At Viracopos Airport in the region of Campinas, a few miles from Piracicaba, Otto waited for the call to board the British Airways flight that would take him to London. It was the 26th. Keeping him company was Diva, who would not accompany Otto this time around. Otto would spend five months at the Department of Botany and Microbiology, at the University College London, where, as a Fellow Researcher, funded by the British Council and the Foundation for Research Support of the State of Sao Paulo (FAPESP), he would develop scientific work with Leslie Fowden. Diva would visit him in London during late February 1969, and after their visit to several countries in Europe, they would return to Brazil in late March.

Celso Rossi and his wife Sirley were also at the airport to say goodbye to Otto. Celso was one of his three assistants at the Faculty of Medical and Biological Sciences of Botucatu, in the State of Sao Paulo, where he was Regent of the Biochemistry Department in the period of 1968-1969. Later, Celso Rossi and Oswaldo Galvão Brazil, his other assistant, would conquer the titles of Master, Doctor and Associate Professor of Biochemistry. Both became Full Professors of that University in February 1987.

Celso Rossi remembers the details of that farewell and, years later, wrote to Otto:

"I never forgot that night at Viracopos Airport. I owned a small 1200 Volkswagen, 6V. The lighting was very poor to travel on a night of heavy fog. I remember the takeoff in the darkness of night; it was 8 or 9 pm. when the aircraft rushed like an arrow into the sky. I thought "what courage has Otto, flying to London in a night like that!" The way back to Piracicaba was difficult. Though, at that time, few cars were on the highway. We arrived safe and sound, or I wouldn't be giving you these details, now."

The third assistant, Luiz Carlos Basso, was subsequently hired by the Department of Chemistry ESALQ, conquering, under Otto's orientation, the Master and Doctor titles. Basso had worked as a Laboratory Assistant in 1963, before being a graduate student at ESALQ in Otto's first scientific project granted by FAPESP. This Foundation was established in 1962 to foster scientific research in the State of São Paulo Throughout his scientific career, Otto has received numerous academic research grants and fellowships from FAPESP, having Basso in several of them as one of his most important collaborators, particularly in research on mineral metabolism and polyamines in plants, whose results were published in international journals. Basso would in the future become one of the best teachers and researchers in biochemistry of plants and microorganisms at ESALQ, participating in public-private projects in the field of bioenergetics, researching enzyme activity during fermentation.

On May, 22 of that year, his father, João, died at the age of 73, after six months of suffering due to a stroke, paralyzed and unable to communicate except by signs, under the care of two of his daughters, Maria and Teresa.

In London, he stayed in the "House of Brazil," close to the Lancaster Gate Underground, which took him every morning, with a transfer at the Tottenham Court Road Underground Station, to the London University College on Gower Street. Very soon he came to admire the capital of England, not only in the context of its participation in the history of Western and Mid-Eastern civilizations, but also, and particularly, for its contribution to the development of science. Now, at the age of 36, Otto saw new horizons to be reached. His continuous search for the essence and the reason of his existence, and of the nature around him. This was characteristic of his ceaseless curiosity, found fertile ground in the science developed in the laboratory of Leslie Fowden and his participation in conferences at the College. He let himself be involved by the European culture and science, just as he previously experienced the benefits of scientific development in the United States. Otto could not find complete answers to his philosophical concerns, but to his intellectual baggage was added now another vision of the world and of the human behavior that, in the future, would greatly contribute much to his conduct in science and his relationship with his fellowmen.

The cosmopolitan and cultural London attracted him. He was in the country of the Beatles, almost at the same time that the participants of the band broke up. But Carnaby Street, which begins in the long and curved and elegant Regent Street, was and still is there, evoking and perpetuating the sounds of their guitars and their voices.

His presence, at weekends, at musical concerts and theater performances left him fascinated.

Carlos, the "Times" is advertising the play "Mousetrap." Have you seen it?" asked Otto.

Yes, they have been showing it in a small theater in West End, the New Ambassador Theatre, since its debut on November 25, 1952, and it is worth watching. It was written by Agatha Christie, said Carlos a young man from Piracicaba, tall, slender, hired by the Brazilian Embassy in London to teach Portuguese in England.

Carlos Toledo Vollet Sachs, who was also housed in the "House of Brazil," was the nephew of Salvador de Toledo Piza Jr., Emeritus Professor of ESALQ, of whom Otto had been a student, both maintaining an excellent intellectual relationship. Years later, Carlos would live in Manchester, England, where he died in 2011.

For a long time I've been a fan of Agatha Christie. The style she used to show the roles of the characters in the unfolding of the plot, and all the arguments she elaborates to unravel the mysteries of police cases, makes the imagination work and helps me in the interpretation of the everyday events that happen around me, Otto said, thoughtfully.

We can go on Saturday, invited Carlos. In fact, I wish we were also seeing the musical "Fiddler on the Roof," which I've seen seven times. Carlos laughed to himself, making his confession as an unconditional fan of the play, and added euphorically: the Jerry Sock's soundtrack and Sheldon Harnick's text are excellent, both are Americans, holding several awards."

The play was being presented at Her Majesty's Theatre in Haymarket, near Trafalgar Square. In Trafalgar Square, dominated by four huge black statues of lions, fountains and the grand column of Nelson, honoring England's great hero Admiral Lord Horatio Nelson, the winner of the battle of Waterloo against Napoleon Bonaparte in the early 19th century, is The National Gallery. This was the favorite place for Otto to spend a good part of his Sunday afternoons in the winter of London. The paintings of the exponents of the art from the XIII to the XIX centuries and the occasional exhibitions of modern and contemporary artists, had his mind enraptured by so much beauty and richness of detail and color harmony-from the vibrant colors of Raphael to the spiritual simplicity of the "cartoon" of Leonardo Da Vinci, "the Virgin and Child with St. Anne and St. John the Baptist," singly exposed in a small acclimatized room, under indirect light. Often, sitting in front of this "sketch" of Da Vinci, he was, in thought, transported to deserted beaches where white sand were kissed by surrounded waves of gentle ripples. Bucolic scenes of unpaved roads that led to the top of hills, where his eyes could see a serene and immense valley, with animals grazing on the green grass, and dotted here and there with cypresses and pines surrounding pointed red roofs houses, visited his mind now enraptured with the chords of the Beethoven's "Pastoral" symphony.

Again, as always, he surprised himself while he pondered the value of being alive and over his future as he had done in the dark pine forest in Viçosa, in August, 1956.

The visits to the Natural History Museum, the British Museum and others, and watching classical music concerts at the Royal Opera House in Covent Garden, or the Royal Albert Hall, Kensington Gore, were moments of extreme beauty, even playful, for the young man eager for new knowledge and new life experiences. It was only the beginning of a series of scientific and cultural visits to the capital of England, in addition to the many other journeys to various countries in Europe.

His childhood in Piracicaba was constantly in his memories. Sitting on a bench in front of the Prince Albert Memorial, in a January morning of intense winter, he remembered speaking to his mother Teresa:

"… When I 'get big' I will see London, Paris and Curitiba." Enchanted with history and geography, he was exploring the maps of Brazil and Europe, sitting at the kitchen table in his parents' house, in Piracicaba, at the age of 10.

In Curitiba, the capital of Parana State, in southern Brazil, he had already been several times before, teaching biochemistry of plants in courses organized by Metry Bacila. Paris would come a little later. But now he was in London! And he was right before the extraordinary monument of love-celebration of Queen Victoria to her husband Albert, in Kensington Gardens, midway between Lancaster Gate, where he lived, and the Royal Albert Hall. This was the path that he often made surrounding the Serpentine Lake in Hyde Park.

By then, he had participated in the Congresses of the Brazilian Society of Biochemistry, which he was a founding partner in 1965. The biannual meetings took place at the Grand Hotel in Caxambu, a hydro mineral spa, in the State of Minas Gerais. Routinely, most of the presentations were panels on biochemistry of animals or of microorganisms. Only he and one or another researcher from institutions, other than ESALQ, exposed work in biochemistry of plants. This situation began to change when in the 1970s and 1980s he was invited to make oral presentations of his works and organize symposia on biochemistry of plants in these Congresses. Other groups were also encouraged, such as the researchers Ladaslav Sodek, UNICAMP, Sonia Dietrich, from the Institute of Botany of Sao Paulo, Walter Handro, from the Institute of Biosciences of USP in Sao Paulo campus. L. Sodek, previously, had been hired by CENA and conducted research in Otto's laboratories. From 1961 to about 1979, Otto toured Brazil from north to south, giving courses and lectures at several institutions of research and teaching on the metabolism of nitrogen and potassium in plants and also the use of radioisotopes to monitor physiological and biochemical processes in plants. He often transported with him his laboratory equipment, including the radioisotopes, to demonstrate the phenomenon of photosynthesis.

VI
Experience with International Projects

1978

Munich, Germany. He got off the streetcar near Marianplatz and walked to the famous square in the city of Munich in a rich region of Germany. He joined the people that looked at the tower of the beautiful building which houses the City Hall. Everyone was waiting for the 12 o'clock chimes of the magnificent clock embedded in the tower. When the hour and minute met, sculptural human figures, one after the other, emerged from within the structure of the tower, some mounted on horses, representing medieval knights, perhaps seeking the release of their beloved ones, he thought. The majestic spectacle, with the statues disappearing inside the tower, came to an end, and despite the cold morning of a September day, the sun was present, and then he continued his walk through the streets of Munich remembering why he was there in that festive and lively Bavarian City.

In Munich he was participating in the 5th and final Meeting of the FAQ/ IAEAIGSF Research Coordination Meeting of Seed Protein Improvement Program. At that meeting he presented the final results of his work on improving the content and quality of storage proteins in seeds of beans (Phaseolus vulgaris) using mutagenic treatments, in collaboration with Augusto Tulmann and Donald Boulter.

In late 1972, Ademar Cervellini, CENA Director, invited Otto to participate in the Program for Improvement of Proteins of Legume Seeds sponsored by the Food Agriculture Organization (FAQ, Italy), the International Atomic Energy Agency (IAEA, Vienna, Austria), and by the German Federal Government. The project led by him would be developed in the Section of Plant Biochemistry of CENA, in collaboration with Augusto Tulmann Neto, of Section of Plant Breeding and an expert in the use of nuclear techniques for inducing mutation in plants. For five years, this program was developed at CENA.

The first results were presented by him in the 2nd Meeting of the Coordination Program, held at the International Institute of Tropical Agriculture (llTA) in Ibadan, Nigeria, Africa, on December 1973.

At this point in his memories, still walking in the streets of Munich, he saw himself in Ibadan, Nigeria, living a unique experience in a country that had been a colony of England, rich in oil and with most of its population having no access to socio-economic resources. In Africa, the Institute of Tropical Agriculture was a leader, and still is, in search to find solutions to fight hunger, malnutrition and poverty of the population of Sub-Saharan Africa.

December, 1973. Before landing at the airport in Lagos, Nigeria, heading to Ibadan, he was in Switzerland, in the winter.

It was night when the plane landed at the airport in Geneva. The sight of snow on the houses from the bus that took him through the silent streets to the hotel in downtown where he would be lodged, brought him an unexpected tranquility. He felt involved by the atmosphere of a secular civilization where everything seemed to be ready, but he was aware that it was not easy for the Swiss people to conquer this!

When the radio in the hotel room played the chords of the Fifth Concerto for Piano and Orchestra, "The Emperor" by Beethoven, one of the classics he most admired, Otto remembered of his wife, Diva, and his five children now, because Daniel was born that year, 1973, on April 28, in Piracicaba. All of them, in their first years of life fell asleep to the sounds of Mozart and Beethoven. The chords of "The Emperor" followed them through adolescence. "They might as well here with me," he thought while contemplating the beautiful photos of the Swiss Alps in his hands. Soon he would actually get to know them when, in Geneva, he would take the train that would take him to Zurich, and climb the mountains, among the snow-covered trees in tortuous paths. In this atmosphere once again, he pondered the meaning of his life, similar to previous experiences he had had, almost mystical.

The third meeting of the Program was held in May, 1975 in Hahnenklee, Germany. After the meeting, he visited the facilities of the Institut for Strahlenbotanik in Hannover. He was exposed to rapid methods for extraction of proteins of legume seeds on devices that protected them from the action of heat, significantly reducing the loss of its content which he would apply in his research in Piracicaba. In Hannover, at a dinner in the basement of a restaurant in a building restored after being bombed during the Second World War, he first heard the song "When a Child is Born," soloed by Cyrus Dammicco. The music made him recall his five children, while they still were very small, each in his own time, just touching his heart. At a store in Hannover he purchased the "single" of the song, taking it with him to Piracicaba. Now, Marco Augusto was 11, Adolfo Egídio 10, Maria Paula 8, Carla Maisa 6, and Daniel 2 years old! The memory of that day and of that place remained forever in his heart.

In late March and early April of that year, 1975, before traveling to Germany, at the age of 42, Otto undertook two days of examinations to be Full Professor of Biochemistry at the Department of Chemistry in ESALQ. The Committee, who evaluated and approved him, was formed by Euripides Malavolta, Renato Catani, Eugenio Acquarone, Henrique Tastaldi and Walter Borzani, all Full Professors of the University of Sao Paulo. He reached then the highest degree of the academic career of USP. At that time, it came to his mind that that dream he had dreamed in his sleep as a child was becoming reality: in the dream he found himself living in a large house among trees and flowers and alleys, at the sides of which is "his" School of Agriculture.

From 1971 on during alternate four year appointments, he served as Head and Deputy Head of the Department of Chemistry, ESALQ until 1989.

In late March, 1977, the bucolic Baden, south of Vienna, Austria, welcomed members of the 4th Meeting of the Program. A streetcar transported him from the bus station in Vienna to Baden, through streets lined with trees whose branches intertwined to form a tunnel and a cozy living shadow. Everything seemed to breathe peace at the spa, located in the vicinity of "the woods" of Vienna. Sitting on a bench before the statue of Strauss, he recalled dancing with one of his sisters, Lidia, in the ball of his graduation of the College "O Piracicabano" in December, 1949. The music was "Tales from the Vienna Woods," a waltz by Johann Strauss, that permeated most of his adolescence and youth.

It was during the various meetings of the Program that he met Donald Boulter, Head of the Department of Botany, University of Durham, Durham, in northern England. In this Department, which later would be called Department of Biological and Biomedical Sciences, he was a visiting professor during the year of 1976. All his family accompanied him. They lived in a large two-story house, owned by the University, located in solitude in the midst of a private park called Elvet Garth, South Road, the same street where the University is located. From the high windows of his room, on the first floor, along with Diva, he found himself many times contemplating the park, admiring the beauty of its landscape: a veritable garden where, in different seasons, bloomed several species of ornamental plants, from "snow drops" in the early spring to roses in the summer. Tall rhododendron bushes surrounded the house, buds blooming in succession, pink, blue, violet and white flowers over the months from June to August. Several species of trees formed corners on the sprawling park, suitable for the games of his children. In Durham, their daughters, Maria Paula and Carla Maisa, attended St. Godrics Primary School, next to "Our Lady of Mercy & St. Godrics Catholic Church." Their older sons, Marco Augusto and Adolfo Egídio, attended Whinney Hill Grammar School. Little Daniel, who was 3 years old, attended a private nursery school.

The imposing Durham Cathedral in Norman style, built between 1093 and 1113, dedicated to St. Cuthbert, who died in 687 AD, and which holds the remains of the Venerable Bede, a Benedictine monk, the author of numerous works on the history of the Church in England, was visited by Otto's family, especially in concert performances of classical music. The Beethoven's Ninth Symphony "Choral" was one of those presentations. In the crowded Cathedral, next to a silent audience, he and Diva embraced the chords sometimes soft sometimes majestic of the melody of Beethoven, which, in a crescendo, exploded in the voices of the Choir on the 4th movement. Overwhelmed by the beauty of the masterpiece, albeit familiar to them, they left the Cathedral and, in the square that surrounded it, they allowed themselves to stand under the moonlight that insisted on appearing from the mist of that cold November night.

Next to the cathedral, there is also the secular Castle, former residence of the Bishops of Durham. These two imposing buildings are located on a hilltop, with access from the Market Square where are the City Hall of Durham, the Church of St. Nicholas, and the statues of Neptune and of the 3rd Marquis of Londonderry, a historic personage of Durham's economy.

In August of that year, just as he had promised himself, he took his whole family to visit several countries in the European continent. At that time he visited and established scientific contacts with protein laboratories in the Faculté de Sciences Agronomiques in Gembloux, Belgium, and with the Department of Plant Biology, University of Geneva, Switzerland, where later his assistant, Maria Tereza Vitral de Carvalho, developed part of her Doctor Thesis on plant proteins. On their way back to England, they visited Edinburgh, in Scotland, on a train ride that enchanted Daniel.

At Boulter laboratories he developed research on pea storage proteins in collaboration with Eric Derbyshire, who was later hired by him as a researcher at Plant Biochemistry Section of CENA. Derbyshire married Maria Teresa V. de Carvalho in Sao Paulo. Maria Tereza died in 2012 some years after the death of her husband.

In September, 1977, Donald Boulter presented a scientific work on plant tissue culture of legumes, with Otto's collaboration, during the 4th Annual College of Biological Sciences Colloquium, organized by Rod Sharp at The Ohio State University, Columbus, Ohio, USA.

In other instances, Otto returned to his always well-remembered Durham. In June 1980, after ministering a course on plant metabolism of amino acids and proteins at the University of Bologna, Italy, and discussing the results of his project on legume proteins in the Department of Biotechnology at the Carlsberg Research Institute, in Copenhagen, Denmark, he visited the laboratories of Donald Boulter, University of Durham. Also, in October and November, 2003, along with Diva and their daughter Maria Paula, then a Masters student in London, they were once more in Durham, when they stayed with George and Breda Gallagher's home, their very good friends since 1976. George died a few years later. Boulter is a Professor Emeritus at the University of Durham. On that occasion, he and Diva revived their beloved London, revisited the museums, theaters, made tours to the parks and once again, as they used to do in other times, attended concerts of classical and contemporary music in the early afternoons on Wednesday Performances at St. Martin in the Fields, by Trafalgar Square, followed by lunch at the "Café in the Crypt" in the basement of that Anglican Church.

In his long walk through the streets of Munich, those memories made his heart beat more strongly with longing, perhaps, but also contentment. There, in the city of the clock tower with moving figures, in Marianplatz, another phase of his scientific life was coming to an end. He took the tram passing by and returned to his hotel.

These recent facts were all in his memory. His colleagues in the Program for Legume Seed Proteins were Augusto Tulmann Neto, S. Blixt, K. Mikaelsen, L. Sodek and Gerald Lee-Sheng Tseng at CENA, and Donald Boulter, from England. Under his guidance, Gerald Lee developed his Doctor thesis, defending it in the Biochemistry Department of the Institute of Chemistry, University of Sao Paulo, in Sao Paulo, in 1978. The seeds of a mutant of the variety Carioca, obtained by Tulmann during the development of the Program at CENA, were the material used by Gerald Lee in his thesis on beans storage prot

VII
Rod Sharp and Otto Meet

1971. In the introductory classes in sciences, in his early days as a high school student, during the forties of the last century, Otto would become fascinated with the description of the cells. These tiny beings, as to him they seemed to be, were inside his inquiring mind: the drawings of cells in the books and those with whom he played, barely drawing them in his notebooks, were to him like living creatures. In his imagination, they were the ones that rocked the flowers and leaves of the trees in the park of the "Escola Agricola" often visited by him, and they were the ones to allow the legs and arms of he and of his brothers and sisters and friends to walk and gesticulate. He had never seen the cells. He saw them for the first time at the optical microscope in the residence of João, one of his colleagues, whose father, Canuto Marmo, was a professor at the "Escola Agricola." In the examination plate were those "beings" going from side to side as if dancing in endless swings. It was true, cells existed! They were not beings, they had no legs or hands! But, as he had imagined, they moved and danced as if they were bodies, weaving, multiplying, and creating life. He was ecstatic! And even more amazed when, years later, he came across the works of Robert Hooke made in the XVII century. He learned that Hooke observed a piece of cork under the microscope and described it as resembling a honeycomb. Hook then called these tinny compartments cells. Hooke published his findings in a paper entitled "Micrographia" on March 20, 1665! Almost two centuries before Scheleiden and Schwann (in 1839) confirmed that cells were the fundamental particles of plants and animals tissues. As Otto would later learn, these two scientists demonstrated the totipotency of plant cells, which during his scientific life, he and William Rod Sharp, and his other colleagues, would use in his walks through the world of plant cells cultivated in vitro

But he was now in the 20th century. Electron microscopes, far more powerful than optical microscopes were used to unravel the "mysteries" of animal, vegetable and mineral natures with greater precision. The cells deciphered, appeared to his eyes with all the importance of their functions: from the stem cells to the most specialized ones. Now he knew: biochemical reactions occur only because the cell exists; at the same time, biochemical reactions are responsible for the existence of the cells, he concluded with wonder, once again, before the complexity and beauty of nature.

All these memories and the constant conjecture about the cells and their function were very much alive in his memory when Malavolta entered his room at CENA in early 1971. This was a moment that would forever mark his scientific life. A young American scientist obtained financial support to develop research on plant cells cultured in vitro for a short period of time in Brazil. Following the suggestion by Henrique Vianna de Amorim, who was studying in the United States, that scientist decided to come to ESALQ. Malavolta asked Otto to receive him. He immediately felt his own eyes shining with excitement. It was everything he expected to happen. The young scientist was William Rod Sharp, a cell biologist, professor at The Ohio State University, USA.

Working with plant cells in vitro was what he had tried so far without positive results. At ESALQ, Darcy Martins da Silva, his colleague at the Chemistry Department, had brought with him from the Netherlands a few glass vials with aseptic cultures of plant cells. Gustav F. Brieger, Professor of Genetics at ESALQ, cultivated orchid seeds in vitro. Handro Walter, of the Department of Botany, Institute of Biosciences, São Paulo campus of the University of Sao Paulo, was in France, at Colette Nitsch laboratory, becoming familiar with the techniques of in vitro culture of anthers, which he subsequently introduced in his laboratory. In that same Department, at the beginning of 1960 Maguro, Hell and Gilberto Karbauy were already taking the first steps in the use of the techniques of plant cell and tissue culture.

Otto enthusiastically endorsed the opportunity to enter the realm of the plant cell science and bring to light the sequences of biochemical reactions, and thereby the possibility to control them. Moreover, he foresaw, in collaboration with Sharp, broader perspectives: the use of the techniques of plant cell and tissue culture as an important auxiliary tool in plant breeding and the production of somaclonal variants and clones. It was his pioneering spirit and that of Rod Sharp's, at CENA and ESALQ, while introducing these techniques in a systematic way, which launched the beginnings of plant biotechnology in Brazil: the application of these in vitro techniques in agriculture, as Otto had always dreamed.

1971 – 1981

December, 1981. On the morning of the 6th, he left his room at CENA and headed to his room as the Head of the Department of Chemistry in ESALQ. He was preparing for the opening of the "International Symposium on Biotechnology for Genetic Engineering" which would take place at the Building of Agricultural Engineering. While looking out the window of the room, once again he admired the beautiful leafy trees in the park of the "Escola", designed by the Belgian Arsenio Puttemans and deployed in the early years of the twentieth century. Its expansion and maintenance over the years were the result of the efforts of many ESALQ professors, mainly Philippe Westin Cabral de Vasconcelos, whose name would be perpetuated as the official name of the park in 1986. At that moment of contemplation, he turned to the past and images, as living entities, took shape in his memory.

Ten years had passed since Rod Sharp, who would later become his great friend and scientific partner, got off the plane which that brought him for the first time to Brazil in June, 1971. Otto, being short, with a swarthy complexion and black hair still, marveled when he saw that tall, blond young man, his antithesis! Then both in broad gestures gave each other a first hug, as if signaling the beginning of a friendship that would last for many years.

Rod Sharp was accompanied by a lady - she was his maternal grandmother. Her natural white hair gleamed in the sunlight that bathed the Congonhas Airport in that mild cold morning in the city of São Paulo. Her kindness and gentleness of gestures soon conquered him and thereafter Diva and those who met her in Piracicaba. Mary Beatrice was her name. Her parents were of Dutch origin, and she was born in Mansfield, Ohio, USA, where she graduated at Ashland College. Mother of two daughters, the eldest of them Rod's mother, with great life experience, had been married twice. She was resourceful, sharing with Rod the routine adaptation to living in a country with customs quite different from hers. Every time she returned to Piracicaba, and those were many, Mary Beatrice delighted her guests with her companionship and her delicious apple pie! (Photo 2).

In 1971, despite the fact that Brazil was under military dictatorship, CENA always created conditions for new scientific advances. Otto's partnership with Rod Sharp was enthusiastically encouraged by Ademar Cervellini, Director of CENA.

The empathy that emerged between the two collaborators facilitated the adaptation of the young Rod. He and Otto bought wood, glass and towels and built the first aseptic chamber, located in the Plant Biochemistry Section of CENA. Another room was adapted to maintain the flasks with cultures of various plant species under light and temperature control. Thus began the intensive learning of these techniques by the technicians, undergraduate and graduate students at ESALQ. Linda S. Caldas, mentored by Rod at The Ohio State University, Columbus, USA, wife of Ruy de Araujo Caldas, Malavolta's Assistant, who had developed his Ph. D. in Biochemistry also at Ohio State University, joined the group at CENA. Some years later, Linda and Ruy were hired by the University of Brasilia, the capital of Brazil, where they developed a productive program of teaching and research in Biology and Biochemistry of plants.

From 1971 to 1981, the Biochemistry of Plants Section at CENA experienced intense research activities in both areas of biochemistry of nutrients, amino acids and proteins and of cell biology in plants. The Rod Sharp's annual visits to the laboratory of CENA and Otto's visits to Rod's laboratory in the United States and several others ones in European countries, fostered many opportunities for both as well as for their students to develop their research works using in vitro techniques with various plant species: sesame, tomato, peanuts, beans, pineapple, coffee, peas, eucalyptus, pine, ornamental plants, medicinal plants, and many others.

In that decade, encouraged by Rod Sharp, on July 1974 Otto participated in the Third International Congress on Cell and Tissue Culture Plants at the University of Leicester, Leicester, England, organized by the International Association of Plant Tissue Culture (IAPTC). In this Congress he presented the results of his research on in vitro culture of bean tissues in collaboration with Rod and Maria Teresa V. de Carvalho. He was a representative of the Association in Brazil for 12 years.

One major research developed by Otto and Rod was the tissue culture of sugar cane (Sacharum spp), of great economic importance to Brazil. Tissues and cells of this species were used for studies of somaclonal variation, aiming to select tolerance to the herbicide ametrin, an herbicide used to control weeds, which can be absorbed through plant leaves and roots. Neftali Ochoa Alejo, of Mexican origin, oriented by Otto and aided by Enio, developed this research in his Doctor thesis in biochemistry in the

Institute of Chemistry, at the Sao Paulo campus of the University of Sao Paulo in 1983. The results of this work were presented by Otto at the International Symposium on Applications of Cell and Tissue Culture in Plant Breeding, at the Academy of Sciences in Olomouc, in the former Czechoslovakia, then part of the Soviet Union, in September 1984. On that occasion, he and David Evans exchanged experiences in the field of cell biology. David Evans, an associate of Rod's collaborator, has visited many times at Otto's laboratory at CENA, greatly contributing to the research developed in there.

The cultivation of sugar cane in Brazil began in 1533 with the first seedlings brought by Martim Afonso de Souza, from the Madeira Island, the main island of the Madeira Archipelago, situated in the Atlantic Ocean southwest of the Portuguese coast. Afonso de Souza was the first colonizer of Brazil, sent by Portugal. This culture came to meet the urgent need for colonization and exploitation of a vast territory, having no economic impact on Portugal. Although the first sugar mill to process sugarcane was installed in Sao Vicente, on the coast of what would later become the State of Sao Paulo, it was on the coast of northeastern Brazil that the cultivation of sugarcane prospered, especially in regions that in the future would be the States of Pernambuco and Bahia. The soil, dark, almost black, rich in clay and humus, present in this region and then called "massapé," was much explored during the Portuguese colonization for the cultivation of this grass. The sugar produced in Brazilian mills, exported to Europe with a high profit margin, was the backbone of the economy of the Portuguese colony between the sixteenth and seventeenth centuries. In fact, Brazil is today the largest exporter of sugar extracted from sugarcane in the world.

In Brazil, the use of ethanol, produced from the fermentation of sugarcane as fuel for vehicles, had been known for many decades. However, it was only in 1970's, during the global crisis of fossil fuels, that the military government, under the presidency of General Ernesto Geisel, decided that alcohol, considered up to then a byproduct of sugarcane would play a major role in the Brazilian economy. The National Alcohol Program (Proalcool), created by government decree in 1975, led to the immediate effect in the production of cars powered by ethanol alone or a mixture of ethanol and gasoline. In the first decade of the 21st century, this practice was resumed with greater intensity with excellent results. Since then, the emissions of carbon monoxide by burning fossil fuel have been greatly reduced, bringing down the degree of pollution in the streets and highways of Brazil.

One of the byproducts derived from the alcoholic fermentation of sugarcane is the vinasse, a residual liquid that has various properties. Currently, its use has been advocated as feedstock for the production of biofuels. Another possibility, studied since the decades of 1940/1950 by the Jaime Rocha de Almeida Group, at ESALQ, is its use as a fertilizer. In this context, one of Otto's graduate student, Joaquim Albenfsio Silveira, developed in 1985 his Doctor thesis on the relationship between nitrogen fertilization and plant growth of sugarcane in field conditions, using the vinasse. Albenfsio, born in the state of Ceara in northeastern Brazil, was hired as an assistant by Otto at the Agricultural Biotechnology Center - CEBTEC in ESALQ. Years later he returned to Ceará.

The identification of hundreds of varieties and hybrids of sugarcane is quite difficult. For taxonomic classification one of the most appropriate methods is based on biochemical differences, which is added to the traditional morphological methods of the various parts of the plants. Although not all of the biochemical differences are reflected in their morphology, they exist and are important in the taxonomy of sugarcane.

Professor, I am interested in sugarcane and would like to develop my Master's thesis exploring the biochemical potential of the cells to determine an analytical tool for the identification of varieties, Marcilio said entering the Otto's room at CEBTEC.

Great, said Otto. "Let's look into this issue," he continued.

Marcilio de Almeida had looked for him in late 1970s while still a graduate student in biology at the University of the State of Sao Paulo "Julio Mesquita" (UNESP) in the city of Rio Claro, near Piracicaba, for an internship at CENA, under Otto's guidance. He stood out among Otto's students. After his graduation he continued under his supervision in his Master's Program at ESALQ under Otto's supervision.

It is important that you work with 10 varieties of sugarcane, the most commonly used by farmers, Otto suggested. "Besides the organographic characterization of leaves and stems, which is your specialty as a biologist, I propose that we also review some biochemical parameters", he continued.

What are these parameters? Asked Marcilio.

You can analyze the activity of peroxidase and esterase isozymes, the total soluble protein content and the level of soluble solids, which will provide you with a great amount of data to draw up a key, which is your main goal, explained Otto and continued: "Enio could assist you in these analyzes."

Marcilio defended his Master's Thesis in 1986, having been hired by the then Department of Botany, ESALQ, and today known as the Department of Biological Sciences. Later, in the 1990s, under the guidance of Gilberto Kerbahuy, Marcilio developed his doctoral program in the Department of Botany in the Institute of Plant Sciences on the Sao Paulo campus of USP.

December 6, 1981. - Professor, the opening ceremony for the Symposium will begin soon. We should head for the Amphitheatre of the Engineering Building, said Enio entering the office of the Head of the Chemistry Department.

I know, said Otto still absorbed in his reveries, his eyes marveled at the beauty of the ESALQ Park. We still have some time.

Has Professor Sharp arrived? He asked, turning to his technician and friend

The driver is picking him up at the hotel, along with the other invited foreign scientists and Professor Tavares, answered Enio.

Otto went back to his memories and saw in his mind Enio setting foot in his lab at CENA at the beginning of summer 1977, as he had just done now.

Enio Tiago de Oliveira was seeking a training post for a few months in the Section of Plant Biochemistry, which was mandatory for all students of Technical Chemistry Course. He was enrolled at Dom Bosco College in Piracicaba. A skinny young man, of small stature, at 18 years of age, Enio appeared before Otto, who accepted his request, since he never rejected anyone wanting to work in his labs in order to become familiar with the techniques of biochemistry and cell biology. Enio soon undertook the laboratory techniques. Under the eyes of Otto and his assistant Maria Teresa, Enio showed his ability and responsibility in the use of in vitro techniques and biochemical analyzes. His training period was extended, and as time went by, Enio exceeded all expectations (photo 3).

Enio was born in a small town in the Vale of Jequitinhonha in northern Minas Gerais state, a region characterized by Guimarães Rosa in his novels, one of the greatest Brazilian writers, passionate about the interior of Brazil. At age 11, Enio left the Taiobeiras city with his parents and eight brothers and sisters. He was the fifth son of Jones and Olivia. They moved to Piracicaba where he worked, still very young, studied and married Joana d´Arc, with whom he has two sons, Marcele and Tiago. He was hired by CENA and later transferred to the Department of Chemistry at ESALQ where, as Senior Technician, joined the scientific staff of the Center for Agricultural Biotechnology-CEBTEC.

Intuition and an unusual curiosity were the predicates that led this young man, coming from one of the poorest regions of

Brazil, to become Otto's right-hand man, participating in a great deal of research, helping students in developing experimental works for their Master and Doctoral Theses. Over time, Enio became responsible for the introduction of cells and tissue culture techniques for the establishment of sugarcane somaclonal variation. In 1979, he actively participated in the five-year Agreement signed by CEBTEC with Planalsucar to Install on their premises in the city of Araras, in the State of Sao Paulo, a tissue culture lab and train their staff in technical skills to obtain somaclonal variants as an auxiliary tool for sugarcane improvement projects.

Planalsucar. - The National Program for Sugarcane Improvement, a program developed by the former Institute of Sugar and Alcohol (IAA), aimed at the continuous renewal of the Brazilian and imported sugarcane varieties allotments, using conventional improvement techniques. This Program was responsible to stock the raw material for sugar and ethanol production in Brazil. Several Planalsucar Experimental Stations developed this work, including the one of Araras, in the State of Sao Paulo, where Otto's team installed laboratories for culturing cells and tissues of sugarcane. After the extinction of IAA, along with Planalsucar in 1990, the Network Inter-Institutional Development of Sugar and Alcohol Sector-RIDESA was created involving several federal universities, including the Federal University of Sao Carlos. Later the Araras unit developed skills of in vitro techniques, under the leadership of Tseng-Sheng Gerald Lee, Otto's former graduate student.

Presently a Biologist at the Department of Biological Sciences in ESALQ, Enio finished his course on Biology in 1987 at the Methodist University of Piracicaba UNIMEP. Under Otto's orientation he took his Master's Degree in Agricultural Sciences in 1993 and, in 2010, his Doctor's degree under the orientation of Luiz Antonio Gallo, both at ESALQ.

A singular moment happened in 2009 when Enio entered Otto's office at CEBTEC and, not hiding his satisfaction, said:

Professor, I received a request from the Editors of the Hortscience Journal to authorize the publication of a photo of our article on Aloe vera on the cover of this coming October issue. That piece of work was part of his Doctor thesis. Enio added: "we need to respond."

With joy, Otto instantly replied, "Congratulations, let's respond right now, saying yes."

The "Symposium on Biotechnology for Genetic Engineering" had been organized by Otto in collaboration with Flavio C. A. Tavares, from Department of Genetics at ESALQ, and Decio Sodrzeieski, from the Department of Industry and Commerce of the State of Sao Paulo. It was the crowning of the research activities of his group with the collaboration of other research institutions in Brazil and abroad, which began when Rod Sharp joined him in June, 1971.

Otto and Enio left the office of the Head of the Department of Chemistry and strode across the extensive park in front of the beautiful ESALQ Administration building, opened in its original condition on May 14, 1907. They both headed to the Engineering Building which, with its imposing dome, dominated the entire border area of a large lake where geese and ducks glided silently, as if dancing, indifferent to the scientific event that was about to begin. The residence of the Director of ESALQ, in architectural style of the southern United States, with its tall white columns, completed the dream scenario in that early sunny morning.

The Brazilian and foreign scientists, who would attend the conferences were waiting for him and Flavio Tavares in the lobby of the building. The opening ceremony chaired by Aristeu Mendes Peixoto, Director of ESALQ, had taken place the night before, in the Great Hall of the Administration Building. Besides

Paulo Fernando Cidade de Araujo and Joaquim Jose de Camargo Engler, also present was Salvador de Toledo Piza, ESALQ renowned scientist, devoted to anatomy and physiology and an expert taxonomist, especially of insects, with a twist: his works were published in Latin. He is the author of the "Ode to ESALQ," which praises the work of Luiz Vicente de Souza Queiroz: "the School is your monument."

The floor of the Amphitheatre of in the Engineering Building was filled to its full capacity with teachers, students and scientists from several countries. The topics presented and discussed during the five days of the Symposium were about the use of genetic engineering and tissue culture cells in energy, agriculture, and animal systems, in microbiology and human cells. The lectures were presented by scientists from Argentina, Brazil, Canada, Scotland, the United States of North America, France and England, as follows: N. Alexander (Dept. Agriculture, USA), J. R. Johnston (University of Strathclyde, Scotland), C. J. Panchal (Labatt Brewing Co. LTD., Canada), D. Boulter (Durham University, England), A.M. Chakrabarty (University of Illinois, Medical Center, USA), I. Roitman (University of Brasilia, Brazil), M. Mares-Guia (Bioferm - Research and Development SA, Brazil), F. G. Nobrega (University of Sao Paulo), Heslot, H. (National Institute Agronomique, France), C. M. Morel (Oswaldo Cruz Foundation, Brazil), F. J. S. Lara (University of Sao Paulo, Brazil), D. M. Glover (Imperial College of Science and Technology, England), D. M. Silva (ESALQ, USP, Brazil), Maragarida L. R. Aguiar-Perecin (ESALQ, USP, Brazil), Alaides P. Ruschel (CENA, USP, Brazil), E. Chartone Souza (Federal University of Minas Gerais, Brazil), R. R. Brentani, University of Sao Paulo, Brazil), C. D. Denoya (Instituto Sidus, Argentina), W. R. Sharp (DNA Plant Technology Corp.., USA), and Joao Lucio de Azevedo, Flavio C. A. Tavares and Otto J. Crocomo, from ESALQ, USP.

The Symposium was one of the most important scientific events in the early programs of biotechnology and genetic engineering in Brazil. The Federal Government had launched the PRONAB - National Biotechnology Program, discussed throughout the year 1981 by a committee of Brazilian researchers, among them, Otto, Flavio Tavares and Joao Lucio de Azevedo. A few days after the Symposium, the text of PRONAB was delivered to the then Finance Minister Delfim Neto, in Brasilia, the capital of Brazil.

VIII
The creation of the Center for Agricultural Biotechnology-CEBTEC

May-July, 1981. On that Brazilian autumn morning in May, Otto and Flavio Tavares were sitting at the boardroom table at the office of the Head of the Section of Plant Biochemistry at CENA. During that meeting, they discussed the latest resolutions of the PRONAB Committee which was held that month at CNPq, Rio de Janeiro.

CNPq - National Council for Scientific and Technological Development, a foundation linked to the Ministry of Science and Technology of the Federal Government of Brazil has, since its creation in 1951, one of the leading and more solid public structures to foster science, technology and its innovations. It is directly connected to the training of teachers, doctors and scientific researchers in Brazil and abroad. Another public company is FINEP - Public Finance for the Development of Studies and Projects, also under the Ministry of Science and Technology, founded in July 24, 1967. FINEP operates across the technology innovation chain, focused on strategic and structural actions and their impact on the economic and social development of Brazil, funding projects that involves public and private sectors. Another institute for scientific research and development is FAPESP - Foundation for Research Support of the State of Sao Paulo. Founded in 1962 by the Government of the State of Sao Paulo, FAPESP has been responsible for much of the scientific and technological development of the Universities and Research Institutes, also financing the public sector partnerships with the private sector in the State of Sao Paulo. Due to the Genome Project, financed by FAPESP in the late twentieth century, Brazil emerged as a leader in the international scientific community in this field. Previously, the program Bioq-FAPESP drove the research and teaching in biochemistry in the Universities of the State of Sao Paulo in the 1970s. FAPESP has been a model for the creation of foundations with the same goals in other Brazilian states.

Over the past 20 years I have had projects approved by CNPq, FINEP, FAPESP and The National Energy Commission (CNEN). If it were not for these financial resources, I could never have done my research here at CENA, Otto said as he got up to look for a package containing reports of his activities.

At the Department of Genetics the Head, Ernesto Paterniani, has given me plenty of support for the development of scientific work funded by several donors, said Flavio Tavares.

- It's interesting that while Professor Cervellini was the Director of CENA my initiatives never lacked support, said Otto. "Unfortunately, he retired," he added.

But isn't he taking over an important administrative position at CNEN, in Rio de Janeiro? Asked Flavio.

Yes, Otto answered, which helps us a lot. But it is not the same situation as before. I feel that I badly need to do something else. Discussions at PRONAB Meetings leave me uneasy. The suggestions that Carlos Morel, Mares-Guia and Francisco Lara are offering make me think about launching some new initiatives, in addition to the ones I develop here at CENA, including our relationships with private companies. That was a reference to Morel, a Brazilian scientist from Oswaldo Cruz Foundation in Rio de Janeiro and Mares-Guia, from Bioferm, a biotech Company at the State of Minas Gerais. With Francisco Jeronymo Salles Lara, Full Professor of Biochemistry, Otto had an old relationship since the 1950s, when he attended his laboratories of biochemistry at USP in Sao Paulo and in Ribeirão Preto.

-I know that the Planalsucar Project can open up many other opportunities for research; nevertheless, I would like to do much more using our techniques of plant cells and tissue culture for agricultural production, continued Otto.

Flavio listened intently. Both had a very good relationship and both were interested in developing biotechnology projects focused on agriculture. In that sense Otto's labs at CENA were relatively well-equipped to develop techniques of plant biotechnology in a multitude of crops.

-Our ten-year partnership with Rod Sharp has allowed us to establish the foundations for a plant biotechnology, Otto emphasized, adding: "Flavio, would you be willing to collaborate in creating a biotechnology center?"

It seemed to Otto that Flavio eyes shone with his proposal. The idea of a biotechnology center was launched. Its implementation would require considerable effort, and Otto predicted the steps that had to be immediately taken.

At the end of the last week of May, Otto met with the Director Aristeu Mendes Peixoto at the beautiful home of the Directors of ESALQ, on campus. Throughout that morning they discussed the proposal to create a Biotechnology Center in ESALQ. With the approval of the Director Peixoto, Otto met with Joaquim Jose de Camargo Engler, one of the Directors of the Foundation for Agronomic Studies Luiz de Queiroz - FEALQ. Engler approved the idea.

FEALQ. It was Paulo Fernando Cidade de Araujo, professor of agricultural economics at ESALQ, who had the idea of creating a School Foundation to launch research in technological and scientific fields with resources outside the USP. With support from the then Director of ESALQ, Salim Simao, the Foundation for Agronomic Studies Luiz de Queiroz - FEALQ was established in December 30, 1976, as a private entity, nonprofit organization.

The first Board of FEALQ was formed by Paulo Cidade as its CEO, Joaquim José de Camargo Engler and lby A. Pedroso. The Board of Trustees was formed by Salim Simao (Chairman), Roberto Cano de Arruda, Dovilio Ometto, Aristeu Mendes Peixoto, Urgel de Almeida Lima, Humberto de Campos and Rubens Valentini.

The aim of FEALQ is to support and administer technological, scientific, economic and social projects from ESALQ, CENA and other units of USP and public and private institutions. FEALQ manages courses, conferences, symposia, seminars and other events and also publishes books and journals for the dissemination of technology. It offers scholarships for students who participate in projects financed by FEALQ and collaborate on social development programs of ESALQ campus communities.

At that time, the headquarters of FEALQ stood on campus in one of the homes previously occupied by ESALQ faculty and staff members. Sitting at the table, Otto expressed his ideas to Engler.

- As you know, there is a committee formed by Brazilian scientists working on a project to develop biotechnology, with support from CNPq and FINEP. Engler, as usual, listened intently. "It is the National Biotechnology Program, the PRONAB, to be presented to the Brazilian Government later this year", Otto continued.

Is it the Committee which you, Tavares and João Lucio de Azevedo belong, is it not? Engler asked.

Yes, we have met quite often, Otto confirmed, adding: "I have high expectations of the resources that can be allocated to develop technological and scientific projects. But I would like to go further: use these in vitro techniques in projects involving agricultural businesses. Indeed, in the last ten years, since the coming of Rod Sharp to the Plant Biochemistry Section at CENA, in June 1971, we have been preparing to definitively undertake public-private projects."

One example is the project you have with Planalsucar, said Engler.

Precisely. This is a project under my responsibility and, as you know, has been administered by FEALQ since 1979, Otto continued, explaining: "the project is underway with good scientific results and personnel training. Our relationship with this company is very good. Osny Bacchi, as a Planalsucar researcher has helped us greatly, as well as Enio, and undergraduate and graduate students as well in CENA. "Noticing Engler's interest, Otto explained: "the possibility of signing an Agreement with Planalsucar came when I presented the results of our studies on cells and tissues of sugarcane cultures during the Sugarcane Producers Congress held in Maceió, the state capital of Alagoas, in January 1979" and continued: "the private sector has been receptive to the idea of using new technologies."

I would like to propose to FEALQ the creation of a Biotechnology Center at ESALQ, which would be administered financially by FEALQ, Otto continued, saying: "I have already gotten the approval of this idea by the Director of ESALQ, Mendes Peixoto."

Motivated by Engler's support, on June 25, he, Flavio Tavares and Ernesto Paterniani, Head of the Department of Genetics, ESALQ, submitted the project to create the Center to Aristeu Mendes Peixoto who officially approved on behalf of the ESALQ Administrative Board.

At the meeting of the FEALQ Board of Trustees, on July 2, 1981, the project for the creation of the Center of Biotechnology was presented and discussed. It has the "main purpose of conducting research based on cellular systems for technological purposes, with the participation of researchers of ESALQ and other institutions interested in issues related to Biotechnology and Genetic Engineering."

The creation of the Center was approved by the Board and, as stated in its minutes, "the FEALQ Board of Trustee will be in charge of the arrangements for its implementation." In addition to Otto, also present at that meeting were: Urgel de Almeida Lima (Chairman of the Board), Pedro Tessinari Filho, Joao Ribas Fleury and Jose Roberto Mendonça de Barros. As guests, Aristeu Mendes Peixoto, Salim Simao, former Director of ESALQ, and Paulo Fernando Cidade de Araujo attended the meeting.

Being one of the first to receive the news, Rod Sharp responded enthusiastically about the approval of the creation of the Center. He was a key part in partnership with Otto in this scientific adventure in a country that was in its last years of military dictatorship.

But, now it was December 6, 1981 and these events were present in his thoughts as he wandered the Amphitheater and greeted the participants of the "International Symposium on Biotechnology for Genetic Engineering," whose implementation had also been approved at the same meeting of the FEALQ Board of Trustees in that afternoon of July 2, 1981.

The Biotechnology Center, which later would be called Center for Agricultural Biotechnology-CEBTEC, was created. Otto was appointed as its Scientific Coordinator.

This was only the beginning of a journey to the achievement of the Center and, as he knew, would be a quite long and laborious task.

Besides him, were part of the first Scientific-Technical Council of CEBTEC: Flavio Cesar Almeida Tavares, Henrique Vianna de Amorim, Luiz Carlos Basso, Caio Cardoso, Elke Cardoso, Wilson Roberto Soares Mattos, Luiz Eduardo Gutierrez and Aline Aparecida Pizzirani Kleiner.

Since 1964, Otto continually received financial support for conducting research in the field of plant biochemistry as well as in cell biology-FAPESP being one of the leading official entities to give him financial support, which was repeated over the subsequent years. The creation of CEBTEC opened broad prospects for research contracts and technology development (R&D) using in vitro techniques supported by Brazilian and foreign Governments. Among the Brazilians, besides FAPESP, are: CNPq; FINEP; CAPES (Coordination for Improvement of Higher Education Personnel, of the Brazilian Ministry of Education); EMBRAPA (Brazilian Agricultural Research Corporation, of the Brazilian Ministry of Agriculture); CNEN (National Nuclear Energy Commission). Among the foreign: The British Council; The OAS (Organization of American States); Fulbright Commission; UNDP (United Nations Development Programme); IAEA (International Atomic Energy Agency, Vienna, Austria); French Government; DADD (Germany Government); Scientific Commission of the European Communities; USA Government I Blue Ribbon Project; Corporacion Andina de Fomento (CAF, Venezuela); National Institute of Higher Education (NIHERST), Trinidad.

IX
The Early Years of CEBTEC

Early 1982, the interest in the fledgling CEBTEC divided their participants. A common physical location would be important to realize the idea.

CEBTEC laboratories must be installed on the campus of ESALQ, said Otto, addressing Luiz Carlos Basso.

But where? Various departments are already using all viable places. Basso showed the concern that he shared with Otto, to make the idea of CEBTEC a physical reality.

The answer which Otto gave served to reassure Basso:

I know the difficulties we are facing, but I have an idea. I will propose to FEALQ that they finance an upgrade of the laboratories of Biochemistry, at the Department of Chemistry, Otto emphasized, continuing: "I just signed a contract with Johnson & Johnson to conduct research and development in Eucalyptus. As you know, Basso, this multinational company, based in Sao Paulo, is interested in the length of eucalyptus fibers. The proposal is to use in vitro cultures to reach this objective. Moreover, Antonio Natal Gonçalves is working with that fast grown species in the Department of Forestry, and will be our collaborator on this project. At our labs at CENA, Enio has begun the cultivation of Eucalyptus using multiple protocols."

I believe that the Board of our Department will approve this proposal. I will also consult with Paulo Cidade and Aristeu Mendes, Otto finished.

Antonio Natal Gonçalves, an expert in the physiology of trees, completed his Master's program under the guidance of Rod Sharp at The Ohio State University, in Columbus, Ohio, in the late 1970s. On his return to ESALQ, he was under Otto´s supervision for his Doctorate Program and defended his thesis on juvenility and cloning of Eucalyptus urophyla in vitro in 1983.

On November 29, 1982, the premises of CEBTEC new laboratories were inaugurated in the basement of the Chemistry Building, remodeled with financial support of FEALQ. A Scientific Seminar, attended by Antonio Vieira Guerra, Rector of the University of Sao Paulo and the Directors of ESALQ and FEALQ, opened the ceremonies. On this occasion, Jose Dall Pozzo Arzolla was honored as Otto's former co-supervisor in his lab in organic and inorganic chemistry during his graduation at ESALQ and in the experiments of his "Livre Docencia" Thesis in 1957/1958, much of it developed in the laboratories of the Institute Zimotecnico, which does not exist anymore. The entrance hall to the new CEBTEC laboratories has a plaque to honor Professor Arzolla.

Jose Dall Pozzo Arzolla and Domingos Pellegrino, along with Demosthenes Santos Correa and Euripides Malavolta, were the ones who most influenced Otto in the search for his scientific identity in the 1950s of the last century. During high school with Demosthenes, he entered the realm of chemical reactions, laboratory techniques, which involved long hours, and his participation in debates of Chemistry, during the three years of his high school course. At the University, with Malavolta, he became familiar with the interpretation of scientific articles on biochemistry: at that time, books and journals of biochemistry in English, French and Spanish took a great deal of time to shipment to Brazil. The Internet would only be available many years later. He became a "bookworm." self-taught on biochemistry. After graduating from ESALQ, and already employed by USP; his training was completed in 1964/1965 with Conn and Stumpf, at University of California, Davis campus. With Pellegrino, his professor of analytical chemistry, he learned to be meticulous in chemical analysis and the interpretation of results; he became interested in statistical analysis of data and attended extracurricular experimental statistics courses with Professors Friedrich Gustav Brieger, in the Genetics Department, and Frederico Pimentel Gomes, in the Mathematics and Statistics Department, at ESALQ. Through Arzolla, his professor of Organic Chemistry, Otto got in touch with the dynamism of theoretical and laboratory classes, developing refined procedures for conducting experiments. He and Arzolla built a strong and beautiful friendship over the years, until his untimely death in 1970.

While walking through the new laboratories, some memories came back to him. In those same counters, now remodeled, after his return from London, where he worked with Fowden, he, Luiz Carlos Basso and Oswaldo Brazil, in the years 1969-1970, set up experiments that led to the discovery of the enzyme N carbamilputrecine decarboxylase, essential to putrecine formation in the ornithine-urea cycle, when Sesamun plants are subjected to potassium stress. So many repetitions of the tests, so many discussions in the preparation of the scientific paper reporting the experiments, so much excitement at the news that it had been accepted for publication in the Phytochemistry journal!

The search for the biochemical basis of cellular phenomena responsible for plant development has always been present in his mind. He constantly repeated to himself that this was the reason why he wished to penetrate into the realm of cell biology. And so, the partnership with Rod Sharp became increasingly timely. However, at that moment he had a premonition that both, far beyond entering the heart of the plant cell, could also contribute to the advancement of agriculture, using in vitro techniques. After all, the physical conditions existed: the premises of the CEBTEC at ESALQ, through which he was now walking, plus his laboratories in the Section of Plant Biochemistry at CENA, would be suitable for the development of this type of project. Indeed, soon, in early December of 1982, he would discuss the potentiality of the techniques of plant cell and tissue culture for the improvement of agricultural crops during the International Conference on New Frontiers of Chemistry and the Relationship between Supply World Food, organized by the International Union of Pure and Applied Chemistry, in Manila, Philippines.

Are you talking to me, Professor? Enio asked entering the laboratory, carrying glass flasks containing cultures of bean embryos.

No, Otto answered, caught off guard: I was talking to myself! "I was just thinking out loud on a project to rescue hybrid embryos that result, for example, from the interspecific cross of two beans species," he said.

But the embryos will not be viable, said Enio.

If these embryos are grown in proper conditions, as we can establish in vitro, we could recover viable hybrid embryos that could develop into adult plants, Otto explained.

It would be a very ambitious project, said Enio.

- Ambitious but not impossible, dear Enio, Otto said. "It's what attracts me in the research. We have already conducted research on storage proteins in bean seeds. Remember the international project on storage proteins of legumes that I was part of? We also have the results that I presented at the Symposium on plant proteins in Las Vegas in August 1980," said Otto.

Yes, professor. Always enthusiastic about Otto's projects, Enio continued: "and, before that trip last June, you were in East Germany presenting the results on the metabolism of storage proteins during the development of bean embryos cultured in vitro."

June, 1980! At that moment, as if by magic, he found himself in an open-air airport, being remodeled, awaiting the arrival of an aircraft that would take him to the interior of East Germany. His destination was the city of Gatersleben. At the Academy of Sciences of the German Democratic Republic he would present the results of experiments on protein metabolism during the development of bean embryos.

In the middle of the "Cold War," he arrived in West Berlin. He headed to Friedrichstrasse train station which makes the connection with the eastern part of town. The station was one of the main border crossings between the two sides of Berlin. In a narrow corridor that allowed the passage of a single person at a time, he presented his documents to the police and then entered the station on the eastern side. In the train, which moved slowly, he tried to observe the city. He felt sad: contrary to what could be seen in West Berlin, buzzing in progress, the streets seemed to him deserted with unkempt buildings, giving him a sense of precariousness. "Poor people," he sighed.

At the heart of East Germany in an airport in the open air under construction, deep in thought and with this sense of sadness in his heart, he walked in back and forth along the wide sidewalk while waiting for the aircraft that would lead him to his final destination.

Suddenly, the kiosks placed on the sidewalk along which he was walking, opened their windows. Flowers in small vessels appeared before his eyes: violets, petunias, geraniums, primroses, and many others, filled with joy that place that seemed harsh to him. The magic of that moment gave way to a show, unusual for him: a bus had just parked and young men and women descended in a row, as if obeying an atavistic ritual, handled those small vessels, carrying them with them, the air leaving the promise that all was not lost. This scene, in the harsh environment of the then communist dictatorship, with throbbing hearts of young people in search of beauty, embodied in the colorful flowers, was the testimony that hope was being renewed once again. And that, the emotional and intellectual freedom of those young people, which are inherent to human beings, would never be separated. "Really, not all is lost," he sighed. Hope will never be lost!

And so he came to Quedlinburg, which would be included as World Heritage by UNESCO in 1994. But now, in June 1980, still owned by East Germany, in Saxony, this incredibly beautiful small town won his heart. "A Saxon gem" thought the now not so young Otto. He, who had lived in Durham, a medieval town in northern England, and felt the full weight of a cultural epoch so distant, immediately felt attracted to the towers and pointed roofs of the houses with half-timbered style facades. They were more than 600 years old! And many of them needed restoration. Many of the monuments of the city were threatened with demolition. "They neither don't have nor can have a modern heating system," said one of the assistants of the hotel where he stayed." And continued, "The vast majority of the monuments are being restored by Polish artisans, who are experts in that kind of architecture."

Otto stayed for a week in Quedlinburg, enough time for him to never forget this town. During that time, he daily rode the bus to the neighboring Gatersleben to participate in discussions on plant proteins in the Academy of Sciences.

X
The Training of Scientific Personnel

- Professor, your student Cabral is here for the meeting, said Solizete while speaking on the phone with Otto. Maria Solizete was the outstanding secretary of the Section of Plant Biochemistry of CENA and Otto's right-hand assistant.

With his unmistakable and pleasant accent from northeastern Brazil, Jose Barbosa Cabral, from Recife, capital of the State of Pernambuco, introduced himself and made himself available to discuss the subject of his dissertation. It would be the rescue of embryos resulting from crosses between three species of bean, Phaseolus vulgaris, Paseos acutifolius and Phaseous lunatus. The viability problem of embryos derived from interspecific crosses of beans was well documented in the literature. Moreover, at CEBTEC, the technique of in vitro culture of bean embryos was routine. The earlier dialogue between Otto and Enio could now yield practical results. Cabral developed research responsibly, rescuing embryos from crosses between those beans species, regenerating fertile adult plants. He proved to be an excellent researcher in cell biology. He defended his dissertation in 1983. Later, he created a plant propagation company for in vitro propagation of economical important plant species, in Recife.

The bean (Phaseolus vulgaris L.) is a legume widely used in Brazil as nourishment, rich in protein, and mineral nutrients like iron, potassium, phosphorus and calcium, plus fiber and vitamins. Besides beans, Brazilians consume rice (Orysa sativa L.), a cereal rich in carbohydrates and essential amino acids. Thus, as a nutritional source, the legume, which is rich in protein and the amino acid lysine, is complemented with the cereal rich in methionine, as a daily meal in Brazil.

Professor, would you see a professor from the State University of Rio de Janeiro (UERJ) after your talk? Asked a young woman turning to Otto, who was up on stage at the amphitheater of the Federal University of Rio de Janeiro (UFRJ), in Rio de Janeiro in October of 1982.

Yes, Otto said, before beginning his lecture on the application of techniques of tissue culture plants in agriculture, at the invitation of Antonio Paes de Carvalho, Director of the Institute of Biophysics, Federal University of Rio de Janeiro (UFRJ).

At that time, Antonio Paes de Carvalo, now Professor Emeritus, was interested in forming a working group to develop the in vitro techniques at his university. This was the reason for Otto's talk. In 1983, Paes de Carvalho led a Biotechnology Program, along with Affonso do Prado Seabra, Maria Apparecida Esquibel and Antonio Rodrigues Carneiro. Later, he founded Biomatrix, which was the first Brazilian plant biotechnology company. Otto and Antonio Natal Goncalves were advisers in the development of eucalyptus clones to the company. Biomatrix technicians were trained in their laboratories in Piracicaba; several meetings with Paes de Carvalho's staff were made at the residence of Otto in Piracicaba and also in New York, USA. In late 1980, Biomatrix crashed. In 1986, Paes de Carvalho founded the Brazilian Association of Biotechnology Companies (ABRABI), and in 1988, was one of the founders of the Bio-Rio Foundation, dedicated to science-industry relationships.

After meeting with Leonardo Carneiro Alves, Professor of Cell Biology, Otto quickly realized his interest in familiarizing himself with the in vitro techniques and explained to him the project on bean embryos which was in development in the laboratories of CEBTEC and CENA:

- Enio and some of my students are conducting studies with cultures of bean embryos, whether to rescue of viable interspecific embryos or to monitor the activity of enzymes committed to the nitrogen metabolism during embryo development in vitro.

Leonardo was enthusiastic. In early 1984 he moved to Piracicaba with his wife Ana Maria and daughters. He began his Master's program in Soil and Plant Nutrition in ESALQ, defending his dissertation in 1990 on the activity of nitrate reductase in bean embryos cultured in vitro. This enzyme is responsible for the initial reactions of nitrogen assimilation when plants are exposed to nitrate. Later, at UERJ, he became interested in ornamental plants, and upon returning to ESALQ in 1994, he enrolled in the Doctoral Program. In 1997, he received a doctor's degree in Agronomy from ESALQ, after defending his thesis on the in vitro control of morphogenesis of endemic bromeliads in southeastern Brazil, under the guidance of Otto-the first study conducted in Brazil with bromeliads.

In 1980, while addressing a problem in his laboratory at CENA, Otto was interrupted by a young woman who introduced herself as a graduate student in the second year at ESALQ. She was also part of a group of ten students selected to participate in a program of specialization in nuclear energy offered by CENA.

The program requires that participants be also trained in the techniques being developed here at the Center, explained Helaine. "I would like to become familiar with plant tissue culture techniques", she continued.

Her hair tresses moved when her light skinned face leaned toward Otto, who was sitting on a bench near one of the lab benches. Helaine was restlessly waiting for a positive response to her application for a trainee position.

Here in our section we have several students from ESALQ and other institutions, including researchers from abroad. You can start your training with them and my technician Enio, who has mastered these techniques, Otto said.

As a trainee, Helaine outdid herself in assimilating the in vi-tro techniques. After completing her undergraduate degree in 1983, she developed her entire Master's Program at ESALQ, under Otto's leadership, completing it in 1988 with a dissertation on the control of tissue morphogenesis of papaya (Carica papaya) in vitro.

Helaine, called Otto, "our group needs to become familiar with new techniques, including the cultivation of anthers for the production of haploid cell that are important tools in plant breeding, because they facilitate the identification and selection of recessive hybrids in both cells and plants. In addition, haploid could accelerate screening in crossings programs."

- Would it be a complement to the program of beans interspe-cific crosses that are being conducted here at your laboratories to obtain viable hybrid embryos? Helaine asked.

- Not exactly, we could have one more method to be available to plant breeders in their plant selection programs, said Otto.

- And what would my contribution be? Helaine was confident that she could be an efficient collaborator in the programs of cell biology.

- You know that CEBTEC needs qualified staff of investigators. You're a strong candidate to further increase your scientific hori-zons. You could become familiar with new techniques in Sharp's laboratories of DNA Plant Technology, in New Jersey, working with Philip Ammirato. Would you agree? Helaine's eyes flashed when she accepted the proposal.

Helaine was hired in 1986 as a researcher in the Department of Chemistry, ESALQ, to provide services in CEBTEC. The following year she became Assistant-Professor of the same Department by public competition. In 1988, she started working with Ammirato at DNA Plant Technology Corp. in New Jersey.

Because of her familiarity with the techniques of cell and plant tissue culture and to her proven record of conducting laboratory work, encouraged by Otto and Rod Sharp, Helaine accepted an invitation from Pal Maliga to join a doctoral program in plant molecular biology. In 1989 she began her studies at the Waksman Institute, Rutgers - The State University of New Jersey. In 1994 she obtained the title of Ph. D. in Plant Biology with a thesis on the expression of bacterial genes in plastids of tobacco.

In 1975, after finishing his undergraduate course in biology at the Universidade Estadual Paulista "Julio Mesquita" (UNESP), Rio Claro, State of Sao Paulo, Luiz Antonio Gallo began his training in biochemistry at CENA

I have observed your dedication in the lab, said Otto addressing Gallo, as Luiz Antonio used to be called by his colleagues. "I think you could enroll in the Master of Science program here at CENA", he added.

I am well acquainted with enzymatic techniques, Gallo said. "I would like to develop my thesis on the enzymes of nitrogen metabolism in beans," he added

I suggest you follow the nitrate assimilation during the growth of bean plants kept in different concentrations of nitrate, Otto said.

That was the theme of the dissertation Gallo defended in 1979, obtaining the title of Master of Science at CENA. In 1987 he was accepted in the Doctor's Program in Biochemistry at the Chemical Institute of USP, São Paulo campus, under Otto's supervision. He continued his research on nitrogen metabolism of bean embryos cultured in vitro and obtained his doctor's degree in 1994. In 1988 he was hired as an Assistant Professor in the Department of Chemistry ESALQ to contribute to the research programs at CEBTEC. In 2002 he was approved in a State Exam Contest for promotion to Associate Professor. At CEBTEC, he has been an important collaborator in projects transferring in vitro technology to private sector companies.

Among Otto's many students, Fernando Broetto stood out because of his great interest for understanding the biochemical phenomena in plant cells. In 1986, after finishing his course in agricultural engineering at the Federal University of Espirito Santo State, in the capital Victoria, Fernando began his Master of Science program at CENA, under Otto's supervision. His thesis was partly carried out in the laboratories of Prof. Ulrich Zeitz, University of Tubingen, Germany, 1988-1989. In 1988, after attending the Symposium of Food and Agriculture Organization (FAO) on Biotechnology for Developing Countries, in Luxembourg, Otto met with Zeitz and Broetto in Tubingen, to discuss the use of biotechnological processes for studies of secondary compounds in plants. In 1991, Broetto earned a master's degree at ESALQ, and in 1995 the title of Doctor of Science from CENA, under the orientation of Euripides Malavolta. At that time, Fernando was already a professor of Biochemistry at the Faculty of Agronomy of the University of the State of São Paulo "Julio Mesquita" (UNESP) in Botucatu, where he is currently Associate Professor and develops intense research on the biochemistry and physiology of stress in plants, the induction of saline resistance in higher plants, and on medicinal plants.

During that same period, a gracious, short young woman, with brown skin peculiar to those living in northern Brazil, was finishing her doctor's degree under Otto's supervision, at CENA. She was born in Belem, the capital of the State of Para, holding a degree in Pharmacy and Biochemistry. Irenice Maria Santos Vieria began graduate work in 1980 and completed the master's degree in 1983. She possessed admirable skills for laboratory work on the influence of potassium levels on sugar concentration and invertase enzyme activities in the leaves of sugarcane plants grown in nutrient solutions under greenhouse conditions. After completion of the master's degree, she continued the work with young sugarcane plants by studying the metabolism of different sugar molecules in four varieties of sugarcane for her Doctor thesis, which she successfully defended in 1988. Then she returned to Belem, resuming her activities in the Federal Rural University of Amazonia and retiring in 2009. In 2011 she became part of the Group of Plant Biotechnology at the Federal University of Minas Gerais, in Belo Horizonte.

Jaime A. Cury, with academic training in dentistry, after his Masters course at the Federal University of Parana in Curitiba, completed his Doctor's Degree in Biochemistry from the Institute of Chemistry, University of Sao Paulo, Sao Paulo campus, under Otto's mentorship. He is currently a professor at the Faculty of Dentistry of Piracicaba and is internationally known for his work on the use of fluoride in dental protection. Also oriented by Otto on his Doctor's Degree in Biochemistry at USP, Luiz Eduardo Gutierrez spent his entire academic career at ESALQ, retiring as Full Professor of Biochemistry.

Over the years many undergraduate and graduate students were instructed by Otto, and many of them stood out in their Master's or Doctor's programs under his guidance. Among them, Maro Ran-ir Sondahl, Guilherme Guimaraes, Maria Aparecida Schiavuzzo, Marina Yukie Murayama, Roberval Salvador de Cassia Ribeiro, Sandra Aparecida Tabai, Carlos Eduardo Corsato, Geraldo Eduardo Cuzzuol, Joao Chadad Junior, Daniela Marques Argollo, Fatima Odahara Kajiki, Antonio Barioni Gusman, Isaac Stringueta Machado, Joao Suzuki, and Sidival Lourengo (his first advisee in a graduate program-in 1967).

XI
The Seminars on Agricultural Biotechnology – SEBIAGRI

1983. Leaving his labs at CENA, Otto took the shortcut road that interconnects the Center to ESALQ. Amid tall palms he inhaled the fresh air of a winter, which stubbornly refused to emerge in that morning of June.

One thought came to his mind: how to establish a productive CEBTEC in the scientific world. The Symposium in December

1981 was an opportunity for him to show to his peers in Brazil and abroad that CEBTEC was created. But that was not enough. The scientific world is demanding - and he agreed that it had to be like that. Skirting the buildings of the Department of Animal Science along his way, roaming through the trees, he came down the slope that, to the left, bordered experimental fields of the Department of Genetics, and finally arrived at building of the Institute of Genetics.

Is Dr. Aline in her office? He asked the receptionist. Upon receiving a positive response, he entered the hallway to his right and was received by Aline.

Aline, I would like to ask you to help me organize a series of lectures focusing on the use of biotechnological processes aimed at agricultural productivity, said Otto.

It is a very good idea, said Aline, adding: "I am ready to work with you; what should be done so that this idea comes to life?"

Aline Aparecida Pizzirani-Kleiner, Assistant Professor of the Department of Genetics, was enthusiastic about the work undertaken by CEBTEC. She was one of his collaborators on the Board of the Center.

I am thinking of inviting Wilson Mattos and Antonio Roque Dechen to collaborate with us in organizing the talks, said Otto.

Wilson Roberto Mattos, an active member of the Department of Animal Sciences, has developed excellent work in his field of research and teaching. Since 2007 he has been Coordinator and then Mayor of the campus "Luiz de Queiroz. "In 2013 he was elected CEO of FEALQ. Antonio Roque Dechen, Full Professor of Plant Nutrition at the Department of Chemistry, was a tireless contributor to the series of lectures and symposia organized by CEBTEC. During 2007-2010 he was Director of ESALQ and in 2010-2011 Vice Dean of Administration at USP.

How will these series of lectures will be called? Aline asked

We could call them "Seminars on Agricultural Biotechnology," with the acronym SEBIAGRI, Otto suggested.

What would be the best date for the first SEBIAGRI? Obviously, Aline was interested in these seminars.

The best date would be next October. Earlier in September, I will attend the National Meeting of Plant Tissue Culture, at the Federal University of Vicosa, during the Ninth Meeting of the Brazilian Society of Plant Physiology, Otto said.

Viçosa, in Minas Gerais, in central-western Brazil! He immediately remembered the first time he was in Viçosa, in 1956, participating in the Congress of Students of Agronomy! The pine forest! The dreams of a young student eager to start in the academic world! And now, in June 1983, he was making plans to return to Vicosa. And he indeed returned to Viçosa, for the "Meeting on Plant Tissue Culture." During that meeting, the Brazilian Association of Tissue Culture ABCTP-was created. He was acclaimed its first President. The Board was completed with Walter Handro, Secretary, Gilberto B. Kerbauhy, Assistant Secretary and Rolf Dieter lllg, Treasurer. Otto held the presidency of the Association until 1987. On October 3, 2003 he was honored by ABCTP, now under the presidency of Renato Paiva. In Vicosa, again!

SEBIAGRI I. took place on October 11, 1983, discussing the theme "Agricultural productivity and animal productivity." The organizing committee consisted of Otto, Aline, Dechen and Mattos.

Other seminars followed. SEBIAGRI II took place on the 29th and 30th of August 1984 on the theme "Biotechnology of Sugarcane: from the cell to the plant", discussing results of the CEBTEC research group on biotechnology of sugarcane, among other lectures. Otto, Dechen, Cristina Gonçalves and Joaquim Albenisio Silveira formed the Organizing Committee.

SEBIAGRI Ill took place in September 1985, from the 10th to the 13th on the theme "Physiological Basis of Plant and Animal Productivities" Otto, Mattos, Dechen, Antonio Natal Gonçalves, Joaquim Albenisio Silveira and Raul Machado Neto formed the organizing committee.

On the 29th and 30th of November 1986, "Forest Biotechnology: prospects and applications" was the theme of SEBIAGRI IV, edited by Otto, Dechen, Natal Gonçalves and Albenisio, in collaboration with Mario Ferreira, Luiz E. G. Barrichello, John W. Simões, Paulo C. T. Carvalho, Angelo DiCiero Neto and N. Nicoliello, the last two representatives of private enterprises.

In April of that same year of 1986, Otto attended a scientific meeting sponsored by the Ministry of Education of Germany in Bonn, along with other professors of the University of Sao Paulo, led by Dean José Goldenberg. On this occasion, he presented results of his work in the area of plant biotechnology.

With the theme "Carbon and nitrogen in plant productivity," SEBIAGRI V took place from the 28th to 30th of September 1987, organized by Otto and Albenisio. Years later, on May 23, 1994, Helaine, back from her studies in the United States, collaborated with Otto in the organization of SEBIAGRI VI on the theme "Molecular biology of plants", at the Room "William R. Sharp" in CEBTEC building.

As soon as she returned from the United States in the mid-90s, Helaine became responsible for the projects involving molecular biology of plants. In the Genome Project, financed by FAPESP, with the help of Enio and Valentina Fatima de Martin, and several graduate students, she provided leadership at CEBTEC for the sequencing of Xylella fastidiosa, a bacterium that causes diseases such as the yellowing plague that affects orange trees, an economically important plant species. Since then, her talent for scientific research has developed tremendously. Her tireless energy to work in her labs, advising students, and her international scientific network, ensure that the activities at the research center that Otto and Flavio Tavares created in 1981 will continue.

XII
The New CEBTEC Building

February, 1988. Enio and Solizete had just been transferred from CENA to the Department of Chemistry at ESALQ to work for CEBTEC: Maria Solizete Granziol as Secretary and Enio as Senior Technician. Enio graduated in Biology from the Methodist University of Piracicaba (UNIMEP) the previous year.

That month began the preparations for the transfer of CEBTEC to the new building still under construction in an area donated by the Board of ESALQ and the Mayor of the campus "Luiz de Queiroz." Students, technicians, and trainees from other institutions of education and research had been taking turns using the laboratories in the Chemistry Building, now too scarce to accommodate everyone properly. So, this transfer was urgent. Among the technicians hired were Pharmacist-Biochemist Antonio Francisco Campos do Amaral, Agronomist João Chadad, and Romeo Aparecido Rocha. Among the trainees was Egidio Miguel Campagnol.

September 1985. Joaquim José de Camargo Engler, then Director of ESALQ and Antonio Sanches de Oliveira, then Mayor of the campus were greeted by Otto in his office at CEBTEC in the Chemistry building.

We have come to you to propose the construction of the CEBTEC building in an area of 9000 m2, here at ESALQ, Engler said.

Sanchez de Oliveira added, "It is a privileged area, near the facilities of FEALQ."

With an emotional expression, Otto readily accepted the offer and, accompanied by Engler and Sanches de Oliveira, headed to the area that bordered with Avenida Centenario. On January

6, 1986, the cornerstone of the building was released within the presence of Antonio Guerra Vieira-at the end of his term as Dean of USP, Engler, Paulo Fernando Cidade de Araujo, Antonio Roque Dechen, Enio, Basso, and other ESALQ professors and technicians.

The construction of the building was made with the resources of Otto's projects that he was developing, under FEALQ administration. Justo Moretti Filho, Full Professor of Agricultural Engineering at ESALQ, designed the layout of the building. Years later, as CEO of FEALQ, Justo Moretti gave special attention to the CEBTEC projects. Former professor of Otto's, he enthusiastically shared with him all the steps in the construction of the new building. He is also the designer of the center's logo which is placed high on the building facade: a plastid enclosed by a circle as if it were an organelle. Justinho, as his friends and colleagues used to call him, passed away in September, 2011.

The facade of the building was designed by Antonio Natal Gonçalves, an expert in tree physiology and also a renowned artist. The three columns in Roman arch, simulating inverted test tubes, receive visitors in the entrance portico of the building.

The same architectural metaphor is present in the shape of the windows in the office of the Coordinator of the Center, to the left of the building. On the right, an auditorium with audiovisual resources has its windows surrounded by embossed circumferences simulating Petri dishes. Biochemistry and Cell Biology are symbolically represented on the facade of the new CEBTEC! (Photo).

October, 1988. Otto was involved with the transfer of scientific equipment and supplies which had been purchased with funds from his scientific projects from his laboratory at CENA to the new building. Enio and Leonardo Carneiro were in charge of the logistics of transportation. Maria Teresa and Eric Derbyshire no longer belonged to his team at CENA, since January 1982. Helaine was at Rutgers, on the eve of starting her Ph.D. program.

She, Murilo Mello, Ph.D. in Plant Physiology, and Luiz Antonio Gallo, M.S., were hired as assistant-professors of the Chemistry Department to work at CEBTEC.

- Professor Otto, here's the folder with all the materials for the Symposium on Plant Biotechnology, said Solizete, entering the room of the CEBTEC Head, in the Chemistry Building.

- You've done an excellent job, I do not know what I'd do without your help in the office, especially now that we are on the eve of the inauguration of the new building and the opening of the symposium, said Otto. For him, both Solizete and Enio were essential for the proper development of CEBTEC.

While reading the contents of the folder once again, he returned to the past. Leaning back in his chair, he was rocked by the memories of his recent past. Seven years had passed since the creation of CEBTEC, in June 1981, and the realization of the Symposium on Genetic Engineering in December of the same year.

October 1984. On day 19, in a one-day meeting, he participated along with six other scientists from different countries, in the "Tokyo Meeting of Plant Biotechnology," organized by Plantech Research Institute. Then, he flew from Japan to Beijing in China, where he discussed the results of his research team at the "International Symposium on Genetic Manipulation in Crops" in Academia Sinica. In the Chinese capital he visited the Great Wall of China, walked through the great Square of Peace and was surrounded by the hub of people in the shops.

- Professor, may we bring the reports of your scientific travels around the world to your office in the new building? Enio asked, interrupting Otto's daydreams.

- Please, they are on that table, Otto said pointing to a stack of books, journals and reports of his research.

Suddenly he remembered that he had made many visits to the laboratories of Rod Sharp, in Ohio State University, The Pioneer

Research Laboratory, and The Campbell Institute for Agricultural Research, The Campbell Soup Company in Cinnaminson, New Jersey and at the DNA Plant Technology in Cinnaminson, New Jersey. In one of those visits, during the luncheon with the participants of the "Annual Conference Scientific Advisory Board of DNAP" in Philadelphia in September 1986, Rod Sharp and David Evans surprised him offering him the "1986-DNAP Biotechnology Award." Wearing only a long-sleeved shirt, while everyone else was in suits and ties, Otto received the honor from the hands of Norman Borlaug, holder of the Nobel Peace Prize.

You seem to be deep in thought, Professor, said Enio heading out the door.

I'm recalling the many times that Rod and I were together at scientific meetings in Brazil and in the United States, said Otto and continued: "right now I'm remembering when, in 1982, Rod invited me to attend a workshop on priorities in biotechnology research for international development in Washington and Berkeley Springs, West Virginia. This event was sponsored by the National Academy of Sciences of the United States. I stayed in a cottage in the middle of a forest. I remember how that place was special for my morning meditation. As you might know Enio, I have always liked to meditate in the midst of trees." Once again, Otto recollected his contemplative experiences in moments of silence.

To the memories of that workshop it was added the Symposium entitled "Biotechnology in the Americas: prospects for developing countries," held in San Jose, Costa Rica, in 1983, whereupon he and Rod Sharp discussed the importance of using techniques of plant cell and tissue culture for the advancement of agriculture. In 1985 he participated in roundtable discussions about the emerging mobile technologies and the "scale up" in biotechnology during the DNAP Scientific Meeting of its Advisory Board in New York.

What about the book that you and Professor Sharp edited? Enio asked.

Ah! That's the book "Biotechnology of Plants and Microorganism," edited in collaboration with David Evans, J. E. Bravo, Flavio Tavares and E. F. Paddock. It was released on May 1987 in Columbus, Ohio, during the commemoration of 20 years of the Agreement between ESALQ and The Ohio State University. Humberto de Campos, who at that time was Director of ESALQ, was also presented at the ceremony.

This dialogue with Enio seemed to Otto to be an "accountability session." But now it was October 1988. A few months earlier, in March, he was on the Island of Trinidad, in the Caribbean, and in Caracas, Venezuela. In August of that year he was in Toronto, Canada, where he spoke, as a guest of A M. Chakrabarty, about his works at the XVI International Congress of Genetics, and when once again Otto went on to meet with Rod Sharp in the United States.

Finally walking away from his recollections, Otto headed out to the new CEBTEC building along with Enio and Solizete.

XIII
1988. International Symposium on Plant Biotechnology

The International Symposium on Plant Biotechnology was held in the amphitheater of the Engineering Building in ESALQ from the 26th to 29th of October, organized by Rod Sharp, Otto and Murilo Melo. In the opening ceremony, chaired by Paulo Fernando Cidade de Araujo, CEO of FEALQ, Rod Sharp delivered the opening speech.

The inauguration of the new CEBTEC building was on the evening of the 27th. Otto would never forget that moment, unique in his academic life. It was eight o'clock in the evening when the lights of the building gradually lit up as the guests moved toward it. Emotion was in the air as they drew near the entrance porch, greeted by the sound of the opening chords of Beethoven's 5th Symphony. During the moments that followed, a group of seven young musicians from the School of Music of Piracicaba "Maestro Ernst Mahler" conducted by Antonio Ribeiro, positioned along the porch, saluted the guests with pieces by Bach and Mozart as they entered the room where a plaque, celebrating the occasion, was inaugurated by Paulo Cidade and Otto.

Among the guests present were P.V. Ammirato (Columbia University, New York, USA), P.A. Arruda (UNICAMP), J.A. Betti (Agronomic Institute, Campinas), J. Daussant (Laboratoire de Physiologie, Meudon, France), P.C. Debergh (State University Gent, Belgium), D.A. Evans (DNAP, USA), J.E. Feldeberg (Suzano Pulp, Brazil), E.I.S. Floh (USP, Sao Paulo), E. Gander (CENARGEN, Brasilia), W. Handro (USP, Sao Paulo), R.D. Ilg (UNICAMP), H.W. Janes (Rutgers, USA), A. Milk (UNICAMP), S.A. Malacrida (UNICAMP), C. Costa (UNESP, Botucatu), R.C. S. Ribeiro (UNESP, llha Solteira, Brazil), J.A Sampaio (CENARGEN, GMT), H.E. Sommer (University of Georgia, USA), J.B. Teixeira (EMBRAPA), M.R. Sondahl (DNAP, USA), M.J.0. Zimmermann (EMBRAPA, Goiania), A. Natal Goncalves, and Jose R. Postali Parra, who later in the future would be elected as ESALQ Director.

A surprise awaited Rod Sharp. A plaque, dedicated to him, in the lobby of the auditorium, claiming it to perpetually be "William R. Sharp" Room (photo 5). Excitement hung in the air when Rod and Otto embraced - the same embracement as they had when they first met at Congonhas Airport in Sao Paulo, on that mild-winter morning in June, 1971!

Later that evening, Rod Sharp, Otto, and guests were welcomed by the Director of ESALQ, Humberto de Campos, and his wife Wanda, in the beautiful residence of the Director on the ESALQ Campus.

November 29, 1990: In the afternoon, on the stage of the auditorium of the building of the Social Service of Commerce of Piracicaba (SESC), a private enterprise, Otto, Murilo Melo, Rod Sharp, Eurípedes Malavolta and Flavio Fava de Moraes, Scientific Director of FAPESP and former Dean of USP, sat at the table. They directed the work to launch the book "Biotechnology for Crop Production."

Before an audience of professors, technicians, students and guests, Otto presented the book containing the lectures delivered at the International Symposium on Plant Biotechnology, and chapters written by invited scientists. With Preface and Introduction by Otto and Rod Sharp, the book was dedicated by Otto for to his first mentor in the intricacies of chemistry, Demosthenes Santos Correa, to Euripedes Malavolta, whom he once was an Assistant at ESALQ, and to his close friend Andre Martin Louis Neptune, who would pass away in July of the following year.

XIV
The International Congresses of Plant Tissue Culture

June 8, 1994. In the lobby of the International Airport of Guarulhos, in the Greater São Paulo, Diva and Otto were waiting to be called to take their seats on the aircraft that would take them to Italy. Otto would attend the International Congress of Plant Tissue Culture, in Florence, Italy.

Dad, Maria Paula finally arrived to say goodbye, Daniel said turning to Otto. Daniel was the one who brought his parents to the airport from Piracicaba.

Diva smiled as she saw her daughter hurriedly walk towards them. She was late due to the heavy traffic conditions on the roads of the state capital. "I am so glad you are here with us", Diva said.

Unfortunately Adolfo Egidio could not come, Maria Paula said. "He is on duty in the hospital", she added.

Carla Maisa, with long hair and swarthy complexion like her father, majored in English teaching, and with pressing engagements in Piracicaba also could not be there to send them off to Italy.

Daniel, the youngest son, with brown hair and light complexion, like his mother, and Maria Paula, dark hair and light complexion, like Adolfo Egidio, joined their parents in an embrace. At that time, Daniel was attending the School of Tourism in Sao Paulo and Maria Paula was already a bilingual translator. They were two of the four children left for Diva and Otto, after the death of their firstborn Marco Augusto on June 16, 1991.

The death of "Maguto" as his brothers called him is constantly reminded by Otto, Diva and his brothers and sisters. His memory, as if he were alive, is always present at every moment of their lives: at least they lived with him for 29 years.

It was June 25, 1991. In the Auditorium of the Chemistry Building, João Lucio de Azevedo, Director of ESALQ, opened the Scientific Session to celebrate the 10th Anniversary of CEBTEC. Paulo Fernando Cidade de Araujo, Rod Sharp, Otto and Philip Ammirato were the speakers. Jeffrey Sharp, Rod's son, was present and while addressing Otto offered a donation in Memory of Marco Augusto. The donation was used by FEALQ to build part of a laboratory attached to the main building of CEBTEC. A plaque in the small entrance hall of the laboratory shows Otto's verses honoring his son Marco Augusto Lovadino Crocomo memory...

"Fly ... fly... I to the stars I and stay there I contemplating the universe..."

1998-1999. "Luiz de Queiroz Medal" Award.

Julio Marcos and Antonio Roque Dechen, then Director and Deputy Director of ESALQ, respectively, supported Otto's initiative to offer Rod Sharp the "Luiz de Queiroz Medal Award." In a ceremony held at ESALQ on June 3, 1999, William Rod Sharp received the Medal Award.

On the same occasion, while there was abundant joy around in celebrating Rod Sharp's Award, Otto's family was shocked by the death of their dear friend orthopedist Caiuby de Souza Arruda. Caiuby and his wife, the poet Jahyra Boucault de Arruda, were very close to Otto, Diva and their children and are Daniel's godparents in Baptism. Rod, accompanied by Otto, visited Caiuby in the hospital as if it were a "good-bye" to a good friend; both had an excellent relationship since the early days when Rod arrived in Piracicaba.

In the Alitalia plane that would take them to Milan, Otto and Diva talked about their children.

-" Maria Paula will feel alone when her brothers and sister travel to Switzerland to meet us after the Congress in Florence, said Otto

- She's just got her first job after completing her course in English Translation, said Diva, with her consistent practical sense of life. "She's a responsible young woman", she added.

Adolfo Egidio, Carla Maisa and Daniel would meet their parents at the airport of Geneva, Switzerland, on June 21, and from there they would together drive to several countries in Europe.

The aircraft landed at the airport of Milan. Together, Otto and Diva visited the city. They admired the grandeur of the "Teatro Alla Scala Milano" and the Cathedral, the "Duomo" in Milan, with its imposing Gothic style. By train they departed for Florence.

Florence, Italy, June 12-19. The place of Michelangelo's "David", of Machiavelli and Leonardo da Vinci and Dante Alighieri, of the city of storytelling streets and whistling jewelers on "Ponte Vecchio" over the Arno River, was where Otto participated in the VII International Congress of Plant Tissue Culture. He was a member of the Scientific Committee of this scientific meeting organized by the International Association of Plant Tissue Culture – IAPTC. In the Congress, he presented results of his research on natural products in plants. He was also the Moderator of the Symposium "Juvenilization and Maturation", which he organized as part of the program.

In 1978, he participated in the IV IAPTC Congress of in Calgary in Canada, in the last days of August. Delighted with the country, he returned there several times in scientific missions.

During that time, the pilots of the Canadian airline were on strike. Otto had made commitment to attend the Meeting of the Project on Legume Proteins that would happen in Munich, Germany in early September of that same year. Luckily, at the end of the IAPTC Congress, the airline company authorized an overnight flight to Toronto and then to Paris, from where he would fly to Munich, after a few days. After leaving Charles de Gaulle Airport, the clock marked 8 o'clock in the morning when Otto entered his room at a hotel in Paris.

Otto was too tired and sleepy to see Paris once again by daylight. He woke up at 7 in the evening, quickly showered, went downstairs, took the subway, and headed to the Arc de Triomphe. Dazzled, he saw the lights go on along Avenue des Champs-Elysees. At that moment, which seemed magic to him, he remembered March 1969 when he walked down that Avenue with Diva from the Arc de Triomphe to the Louvre, through the Jardin de Tulleries. That walk he would repeat several times, including on August 1976 with all their five children and in other times he visited Paris in the future.

Otto was always impressed by the architectural beauty of the Hotel de Ville, near the majestic Cathedral of Notre Dame de Paris, considered one of the best examples of the Gothic style. Years later he was delighted while reading the accounts of the cathedral's construction, which was handmade by art masters from the middle Ages, and even before, in novels set around historical events, such as those by the English writer Ken Follet.

On July 10, 1982, Otto left the Japan Airlines plane that landed at Narita Airport in Tokyo. The trip was long: from Sao Paulo in Brazil to Los Angeles in the United States and then Tokyo. Once settled in his seat on the aircraft that would take him to Tokyo, he started reading the novel "Adrian" by Marguerite Yourcenar about the trajectory of the great emperor conquering new lands for Rome. Having lived between the first and second centuries of our era, Adrian's romanticized life is masterly described by the great Belgian writer, naturalized American and a Member of the French Academy of Letters.

From a young age, Otto admired the courage and skillfulness of conquerors in the strategies of their struggles. As he developed and reached maturity, he also started to admire the great industrialists, great engineers and architects, philosophers, writers, teachers, and scientists who do not let themselves down by the setbacks in their aims. Otto was always searching for the biographies of the human figures that have helped build the history of civilization. What are the reasons, biological or circumstantial, that caused them to be who they are? Also fond of classical music, he chose, as a gift to himself, at the end of his High School years, a biography of Beethoven. It was December 1949. He was then 17 years old.

At the V Congress of IAPTC, which took place in a convention center at the foot of Mount Fuji, he presented the results of his research on irradiated sugarcane cells growing in vitro. The exchange of his experiences with Japanese scientists during the daily comings and goings on the bus between the hotel, where he was staying, and the convention center, and during the discussion sessions, opened new prospects in his life as a researcher.

The possibility of CEBTEC to participate in upcoming scientific meetings was promising. In August 1986, Otto assumed leadership for a group of young researchers from private Brazilian producers of pulp and paper, to participate in the VI Congress of IAPTC in Minneapolis, Minnesota, USA.

In this Congress he presented papers on the hybrid embryo rescue of beans, the in vitro control of Pinus morphogenesis, and the juvenilization reversion during Eucalyptus micropropagation. After the Congress, the group met with Rod Sharp and David Evans at DNAP in Cinnaminson, New Jersey, where they discussed the state of the art of the research being done at CEBTEC.

It was in a morning of June 1990 when Otto left the aircraft of Air France and once again walked the corridors of Charles de Gaulle Airport in Paris, from where he soon would head to the Netherlands. The VI I Congress of IAPTC would take place from the 24 to the 29 of June in Amsterdam. There he presented the results of his research on the control of the formation of somatic embryos of the Macauba Palm (Acrocomia acu/eata), characterized by high oil content and potential use as an alternative fuel -a national strategic interest for Brazil. In the beginning of 21 century these kinds of fuels would be called biofuels. As always, during his scientific travels, his interest in universal culture allowed him to appreciate the works of great painters at the Van Gogh Museum, in Amsterdam, in the company of Philip Ammirato, then Rod Sharp's collaborator. Also on that trip, he visited several laboratories of biotechnology in Italy, Germany, Switzerland and Belgium.

XV
The Application of In Vitro Techniques: Plant Biotechnology

1979-2001

The 1980s were emblematic for Otto: the "boom" of biotechnology, in Brazil and abroad, which became evident at the end of the previous decade, now showed its total capacity for introducing technological innovations in agriculture. At the beginning of this decade, the Planalsucar Project, under his responsibility was in full development.

In his office, talking with Enio about **CEBTEC** achievements, he added: "this project was born when I received an invitation to present our results on plant regeneration from sugar cane cells grown in vitro in the First Brazilian Congress of Sugarcane Technology, held in Maceio, the capital of Alagoas State, in northeastern Brazil, on January 1979." He recalled: "The Planalsucar technicians were excited when they saw the potential for the technology to complement their sugarcane breeding program."

That was the first collaborative research and development project accomplished with a private sector company. In collaboration with Enio and Neftali Ochoa-Alejo, Otto presented results of their research on the tolerance of sugarcane cells to the herbicide Ametrin, in the II Brazilian Congress of Sugarcane Technology, in Rio de Janeiro, on August 1981.

Otto also presented other research results on sugarcane cells experiments in successive international meetings during the year of 1984. In September, he attended the "International Symposium on Plant Tissue and Cell Culture: Application to Crop Improvement" in Olomouc, in the former Czechoslovakia, then under communist rule.

That was a moment of remembrance. He remembered his crusade in spreading the application of in vitro techniques in agriculture. Before an audience of agricultural producers, who mostly did not accept the use of these techniques, in the "Symposium on Fruit Biotechnology" held in Campinas, in Sao Paulo in 1986, he emphasized: "the time will come when all of you will use these techniques to get better and larger quantities products."

And So It Happened.

From 1983 to 1988, Brazilian enterprises, such as Duratex, part of the ltaú Bank Group, and Duraflora, funded the project "Cloning and Selection of Eucalyptus Clones" under Otto's responsibility, in collaboration with Antonio Natal Goncalves and the invaluable collaboration of Angelo DiCiero and Antonio S. Rensi Coelho. During the development of the project, staff from these private companies were trained in in vitro techniques. The clones provided by CEBTEC were kept in greenhouses at the companies premises.

April 1986. In Bonn, Germany, he presented results of his research on various plant species, and in Brussels, Belgium, along with Paulo Cidade, he successfully defended the continuation of the project sponsored by the Scientific Commission of the European Community that started in 1983. After his return to Piracicaba, he was approached by investors to form a public private partnership for development of in vitro techniques for fruit production. After several meetings, FEALQ signed with Brazil Venture Capital the project "Engevit": research and development for large scale production of virus-free strawberries, to be conducted in CEBTEC. Acclimatized in a greenhouse, the seedlings obtained in vitro through meristem culture were transferred to the field on a farm located in Itatiba, a mountainous region in the State of Sao Paulo, where no incidences of virus were recorded, and commercialized.

On October 1989, accompanied by Walter Traldi, one of the sponsors of the project "Engevit," Otto visited several laboratories of plant biotechnology in Italy, France, Belgium, and Spain. At that time he presented results of his research using the techniques of in vitro culture of plants during the "World Food Day: Biotechnology for the Third World" at the Deutsches Institut Ubersee, in Hamburg, Germany.

Other investors also looked for him to produce virus-free strawberries. From 1991 to 1992, the seedlings grown in vitro and acclimatized in CEBTEC were planted on a farm in the Serra de São Pedro, 700 meters high, in a climate that was suitable for this type of cultivation. That was the Project ENGEVEG, with the participation of Enio and Vitor S. Sheffer, Otto's graduate student.

Throughout the 1990s, given the intense demand for disease-free banana seedlings, CEBTEC installed an appropriate structure for their production. Several companies producing different varieties of banana had signed agreements with the Center through FEALQ; these projects also provided staff training. Many students developed scientific projects using these plants, under Otto's guidance, including Humberto Zaidan. Humberto, with the aid of Enio, studied the metabolism of polyamines in banana plantlets. The detection of these substances were made in the laboratories of Fernando Broetto at UNESP in Botucatu. Humberto is currently an Associate Professor at Federal University of Campina Grande, State of Paraiba, in northeastern Brazil.

1989-1995: The Citrus Project

February 1989. When answering a phone call, Otto was surprised by the invitation made to him by one of the Managing Directors of the Votorantin Group: to explore the advantages of using in vitro techniques for plant propagation.

- Where and when will be the meeting? He asked, adding: "in the coming days I should go to Lima, Peru, to attend the 'Peruvian Encuentro de Cultivo de Tejidos Vegetales', could we meet before?"

The Scientific Meeting in Lima, in which he participated, was held on February 23- 25, to discuss strategies to increase research in plant biotechnology in the Andean Countries. He proposed that the meeting with the Votorantin Group Staff be held on the morning of February 21st In the Votorantim offices in Sao Miguel Paulista, in the Greater Sao Paulo, he was received by Ricardo Ermírio Morais, responsible for the Company "Citrovita." After the presentation of scientific data, a discussion occurred about the participation of CEBTEC in a university/private company partnership for the production of orange plants free of the tristeza virus which, when present, decimates orange plantations. The collaboration of Ary A Salibe, an internationally renowned expert on plant viruses, was important to Otto's realization of the Project. At that time, Salibe was full professor of Fruit Horticulture at UNESP in Botucatu. He and Otto were friend since 1953, when both began their graduate studies at ESALQ,

The Project "Citrus Micrografting: the production of oranges free of the tristeza virus and viroids" progressed from 1989 to 1995. In the laboratories of CEBTEC were produced orange pear micrografts, obtained from the meristems of the best citrus trees selected by Salibe, who toured farms in the States of Sao Paulo and Bahia. The orange seedlings obtained from the micrografts were maintained in clean insect-free greenhouse environments at CEBTEC and then transferred to the UNESP laboratory, in Botucatu, where they were indexed by Salibe to the tristeza virus, in greenhouses. The selected plants were transferred to greenhouses in the farm of Citrovita in the city of ltapetininga, in the south of the State of Sao Paulo and, after being acclimatized, they were maintained in the field. All tests confirmed the non-presence of the tristeza virus in Pera Sweet Orange Trees, which also have a high-yield production. At CEBTEC, the project had the collaboration of Enio and Gallo, and at UNESP the collaboration of Antonio Tubelis.

XVI
The Living Memory of That Time That Was Not Lost

2011-2012

In that morning of a sub-tropical spring, Otto opened the door of his living room and slowly began to descend the stairs that lead to the recreational area of his home. On the wide terraces along the steps, at the sounds of bird songs, he appreciated the potted plants that flank the stairs, and over the enclosing wall, he glanced with admiration at the beauty of the flowers and shrubs outside the houses across the street. On a corner at the top of the hill, the house offers him a broad view of the two streets that meet. Many times, as he walked around the neighborhood, he caught himself admiring the trees along the sidewalks-some tall, others short -- the palm trees, and the flowers that decorate the entrances of the houses as if to welcome guests. Joining them were an array of tall Thujas and Cypresses that surround the exterior of his own house. A true garden in his eyes. He left his house where he has lived for about 40 years and, slowly, walked toward the nearby cul-de-sac. Trees, with bunches of drooping yellow flowers that look at the ground on both sides of the street, formed a live tunnel sheltering from the above the green bushes and blue flowers and red roses that surround the walls of the neighboring houses. Suddenly he found himself in this same place, playing ball with Rod Sharp and his sons Marco Augusto and Adolfo Egidio, still little, in the early 1970s. The houses and trees in that alley did not exist at that time. Daniel, his youngest son, was taking his first steps; the girls, Carla Maisa and Maria Paula, played with dolls. Everything was different, he sighed. It was a time when the first fruits of the coming-to-be were chiseled in his heart and mind, paving his way to the scientific achievements that would follow over the coming years.

His admiration for beauty, once again a magical moment, for the lush of the green and the variegated flowers and their myriad of shapes, was at the root of his inquisitive mind for why things are the way they are. Oh! The biochemical control of the forms of things: the purpose of the allosteric enzymes in the regulation of biochemical reactions! How crucial was the development of the research by Jean-Pierre Changeux in that field in the early 1960, he remembered. This and other questions were always present in his mind while observing nature.

This unrest was the result of his deep interest in the world of philosophical ideas that since his youth he has learned to dig, as if it were in pursuit of gold and diamonds, in an attempt to explain life. The company of philosophers from the Greek to the contemporary, through the Middle Ages, the Renaissance and Enlightenment eras, was fundamental for him, either by interpreting data from his research or imparting knowledge with his students, in the creation of his scientific mindset. Many years had passed since he entered the field of plant cells where he searched for the biochemical reactions that are at the base of life. This mind wandering around the "world of the invisible" enabled him to experience the marriage of cell biology and molecular biology in his life, as a professor and researcher entering the wonderful world of plant biotechnology. From the plant cell to the whole plant cultivated in the agricultural field! In the early 80s, the transfer of cellular technology to the private sector was in the first step and this, and many others, was taken by him and his team over the last thirty years.

Leaving the dead-end alley, he walked toward the avenue, one block away from his home, and began the descent of the hill that would lead to another large and busy avenue. To his right was the Clube de Campo, a wide area, once a farm owned by Rodolfo Lara Campos - the Count Lara, an elegant man in his dress and his manners, who would ride around town in his carriage drawn by black horses, as Otto remembered him from his childhood. In the early 1950s, a group of Piracicaba residents bought the area and on one fraction of it, the Club was established; the rest was divided into lots, one of which Otto built his house on. The swimming pools in the Club and the one in his own home! Walking along the wall surrounding the club, he remembered it well: in one of his visits to Piracicaba, on the deck, as he got out of the pool in his house, Rod Sharp slipped and fell injuring his elbow. It was a rush - Rod was rescued taken to the hospital for treatment. It was only a scare!

He crossed the Armando de Salles Oliveira Avenue and after a long walk went to the left-hand bank of the Rio Piracicaba, decanted in prose and verse by the residents of Piracicaba, from the fishermen to the intellectuals. Down by the curves of Beira Rio Avenue, he appreciated the spectacle of water sliding down the huge basalt stones arranged like irregular steps and stairs, seeming to imitate a waterfall, the "Leap of the Piracicaba River." Below, the waters, plentiful, were now a majestic sheet calmly following the fate of every river on their way to the sea.

Through the ups and downs of the avenue, he came upon a building dating from the mid nineteenth century: the so-called "settler's house," in honor of Captain Antonio Correa Barbosa who in 1767 founded the village that eventually became the city of Piracicaba, with its current economic strength. Despite not having been built by Captain Barbosa, the house is considered as if it were, and as such a symbol of historical memory. It is a tourist landmark, with frequent cultural events and art exhibitions. Made of wood and cloth by Elias Rocha, an Indian descent fisherman, and a resident on the edge of the river who, during his life, kept a respectful coexistence with the river, the "figures of Elias" are scattered around the house and the River edges representing fishermen as guardians of historical reminiscences.

Enjoying nature along his walk, Otto stopped at the "Lagos dos Pescadores." Surrounded by small houses that sell goods for tourists, the square is the scene of musical meetings, the "night of serenades." Singers, the serenaders, sang sentimental songs, with lyrics that celebrated the wonders of love and often, the pain of unrequited love. Looking around, he came upon a group of serenaders preparing to sing the Hymn of Piracicaba, written by Newton Almeida Mello. "Piracicaba which I love much, full of flowers, full of charms ..." the verse is sung by all who live in Piracicaba. In a flash he saw himself still as a teenager in the late 1940's, in the company of his father João, after the death of his mother Teresa, when they often entered the spacious home of the DiGiaimo's, of Italian Neapolitan origin. Their house is no longer there in that square. One of the daughters of the DiGiaimo's, Ada, married Francisco, Otto's brother. The memory of the sounds of Italian music, which they sang in chorus during those visits, played in his mind and his heart, mingling with the strains of the Piracicaba serenades, that now he heard.

At the sound of the murmuring river, he was seeking for the past. A few dreams and desires were not realized, but now reaching 80 years of age, they had no more importance. They sit on the depths of his being and there they will be forever. No need to rummage around for them, exposing them. The ups and downs of human contact during his life in the academy had taught him to fight for something that seemed important. The achievements in the technical scientific world that to him did not seem to be as great as they are considered to be by his peers and friends, now belong to a prominent place in his accomplishments, in spite of everything and everyone. To paraphrase Marcel Proust, his time was not lost!

Continuing to walk, he found himself entering the "Rua do Porto Park." Twenty thousand square meters of green area of historic preservation. Again and again he would walk for his physical exercises along these alleys lined with leafy trees that winded the extensive lake where canoeing championships are held. Ahead of the lake, across the boardwalk and the promenade are bars, cold coconut kiosks and typical sea and freshwater restaurants.

At a bend in the road where he slowly walked, he felt in his skin the air that wrapped the trees under which now he rested on a wooden bench respiring the oxygen coming from the leaves, a result of the "miracle" of photosynthesis! And he breathed. How magnificent and wise nature is and the one who made it! The biochemical reactions that take place silently behind what our senses experience and we do not see: are always present in his mind!

A few kilometers away is CEBTEC. His heart beat strongly. Since 2001 he has no longer been its Scientific Director, but his office is still there, with all his memories and frequent presence. How many students discussed with him the results of their research in that room!

The contact not only with Enio, Helaine, Gallo and the many students who pass through the CEBTEC laboratories, but also with his colleagues from the Association of Retired Professors of ESALQ, his beloved ADAE, of which he is currently the President, has given him continuous encouragement to keep alive his desire to continue studying the nature around him. There's always time to learn!

The branches of the trees swayed. He felt the touch of the breeze that wrapped him, softening the heat that was coming along in that late morning of recollections. The breeze...the wind..."Sometimes I hear the wind passing by; and just by hearing the wind pass, it is worth being born." verses by the Portuguese poet Fernando Pessoa playing in his thoughts. There, sitting on that wooden bench, his thoughts, as in a flight, made him remember: on March 1998, Pedro Augusto was born, his only grandson, son of Daniel and Fernanda. Tall and slender, with lively intelligence, he is the joy of the whole family, confident in his future.

- Many years passed by, I'm in 2011! He was dazzled by this thought. And then he exulted: "CEBTEC is celebrating 30 years! A few weeks ago, in early October, ESALQ and the Association of Former Students of ESALQ (ADEALQ) gave me a plaque in honor of that date, full of meaning for me. I got it from the hands of Helaine and Roque Dechen, during the ceremony of the "54th Week Luiz de Queiroz." The recognition of his School would never be forgotten, he was sure.

During that ceremony he and his colleagues of ESALQ Agronomy Course were honored for their 55th anniversary of graduation. Otto was offered another card plaque celebrating his appointment to receive the "Medal Fernando Costa" that will be handed by the Agricultural Engineers Association of the State of Sao Paulo (AEASP) in a ceremony to be held on June 15, 2012, in Sao Paulo (photos 6 and 7).

He walked to Beira Rio Avenue and sat on the bench sipping ice-cold coconut water. In one of the bars, in front of him, the clock reminded him that it was already 12 noon. He stood up. "Helaine and Enio are waiting for me in CEBTEC," he said to himself, quickening his steps. "Ten years ago, in June, Helaine and Raul Machado Neto came to my room to announce that Rutgers University would honor me", he suddenly remembered.

Rutgers, 2001

Like a flash, he saw himself beside Diva and his daughter Carla Maisa, who came from San Francisco, California, where she was studying, to New York to meet her parents, all embracing Rod Sharp at John F. Kennedy International Airport. It was 7 o'clock in the morning of August 4.

On day 8, with Rod, they left for New Brunswick. In the afternoon of that day, in the presence of the President and Professors of Rutgers University, Helaine, Raul Machado, and Carlos Alberto Labate, and ESALQ graduate students, he was honored by the Cook College & the New Jersey Agricultural Experiment Station, and Rutgers - The State University of New Jersey.

Absorbed by in his memories while walking in the promenade of the park, Otto did not realize that a car was parked along the curb of the Beira Rio Avenue. When approaching it, he was surprised by a familiar voice calling him: "Dad, we're here." That was Daniel who was leaving his car to greet Otto. Daniel, who owns a powerful, pleasant tone of voice, inherited from his mother Diva, is the vocalist of the rock band "Royales."

A young, still almost adolescent boy got out of the car and ran toward him. "they're waiting for us", said Pedro Augusto (Photos 8 and 9).

In the car, which traveled the reverse way he had done in that morning when he went out of his house toward the river, his eyes, which reflected a future longing of those curves, climbs and descents, was looking for the sliding waters, visible among the tops of the trees that, while surrounding the edges of the river, cast a warm shadow left behind by the car driven by Daniel.

Finally they arrived at CEBTEC. His eyes met the eyes of his friends. Along with Diva, his children and grandson, he heard the news transmitted by Helaine, seconded by Enio: Jose Vicente Caixeta, ESALQ Director, informed that the Congregation of the College had approved the request of the Department of Biological Sciences, made through its Chief Ricardo Ferraz de Oliveira. The building of CEBTEC would have his name! (Photo 10).

Cheerful handclaps were heard but he did not listen to them. His mind was filled with memories and emotions of his childhood on Rangel Pestana Street, in Piracicaba. He recalled once again. He was sleeping. He found himself playing in a big house, surrounded by flowers and trees. He awaked. It had only been a dream: he was at the home of his parents! Over the years, as time passed by, that huge house would be where he would study and work, but still was not his home.

Embraced by Enio, Otto whispered in his ears: "now I'm also in my other home." He paraphrased Shakespeare:

"We are made of the matter of our dreams"

Acknowledgements

I would like to express my great appreciation to the comments made by Cecilio Elias Neto, Plinio Montagner, Celso Rossi, Enio Tiago de Oliveira Leonardo Alves Carneiro and Maria Helena

Aguiar Corazza in the process of writing this chapter. I offer special thanks for the invaluable support of my wife, Diva, and my children, Adolfo Egidio, Maria Paula, and Daniel. I am particularly grateful for the assistance given by my daughter, Carla Maisa, in the editing steps of the English version of this chapter. I extend my deep gratitude to all that shared the adventures on my scientific life.

PHOTO 1:
OTTO AND HIS WIFE DIVA IN EARLY 1980

PHOTO 2
DIVA, MARY BEATRICE, ROD´S GRANDMA AND ROD.
GARDEN OF OTTO´S HOME. PIRACICABA. 1972

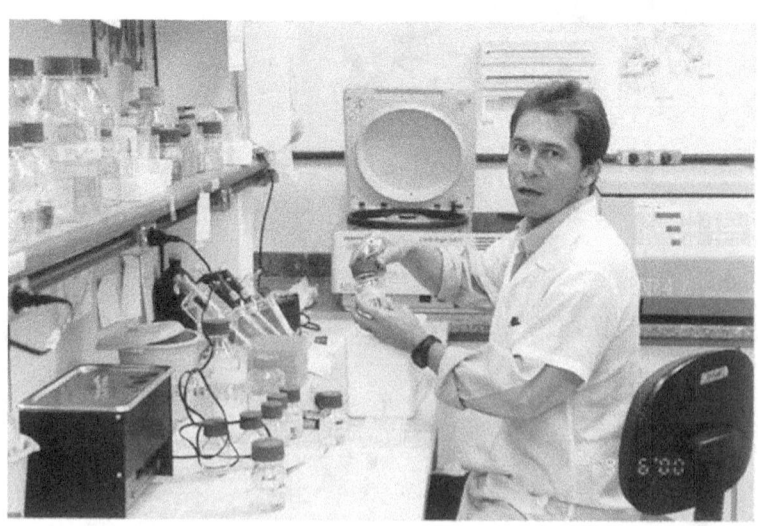

PHOTO 3
ENIO TIAGO DE OLIVEIRA AT CEBTEC´S LABORATORY (2001)

PHOTO 4
CEBTEC PREMISES (AERIAL VIEW) AT
ESALQ, PIRACICABA, SP, BRASIL

PHOTO 5
INAUGURATION OF THE "WILLIAM R. SHARP" ROOM AT
CEBTEC BUILDING. FROM LEFT TO RIGHT: ROD SHARP, OTTO
AND PAULO F. CIDADE DE ARAUJO (OCTOBER 27, 1988)

PHOTO 6
OTTO´S ADDRESS THE FERNANDO COSTA MEDAL (JUNE, 2012)

PHOTO 7
OTTO AND HIS WIFE DIVA WITH THE
FERNANDO COSTA MEDAL (JUNE, 2012)

PHOTO 8
OTTO´S FAMILY GATHERING AT SOFITEL HOTEL, GUARUJA
BEACH, BRASIL, ON THE OCCASSION OF THE CELEBRATION
OF THEIR 50 YEARS MARRIAGE. FROM LEFT TO RIGHT:
DIVA, ADOLFO EFIDIO, DANIEL, MARIA PAULA, CARLA
MAISA, PEDRO AUGUSTO AND OTTO (MARCH, 2011).

PHOTO 9
CELEBRATION OF ROD SHARP AND OTTO ANIVERSARIES
IN PIRACICABA, SP, BRAZIL (SEPTEMBER 23, 2013)

PHOTO 10
TOTEM IN FRONT OF CEBTEC BUILDING AT
ESALQ, USP, PIRACICABA, SP, BRAZIL

CHAPTER 6.

O Encontro da Quimica com a Biologia Celular e a Busca pelo Significado do Existir

Otto J. Crocomo

I
Em busca da identidade científica

1956

O trem movia-se com a lentidão própria dos trilhos de bitola estreita, rasgando as entranhas das montanhas que ao jovem de 23 anos lhe pareciam nuas, sem vegetação alguma. Ele tomara o trem às 6 horas da manhã de um dia do mês de agosto de 1956 na Estação Ferroviária Leopoldina, no Rio de Janeiro, com destino a Viçosa, Estado de Minas Gerais, no centro-oeste do Brasil, quando então participaria do III Congresso de Estudantes de Agronomia na Universidade Federal de Viçosa, apresentando dois trabalhos científicos que realizara sob a orientação de Eurípedes Malavolta e José Dall Pozzo Arzolla e em colaboração com o seu colega Ary Aparecido Salibe. O trem era apelidado de "Maria Fumaça" devido à densa nuvem de vapor e fuligem, expelida pela locomotiva que o conduzia, e que sujava as roupas dos passageiros em todos os vagões. A Estação Ferroviária recebeu o nome de Leopoldina em homenagem à primeira esposa do primeiro Imperador do Brasil, D. Pedro I, o qual, em 1822, proclamou a independência do Brasil de Portugal. A chegada em Viçosa estava prevista para 22 horas.

A lentidão do trem para avançar no percurso, o calor, a fumaça constante, proporcionaram ao jovem devaneios e recorrentes visitas ao seu passado em busca do porque estava ele ali sentado em um desconfortável banco de madeira. Ele cursava, naquele então, o quarto ano do Curso de Agronomia da Escola Superior de Agricultura "Luiz de Queiroz" (ESALQ) da Universidade de São Paulo (USP). A Universidade fora criada em 1934, mas a sua Escola, a ESALQ, era mais antiga: fora criada em 1900 e inaugurada em 1901, com início das aulas no dia 3 de junho desse ano. A ESALQ nasceu do desejo de um visionário, Luiz Vicente de Souza Queiroz, com formação técnica em agricultura na França e na Suiça, e que em 1892 doou ao Estado de São Paulo uma fazenda no entorno de Piracicaba para que nela fosse instalada uma "Escola Agrícola". A ESALQ foi uma das primeiras faculdades de ensino superior do Estado de São Paulo a integrar-se à USP em 1934.

Em um determinado momento dos seus devaneios, veio à memória do jovem um sonho que tivera quando criança durante o sono: ele, que morava no centro da cidade de Piracicaba, no Estado de São Paulo, desde que nascera, viu-se dentro de uma casa enorme no meio de árvores ladeando as aléias que a ela conduziam, lá pelos lados onde agora situava-se a "sua" Escola. Na sua infância, tomando o bonde que serpenteava as ruas de sua cidade, constantemente visitava a Escola Agrícola, na companhia de sua mãe ou de um de seus sete irmãos. Pareceu-lhe, então, ter sido aquele sonho uma antevisão do futuro: o sonho estava tornando-se realidade desde o dia em que, tomando esse mesmo bonde, foi à Escola para submeter-se aos exames de admissão, ser aprovado e adentrar a esse templo sagrado do saber agronômico.

As lembranças do jovem, entremeadas de poucas horas de sono e de paradas para almoço e jantar nas cidades que ladeavam a estrada de ferro, foram interrompidas quando o trem finalmente chegou na Estação Ferroviária de Viçosa. Porém, não terminaram aí. Dono de uma mente inquisitiva, após a apresentação de seus trabalhos científicos estava decidido a fazer o reconhecimento do *campus* universitário e da cidade. Um bosque de altos pinheiros que, com seus galhos entrelaçados, lançavam uma sombra acolhedora, o seduziu e foi o reduto que escolheu para meditar durante horas seguidas sobre o sentido de sua vida e de como seria o seu futuro no mundo científico. Aliás, essa era uma característica do jovem: ser capaz de isolar-se do mundo, como naquele momento, e também quando em meio a uma multidão.

1932 – 1956. Filho de imigrantes italianos, ambos do sul da Itália, sua mãe Tereza, da região da Calábria e seu pai João, originário de Ravello, na região de Napoles, o jovem foi o último a nascer. Apresentou-se ao mundo em Piracicaba, às 8 horas do dia 23 de setembro de 1932, pelas mãos de um médico, na residência de seus pais. Por ser o oitavo filho, seu pai deu-lhe o nome de Otto (oito, no idioma italiano). Seus outros sete irmãos eram: Tereza, Salvador, Maria, Letícia, Francisco, Lídia e Ada. Uma grande família. Maria, uma de suas irmãs, logo dele afeiçoou-se e passou a auxiliar sua mãe nos cuidados para com ele, substituindo-a após o seu falecimento em 2 de agosto de 1944, quando Otto tinha 12 anos de idade. Otto dedicou seus cuidados a ela até que falecesse aos 81 anos de idade. Ele a considerava sua "segunda mãe". Seu pai João, um industrial, fabricava grandes aparelhos de cobre para produção de álcool e aguardente de cana-de-açucar. A Prefeitura de Piracicaba homenageou-o denominando uma das ruas da cidade com seu nome. A mãe, descendente de família abastada, influenciou definitivamente na sua formação intelectual: aos 11 anos, ele escreveu um romance baseado nos contos que sua mãe lhe narrava sentados na porta de sua casa, ao anoitecer. Antes mesmo de frequentar a escola primária, já sabia ler e escrever. Na adolescência, não só as letras lhe encantavam: nos primeiros anos na escola secundária a matemática, a geografia e a história eram suas favoritas, como também, ao longo desse período escolar, as ciências biológicas, a química e a física; estas foram essenciais para a sua visão e compreensão do mundo físico e definição de sua carreira científica.

1950 – 1952. No trem, que engolia os trilhos de ferro à medida que se distanciava do Rio de Janeiro, onde chegara à meia-noite vindo de Piracicaba via São Paulo, teve vislumbres de suas atividades como estudante secundário e seu interesse pelas aulas teóricas e práticas de Química, ministradas com maestria por Demósthenes Santos Correia no Colégio "O Piracicabano", de orientação Metodista, nos anos 1950 a 1952. Os cinco debates de química, realizados nesse período, enfrentando o Colégio Estadual "Sud Mennucci", ou interclasses do próprio Colégio onde estudava, para os quais fora convocado pelo seu professor para participar e ser líder de equipe, marcaram indelevelmente o seu futuro.

1956. Naquele momento, no trem, mais do que em outros, como no bosque de pinheiros no qual ele, naquele momento, sem saber, meditaria, teve a certeza de que, muito além do fato de suas equipes terem sempre vencido os debates, a semente plantada em sua mente nessas ocasiões tensas para um adolescente, fazia aumentar, mais e mais, o desejo de saber a razão da sua existência e das coisas que o cercavam, como e para que são feitas, o que as faz serem como são, quais seriam os controles das reações químicas que acontecem dentro das células das plantas e que levam ao produto final, muitos deles sensíveis aos nossos sentidos quando extraídos das células. "O progresso cientifico deve-se à curiosidade do homem na busca da verdade": foram essas as palavras com que iniciou a sua participação no seu primeiro "Debate de Química", no Salão Nobre do Colégio "O Piracicabano" na manhã do dia 19 de setembro de 1950. Tinha então 18 anos de idade e em sua mente se desenhava seu profundo interesse sobre o porque da existência do homem e dos seres que em torno dele vivem.

Esse desejo, fruto de sua mente inquieta, era, pensou, a razão pela qual estava ele ali, sentado em um duro banco de madeira de um trem que vomitava fumaça, rumo à conquista de novos conhecimentos científicos que o Congresso de Viçosa poderia lhe proporcionar. Esse seria o terceiro e último congresso de estudantes no qual participaria, antes de se graduar; o primeiro fora realizado na ESALQ, em *1954* e o segundo na Escola Nacional de Agronomia (ENA) no Rio de Janeiro, em *1955*, quando então encontrou-se pela primeira vez com Paulo Fernando Cidade de Araújo que, anos mais tarde, se tornaria seu grande parceiro na administração de seus projetos científicos. Todos esses congressos tiveram a participação do Centro de Debates e Estudos Agronômicos que ele ajudara a criar no Centro Acadêmico "Luiz de Queiroz" (CALQ), dos alunos da ESALQ, *em 1954.*

A lembrança de acontecimentos recentes lhe vieram à mente. Como estudante, ainda no terceiro e também no atual último ano do curso de agronomia, ele se viu tomando parte ativamente de reuniões semanais em que eram apresentados e discutidos trabalhos científicos no então Instituto Zimotécnico, sob a coordenação de Jaime Rocha de Almeida. Recordou-se que era o único aluno de graduação aceito para participar dessas reuniões. Nesse momento de suas divagações tomou novo alento ao lembrar-se que, ainda como estudande de graduação, participou do Curso de Enzimologia, nesse mesmo Instituto, ministrado por Metry Bacila, professor da Universidade do Paraná, Estado do Sul do Brasil, um dos nomes mais respeitados em bioquímica de microorganismos, naquela época.

Estava a poucos meses do término de seu curso na ESALQ, era quase um Engenheiro Agrônomo. O trabalho que apresentaria em

Viçosa seria o seu último como estudante de graduação. Sentia-se mais seguro de si, com planos para o futuro próximo: queria dedicar-se ao estudo da química tão logo terminasse o curso. E esse, para ele, terminou no dia *30 de novembro de 1956*. A solenidade de graduação aconteceria quatro meses mais tarde, em *16 de março de 1957*, quando então recebeu o Prêmio MANAH pelo seu desempenho nas disciplinas de Química durante o seu Curso de Graduação. Seria então, de fato, um Engenheiro Agrônomo.

II
As experiências acadêmicas

1957 – 1962

Outubro de 1961. O avião, um Constalation da PANAIR, estava pronto para decolar do aeroporto de Congonhas, na cidade de São Paulo, com destino à Venezuela. Nele, estavam Otto e sua jovem esposa Diva. Haviam se despedido de seus familiares: seu pai e duas de suas irmãs, Tereza e Maria. Estavam também presentes os pais de Diva, Gelindo e Maria, ambos filhos de imigrantes do Norte da Itália. A despedida tinha sido ao mesmo tempo triste e alegre. Triste porque deixariam os seus entes queridos pelo período de 14 meses e alegre porque ele havia sido contratado como professor de Bioquímica e Química Agrícola pela Facultad de Agronomia da Universidad del Zulia na cidade de Maracaibo, às margens do Lago de mesmo nome, em uma região petrolífera da Venezuela.

Otto e Diva conheceram-se em *9 de novembro de 1957* (foto 1). O sucesso musical nessa época era "Unchained Melody" de Alex North e Hy Zaret que elegeram como "sua música" e repetidas vezes cantada por Diva, dona de um timbre de voz único, com a sonoridade de uma excelente soprano que é. Quase 4 anos depois, o casamento civil ocorreu em *25 de março de 1961*; um dos padrinhos foi o grande amigo e colaborador de Otto em vários de seus trabalhos científicos, André Martin Louis Neptune, de origem haitiana, naturalizado brasileiro, e sua esposa Nair. O casamento religioso, celebrado por D. Aniger Melillo, Bispo da Diocese de Piracicaba, aconteceu em *23 de maio do mesmo ano*. Eurípedes Malavolta, de quem Otto era assistente, e sua esposa Leila, e Bem-Hur Carvalhaes de Paiva e sua esposa Maria Helena, estavam entre os padrinhos. Maria Helena fora colega de Otto durante o Curso Secundário no Colégio "O Piracicabano".

Junho de 1958. Na manhã do dia 22, como de costume, Otto desceu do bonde que o levava do centro de Piracicaba até a ESALQ. Contornou o Prédio de Química e subiu a alta escada com degraus de ladrilhos vermelhos, situada na parte de trás do Prédio, e adentrou no Laboratório de Química Agrícola. Euripedes Malavolta estava sentado em uma banqueta, debruçado sobre livros e revistas distribuídos sobre uma bancada. Preparava-se para o concurso de Professor Catedrático que teria lugar nos próximos meses.

- Otto, tenho um convite para fazer-lhe, disse Malavolta.

Admirado, aquiesceu com a cabeça e aproximou-se de Malavolta.

- Eu gostaria que você se submetesse ao concurso de Livre-Docente, disse Malavolta olhando diretamente para o jovem que, surpreso, balbuciou: "mas, como? eu sou recém-formado, e o Professor Arzolla é quem está se preparando para esse concurso".

- Voces dois prestarão o concurso na mesma época, respondeu Malavolta. E continuou: "eu gostei muito do seu seminário no dia 8, no Instituto Zimotécnico, aqui na ESALQ, sobre os resultados de seu trabalho com ureia-^{14}C. Voce impressionou os meus colegas presentes".

- Mas, são resultados de minha Tese de Doutorado em Agronomia, respondeu Otto, deixando no ar uma pergunta: "será...?"

- Voce poderá utilizá-los como Tese de Livre-Docência, disse Malavolta. Voltando para seus livros, acrescentou: "confio em você e dou-lhe alguns dias para decidir".

Setembro de 1959. Tendo aceitado o convite de Malavolta, submeteu-se, durante 4 dias, de 19 a 22, à provas teóricas e práticas do Concurso de Livre-Docência em Química Orgânica e Química Biológica, perante uma Comissão formada por cinco professores: Euripedes Malavolta, Tufi Cury, ambos da ESALQ, e Veronica Rapp de Eston e Henrique Tastaldi, do *campus* da USP em São Paulo, e Metry Bacila, da Universidade Federal do Paraná, em Curitiba. Uma das 5 provas foi a defesa pública de sua Tese, a primeira na ESALQ, no campo biológico, a utilizar radioisótopos: "metabolismo de ureia-^{14}C em folhas de cafeeiro", contribuindo para evidenciar a presença do ciclo da uréia em vegetais superiores. Transformações metabólicas: as constantes perguntas que ele se fazia sobre o funcionamento das reações químicas dentro das células vivas, e como são controladas, estavam começando a ser-lhe respondidas! Compreender as reações químicas no contexto da célula vegetal seria o objetivo de seu interesse científico no futuro. A bioquímica e Otto estabeleceram nesse momento uma feliz e duradoura união.

A sua contratação pela Universidade de São Paulo como Professor Assistente Doutor da Cadeira de Química Orgânica e Química Biológica da sua Escola, a ESALQ, ocorreu em *16 de abril de 1960*. Nesse mesmo ano, o título de Doutor em Agronomia foilhe concedido pela Universidade de São Paulo em consequência de já ter obtido o Título de Livre-Docente pela mesma Universidade. A ele não foi exigido o período probatório de 5 anos, que normalmente o era para Professores-Assistentes contratados em tempo integral. Finalmente, poderia ministrar aulas de bioquímica para os alunos da ESALQ.

Ao longo de sua vida, muitas vezes ele refletiu sobre a razão de seu profundo interesse pelas reações químicas que acontecem no interior das células. Era fruto de sua incessante procura pela verdade científica. Mas, como professor de bioquímica, a sua preocupação constante era como explicar aos seus alunos as equações químicas que representam, no papel e no quadro negro, as reações químicas? Talvez fosse algo relacionado ao seu incessante debruçar sobre as minúcias das equações matemáticas, na resolução de problemas de álgebra superior. Tivera excelentes professores de matemática. Sempre teve consciência de que, apesar de não perceptíveis aos nossos sentidos, verdades matemáticas existem nas mentes humanas e se materializam e são palpáveis nas grandes realizações que tem promovido o desenvolvimento da humanidade. Basta um exemplo: as teorias de Albert Einstein, frutos de sua profunda intuição, tornadas vivas pelas equações matemáticas! O mundo moderno é o que é hoje devido à aplicação dos princípios matemáticos. Informática e viagens interplanetárias sempre são manchetes. Como "manusear" as fórmulas matemáticas? Assim também perguntava a sua mente inquisitiva para o "manuseio" das fórmulas químicas. Fora treinado na resolução de problemas matemáticos, o que lhe facilitou a transmissão dos acontecimentos bioquímicos a seus alunos, manipulando no quadro negro as fórmulas dos componentes químicos das reações bioquímicas. As fórmulas e as equações químicas são modelos que servem ao propósito de facilitar a compreensão do que acontece no mundo microscópico da célula, que não vemos. Ele, disso, sempre teve consciência.

No avião que os conduzia à Venezuela, recordava-se e refletia sobre isso tudo enquanto Diva tomava notas para uma carta que, mais tarde, enviaria a seus pais. Na carta, Diva descrevia a maravilha que era sobrevoar a selva amazônica: aquele tapete verde que parecia estar aos seus pés. O percurso aéreo estava sendo feito a baixa altitude.

Maracaibo, Venezuela. O convite para lecionar na Venezuela fora feito pelo Decano da Facultad de Agronomia, José Gonzalez

Mateus, intermediado por André M. Louis Neptune, que lá estivera anteriormente. No Aeroporto Internacional de Maiquetia "Simon Bolivar", em Caracas, foram recebidos por Felix Taborda, ex-aluno da ESALQ. À noite encontraram-se com Gonzalez Mateus que logo colocou à sua disposição todas as facilidades para que ele, além de responsabilizar-se pelas Disciplinas de Bioquímica e Química Agrícola, também instalasse um Laboratório de Radioisótopos.

Devido ao uso de textos e livros em idioma espanhol, nos quais estudara química inorgânica, e seu interesse pelos ritmos caribenhos, como boleros, salsas e cumbias, sentiu-se familiarizado no novo ambiente e, desse modo, a sua adaptação, e de sua esposa Diva, ao novo estilo de vida venezuelano, foi muito facilitado. Os professores da Facultad de Agronomia e o pessoal administrativo lhe deram todo apoio. Reuniões festivas, próprias do espírito alegre e afetuoso dos venezuelanos, eram frequentes. Talvez por ser um professor estrangeiro os alunos logo se entusiasmaram com suas aulas. Dois deles, em particular, interessaram-se em ser seus monitores nas aulas práticas. Um deles, J. Villasmill, era o mais dedicado, excelente aluno, que mais tarde se envolveria na administração da Facultad. O outro, Tibério Perozo Yori, que não era considerado bom aluno, interessou-se pelos seus ensinamentos, auxiliando-o inclusive na instalação e acompanhamento de um experimento a campo sobre nutrição mineral de milho (*Zea mais*), em uma das fazendas de seu pai. A sua dedicação aos estudos levou-o a apresentar com sucesso, sob a orientação de Otto, um seminário, perante seus colegas e professores da Facultad, com a presença do Decano. Em janeiro e fevereiro de 1964, Tibério estagiou nos laboratórios de Otto no CENA, em Piracicaba. Em agosto desse mesmo ano, Otto e Diva, a caminho dos Estados Unidos, visitaram em Maracaibo a família de Tibério, cujo pai considerava Otto como o "segundo pai de Tibério". Em *fevereiro de 1972*, Tibério estava desenvolvendo seu programa de

Mestrado no Instituto Interamericano de Ciências Agrícolas (IICA), em Turrialba, Costa Rica. Nessa ocasião, encontrou-se com Otto que estava ministrando um Curso sobre "Bioquímica da Nutrição Mineral de Plantas" nesse Instituto. Tibério faleceria, poucos anos mais tarde, ainda bem jovem.

Costa Rica, 1972. O carro esperava Otto à porta da casa de hóspedes na qual se hospedara durante o Curso que ministrara no Instituto Interamericano, em Turrialba. Partiram em direção a San José. Eram os últimos dias do mês de fevereiro. O sol teimava em aparecer entre as nuvens que emolduravam as curvas da estrada, o asfalto sendo engolido pelas rodas do carro, enquanto ele se encantava com a magnífica vista do vale lá embaixo onde, em algum ponto, estava Turrialba, na província de Cartago, conhecida como "um pedaço do paraíso do Caribe, as portas da Costa Rica para o oceano Atlântico". Suspirou ao ouvir o som de Guantanamera vindo do rádio do carro. Uma das mais preferidas por ele dentre as que formavam a sua coleção de músicas com letras em castelhano, de autoria do intelectual e professor cubano José Mautas Martí, e música de Josito Fernandez. Sim, Martí, importante participante das lutas pela independência de Cuba do domínio espanhol, no final do século XIX! Mal sabia ele que, *21 anos mais tarde, em maio de 1993*, acompanhado por Diva, iria em missão científica, como perito das Nações Unidas, a Havana, para acompanhar o desenvolvimento de pesquisas sobre cana-de-açúcar no Instituto de Biotecnologia e conviver com o generoso e sofrido povo cubano, que canta esses versos de José Martí, poeta que venera.

O carro continuou a serpentear as encostas montanhosas e ele chegou ao aeroporto de San José, capital de Costa Rica, situada a mais de mil metros de altitude. No Brasil, a sua família e a ESALQ o esperavam.

Maio de 1962. Marco Augusto, o seu primeiro filho nasceu no dia 31, no Hospital Nossa Senhora de Coromoto. Por ter nascido em Maracaibo recebeu o apelido de "el Maracucho" pelos amigos venezuelanos, entre eles Felix Taborda e Hiram Reyes-Zumeta, ambos ex-alunos da ESALQ e colegas de Otto na Facultad, e Emiro Gomez também ex-aluno da ESALQ. Emiro e sua esposa Maria José, piracicabana, foram padrinhos de batismo de Marco Augusto.

A experiência de ser professor em um país que não era o seu, aos 29 anos de idade, reger duas cátedras, instalar laboratório de radioisótopos, coordenar um Simpósio sobre Uso de Radioisótopos em Agricultura, realizar experimento a campo, falando um idioma que também não era o seu, foi essencial para o desenvolvimento de sua vida acadêmica, pois estava sozinho para tomar decisões. Não havia tutores ou colegas que pudessem auxiliá-lo em suas autocríticas. Passou a olhar o mundo com novas perspectivas, com desejo de novas realizações. Essa experiência estava presente na sua mente e no seu coração quando, em companhia de Diva e de Marco Augusto, deixou Maracaibo com destino a Caracas e, em seguida, a São Paulo onde, no Aeroporto de Congonhas, abraçou seus familiares. Era a manhã de *19 de dezembro de 1962.*

Já havia tido uma experiência similar, se bem que não tão intensa quanto aquela vivida na Venezuela. Fora convidado pelo cardiologista Ben-Hur Carvalhaes de Paiva para ser seu Assistente e Responsável pela Disciplina Bioquímica na então Faculdade de Farmácia e Odontologia de Piracicaba, nos *anos 1959-1960.* Este foi o seu primeiro emprego após ter-se formado na ESALQ. Bem-Hur era, então, professor de Fisiologia Animal. Essa Faculdade foi criada em 1955 como parte dos Institutos Isolados do Conselho de Ensino Superior do Estado de São Paulo e em 1967 transformada em Faculdade de Odontologia de Piracicaba, passando a fazer parte da Universidade Estadual de Campinas (UNICAMP). Nessa época ele teve vários alunos muito bons, como Antonio Carlos Neder, seu monitor nas aulas práticas, o qual, segundo ele

próprio, dedicou-se à farmacologia devido a influência das aulas de bioquímica. Neder se tornaria, no futuro, Diretor daquela
Faculdade e um nome respeitado internacionalmente no campo
da farmacologia odontológica, a qual introduziu como disciplina
nos Cursos de Odontologia no Brasil.

III
A contínua busca de conhecimento científico

1964

A sua natureza inquisitiva continuava presente em todas as suas
atividades pessoais e profissionais. Ao lado das suas indagações
filosóficas e suas profundas e contínuas inquietações transcen-
dentais, que frequentemente discutia com seu amigo o jornalista
Cecílio Elias Neto, brilhante intelectual piracicabano, o desejo de
aprofundar-se cada vez mais no estudo da bioquímica de plantas
tornava-o ainda mais inquieto. Na Venezuela, onde publicara seu
primeiro livro "Bioquímica de la Respiración Celular", editado
pela Facultad de Agronomia, em Maracaibo (1962), arquitetara
planos que poderiam levá-lo a estudar nos Estados Unidos: esta
era uma das razões pelas quais não aceitou a renovação do seu
contrato junto à Facultad de Agronomia, somando-se ao fato de
que a ESALQ desejava-o de volta à Piracicaba.

Agosto. Otto, de mãos dadas com seu filho Marco Augusto, e
Diva, levando ao colo o filho Adolfo Egídio, que nascera em *6 de
setembro de 1963*, em Piracicaba, subiram a alta escada do avião
DC-8 da PANAM, no Aeroporto de Congonhas, em São Paulo.
Essa segunda viagem internacional do casal o levaria aos Estados
Unidos. Como Fellow Researcher da Rockefeller Foundation, iria
desenvolver trabalhos científicos no Departamento de Biofísica e
Bioquímica da Universidade da Califórnia, *campus* de Davis.

Em seu caminho para os Estados Unidos, fez uma série de
palestras na Facultad de Agronomia da Universidad del Zulia. As

palestras foram publicadas pela Facultad no livro "Absorción de iones por las plantas", com a colaboração de André Martin Louis Neptune e Hiram Reyes-Zumeta.

Em um apartamento no condomínio *Orchard Park*, em Davis, ele e Diva conviviam com famílias de estudantes vindos de vários países europeus, africanos e orientais. Frequentemente reuniam-se com amigos brasileiros de diferentes regiões do Brasil, com seus filhos ainda pequenos que se tornaram companheiros de Marco Augusto. O segundo filho, Adolfo Egídio, aprendeu a caminhar poucos meses após a chegada em Davis. Laços de amizade foram estabelecidos e outros estreitados com brasileiros que desenvol-viam seus programas de doutorado no *campus*, entre eles Lourival Carmo Mônaco e sua esposa Mercedes. Lourival posteriormente desenvolveu intenso e importante trabalho em genética de plan-tas no Instituto Agronômico de Campinas. Ele e Otto haviam sido colegas nos tempos da escola secundária e durante o Curso de graduação na ESALQ.

Em Davis, além de desenvolver pesquisa com microorganis-mos com C. C. Delwiche, frequentou curso diurno de bioquímica teórica e curso noturno de laboratório, ministrados por E. E. Conn e P. K. Stumpf e seus Assistentes, e também o curso de Físico-Química Conceitual e o curso de Princípios de Nutrição Mineral de Plantas, este último ministrado por E. Epstein. Nessa oportunidade aprendeu a linguagem computacional Fortran, ne-cessária para a elaboração de dados obtidos nos seus trabalhos de laboratório. Sua passagem pela Universidade da Califórnia foi essencial para abrir-lhe novas perspectivas, preparando-o para assumir, imediatamente após sua volta ao Brasil, a Disciplina de Pós-Graduação "Bioquímica de Plantas", na ESALQ, a qual lecionou durante 23 anos, também orientando alunos de Mestrado e Doutorado. Em colaboração com Euripedes Malavolta e Darcy Martins da Silva, traduziu para o Português o livro "Outlines of

Biochemistry" de E. E. Conn e P. K. Stumpf, edição de 1972, John Wiley & Sons Inc., com várias edições no Brasil.

A revolução militar. O *ano de 1964* marcou indelevelmente a história do Brasil. Em 31 de março, militares, de vários estados brasileiros, principalmente de Minas Gerais e São Paulo, rebelaram-se contra o que se considerava desmandos autoritários do então Presidente da República do Brasil, João Goulart. Desde 1961, quando assumira a Presidência, como consequência da renúncia de Jânio Quadros, do qual era Vice-Presidente, os militares sempre fizeram-lhe oposição. Só foi-lhe permitido governar o Brasil dentro de um regime parlamentarista estabelecido pelas forças que se lhe opunham. Goulart, então, passou a governar tendo a máquina da previdência social em suas mãos e fazendo alianças com a esquerda no controle dos sindicatos.

Foi nesse cenário político que, em *21 de agosto de 1965*, logo após o regresso de Otto e família ao Brasil, vindos dos Estados Unidos, nasceu Maria Paula, a sua terceira filha. Nesse mesmo ano, de outubro a dezembro, ele ministrou aulas teóricas e práticas sobre "Aplicação de Radioisótopos nas Relações Solo/Planta" no Instituto de Asuntos Nucleares, em Bogotá, Colombia.

Como consequência da rebelião dos militares, estabeleceu-se no Brasil um Regime Militar com amplos poderes ditatoriais, principalmente a censura da imprensa, perseguição política a estudantes, líderes sindicais, músicos e intelectuais. Entre estes, muitos eram professores da Universidade de São Paulo que se exilaram em países estrangeiros, como Fernando Henrique Cardoso. O líder sindicalista Luis Inácio Lula da Silva foi preso por alguns dias. Quando o Regime Militar terminou e a democracia foi restabelecida no Brasil, o que ocorreu em 1985, Fernando Henrique foi o quarto Presidente eleito direta e livremente pelo povo (1994-2002) e responsável pela implementação do Plano Real, criado pelo seu antecessor, Itamar Franco. O Plano Real valorizou a moeda brasileira, estabilizando a economia. Foi criada a Lei de Responsabilidade Fiscal, a que todo governante deve obedecer; a inflação passou a ser controlada. Luiz Inácio Lula da Silva foi eleito o quinto Presidente e governou o Brasil por duas legislaturas (2002-2010), com forte apelo popular. Lula apoiou a candidatura de Dilma Roussef, ex-combatente contra a ditadura militar, para a Presidência, tornando-se a primeira Presidente mulher do Brasil (2011-).

IV
Os Primeiros Anos no CENA

Dezembro de 1973. Ele saiu de sua sala e, a passos curtos, percorreu o corredor da Seção de Bioquímica. de Plantas, da qual era o Responsável, até a porta que dava para o parque onde se situa o Centro de Energia Nuclear na Agricultura (CENA). Era uma manhã de sábado, em princípios de dezembro. Apesar do verão, uma intensa névoa cobria as árvores e os arbustos que emolduravam os poucos edifícios que ainda podiam ser vistos. Névoa adentro, caminhou pelas aléias, recebendo o ar úmido. De imediato, recordou a recente viagem de trem que fizera em meio à neve que revestia as árvores, no percurso entre Genève e Zurique, nos Alpes Suiços. A passos lentos, continuou contornando os edifícios e relembrando que as suas viagens científicas ao redor do mundo (que apenas estavam começando) deviam-se principalmente àquele lugar, agora inundado pela névoa inusitada em uma manhã do verão piracicabano. E uma vez mais recordou-se de como iniciara seus trabalhos científicos no CENA.

Em 1967, já Professor Associado desde 1966, estava trabalhando nos laboratórios de bioquímica no Pavilhão de Química na ESALQ, com três de seus orientados, Luiz Carlos Basso, Oswaldo Galvão Brasil e Celso Rossi. Malavolta aproximou-se e pediu que eles e seus alunos, e todo material que estavam usando, se transferissem para um novo laboratório no CENA. Assim recomeçara a sua vida científica uma vez mais em outro local. Sim, uma vez mais, porque anteriormente utilizara os laboratórios do Instituto Zimotécnico, também na ESALQ, por longo período para conduzir os experimentos de sua Tese de Livre Docência nos anos 1950. E mais uma mudança na sua vida acontecera: nesse mesmo ano de *1967* nasceu a quarta filha, Carla Maisa, no *dia 25 de setembro*, em Piracicaba.

À medida que caminhava pelas aléias do parque do CENA, absorto em suas lembranças, a névoa paulatinamente se dissipava e de repente viu-se envolto em uma luz brilhante, era o sol que o abraçava e lhe anunciava novos tempos de aventuras científicas. Entrou pela porta principal da sua Seção, foi à sua sala e debruçou-se sobre novas ideias.

Naquele momento, entretanto, o seu pensamento dispersouse, voltando para os acontecimentos científicos iniciais que deram nascimento ao CENA.

Os pioneiros. Os passos que levaram à criação do CENA começaram muito antes. *Em 1953*, Eurípedes Malavolta introduziu, na ESALQ, as técnicas do uso de radioisótopos que, como investigador associado na Universidade da Califórnia, em Davis, utilizara em seus estudos de nutrição mineral de plantas com a colaboração de E. R. Stout. Malavolta foi um dos primeiros a publicar no Brasil um trabalho científico sobre o uso de radioisótopos em plantas: absorção de zinco radioativo pelas folhas de orquídeas. *Em 1956*, a vinda à ESALQ de Constant C. Delwiche, do IBEC Research Institute, da Universidade da Califórnia, com quem Otto trabalharia mais tarde (1964-1965), foi decisiva para implementar o uso das técnicas nucleares em vegetais superiores na ESALQ. *Em 1958* os trabalhos de Malavolta sobre a absorção de superfosfato radioativo em cafeeiro, a campo, demonstraram a melhor localização, no solo, do adubo fosfatado para essa cultura agrícola. Em *setembro de1959* Otto defendeu sua Tese de Livre-Docência cujos resultados científicos foram obtidos utilizando radioisótopos.

Esses trabalhos científicos pioneiros, utilizando técnicas radio-sotópicas no campo da nutrição e da bioquímica de plantas, e outros no campo da física realizados por Ademar Cervellini, Anivaldo Pedro Cobra, Eneas Salati e José Goldemberg, foram conduzidos no Laboratório de Isótopos da Cadeira de Física e Meteorologia, e nos laboratórios da Cadeira de Química Orgânica e Química Biológica e do Instituto Zimotécnico, todos na ESALQ, nos anos 1950. Um dos seus colaboradores foi seu grande amigo André Martin Louis Neptune. Devidamente protegidos contra radiação nuclear, ambos e seus alunos, deixavam suas residências, de madrugada, para pincelar folhas de plantas de cafeeiro com solução de enxofre radioativo (^{35}S), com a finalidade de comprovar a absorção foliar de minerais e acompanhar o metabolismo de substâncias sulfuradas em vegetais superiores. Em *janeiro e fevereiro de 1960*, participou, como aluno, do Curso Internacional de Medicina Nuclear, no Centro de Medicina Nuclear da Universidade de São Paulo, em São Paulo. No ano seguinte, *em janeiro de 1961*, ministrou aulas teóricas e práticas sobre o uso de radioisótopos no metabolismo de plantas e métodos de medição de radioatividade em extratos vegetais, nesse mesmo Curso.

Juntamente com Malavolta, Otto foi um dos muitos colaboradores de Cervellini na criação e consolidação do CENA, entre os quais estavam Almiro Blumenschein, Akihiko Ando, André Martin Louis Neptune, Darcy Martins da Silva, Eneas Salati, Epaminondas Sansigolo Ferraz, Frederico Maximiliano Wiendl, Henrique Bergamin Filho, Klaus Reichardt, Renato Amilcare Catani e Valdomiro Corrêa Bittencourt.

Apesar das investidas ditatoriais dos governos militares, cujos Presidentes sucederam-se durante cerca de 20 anos, o Brasil experimentou grandes avanços em várias áreas, como a científica. Foi nessa época, em *22 de setembro de 1966*, que o Centro de Energia Nuclear na Agricultura (CENA) foi criado como Instituto anexo à ESALQ. Em 1968, o Convênio assinado entre a Universidade de São Paulo e a Comissão Nacional de Energia Nuclear (CNEN), entidade estratégica do Governo Federal na área de energia nuclear, foi fundamental para a consolidação da Instituição. A partir de 1985 o CENA passou a integrar, juntamente com a ESALQ, o *campus* "Luiz de Queiroz" e desde 1988, é Instituto Especializado da USP.

V
A experiência científica e cultural

1968

Outubro. No Aeroporto de Viracopos, na região de Campinas, a alguns quilômetros de Piracicaba, Otto aguardava ser chamado para ocupar seu lugar no avião da British Airways que o levaria a Londres. *Era o dia 26.* Com ele estava Diva que não o acompanharia nessa estada de 5 meses no Department of Botany and Microbiology, no University College London, onde, como Fellow Researcher, financiado pelo British Council e pela Fundação de Amparo à Pesquisa do Estado de São Paulo (FAPESP), desenvolveria trabalho cientifico com Leslie Fowden. Diva iria visita-lo em Londres em final de fevereiro de 1969 quando então, após visita a vários países do continente europeu, regressariam ao Brasil em finais de março.

Celso Rossi e sua esposa Sirley foram despedir-se dele. Celso era um de seus três Assistentes na Faculdade de Ciências Médicas e Biológicas de Botucatú, no interior do Estado de São Paulo, onde ele foi Regente da Disciplina Bioquímica no *período 1968-1969.* Posteriormente, Celso Rossi e Oswaldo Galvão Brasil, o outro Assistente, conquistariam os títulos de Mestre, Doutor e Livre-Docente de Bioquímica. Ambos tornaram-se Professores Titulares daquela Faculdade em fevereiro de 1987.

Celso Rossi lembra-se com detalhes dessa despedida e, anos depois, escreveu:

"...jamais me esqueci dessa noite no Aeroporto de Viracopos. Eu possuía um pequeno carro Volkswagen 1200, bateria de 6V. A luminosidade era bastante fraca para viajar numa noite de muita neblina. Lembro-me da decolagem noturna, eram 20 ou 21 h quando a aeronave arremeteu como uma flecha em direção ao céu. Eu pensei "que coragem tem o Otto, partir numa noite dessas em direção a Londres". A volta para Piracicaba foi difícil, apesar de que, naquele tempo, poucos carros transitavam na rodovia, mas chegamos vivos. A prova é que hoje estou lhe contando esses detalhes"

O terceiro Assistente, Luiz Carlos Basso, foi posteriormente contratado pelo Departamento de Química da ESALQ, conquistando, sob orientação de Otto, os títulos de Mestre e Doutor. Basso trabalhara como Auxiliar de Laboratório em 1963, antes de ser aluno de graduação na ESALQ, no primeiro Auxílio Científico concedido a Otto pela Fundação de Amparo à Pesquisa do Estado de São Paulo (FAPESP), criada em 1962 para fomentar a pesquisa científica no Estado. Ao longo de sua carreira científico-acadêmica Otto recebeu inúmeros Auxílios Financeiros da FAPESP, em vários deles tendo Basso como um de seus mais importantes colaboradores, principalmente nas pesquisas sobre metabolismo mineral e de poliaminas em plantas, publicados em revistas científicas internacionais. Basso se tornaria no futuro um dos melhores professores e pesquisadores de bioquímica de plantas e de microorganismos da ESALQ e participante de projetos de parceria público-privada no campo da bioenergética, pesquisando atividades enzimáticas durante a fermentação alcoólica.

Em 22 de maio desse ano, seu pai João, faleceu aos 73 anos de idade, vítima de um derrame cerebral, após seis meses de sofrimento, paraplégico e sem poder comunicar-se a não ser por sinais, sob os cuidados de duas de suas filhas, Maria e Tereza.

Em Londres, hospedou-se na Casa do Brasil, próxima ao metrô Lancaster Gate, que o levava todas as manhãs, com uma troca de metrô na Tottenham Court Road Station, para o University

College, na Gower Street. Logo passou a admirar a capital da Inglaterra, não só no contexto da sua participação na história da civilização ocidental, mas também, e particularmente, pela sua contribuição para o desenvolvimento da ciência. Agora, aos 36 anos de idade, vislumbrou novos horizontes a serem atingidos. A sua contínua procura pela essência e o porque da existência de si mesmo e da natureza da qual era parte, características de sua incessante curiosidade, encontrou um solo fértil na ciência que praticava no Laboratório de Leslie Fowden e na participação em conferências no College. Deixou-se envolver pela cultura e pela ciência européias, do mesmo modo que anteriormente experimentara os benefícios do desenvolvimento científico dos Estados Unidos. Não encontrou respostas completas para suas inquietações filosóficas, mas de sua bagagem intelectual fazia parte, agora, uma outra visão de mundo e de comportamento humano que, no futuro, muito contribuiria para a sua conduta científica e o seu relacionamento com o outro.

A Londres cosmopolita e cultural atraiu-o. Estava no país dos Beatles, quase na mesma época em que os participantes da banda se separaram. Mas a Carnaby Street, que se inicia na longa e curva e elegante Regent Street, estava e ainda está lá, evocando e eternizando os sons de suas guitarras e de suas vozes.

Suas idas, nos finais de semana, a concertos musicais e a peças teatrais deixaram-no fascinado.

- Carlos, no "Times" estão anunciando a peça "Mousetrap". Voce já a viu?, pergunto Otto.

- Sim, está sendo apresentada em um pequeno teatro no West End, o New Ambassador Theatre, desde de sua estréia em 25 de novembro de 1952, vale a pena assistir, respondeu Carlos e continuou: "é de autoria de Agatha Christie". Carlos, um jovem piracicabano, alto, esguio, contratado pela Embaixada Brasileira em Londres, para ensinar o idioma Português na Inglaterra, também residia na Casa do Brasil.

Carlos Toledo Vollet Sachs era sobrinho de Salvador de Toledo Piza Jr., Professor Emérito da ESALQ, de quem Otto havia sido aluno, ambos mantendo um excelente relacionamento intelectual. Anos depois, Carlos iria viver em Manchester, na Inglaterra, onde viria a falecer em 2011.

- Há muito tempo sou fã de Agatha Christie. O seu estilo ao apresentar o envolvimento dos personagens no desenrolar da trama, e toda a argumentação para desvendar os mistérios dos casos policiais, me excita a imaginação e me auxilia na interpretação de fatos cotidianos que acontecem ao meu redor, Otto afirmou, pensativamente.

- Podemos ir no próximo sábado, convidou Carlos. Aliás, gostaria que também fossemos ver o musical "Fidler on the Roof", que já vi 7 vêzes. Carlos riu-se de si mesmo, ao fazer essa confissão de fã incondicional da peça, e acrescentou eufórico: "a trilha musical é excelente, é de Jerry Bock, com texto de Sheldon Harnick, ambos americanos, detentores de vários prêmios". A peça teatral estava sendo apresentada no Her Majesty's Theatre, na Haymarket, próximo a Trafalgar Square.

Na Trafalgar Square, dominada pelas quatro enormes estátuas negras de leões, os chafarizes e a grandiosa Coluna de Nelson, vencedor da batalha de Waterloo contra Napoleão Bonaparte em princípios do século 19, está a The National Gallery. Este era o lugar predileto por ele para passar boa parte das tardes de domingo, no inverno de Londres. Os quadros dos expoentes das artes plásticas dos séculos XIII ao XIX, e as exposições esporádicas de artistas modernos e contemporâneos, deixavam a sua mente extasiada diante de tanta beleza e riqueza de detalhes e de harmonia de cores. Das cores vibrantes de Rafael à singeleza espiritualizada do "cartoon" de Leonardo Da Vinci, "the Virgin and Child with St. Anne and St. John the Baptist", solitariamente exposto em uma saleta aclimatizada, com luz indireta. Com frequência, sentado à frente desse "esboço" de Da Vinci, era, em pensamento, transportado para paragens de praias desertas onde brancas areias deixavam-se beijar pelas ondas de suaves marolas que as envolviam. Cenários bucólicos de caminhos de terra que o conduziam ao topo de morros, de onde seu olhar podia vislumbrar um sereno e imenso vale com animais saboreando a relva verde, salpicado aqui e acolá por ciprestes e pinheiros rodeando casas de pontiagudos tetos vermelhos, visitavam sua mente embalada pelos acordes da sinfonia "Pastoral" de Beethoven. Uma vez mais, e como sempre, surpreendia-se também meditando sobre o valor de estar vivo e sobre seu futuro, como fizera no escuro bosque de pinheiros em Viçosa, em *agosto de 1956*.

As visitas ao Museu de História Natural e ao Museu Britânico e outros e também assistindo aos concertos de música clássica no Royal Opera House, em Covent Garden, ou no Royal Albert Hall, em Kensington Gore, eram momentos de extrema beleza, até mesmo lúdicos, para aquele jovem ávido de novos conhecimentos e de novas experiências vivenciais. Era somente o começo de uma série de visitas científicas e culturais à capital da Inglaterra, além de inúmeras outras para vários países do continente europeu.

A sua infância em Piracicaba estava constantemente em suas recordações. Sentado em um banco diante do Prince Albert Memorial, em uma manhã de janeiro de intenso inverno, recordava-se falando para sua mãe Teresa:

- "...quando eu "ficar grande" quero conhecer Londres, Paris e Curitiba". Encantava-se com história e geografia e, naquele momento, estava explorando os mapas do Brasil e da Europa, sentado à mesa da cozinha na casa de seus pais, em Piracicaba, aos 10 anos de idade.

Em Curitiba, capital do Estado do Paraná, ao Sul do Brasil, já estivera várias vezes, lecionando bioquímica de plantas em cursos de especialização organizados por Metri Bacila. Paris conheceria um pouco mais tarde. Mas, agora estava em Londres! E estava diante do extraordinário monumento-celebração do amor da Rainha Vitória pelo seu esposo Albert, em Kensington Gardens, a meio caminho entre Lancaster Gate, onde morava, e o Royal

Albert Hall. Esse era o caminho que frequentemente fazia circundando o lago Serpentine, no Hyde Park.

Naquela época já participava dos Congressos da Sociedade Brasileira de Bioquímica, da qual fora sócio-fundador em 1965. As reuniões bianuais aconteciam no Grande Hotel, na Estância Hidromineral Caxambú, no Estado de Minas Gerais. Rotineiramente, a maioria das apresentações em painéis era sobre bioquímica de animais ou de microrganismos. Somente ele e um ou outro pesquisador de outras instituições, que não a ESALQ, expunham trabalhos em bioquímica de plantas. Essa situação começou a mudar quando nas décadas de 1970 e 1980 foi convidado para fazer apresentações orais de seus trabalhos e organizar simpósios sobre bioquímica de plantas nesses Congressos. Outros grupos também foram incentivados, como os dos pesquisadores Ladaslav Sodek, da UNICAMP, Sonia Dietrich, do Instituto de Botânica de São Paulo, Walter Handro, do Instituto de Biociências, do *campus* da USP em São Paulo. Sodek, anteriormente, fora contratado pelo CENA e realizou pesquisas nos Laboratórios de Otto. *De 1961 a 1979*, Otto percorreu o Brasil de Norte a Sul, ministrando Cursos e Palestras em várias instituições de pesquisa e de ensino, sobre o metabolismo do nitrogênio e do potássio nas plantas e também o uso de radioisótopos para acompanhar processos fisiológicos e bioquímicos em vegetais. Muitas vezes transportou consigo o material de laboratório, incluindo radioisótopos, para demonstração do fenômeno da fotossíntese.

VI
A experiência com projetos internacionais

1978

Munique, Alemanha. Desceu do bonde perto da Marianplatz e caminhou até essa famosa praça na cidade de Munique, em uma rica região da Alemanha. Juntou-se ao povo que olhava para o alto da torre do belo edifício que abriga a Prefeitura da cidade. Todos estavam esperando as badaladas das 12 horas do magnífico relógio incrustado na torre. Quando os ponteiros das horas e dos minutos se encontraram, figuras humanas, esculturais, surgiram umas após outras de dentro da estrutura da torre, algumas montadas em cavalos, representando cavalheiros medievais, talvez em busca da libertação de suas amadas, imaginou. O espetáculo majestoso, com as estátuas desparecendo no interior da torre, terminou e, apesar da manhã fria de um dia de *setembro de 1978*, o sol estava presente e ele prosseguiu em seu caminhar pelas ruas de Munique lembrando-se do porque estava ali na festiva e alegre cidade da Baviera.

Em Munique, estava participando da 5ª e última Reunião do Programa de Melhoramento de Proteinas de Sementes da FAO/IAEA/GSF (5th Research Coordination Meeting of Seed Protein Improvement Programme). Nessa Reunião, apresentou resultados finais de seu trabalho sobre o melhoramento do teor e qualidade de proteínas de reserva em sementes de feijão (*Phaseolus vulgaris*) utilizando tratamentos mutagênicos, em colaboração com Augusto Tulmann e Donald Boulter.

No final de 1972, Ademar Cervellini, Diretor do CENA, convidou-o para participar do Programa de Melhoramento de Proteinas de Sementes de Leguminosas patrocinado pela Food Agriculture Organization (FAO, Itália), pela Agência Internacional de Energia Atômica (IAEA, Viena, Austria) e pelo Governo da Alemanha. O projeto liderado por ele seria desenvolvido na Seção de Bioquímica de Plantas do CENA, com a colaboração de Augusto Tulmann Neto, da Seção de Melhoramento de Plantas e especialista no uso de técnicas nucleares para indução de mutação em vegetais. Durante cinco anos, desenvolveu esse Programa no CENA.

Os primeiros resultados foram por ele apresentados na 2ª Reunião da Coordenação do Programa, realizada no International

Institute of Tropical Agriculture (IITA), em Ibadan, Nigéria, África, *em dezembro de 1973.*

Nesse momento de suas recordações, ainda caminhando pelas ruas de Munique, viu-se em Ibadan, na Nigéria, vivendo uma experiência peculiar, em um país que havia sido colônia da Inglaterra, riquíssimo em petróleo e com grande parte de sua população não tendo acesso aos recursos sócio-econômicos. Na África, o Instituto de Agricultura Tropical era lider, e ainda o é, na pesquisa para procurar soluções para combater a fome, a mal nutrição e a pobreza da população da região sub-Saariana da África.

Dezembro de 1973. Antes de desembarcar no aeroporto de Lagos, na Nigéria, de onde partiria para Ibadan de carro, ele esteve na Suiça, em pleno inverno europeu. Era noite quando o avião aterrissou no aeroporto de Génève. A visão da neve sobre as casas, o trafegar silencioso do ônibus que o levou pelas ruas silenciosas até o hotel onde se hospedaria no centro da cidade, trouxe-lhe uma tranquilidade inesperada. Sentiu-se envolvido pela atmosfera de uma civilização secular onde tudo parecia estar pronto, mas que ele sabia não ter sido fácil para o povo suíço tê-la conquistado!

Quando o rádio do quarto do hotel apresentou os acordes do Quinto Concerto para Piano e Orquestra, "O Imperador", de Beethoven, um dos clássicos que ele mais admirava, lembrou-se de sua esposa Diva e de seus agora 5 filhos, porque Daniel nascera naquele ano de *1973, no dia 28 de abril*, em Piracicaba. Todos eles, nos seus primeiros anos de vida adormeciam ao som de Mozart e de Beethoven. Os acordes de "O Imperador" os acompanharam até a adolescência. "Eles poderiam também estar aqui junto a mim" pensou extasiado diante da beleza das fotos dos Alpes suíços em suas mãos. Logo mais iria realmente conhecê-los quando, em Génève, tomaria o trem que o levaria a Zurique, e que subia as montanhas, por entre as árvores cobertas de neve, em caminhos tortuosos. Nessa atmosfera lúdica, uma vez mais meditou sobre o significado de sua vida, à semelhança de experiências anteriores, quase místicas.

A 3ª Reunião do Programa foi realizada em *maio de 1975* em Hahnenklee, Alemanha. Após a reunião, visitou as instalações do Institut für Strahlenbotanik, em Hannover. Conheceu então métodos rápidos de extração de proteínas de sementes de leguminosas em aparelhos que as protegiam da ação do calor, diminuindo sensivelmente as perdas do seu conteúdo, e que ele aplicaria em suas pesquisas em Piracicaba. Em Hannover, durante um jantar no subsolo de um restaurante em um prédio restaurado após ter sido bombardeado durante a 2ª Guerra Mundial, ele ouviu pela primeira vez a música "When a Child is Born", originalmente "Soleado" de Ciro Dammicco. A melodia o embalou e a lembrança de seus 5 filhos, quando ainda muito pequenos, cada um a seu tempo, logo aflorou em seu coração. Em uma loja de Hannover comprou o "single" da música, levando-o consigo para Piracicaba. Agora, Marco Augusto tinha 11, Adolfo Egídio 10, Maria Paula 8, Carla Maisa 6 e Daniel, 2 anos de idade! A lembrança desse dia e daquele lugar permaneceu para sempre em sua memória afetiva.

No final de março e início de abril desse mesmo ano de 1975, antes de viajar para a Alemanha, submeteu-se, durante dois dias, a provas públicas tornando-se, aos 42 anos de idade, Professor Titular de Bioquímica do Departamento de Química da ESALQ. A Comissão que o examinou e o aprovou era formada por Eurípedes Malavolta, Renato Catani, Eugênio Acquarone, Henrique Tastaldi, e Walter Borzani, todos professores titulares da Universidade de São Paulo. Atingiu, então, o mais alto grau da carreira acadêmica da USP. Nesse momento, veio-lhe à mente que o sonho que sonhara durante o seu sono, ainda criança, em que se viu morando em uma casa enorme em meio a árvores e flores e aléias, lá pelos lados onde ficava a "sua" Escola Agrícola, estava tornando-se realidade!

A partir de *1971*, com alternância de 4 anos, exerceu a função de Chefe e Vice-Chefe do Departamento de Química da ESALQ, até *1989*.

No final de *março de 1977*, a bucólica Baden, localizada ao sul de Viena, na Austria, acolheu os integrantes da 4ª Reunião do Programa. O bonde que o transportou da estação de ônibus em Viena até Baden, trafegou por ruas ladeadas de árvores cujos ramos se entrelaçavam formando um túnel vivo e uma acolhedora sombra. Tudo parecia respirar tranquilidade nessa estância hidromineral, situada no entorno dos "bosques" de Viena. Em Viena, sentado em um banco diante da estátua de Strauss, viu-se dançando no baile de sua formatura de ginásio do Colégio "O Piracicabano", no Clube Coronel Barbosa, em *dezembro de 1949*, com uma de suas irmãs, Lídia, ao som dos "Contos dos Bosques de Viena", valsa de Johann Strauss, que marcou grande parte de sua adolescência e juventude.

Foi durante as várias Reuniões do Programa que conheceu Donald Boulter, Chefe do Departamento de Botânica da Universidade de Durham, em Durham, no norte da Inglaterra. Nesse Departamento, que no futuro passaria a ser denominado Departamento de Ciências Biológicas e Biomédicas, foi professor visitante durante o *ano de 1976*. Toda sua família o acompanhou. Viveram em uma ampla casa de dois pisos, de propriedade da Universidade, situada solitariamente no meio de um parque privado chamado Elvet Garth, em South Road, a mesma rua onde se localiza a Universidade. Da alta janela de seu quarto, no primeiro andar, ao lado de Diva, contemplava o parque, admirando a beleza do seu paisagismo: um verdadeiro jardim onde, em diferentes épocas do ano, floresciam diversas espécies de plantas ornamentais, de "snow drops", no início da primavera, a rosas no verão. Altos arbustos de rododendros cercavam a casa, seus botões desabrochando-se, sucessivamente, em flores de cor rosa, azul, violeta e branca ao longo dos meses de junho a agosto. Várias espécies de árvores formavam rincões no amplo espaço do parque, apropriado para as brincadeiras de seus filhos. Em Durham, as filhas, Maria Paula e Carla Maisa frequentaram St. Godrics Primary School, próximo a "Our Lady Of Mercy & St. Godrics Catholic Church". Os dois filhos, Marco Augusto e Adolfo Egídio frequentaram Whinney Hill Grammar School. O pequeno Daniel, com 3 anos de idade, frequentou uma escola maternal particular.

A imponente Catedral de Durham em estilo normando, construída entre 1093 e 1113, dedicada a St. Cuthbert, morto em 687 d.c., e que guarda os restos mortais do Venerável Bede, monge beneditino, autor de inúmeros trabalhos sobre a história da Igreja na Inglaterra, era com frequência visitada pela família de Otto, principalmente nas apresentações de concertos de músicas clássicas. A Nona Sinfonia – "Coral" - de Beethoven foi uma dessas apresentações. Na Catedral lotada, junto a uma audiência silenciosa, ele e Diva deixaram-se envolver pelos acordes ora suaves ora majestosos da melodia de Beethoven que, em um crescendo, explodiu nas vozes do Coral, no 4^0 movimento. Inebriados de tanta beleza da obra, que já conheciam, saíram da Catedral e, na praça que a rodeava, deixaram-se ficar à luz da lua que teimava em aparecer entre a névoa da fria noite de novembro.

Próximo à catedral, lá esta o também secular Castelo, antiga residência dos Bispos de Durham. Essas duas imponentes construções situam-se no alto de uma colina, com acesso pela Praça do Mercado onde estão a Prefeitura de Durham, a Igreja de São Nicolau e as estatuas de Netuno e do 3^0 Marquês de Londonderry, importante personagem histórica da vida econômica de Durham.

Em agosto desse ano, como prometera a si mesmo, levou toda a sua família para visitar vários países do continente europeu. Nessa oportunidade visitou e estabeleceu contatos científicos com os laboratórios de proteína da Faculté de Sciences Agronomiques, em Gembloux, na Bélgica, e o Departamento de Biologia Vegetal, na Universidade de Génève, na Suiça, onde posteriormente sua assistente e orientada, Maria Tereza Vitral de Carvalho, desenvolveria parte de sua Tese de Doutorado sobre proteínas vegetais. No retorno à Inglaterra visitaram Edinburgo, na Escócia, em uma viagem de trem que encantou Daniel.

Nos laboratórios de Boulter desenvolveu trabalho sobre proteínas de reserva de ervilha, com a colaboração também de Eric Derbyshire, o qual, posteriormente, foi por ele contratado como pesquisador da Seção de Bioquímica de Plantas do CENA. Derbyshire casou-se com Maria Tereza V. de Carvalho, em São Paulo.

Em setembro de 1977, Donald Boulter apresentou trabalho sobre cultivo de tecidos vegetais de leguminosas, em colaboração com ele, no 4th Annual College of Biological Sciences Colloquium, organizado por Rod Sharp, na The Ohio State Univeristy, Columbus, Ohio, USA.

Outras vezes retornou à sua sempre bem lembrada Durham. *Em junho de 1980,* após ministrar curso sobre metabolismo de aminoácidos e proteínas em plantas na Universidade de Bologna, na Itália, e discutir os resultados de seu projeto sobre proteínas de leguminosas no Departamento de Biotecnologia no Carlsberg Research Institute, em Copenhagem, na Dinamarca, ele esteve nos laboratórios de Donald Boulter, na Universidade de Durham. Também, *em outubro e novembro de 2003,* em companhia de Diva e da filha Maria Paula, estudante de Mestrado em Londres, esteve uma vez mais em Durham, quando então visitaram George e Breda Galagher, com quem estabeleceram fortes laços de amizade desde 1976. George faleceria alguns anos mais tarde.

Nessa ocasião, ele e Diva reviveram a sua sempre lembrada Londres, revisitaram os museus, os teatros, os passeios pelos parques e uma vez mais, como em outros tempos, participaram das apresentações de música clássica e contemporânea no início das tardes de quarta-feiras em St. Martin in the Fields, na vizinhança da Trafalgar Square, seguido de almoço no "Café in the Crypt", no subsolo dessa Igrela Anglicana.

Em seu longo caminhar pelas ruas de Munique, essas lembranças fizeram o seu coração bater mais fortemente de saudades, talvez, mas também de contentamento. Ali, na cidade do relógio com as figuras andantes da torre da Prefeitura, na Marianplatz, terminava mais uma etapa de sua vida científica. Tomou o bonde, que por ali passava, e retornou ao seu hotel.

Esses fatos todos eram recentes na sua memória. Seus colaboradores no Programa de Proteinas de Leguminosas foram Augusto Tulmann, S. Blixt, K. Mikaelsen, L. Sodek e Tseng-Sheng Gerald Lee e, posteriormente, Donald Boulter. Sob sua orientação, Gerald Lee desenvolveu sua Tese de Doutorado, defenden-do-a no Departamento de Química do Instituto de Química da Universidade de São Paulo, em São Paulo, em 1978. Sementes do mutante da variedade Carioca precoce, obtido por Tulmann durante o desenvolvimento do Programa no CENA, foram o material utilizado por Gerald Lee em sua tese sobre proteínas de reserva de feijão.

VII
Rod Sharp e Otto encontram-se

1971. Nas aulas de introdução às ciências, em seus tempos iniciais de estudante secundário, na *década de 1940*, ele ficava deslumbrado com a descrição das células. Esses pequeníssimos seres, como a ele pareciam ser, povoavam sua mente inquisidora: as figuras das células nos livros e aquelas com as quais brincava, mal desenhando-as em seus cadernos, eram para ele criaturas vivas. Na sua imaginação, elas é que faziam balançar as flores e as folhas das árvores do parque da Escola Agrícola que frequentemente visitava, e também faziam os braços e as pernas, suas e de seus irmãos e amigos, gesticularem e caminharem. Ele nunca havia visto as células. Foi vê-las pela primeira vez ao microscópio óptico da residência de João, um de seus colegas, cujo pai, Canuto Marmo, era Professor da Escola "Luiz de Queiroz". Na lâmina examinada estavam aqueles "seres" indo de um lado para outro como que dançando em rodopios sem fim. Era verdade, as células existiam! Não eram seres, não tinham pernas nem mãos! Mas, como imaginara, moviam-se e dançavam como se corpos fossem, entrelaçando-se, multiplicando-se, gerando vida. Ficou extasiado! E mais ainda maravilhou-se quando, anos mais tarde, encontrou-se com os trabalhos de Robert Hooke realizados no século XVII. Ele aprendeu que Hooke, observando ao microscópio um pedaço de cortiça, descreveu–a como semelhante a um favo de mel, dando aos seus pequenos compartimentos o nome de células. E Hooke publicara seus achados em um trabalho intitulado "Micrographia" em *20 de março de 1665*! quase dois séculos antes que Schwann e Scheleiden, em *1839*, confirmassem que as células eram partículas fundamentais dos tecidos de plantas e animais. Como logo aprenderia, esses dois cientistas evidenciaram a totipotência das células vegetais, a qual, no percorrer de sua vida científica, William Rod Sharp e ele, e seus outros colaboradores, utilizariam em suas caminhadas pelo mundo das células de plantas cultivadas *in vitro*.

Ele, porém, estava agora no século XX. Microscópios eletrônicos, muitíssimo mais potentes que os microscópios ópticos, eram utilizados para desvendar os "mistérios" da natureza animal, vegetal e mineral com maior precisão. A célula, decifrada, apareceu aos seus olhos com toda a importância de suas funções: das células troncos até as mais especializadas. Agora ele sabia: reações bioquímicas só acontecem porque a célula existe, mas ao mesmo tempo, as reações bioquímicas são responsáveis pela existência da célula, concluiu com deslumbramento, uma vez mais, diante da beleza da natureza!

Todas essas lembranças e constantes conjecturas a respeito da célula e sua estrutura estavam bem vivas na sua memória quando

Malavolta entrou em sua sala no CENA, *em princípios de 1971.* Esse foi um momento que marcaria a sua vida científica para sempre. Um jovem cientista americano obtera auxílio financeiro para desenvolver pesquisa com células vegetais cultivadas *in vitro* durante um curto período no Brasil e que, informara Malavolta, por sugestão de Henrique Vianna Amorim, que estava estudando nos Estados Unidos, decidira vir à ESALQ. Malavolta solicitou-lhe que o recebesse. Imediatamente Otto sentiu seus próprios olhos brilharem de entusiasmo. Era tudo o que ele esperava acontecer. O jovem cientista era William Rod Sharp, biólogo celular, professor da The Ohio State University, nos Estados Unidos.

Trabalhar com células de plantas *in vitro* era o que ele havia tentado sem resultados positivos até aquele momento. Na ESALQ, Darcy Martins da Silva, seu colega de Departamento, havia estado na Holanda de onde trouxera alguns frascos de vidro contendo culturas assépticas de células de vegetais. Gustav F. Brieger, professor de Genética na ESALQ, cultivava sementes de orquídeas *in vitro*. Walter Handro, do Departamento de Botânica, do Instituto de Biociências, no *campus* de São Paulo da Universidade de São Paulo, estava na França, no laboratório de Colette Nitsch, familiarizando-se com técnicas de cultura *in vitro* de anteras, que, posteriormente, introduziu em seus laboratórios. Nesse mesmo Departamento, Maguro, Hell e Gilberto Karbauy já davam os primeiros passos no uso das técnicas de cultura de células e tecidos de plantas na *década de 1960.*

Para ele, desenhava-se agora a possibilidade de adentrar no reino celular e trazer o conhecimento das sequências de reações bioquímicas para o contexto dos meandros da célula vegetal e, assim, estudar o controle das mesmas. Mais ainda, preconizou, com a colaboração de Sharp, perspectivas mais amplas: a utilização das técnicas de cultura de células e tecidos de plantas como um importante instrumento auxiliar no melhoramento de plantas, produção de clones e variantes somaclonais. O seu pioneirismo e de Rod Sharp, no CENA e na ESALQ, introduzindo as técnicas de cultura de células e tecidos de modo sistemático, lançou o embrião da biotecnologia de plantas no Brasil: a aplicação das técnicas de cultura de células e tecidos de plantas em agricultura, como sempre sonhara.

1971 - 1981

Dezembro, 1981. Na manhã do dia 6, deixou sua sala no CENA e dirigiu-se para a sala da Chefia do Departamento de Química da ESALQ. Estava preparando-se para, em seguida, abrir os trabalhos do "International Symposium on Genetic Engineering for Biotechnology" no Prédio de Engenharia. Ao olhar pela janela da sala, admirou uma vez mais as belas e frondosas árvores do parque da Escola, idealizado pelo paisagista belga Arsenio Puttemans e implantado nos primeiros anos do século XX. Sua expansão e manutenção ao longo dos anos posteriores foram resultados do esforço de muitos professores da ESALQ, destacando-se Philippe Westin Cabral de Vasconcelos, cujo nome seria perpetuado na denominação oficial do parque em 1986. Naquele momento de contemplação do belo, voltou-se para o passado e imagens, como entidades vivas, tomaram corpo na sua memória.

Mais de dez anos haviam se passado desde que Rod Sharp, que se tornaria seu grande amigo e parceiro científico, desembarcara do avião que o trazia pela primeira vez ao Brasil, *em junho de 1971*. Otto, sendo de estatura baixa, tez morena e cabelos ainda pretos, admirou-se quando olhou para aquele jovem alto e louro, a sua antítese! Mas, então, ambos, em gestos largos deram-se um primeiro abraço, como que selando o início de uma forte amizade que iria perdurar por muitos anos.

Uma senhora acompanhava Rod Sharp, era sua avó materna. Seus cabelos brancos reluziam à luz do sol que banhava o aeroporto de Congonhas, naquela manhã de frio ameno, na cidade de São Paulo. A sua simpatia e delicadeza de gestos logo conquistou-o e depois Diva e todos aqueles que passaram a conviver com ela em Piracicaba. Mary Beatrice, esse era seu nome. Seus pais eram de origem holandesa, tendo ela nascido em Mansfield, Ohio, USA, onde graduou-se pelo Ashland College. Mãe de duas filhas, a mais velha delas mãe de Rod, com grande experiência de vida, fora casada duas vezes. Agia com desenvoltura, compartilhando com Rod a rotina da adaptação de viver em um pais com costumes muito diferentes do seu. Todas as vezes que voltava a Piracicaba, e foram várias, Mary Beatrice encantava seus convivas com o seu companheirismo e sua deliciosa torta de maçã! (foto 2).

Em 1971, apesar de o Brasil estar em plena vigência da ditadura militar, o CENA sempre criara condições para novas investidas científicas. A sua parceria com Rod Sharp foi entusiasticamente estimulada por Ademar Cervellini, Diretor do CENA.

A empatia que surgira entre os dois colaboradores facilitou a adaptação do jovem Rod. Ele e Otto compraram madeira, vidro e toalhas e construíram a primeira câmara asséptica, instalada na Seção de Bioquímica de Plantas do CENA. Uma outra sala foi adaptada para manutenção dos frascos de culturas de várias espécies vegetais sob condições de luminosidade e temperatura controladas. Teve início, então, um intenso aprendizado de Técnicos da Seção e de estudantes de graduação e de pós-graduação da ESALQ nessas técnicas. Linda S. Caldas, orientada de Rod na The Ohio State University, Columbus, USA, e esposa de Rui de Araujo Caldas, assistente de Malavolta, e que desenvolvera seu Programa de Ph. D. em Bioquímica também na mesma Universidade, juntou-se ao grupo no CENA. Alguns anos mais tarde, Linda e Rui foram contratados pela Universidade de Brasília, na Capital do Brasil, onde desenvolveram um produtivo programa de ensino e pesquisa em biologia e em bioquímica de plantas.

De 1971 a 1981, a Seção de Bioquímica de Plantas do CENA experimentou intensa atividade tanto na sua área de pesquisa em bioquímica de nutrientes, de aminoácidos e proteinas quanto na área de biologia celular, em plantas. As visitas anuais de Rod Sharp ao laboratório do CENA e as visitas de Otto aos laboratórios de Rod nos Estados Unidos e em vários países europeus, trouxeram muitas oportunidades para que ambos e seus alunos desenvolvessem trabalhos utilizando técnicas *in vitro* com várias espécies vegetais: gergelim, tomate, amendoim, feijão, abacaxi, café, ervilha, eucalipto, pinus, plantas ornamentais, plantas medicinais e tantas outras.

Nessa década, incentivado por Rod Sharp, Otto participou do Terceiro Congresso Internacional de Cultura de Células e Tecidos de Plantas, na University of Leicester, em Leicester, Inglaterra, organizado pela Associação Internacional de Cultura de Tecidos de Plantas (IAPTC), em *julho de 1974*. Nesse Congresso apresentou resultados de pesquisa com tecidos de feijão, em colaboração com Rod e Maria Tereza V. de Carvalho. Foi Representante no Brasil dessa Associação durante 12 anos.

Uma das principais pesquisas desenvolvidas pelo Grupo de Rod e Otto foi a cultura de tecidos de cana-de-açúcar (*Sacharum* spp), de grande importância econômica para o Brasil. Tecidos e células de cana-de-açúcar foram usados para estudos de variação somaclonal, visando tolerância a ametrin, herbicida utilizado para controlar ervas daninhas, que o absorvem através de suas folhas e raízes. Neftali Ochoa Alejo, de origem mexicana, orientado de Otto, foi quem, auxiliado por Enio, desenvolveu essa pesquisa em sua Tese de Doutorado em Bioquímica pelo Instituto de Química da Universidade de São Paulo, *campus* de São Paulo, em 1983. Resultados desse trabalho foram apresentados por Otto no Simpósio Internacional de Aplicações de Cultura de Células e Tecidos no Melhoramento de Plantas, na Academia de Ciências, em Olomuc, na então Czechoslocakia, do então bloco da União Soviética, *em setembro de 1984*. Nessa oportunidade, ele e David Evans trocaram experiências no campo da biologia celular. Em várias oportunidades David Evans, colaborador de Rod, esteve nos laboratórios de Otto, no CENA, contribuindo sobremaneira para as pesquisas ali desenvolvidas.

O cultivo de cana-de-açúcar no Brasil teve início com as primeiras mudas trazidas em 1533 por Martim Afonso de Souza, da Ilha da Madeira, principal ilha do Arquipélago da Madeira, situada no Oceano Atlântico a sudoeste da costa portuguesa. Afonso de Souza foi o primeiro colonizador do território brasileiro, enviado por Portugal. Essa cultura veio suprir a premente necessidade de colonização e exploração de um território imenso, porém sem expressão econômica para Portugal. Apesar de ter sido instalado o primeiro engenho de açúcar, para processar a cana, em São Vicente, no litoral daquele que posteriormente viria a ser o Estado de São Paulo, foi no litoral do nordeste do Brasil que a cultura da cana-de-açúcar prosperou, principalmente nas regiões que, no futuro, seriam os Estados de Pernambuco e da Bahia. O solo, de cor escura, quase preta, rico em argila e humus, presente nessa região, e que era denominado, naquela época, de "massapé", foi muito explorado durante a colonização portuguesa para o cultivo dessa gramínea. O açúcar produzido nos engenhos brasileiros, exportado para a Europa com elevado lucro, era o alicerce da economia da colônia portuguesa entre os séculos XVI e XVII. Aliás, até hoje o Brasil é o maior exportador de açúcar, extraído de cana-de-açúcar, do mundo.

No Brasil, a utilização de etanol, produzido a partir da fermentação do açúcar de cana, como combustível para automotores, era conhecida há muitas décadas. Entretanto, foi somente nos anos 1970, diante da crise mundial de combustíveis fósseis, que o governo militar, sob a presidência do general Ernesto Geisel, decidiu que o álcool, até então considerado um sub-produto do açúcar da cana, passaria a desempenhar papel estratégico na economia brasileira. O Programa Nacional do Alcool (Proalcool), criado por portaria governamental em 1975, teve como efeito imediato a produção de automóveis movidos seja somente a etanol seja com uma mistura de álcool e gasolina. Na primeira década do século XXI o Brasil retomou com maior intensidade essa prática, com excelentes resultados, pois tem sido bastante reduzida a emissão de monóxido de carbono pela combustão do combustível fóssil, diminuindo a poluição nas ruas e rodovias brasileiras.

Um dos subprodutos durante a fermentação alcoólica de cana-de-açucar é a vinhaça, líquido residual que apresenta diferentes propriedades. Atualmente, tem sido preconizada sua utilização como matéria prima na produção de biocombustíveis. Uma outra possibilidade, estudada desde as décadas de 1940/1950 pelo grupo de Jaime Rocha de Almeida, na ESALQ, é o seu uso como fertilizante. Nesse contexto, um dos orientados de Otto, Joaquim Albenísio Silveira, desenvolveu, em 1985, sua Tese de Doutorado sobre a relação entre adubação nitrogenada e crescimento de plantas de cana-de-açucar, em condições de campo, utilizando a vinhaça. Albenísio, natural do Estado do Ceará, nordeste do Brasil, foi contratado como assistente junto ao Centro de Biotecnologia Agrícola – CEBTEC, na ESALQ. Anos depois Albenísio regressou ao Ceará.

A identificação de centenas de variedades e híbridos de cana-de-açucar é bastante difícil. No reconhecimento taxonômico, um dos métodos mais apropriados é o que se baseia em diferenças bioquímicas, o qual se soma aos métodos tradicionais de medições da morfologia das várias partes das plantas. Apesar de nem todas as diferenças bioquímicas se refletirem na morfologia, elas existem e são importantes na taxonomia de cana-de-açucar.

- Professor, estou interessado em cana-de-açucar e gostaria de desenvolver minha Dissertação de Mestrado explorando o potencial bioquímico das células para determinar uma chave analítica para identificação de variedades, disse Marcilio entrando na sala de Otto no CEBTEC.

- Ótimo, respondeu Otto. Vamos nos debruçar sobre esse tema", continuou.

Marcilio de Almeida procurou por ele nos finais da década de 1970, quando ainda estudante de graduação em Biologia na Universidade do Estado de São Paulo "Júlio Mesquita" (UNESP), na cidade de Rio Claro, próxima a Piracicaba, para estagiar no CENA, sob sua orientação. Destacou-se entre os seus outros orientados. Após a graduação continuou como seu orientado em seu Programa de Mestrado na ESALQ.

- Seria importante você trabalhar com 10 variedades de cana, as mais utilizadas pelos agricultores, sugeriu Otto. "Além da caracterização organográfica das folhas e colmos, que é sua especialidade como biólogo, proponho que também analise alguns parâmetros bioquímicos", prosseguiu.

- Quais seriam esses parâmetos?, perguntou Marcílio.

- Voce poderá analisar a atividade das isoenzimas de esterase e peroxidase, o teor de proteína total solúvel e o nível de sólidos solúveis, o que lhe proporcionará um maior número de dados para elaborar a chave analítica, que é o seu principal objetivo, esclareceu Otto e continuou: "o Enio poderá auxiliá-lo nessas análises".

Marcilio defendeu sua Dissertação de Mestrado em 1986, tendo sido contratado pelo então Departamento de Botânica da ESALQ, hoje Departamento de Ciências Biológicas. Posterirormente, na década de 1990, sob orientação de Gilberto Kerbahuy, desenvolveu seu programa de Doutorado no Departamento de Botânica do Instituto de Ciências, no *campus* da USP em São Paulo.

6 de dezembro, 1981. - Professor, a cerimônia de abertura do Simpósio terá início logo mais. Devemos nos dirigir ao Anfiteatro do Prédio de Engenharia, disse Enio entrando na sala da Chefia do Departamento de Química.

- Eu sei, respondeu Otto ainda absorto em seus devaneios, olhos maravilhados diante da beleza do parque da ESALQ. Ainda temos algum tempo. O Prof. Sharp já chegou? perguntou, voltando-se para o seu técnico e amigo Enio.

- O carro foi busca-lo no Hotel, juntamente com os cientistas estrangeiros que o senhor e o Professor Tavares convidaram, respondeu Enio.

Otto retomou suas lembranças e, do mesmo modo que, naquele momento, Enio entrara na sala, viu-o, em pensamento, entrando em seu laboratório no CENA, naquele início do verão piracicabano de *1977*.

Enio Tiago de Oliveira procurara-o para fazer um estágio de poucos meses na Seção de Bioquímica de Plantas, estágio esse obrigatório para todos os alunos do Curso Técnico em Química que ele estava cursando no Colégio Dom Bosco, em Piracicaba. Um rapaz franzino, de estatura baixa, de 18 anos, surgiu diante dele que aceitou o pedido, uma vez que, como de costume, nunca dispensara alguém que desejasse trabalhar em seus laboratórios, familiarizando-se com as técnicas de bioquímica e de biologia celular. Enio logo iniciou-se nas técnicas laboratoriais. Sob os olhares de Otto e de sua assistente Maria Tereza, mostrava seguidamente sua habilidade e responsabilidade no uso das técnicas *in vitro* e análises bioquímicas. O seu estágio prolongou-se e, à medida que o tempo passava, Enio ultrapassava todas as expectativas (foto 3).

Enio nasceu em uma pequena cidade da região do Vale de Jequitinhonha, no Norte do Estado de Minas Gerais, região decantada em prosa por Guimarães Rosa, um dos grandes escritores brasileiros, apaixonado pelo sertão do Brasil. Aos 11 anos, Enio deixou a cidade de Taiobeiras com seus pais e seus 8 irmãos. Era o quinto filho de Jones e Olívia. Mudaram-se para Piracicaba onde trabalhou, estudou e casou-se com Joana d´Arc, com quem teve dois filhos, Marcele e Tiago. Foi contratado pelo CENA e posteriormente transferido para o Departamento de Química da ESALQ onde, como Técnico Superior, passou a integrar a equipe científica do Centro de Biotecnologia Agrícola –CEBTEC.

Intuição e uma inusitada curiosidade foram os predicados que fizeram com que aquele jovem, advindo de uma das regiões mais pobres do Brasil, passasse a ser o "homem de confiança" de Otto, participando de muitas das pesquisas, auxiliando os alunos no desenvolvimento dos trabalhos experimentais de suas Dissertações de Mestrado e Teses de Doutorado. Ao longo do tempo, Enio responsabilizou-se pela introdução de técnicas de cultura de tecidos e células de cana-de-açúcar para estudos de variação somaclonal. Foi peça chave para que, em 1979, fosse assinado um Convênio pelo período de 5 anos com o Planalsucar para instalar em suas dependências, na cidade de Araras, no Estado de São Paulo, um laboratório de cultura de tecidos e treinar seu pessoal nas técnicas para obtenção de variantes somaclonais como coadjuvantes no melhoramento da cana-de-açúcar.

O Planalsucar – O Programa Nacional de Melhoramento da Cana-de Açúcar, pertencente ao antigo Instituto do Açúcar e do Alcool - IAA- tinha como objetivo a renovação contínua de plantas do plantel de variedades de cana brasileiras e importadas de outros países, utilizando técnicas convencionais de melhoramento. Essas matrizes constituíam a matéria prima para a produção sucroalcooleira do Brasil. Várias estações experimentiais do Planalsucar desenvolviam esse trabalho, inclusive a de Araras, no Estado de São Paulo, onde a equipe de Otto instalou os laboratórios para a cultura de células e tecidos de cana-de-açúcar. Após a extinção do IAA, e com ele do PLANALSUCAR, em 1990, criou-se a RIDESA – Rede Interinstitucional de Desenvolvimento do Setor Sucroalcooleiro, englobando várias Universidades Federais, entre elas a Universidade Federal de São Carlos, cuja Unidade de Araras, passou a responsabilizar-se pelas técnicas *in vitro*, sob a coordenação de Tseng-Sheng Gerald Lee, que fora orientado de Otto em sua Tese de Doutorado no Instituto de Química do *campus* da USP, em São Paulo.

Atualmente Biólogo do Departamento de Ciências Biológica da ESALQ, Enio graduou-se em 1987 em Biologia pela Universidade Metodista de Piracicaba–UNIMEP. Na década de 1990, sob orientação de Otto, obteve o título de Mestre e na década de 2000, sob a orientação de Luiz Antonio Gallo, o título de Doutor em Ciências, ambos na ESALQ.

Nas lembranças recentes de Otto um momento ímpar aconteceu em *2009*, quando Enio entrou em sua sala.

- Professor, recebi uma solicitação dos editores da revista *Hortscience* para autorizar a publicação de uma das fotos do nosso artigo sobre *Aloe vera* na capa do número de outubro próximo, disse Enio, não escondendo sua imensa satisfação. O trabalho era parte de sua Tese de Doutoramento. Enio acrescentou: "precisamos responder".

Com alegria, Otto, de imediato, respondeu: "meus parabéns, vamos responder agora".

A realização do "Simpósio de Engenharia Genética para Biotecnologia" que logo mais se iniciaria na ESALQ, e que ele organizara com a colaboração de Flávio C. A. Tavares, do Departamento de Genética da ESALQ e Décio Sodrzeieski, da Secretaria de Indústria e Comércio do Estado de São Paulo, seria a coroação das atividades de pesquisa do seu Grupo, e de colaboração com outras entidades de pesquisa do Brasil e do Exterior, que se iniciara quando Rod Sharp juntou-se a ele em junho de 1971.

Ele e Enio deixaram a sala da Chefia do Departamento de Química. A passos largos atravessaram o extenso parque defronte ao belo Edifício da Administração da ESALQ, inaugurado em sua condição original em 14 de maio de 1907, contornando-o pelas suas aléias. Dirigiram-se ao Prédio de Engenharia que, com sua imponente cúpula, dominava todo o espaço fronteiriço ao grande lago, onde gansos e patos deslizavam silenciosamente, como que dançando, indiferentes ao acontecimento científico que estava prestes a começar. A Residência do Diretor da ESALQ, em estilo arquitetônico do sul dos Estados Unidos, com suas altas colunas brancas, completava o cenário onírico daquele início de ensolarada manhã.

Os cientistas estrangeiros e brasileiros, que fariam as conferências, esperavam por ele e por Flávio Tavares no saguão do Prédio. A cerimônia de Abertura, presidida por Aristeu Mendes Peixoto, Diretor da ESALQ, fora realizada na noite do dia anterior, no Salão Nobre do Prédio da Administração. Além de Paulo Fernando Cidade de Araújo e Joaquim José de Camargo Engler, estava presente Salvador de Toledo Piza, renomado cientista da ESALQ, dedicado à anatomia e fisiologia animal e exímio taxonomista, principalmente de insetos, com uma peculiaridade: publicava seus trabalhos em latim. É ele o autor da "Ode à ESALQ", na qual enaltece a obra de Luiz Vicente de Souza Queiroz: "a escola é o teu monumento".

O plenário do Anfiteatro do Prédio da Engenharia, estava totalmente lotado com professores, estudantes e cientistas vindos de vários estados brasileiros e de alguns países sul americanos. Os tópicos apresentados e discutidos durante os cinco dias do Simpósio versaram sobre o uso das técnicas de engenharia genética e cultura de tecidos e células em energia, em agricultura, em sistemas animais, em microbiologia e em células humanas. As conferencias foram apresentadas por cientistas da Argentina, Brasil, Canadá, Escócia, Estados Unidos da América do Norte, França e Inglaterra.

Estavam presentes: N. Alexander (Dept. Agriculture, USA), J. R. Johnston (University of Strathclyde, Scotland), C. J. Panchal (Labatt Brewing Co. Limt., Canada), D. Boulter (University od Durham, England), A. M. Chakrabarty (University of Illinois, Medical Center, USA), I. Roitman (Universidade de Brasília, Brasil), M. Mares-Guia (Bioferm – Pesquisa e Desenvolvimento SA, Brasil), F. G. Nobrega (Universidade de São Paulo), (ESALQ, USP,

Brasil), Heslot, H. (Institut National Agronomique, França), C. M, Morel (Fundação Oswaldo Cruz, Brasil) F. J. S. Lara (Universidade de São Paulo, Brasil), D. M. Glover (Imperial College of Science and Technology, England), D. M. Silva (ESALQ, USP, Brasil), Maragarida L. R. Aguiar-Perecin (ESALQ, USP, Brasil), Alaides P. Ruschel (CENA, USP, Brasil), E. Chartone Souza (Universidade Federal de Minas Gerais, Brasil) R. R. Brentani, Universidade de São Paulo, Brasil), C. D. Denoya (Instituto SIDUS, Argentina), W. R. Sharp (DNA Plant Technology Corp., USA), além de João Lúcio de Azevedo, Flavio C. A. Tavares e Otto J. Crocomo, os três da ESALQ, USP.

O Simpósio foi um dos mais importantes acontecimentos científicos no início dos programas de biotecnologia e engenharia genética no Brasil. O Governo Federal havia lançado o PRONAB – Programa Nacional de Biotecnologia, discutido durante todo o ano de 1981 por uma Comissão de pesquisadores brasileiros, entre os quais, ele, Flavio Tavares e João Lúcio de Azevedo. Dias após a realização do Simpósio, o texto do PRONAB foi entregue ao então Ministro da Fazenda, Delfim Neto, em Brasília, capital do Brasil.

VIII
A criação do Centro de Biotecnologia Agrícola – CEBTEC

Maio-julho, 1981. Naquela manhã de outono brasileiro no mês de maio, ele e Flávio Tavares estavam sentados à mesa de reuniões na sala da chefia da Seção de Bioquímica de Plantas no CENA. Discutiam as últimas resoluções da Comissão do PRONAB, durante a reunião daquele mês realizada na sede do CNPq, no Rio de Janeiro.

O CNPq, Conselho Nacional de Desenvolvimento Científico e Tecnológico, uma Fundação vinculada ao Ministério da Ciência e Tecnologia do Governo Federal do Brasil é, desde sua criação em 1951, uma das principais e mais sólidas estruturas públicas de fomento à ciência, tecnologia e suas inovações. Está diretamente ligado à formação de mestres, doutores e pesquisadores científicos brasileiros, no Brasil e no Exterior. Outra empresa pública é a FINEP – Financiadora de Estudos e Projetos, também vinculada ao Ministério de Ciência e Tecnologia, fundada em 24 de julho de 1967. A FINEP atua em toda a cadeia de inovação, com foco em ações estratégicas, estruturantes e de impacto para o desenvolvimento econômico e social do Brasil, financiando projetos envolvendo setores públicos e privados. Uma outra instituição de fomento à pesquisa científica é a FAPESP – Fundação de Amparo à Pesquisa do Estado de São Paulo. Fundada em 1962 pelo Governo do Estado de São Paulo, a FAPESP tem sido responsável por grande parte do desenvolvimento científico e tecnológico das Universidades e Institutos de Pesquisa, financiando também parcerias do setor público com o setor privado, no Estado de São Paulo. Devido ao Projeto Genoma, financiado pela FAPESP no final do século XX, é que o Brasil entrou definitivamente no contexto científico internacional da biotecnologia e engenharia genética. Anteriormente, financiara o Programa BIOQ-FAPESP que impulsionou a pesquisa e o ensino da bioquímica nas universidades paulistanas, nos anos 1970. A FAPESP tem sido modelo para a criação de Fundações com os mesmos objetivos, em outros Estados brasileiros.

- Nos últimos 20 anos eu tenho tido projetos aprovados pelo CNPq, pela FINEP e pela FAPESP. Se não fossem esses suportes financeiros eu jamais poderia ter realizado minhas pesquisas aqui no CENA, disse Otto levantando-se para procurar um volume contendo relatórios de suas atividades.

- No Departamento de Genética o atual Chefe, Ernesto Paterniani, tem me dado bastante apoio para desenvolver trabalhos financiados pelas três financiadoras, comentou Flávio.

- Interessante, enquanto o Prof. Cervellini foi Diretor do CENA nunca faltou apoio às minhas iniciativas, disse Otto. "Infelizmente ele aposentou-se", acrescentou.

- Mas ele não está ocupando um importante cargo administrativo na Comissão Nacional de Energia Nuclear, a CNEN, no Rio de Janeiro? perguntou Flávio.

- Sim, respondeu Otto, como sempre a CNEN continua me auxiliando e bastante. Mas não é a mesma situação anterior. Eu sinto que preciso fazer algo mais. As discussões nas reuniões do PRONAB deixam-me inquieto. As sugestões do Carlos Morel, do Mares-Guia e do Francisco Lara estão fazendo com que eu pense em partir para novas iniciativas, diferentes daquelas que desenvolvo aqui no CENA, como, por exemplo, relacionamento com empresas privadas. Era uma referência a Morel, cientista brasileiro da Fundação Oswaldo Cruz, no Rio de Janeiro e a Mares-Guia, da Empresa Bioferm, de Minas Gerais. Com Francisco Jeronymo Salles Lara, professor titular de bioquímica, tinha um relacionamento antigo, desde a década de 1950, quando frequentara os seus laboratórios de bioquímica na USP, em São Paulo.

- Eu sei que o projeto com o PLANALSUCAR pode abrir muitas outras possibilidades de pesquisa, mas eu gostaria de fazer muito mais utilizando as nossas técnicas de cultura de células e tecidos no campo da produção agrícola, acrescentou Otto.

Flávio ouvia atentamente. O relacionamento entre ambos era muito bom. Ambos estavam interessados em desenvolver projetos biotecnológicos voltados para a agricultura. Nesse sentido os laboratórios de Otto no CENA eram relativamente bem equipados para desenvolver técnicas de biotecnologia de plantas.

- Graças à parceria com Rod Sharp, há dez anos, a Seção de Bioquímica de Plantas criara as bases para desenvolver inovação biotecnológica em plantas, enfatizou Otto, acrescentando: "Flávio, você estaria disposto a colaborar na criação de um centro de biotecnologia?"

Pareceu-lhe que os olhos de Flávio brilharam com a sua proposta, quando ele assentiu. A idéia de um centro de biotecnologia estava lançada. A sua concretização demandaria muito esforço, mas Otto previu os próximos passos a tomar de imediato,

No final da última semana de maio, na bela casa dos diretores da ESALQ, no *campus*, encontrou-se com o Diretor Aristeu Mendes Peixoto. Durante toda a manhã daquele dia discutiram a sua proposta de se criar na ESALQ um Centro de Biotecnologia. Com a aprovação, em princípio, do Diretor Peixoto, Otto reuniu-se com Joaquim José de Camargo Engler, um dos Diretores da Fundação de Estudos Agrários Luiz de Queiroz – FEALQ. Engler aprovou a idéia.

FEALQ. Foi Paulo Fernando Cidade de Araújo, professor de economia agrícola da ESALQ, quem teve a idéia da criação na Escola de uma Fundação para agilizar pesquisas nos campos tecnológico e científico com recursos externos à USP. Com apoio do então Diretor da ESALQ, Salim Simão, a Fundação de Estudos Agrários Luiz de Queiroz – FEALQ foi criada em 30 de dezembro de 1976, como uma entidade de direito privado, sem fins lucrativos.

A primeira Diretoria da FEALQ foi formada por Paulo F. Cidade de Araujo, (Diretor Presidente), Joaquim J. de Camargo Engler e Iby A. Pedroso e o seu Conselho Curador por Salim Simão (Presidente), Roberto Cano de Arruda, Dovilio Ometto, Aristeu Mendes Peixoto, Urgel de Almeida Lima, Humberto de Campos e Rubens Valentini.

O objetivo da FEALQ é o de apoiar e administrar projetos tecnológicos, científicos, econômicos e sociais da ESALQ, do CENA e de outras unidades da USP e Instituições públicas e privadas. Gerencia cursos, congressos, simpósios, seminários e outros eventos e também edita livros e publicações de divulgação de tecnologia. De seu portifólio faz parte o oferecimento de bolsas para estudantes que participam de projetos por ela financiados e a colaboração em programas de desenvolvimento social da comunidade do *campus* da USP em Piracicaba.

Naquela época, a sede da FEALQ situava-se no *campus*, em uma das residências anteriormente ocupadas por professores e funcionários da ESALQ. Sentados à mesa, ele expunha suas idéias a Engler.

- Como você sabe, há uma Comissão formada por cientistas brasileiros elaborando um projeto de desenvolvimento de biotecnologia, com apoio do CNPq e da FINEP, explicou para Engler que, como sempre, ouvia atentamente. "É o Programa Nacional de Biotecnologia, o PRONAB, que deverá ser apresentado ao Ministro do Planejamento, Delfim Neto no final deste ano", continuou Otto.

- É a Comissão da qual você, Tavares e João Lúcio fazem parte, não é? perguntou Engler.

- Sim. Temos nos reunido com bastante frequência, confirmou Otto acrescentando: "tenho grandes expectativas quanto aos recursos que poderão ser alocados para desenvolver projetos tecnológicos e científicos. Mas, eu gostaria de ir mais além: utilizar essas técnicas *in vitro* em projetos envolvendo empresas agrícolas. Aliás, nos últimos dez anos, desde a vinda de Rod Sharp para a Seção de Bioquímica de Plantas do CENA, em junho de 1971, que estamos nos preparando para entrar definitivamente em projetos público-privados".

-O exemplo disso é o Projeto que você tem com o PLANALSUCAR, afirmou Engler.

- Justamente. Esse é um projeto sob minha responsabilidade que, como voce sabe, é administrado pela FEALQ desde 1979, prosseguiu Otto, esclarecendo: "O projeto está em andamento, com bons resultados científicos e de treinamento de pessoal. O nosso relacionamento com essa empresa é muito bom. Osny Bacchi, como pesquisador do PLANALSUCAR, tem nos ajudado bastante no projeto, assim como também Enio e alunos de graduação e de pós-graduação no CENA".

Observando o interesse de Engler, Otto explicou: "na realidade a possibilidade de assinatura de um Convênio com o PLANALSUCAR surgiu quando apresentei resultados de nossas pesquisas com cultura de células e tecidos de cana-de-açúcar no Congresso de produtores de cana realizado em Maceió, capital do Estado de Alagoas, em janeiro de 1979" e continuou: "a iniciativa privada tem se mostrado receptiva em utilizar novas tecnologias".

- Gostaria de propor para a FEALQ a criação de um Centro de Biotecnologia na ESALQ, que seria administrado financeiramente pela FEALQ, prosseguiu Otto, afirmando: "tenho já a aprovação dessa idéia, em princípio, pelo Diretor da ESALQ, Mendes Peixoto".

Entusiasmado com o apoio de Engler, no dia 25 de junho, ele, Flávio Tavares e Ernesto Paterniani, Chefe do Departamento de Genética da ESALQ, submeteram o projeto de criação do Centro para Aristeu Mendes Peixoto que o aprovou oficialmente em nome da Diretoria da ESALQ.

Na reunião do Conselho Curador da FEALQ, no dia 2 de julho de 1981, foi apresentado e discutido o projeto de criação do Centro de Biotecnologia que teria a "finalidade precípua de realizar pesquisas fundamentadas em sistemas celulares para fins tecnológicos, com a participação de pesquisadores da ESALQ e de outras instituições, interessados em assuntos relacionados com a Biotecnologia e Engenharia Genética". A criação do Centro foi aprovada pelos Conselheiros e, como consta em sua Ata, "ficando a Diretoria da FEALQ encarregada das providências necessárias para sua concretização". Além de Otto, estavam presentes nessa reunião os também Conselheiros: Urgel de Almeida Lima (Presidente), Pedro Tessinari Filho, João Ribas Fleury e José Roberto Mendonça de Barros. Como convidados, participaram dessa reunião: Aristeu Mendes Peixoto, Salim Simão, ex-Diretor da ESALQ, e Paulo Fernando Cidade de Aráujo.

Um dos primeiros a receber a notícia, Rod Sharp reagiu entusiasticamente com a aprovação da criação do Centro. Ele fora peça chave na parceria com Otto nessa aventura científica em um país que estava nos últimos anos da ditadura militar.

Agora era *6 de dezembro de 1981* e esses acontecimentos estavam presentes em seus pensamentos enquanto ele adentrava o Anfiteatro e cumprimentava os participantes do "Simpósio Internacional de Biotecnologia para Engenharia Genética", cuja realização também fora aprovada na reunião do Conselho Curador da FEALQ naquela tarde do dia 2 de julho de 1981.

O Centro de Biotecnologia, que posteriormente passaria a ser denominado Centro de Biotecnologia Agrícola–CEBTEC, estava criado. Ele foi indicado como seu Coordenador Científico.

Este era apenas o começo de uma jornada para a concretização do Centro que, como ele tinha consciência, seria bastante longa e trabalhosa.

Além dele, fizeram parte do primeiro Conselho Técnico-Científico do CEBTEC: Flávio Cesar Almeida Tavares, Henrique Vianna Amorim, Luiz Carlos Basso, Caio Cardoso, Elke Cardoso, Wilson Roberto Soares Mattos, Luiz Eduardo Gutierrez e Aline Aparecida Pizzirani Kleiner.

Desde 1964, ele continuamente recebera auxílios para realização de pesquisa seja no campo da bioquímica de plantas seja em biologia celular, a FAPESP sendo uma das principais entidades oficiais a dar-lhe suporte financeiro, o que se repetiu ao longo dos anos posteriores. A criação do CEBTEC abriu amplas perspectivas de contratos de pesquisa e desenvolvimento de tecnologias *in vitro* com entidades oficiais, brasileiras e estrangeiras. Entre as brasileiras, além da FAPESP, estão: CNPq; FINEP; CAPES (Coordenação de Aperfeiçoamento de Pessoal de Nível Superior, do Ministério da Educação do Brasil); EMBRAPA

(Empresa Brasileira de Pesquisa Agropecuária, do Ministério de Agricultura, Pecuária e Abastecimento do Brasil); CNEN (Comissão Nacional de Energia Nuclear). Entre as estrangeiras: British Council; OEA (Organização dos Estados Americanos); Fulbright Commission; UNDP (United Nations Development Programme), IAEA (International Atomic Energy Agency, Viena, Austria); French Government; Germany Government (DADD); Scientific Commission of the European Communities; USA Government / Blue Ribbon Project; Corporacion Andina de Fomento (CAF, Venezuela); National Institute of Higher Education (NIHERST, Trinidad).

IX
Os primeiros anos do CEBTEC

Logo no *início de 1982*, o interesse pelo incipiente CEBTEC dividia os seus participantes. Um local físico seria importante para concretizar a idéia.

- Os laboratórios do CEBTEC devem ser instalados no *campus* da ESALQ, disse Otto dirigindo-se a Luiz Carlos Basso.

- Mas, onde? Todos os possíveis lugares que já procuramos estão sendo utilizados pelos vários Departamentos. Basso demonstrava a sua preocupação, que compartilhava com Otto, em tornar realidade física a idéia do CEBTEC.

A resposta que ouviu tranquilizou Basso:

- Eu sei das dificuldades que estamos enfrentando, mas tenho uma idéia. Vou propor à FEALQ para financiar a reforma dos laboratórios das Disciplinas de Bioquímica, aqui no Departamento de Química, enfatizou Otto, prosseguindo:, "logo assinaremos um contrato com a Johnson & Johnson para realização de um projeto de pesquisa e desenvolvimento em eucalipto. Como você sabe, Basso, essa empresa multinacional, com sede em São Paulo, está interessada no comprimento das fibras de eucalipto. A proposta é utilizar culturas *in vitro*. Aliás, Antonio Natal Gonçalves está trabalhando com essas espécies de rápido crescimento no Departamento de Ciências Florestais, aqui da ESALQ, e será nosso colaborador nesse projeto. E nos nossos laboratórios no CENA, o Enio já iniciou as culturas de *Eucalyptus* utilizando vários protocolos".

- Creio que o Conselho do nosso Departamento aprovará. Falarei também com Paulo Cidade e com Aristeu Mendes Peixoto, finalizou.

Antonio Natal Gonçalves, especialista em fisiologia de árvores, fez seu programa de Mestrado sob orientação de Rod Sharp, na The Ohio State University, em Columbus, Ohio, nos últimos anos da década de 1970. Em seu regresso a ESALQ, passou a ser orientado por Otto em seu doutoramento, e defendeu sua Tese sobre juvenilidade e clonagem de *Eucalyptus urophyla in vitro*, em 1983.

Em 29 de novembro de 1982, com financiamento da FEALQ, foi inaugurada a reforma do subsolo do Prédio da Química, abrigando laboratórios do CEBTEC para análises bioquímicas e salas de transferência assépticas e de manutenção de culturas de células e tecidos vegetais. Um Seminário Científico, com a presença de Antonio Guerra Vieira, Reitor da Universidade de São Paulo, dos Diretores da ESALQ e da FEALQ, abriu as comemorações. Nessa ocasião foi lembrada a atuação de José Dall Pozzo Arzolla como seu co-orientador em seus estágios em química inorgânica e orgânica durante sua graduação na ESALQ e participação nos experimentos de sua Tese de Livre-Docência em 1957/1958, grande parte dela desenvolvida nos laboratórios do Instituto Zimotécnico, desativado anos mais tarde. Na entrada desses novos laboratórios do CEBTEC uma Placa homenageia o professor Arzolla.

José Dall Pozzo Arzolla e Domingos Pellegrino foram, juntamente com Demóstenes Santos Corrêa e Eurípedes Malavolta, os que mais o influenciaram na busca por sua identidade científica, *nos anos 1950* do século XX. Com Demóstenes, adentrou o reino das reações químicas: as técnicas de laboratório, que o envolviam horas a fio, e sua participação nos Debates de Química, durante os três anos de seu Curso Colegial. Na Universidade, com Malavolta, familiarizou-se com a interpretação dos artigos científicos sobre bioquímica: naquela época, os livros e revistas de bioquímica em inglês, francês e espanhol, demoravam a chegar ao Brasil. A internet estaria disponível muitos anos mais tarde. Tornou-se "rato de biblioteca". Autodidata em bioquímica, após graduar-se na ESALQ e já contratado pela USP, sua formação foi completada em 1964/1965 com Conn e Stumpf, na Universidade da Califórnia, no *campus* de Davis. Com Pellegrino, seu professor de Química Analítica, aprendeu a ser meticuloso nas análises químicas e na interpretação dos resultados, fazendo com que se interessasse pelas análises estatísticas dos dados e frequentasse cursos extracurriculares de estatística experimental com os professores Friedrich Gustav Brieger, de Genética, e Frederico Pimentel Gomes, de Matemática e Estatística, na ESALQ. Através de Arzolla, seu professor de Química Orgânica, entrou em contato com o dinamismo das aulas teóricas e práticas, desenvolvendo sentido de perspicácia na condução de experimentos. Ele e Arzolla tiveram uma forte e bonita amizade ao longo dos anos, até o seu falecimento prematuro em *1970*.

Caminhando pelos laboratórios recém inaugurados, lembranças afloraram à sua mente. Naqueles mesmos balcões, agora reformados, após seu regresso de Londres, onde trabalhou com Fowden, ele, Luiz Carlos Basso e Oswaldo Brasil, nos *anos de 1969/1970,* montaram os experimentos que levaram à descoberta da atividade da enzima N-carbamilputrecina descarboxilase, essencial para a formação de putrecina no ciclo da ornitina-uréia, quando plantas de *Sesamun* são submetidas ao estresse de potássio. Quantas repetições dos ensaios, quantas discussões no preparo do artigo científico relatando os experimentos, quanta excitação ao receber a notícia de que havia sido aceito para publicação na revista *Phytochemistry*!

A busca pelas bases bioquímicas dos fenômenos celulares responsáveis pelo desenvolvimento vegetal sempre esteve presente em sua mente. Ele constantemente repetia para si mesmo que esta era a razão pela qual desejava penetrar no reino da biologia celular. E, por isso, a parceria com Rod Sharp tornava-se cada vez mais oportuna. Entretanto, naquele momento, ele teve a premonição de que ambos, além desse adentrar no âmago da célula vegetal, poderiam também contribuir para o avanço da agricultura, utilizando as técnicas *in vitro*. Afinal, as condições físicas existiam: as instalações do CEBTEC na ESALQ, pelas quais agora estava caminhando, somadas aos seus laboratórios na Seção de Bioquímica de Plantas no CENA, seriam apropriadas para o desenvolvimento desse tipo de projeto. Aliás, logo mais, em princípios de *dezembro desse ano de 1982*, iria discutir o potencial das técnicas de cultura de células e tecidos vegetais para o melhoramento de culturas agrícolas, na Conferência Internacional sobre as Novas Fronteiras das Relações entre a Química e o Suprimento Mundial de Alimentos, organizada pela União Internacional de Química Pura e Aplicada, em Manila, nas Filipinas.

- O senhor está falando comigo, professor? perguntou Enio ao entrar no laboratório tendo em suas mãos frascos de cultura de embriões de feijão.

- Não, respondeu Otto, admirando-se pelo inusitado do acontecimento: estava dialogando consigo mesmo, em voz alta! "É que eu estava pensando alto em um projeto de resgate de embriões híbridos que resultam, por exemplo, do cruzamento interespecífico de duas espécies de feijão", explicou.

- Mas os embriões não seriam viáveis, afirmou Enio.

- Se esses embriões forem cultivados em condições corretas, como as que podemos estabelecer *in vitro*, poderiamos resgatar embriões híbridos viáveis que se desenvolveriam em plantas adultas, explicou Otto.

- Seria um projeto bastante ambicioso, disse Enio.

- Ambicioso, mas não impossível, meu caro Enio, respondeu Otto. "É o que me seduz na pesquisa. Nós já temos feito algo sobre proteínas de reserva em sementes de feijão. Lembra-se do projeto internacional sobre proteínas de reserva de leguminosas do qual participei? Temos também resultados que apresentei no Simpósio sobre proteínas vegetais, em Las Vegas, *em agosto de 1980*, afirmou Otto.

- Sim, professor. Sempre entusiasta com os projetos de Otto, Enio continuou: "mas, antes dessa apresentação, em junho passado, o senhor esteve na Alemanha Oriental discutindo o metabolismo de proteínas de reserva durante o desenvolvimento de embriões de feijão *in vitro*, não é mesmo?"

Junho de 1980! Nesse momento, como que num passe de mágica, ele viu-se em um aeroporto ao ar livre, em reforma, esperando a chegada de uma aeronave que o levaria ao interior da então Alemanha Oriental. Seu destino era a cidade de Gatersleben. Na Academia de Ciências da República Democrática Alemã apresentaria os resultados dos experimentos sobre o metabolismo de proteínas durante o desenvolvimento de embriões de feijão cultivados *in vitro*.

Em plena "Guerra Fria", chegou a Berlin Ocidental. Dirigiu-se à estação de trem Friedrichstrasse, que fazia a ligação com a parte oriental da cidade. A estação era um dos principais pontos de fronteira entre os dois lados de Berlin. Em um estreito corredor que permitia a passagem de uma única pessoa de cada vez, apresentou seus documentos para os policiais e então adentrou na estação no lado oriental. Do trem que, de início, movimentava-se lentamente, procurava observar a cidade. Sentiu-se triste: ao contrário do que vira em Berlin Ocidental, fervilhante em progresso, as ruas lhe pareceram desertas e os prédios mal cuidados, dando-lhe uma sensação de precariedade. "Pobre povo", suspirou.

No coração da Alemanha Oriental, em um aeroporto a céu aberto, ainda em construção, absorto em seus pensamentos, com a sensação de tristeza presente em seu coração, caminhava, em idas e vindas, ao longo da larga calçada à espera da aeronave que o levaria ao seu destino final.

De repente, os quiosques dispostos na calçada, à frente dos quais caminhava, abriram suas janelas. Flores em pequenos vasos surgiram diante de seus olhos: violetas, petúnias, gerânios, prímulas, e tantas outras, encheram de alegria aquele lugar que lhe parecia inóspito. A magia daquele momento deu lugar a um espetáculo, para ele inusitado: um ônibus acabara de estacionar e dele jovens, homens e mulheres, desceram e, em fila, como que obedecendo a um ritual atávico, sobraçavam os pequenos vasos, levando-os consigo, deixando no ar a promessa de que nem tudo estava perdido. Essa cena, no ambiente austero da ditadura comunista de então, com corações de jovens palpitando em busca da beleza, personificada nas flores coloridas, era o testemunho de que a esperança renovava-se uma vez mais e mais. E que a liberdade emocional e a liberdade intelectual, inerentes ao ser humano, daqueles jovens não se separariam jamais. "Realmente, nem tudo está perdido", suspirou. Desta vez, de esperança.

E assim, chegou a Quedlinburg, que seria incluida como patrimônio cultural da humanidade pela UNESCO, em 1994. Mas agora, em *junho de 1980*, ainda pertencente à Alemanha Oriental, na Saxônia, essa incrivelmente bela pequena cidade conquistou o seu coração. "Uma joia saxônica" pensou o não mais tão jovem Otto. Ele, que vivera em Durham, uma cidade medieval no norte da Inglaterra, e sentira todo o peso cultural de uma época tão distante, de imediato sentiu-se atraído pelas torres e telhados pontiagudos das casas com fachadas decoradas em estilo enxaimel. Tinham elas mais de 600 anos! E muitas delas necessitavam de restauração. Muitos dos monumentos da cidade estavam ameaçados de demolição. "Não possuem e nem podem ter sistema moderno de aquecimento", disse um dos auxiliares do hotel onde se hospedara. E continuou: "a grande maioria está sendo restaurada por artesãos poloneses, que são peritos nessa arquitetura".

Uma semana esteve em Quedlinburg, tempo suficiente para dessa cidade não mais se esquecer. De ônibus, nesse mesmo período, diariamente ele dirigia-se à vizinha Gatersleben para participar das discussões sobre proteínas vegetais na Academia de Ciências.

X
Formação de pessoal científico

- Professor, o seu aluno Cabral está aqui para a reunião, disse Solizete ao telefone, dirigindo-se a Otto. Maria Solizete era a exímia secretária da Seção de Bioquímica de Plantas do CENA e de sua total confiança.

Com seu inconfundível e agradável sotaque do nordeste brasileiro, José Barbosa Cabral, vindo de Recife, capital do Estado de Pernanmbuco, apresentou-se e colocou-se à disposição para discutir o tema de sua Dissertação. Seria o resgate de embriões viáveis resultantes do cruzamento entre três espécies de feijão: *Phaseolus vulgaris, Phaseolus lunatus* e *Phaseolus acutifolius*. A inviabilidade de embriões resultantes de cruzamentos interespecíficos de feijões era muito bem documentada na literatura. Por outro lado, a técnica de cultivo de embriões *in vitro* já era dominada pelo grupo do CEBTEC. O diálogo anteriormente travado entre ele e Enio poderia dar resultados práticos. Cabral desenvolveu a pesquisa com responsabilidade, resgatando embriões viáveis dos cruzamentos entre aquelas espécies de feijão, regenerando plantas adultas e férteis. Revelou-se um excelente pesquisador na área de biologia celular. Defendeu sua Dissertação em 1983. Posteriormente, criou uma empresa de biofábricas, utilizando técnicas de cultura de vegetais *in vitro*, em Recife.

O feijão (*Phaseolus vulgaris* L.) é uma das principais leguminosas utilizadas como alimento no Brasil, rico em proteínas, e nutrientes minerais como ferro, potássio, fósforo e cálcio, além de fibras e vitaminas. Os brasileiros o consomem com arroz (*Orysa sativa* L.), um cereal rico em carboidratos e aminoácidos essenciais. Desse modo, como fonte nutricional, a leguminosa rica em proteinas e no aminoácido lisina, complementa-se com o cereal rico em metionina, nas refeições diárias no Brasil.

- Professor, poderia receber um professor da Universidade Estadual do Rio de Janeiro (UERJ) logo após a sua palestra? perguntou uma jovem dirigindo-se a Otto que estava subindo ao palco do anfiteatro da Universidade Federal do Rio de Janeiro (UFRJ), no Rio de Janeiro, *em outubro de 1982*.

- Sim, respondeu Otto, antes de iniciar sua palestra sobre aplicação das técnicas de cultura de tecidos de plantas em agricultura, a convite de Antonio Paes de Carvalho, Diretor do Instituto de Biofísica da Universidade Federal do Rio de janeiro (UFRJ).

Nessa época, Antonio Paes de Carvalho, atualmente Professor Emérito, estava interessado em formar um grupo de trabalho para desenvolvimento dessas técnicas em sua Universidade. Essa era a razão da palestra de Otto. Em 1983 Paes de Carvalho liderou um Programa de Biotecnologia, juntamente com Affonso do Prado Seabra, Maria Apparecida Esquibel e Antonio Rodrigues Carneiro. Posteriormente, fundou a Biomatrix, que foi a primeira empresa brasileira de biotecnologia vegetal. Otto e Antonio Natal Gonçalves atuaram como seus assessores no desenvolvimento de clones de eucaliptos, para a empresa. Técnicos da Biomatrix foram treinados nos seus laboratórios em Piracicaba; várias reuniões com o pessoal de Paes de Carvalho foram feitas na residência de Otto em Piracicaba e também em New York, USA. No final da década de 1980, a Biomatrix deixou de funcionar. Em 1986 Paes de Carvalho fundou a Associação Brasileira de Empresas de Biotecnologia (ABRABI) e, em 1988, foi um dos fundadores da Fundação Bio-Rio, dedicada às relações Ciência-Indústria.

Ao receber Leonardo Alves Carneiro, professor de Biologia Celular na Universidade Estadual do Rio de janeiro (UERJ), Otto logo percebeu o seu interesse em familiarizar-se com as técnicas *in vitro* e explicou para Leonardo o projeto sobre embriões de feijão que estava desenvolvendo em seus laboratórios no CEBTEC e no CENA:

- Enio e alguns de meus orientados estão realizando trabalhos com culturas de embriões de feijão, seja para o resgate de embriões interespecíficos viáveis seja para acompanhar a atividade de enzimas comprometidas com o metabolismo de nitrogênio durante o desenvolvimento dos embriões *in vitro*.

Leonardo entusiasmou-se. Em *princípios de 1984* mudou-se para Piracicaba, com sua esposa Ana Maria e suas filhas. Iniciou seu programa de Mestrado em Solos e Nutrição de Plantas, na ESALQ, defendendo sua Dissertação em 1990 sobre a atividade de redutase de nitrato em embriões de feijão cultivados *in vitro*. Essa enzima é responsável pelas reações iniciais da assimilação do nitrogênio quando plantas são expostas a nitrato. Posteriormente, na UERJ, interessou-se por plantas ornamentais e, retornando à ESALQ em 1994, inscreveu-se no Programa de Doutorado. Em 1997 recebeu o título de Doutor em Agronomia pela ESALQ, defendendo sua Tese sobre controle da morfogênese *in vitro* de bromélias endêmicas do sudeste brasileiro, sob orientação de Otto: o primeiro trabalho científico realizado no Brasil com Bromeliaceas.

Em 1980, procurando resolver um problema no seu laboratório no CENA, Otto foi interrompido por uma jovem que se apresentou como sendo aluna do segundo ano de graduação da ESALQ. Também fazia parte de um grupo de 10 alunos selecionados para participarem de um programa de especialização em energia nuclear oferecido pelo CENA.

- O programa exige que os participantes também sejam treinados em técnicas que estão sendo desenvolvidas aqui no Centro, explicou Helaine. "Eu gostaria de familiarizar-me com culturas de tecidos de plantas", continuou.

As madeixas de seus cabelos movimentaram-se quando seu rosto, de pele clara, inclinou-se em direção a Otto, que estava sentado em um banco próximo a uma das bancadas do laboratório. Helaine estava inquieta à espera de uma resposta positiva ao pedido de estágio que fizera.

- Aqui na nossa Seção temos vários alunos da ESALQ e de outras instituições, inclusive pesquisadores do exterior. Voce poderá iniciar seu treinamento com eles e com o meu técnico Enio, que domina essas técnicas, Otto explicou.

Como estagiária, Helaine Carrer superou-se ao assimilar as técnicas *in vitro*. Após concluir o seu curso de graduação em 1983, desenvolveu todo seu programa de Mestrado na ESALQ, sob orientação de Otto, concluído-o em 1988 com a Dissertação sobre controle da morfogênese de tecidos de mamão (*Carica papaya*) *in vitro*.

- Helaine, chamou Otto: "o nosso grupo precisa familiarizar-se com novas técnicas, entre elas, o cultivo de anteras para a produção de haploides que são instrumentos celulares importantes no melhoramento de plantas, pois facilitam a identificação e seleção de híbridos recessivos em células e em plantas. Além disso, os programas de seleção podem ser acelerados, após os cruzamentos, pela haploidia".

- Seria um complemento ao programa de cruzamentos interespecíficos de feijão que estão sendo conduzidos aqui nos seus laboratórios para obtenção de embriões híbridos viáveis? perguntou Helaine.

- Não exatamente, teriamos uma metodologia a mais para colocar à disposição dos melhoristas de plantas em seus programas de seleção de plantas, respondeu Otto.

- E qual seria minha contribuição? Helaine estava confiante de que poderia ser uma colaboradora eficiente nos programas de biologia celular.

- Voce sabe que o CEBTEC necessita pessoal qualificado para o seu quadro de pesquisadores. Voce é uma forte candidata e precisa aumentar seus horizontes científicos. Já entrei em contacto com Rod Sharp para que você se familiarize com novas técnicas nos laboratórios da DNA Plant Technology, em New Jersey, com o Philip Ammirato. Voce aceita? Os olhos de Helaine brilharam ao aceitar a proposta.

Em 1986 Helaine foi contratada como pesquisadora do Departamento de Química da ESALQ para prestar serviços no CEBTEC. No ano seguinte passou a ser Docente do mesmo Departamento, após concurso público. Agora, *em 1988*, Helaine passou a trabalhar com Amirato, em New Jersey.

Devido à sua familiarização com as técnicas de cultura de células e tecidos vegetais e à sua grande responsabilidade na condução de trabalhos laboratoriais, e incentivada por Otto e Rod Sharp, aceitou o convite de Pal Maliga para realizar seu programa de doutorado em biologia molecular. Em 1989 iniciou seus estudos no Waksman Institute, The State University of New Jersey, Rutgers. Em 1994 obteve seu título de Ph. D. em Biologia de Plantas com tese sobre a expressão de genes bacterianos em plastídeos de fumo.

Em 1975, após terminar seu Curso de Graduação em Biologia na Universidade Estadual Paulista "Júlio Mesquita" (UNESP), em Rio Claro, no Estado de São Paulo, Luiz Antonio Gallo iniciou seu treinamento em técnicas bioquímicas no CENA.

- Tenho observado sua dedicação no laboratório, disse Otto dirigindo-se a Gallo, como era costume Luiz Antonio ser chamado pelo seus colegas. Creio que você poderia inscrever-se no programa de Mestrado em Ciências aqui do CENA, acrescentou.

- Já estou bem familiarizado com técnicas enzimáticas, disse Gallo. Eu gostaria de desenvolver minha Dissertação sobre enzimas do metabolismo de nitrogênio em feijão, afirmou.

- Sugiro que você acompanhe a assimilação de nitrato durante o crescimento de plantas de feijoeiro mantidas em diferentes doses de nitrato, disse Otto.

Esse foi o tema da Dissertação que Gallo defendeu em 1979, obtendo o título de Mestre em Ciências pelo CENA. Em 1987 foi aceito no Programa de Doutorado em Bioquímica no Instituto de Química da USP, *campus* de São Paulo, sob orientação de Otto, continuando suas pesquisas sobre metabolismo do nitrogênio em embriões de feijão cultivados *in vitro*. Defendeu sua Tese em 1994. Anteriorrmente, em 1988, foi contratado como docente do Departamento de Química da ESALQ para prestar serviços no CEBTEC. Em 2002 foi aprovado em concurso público para Livre Docência, tornando-se professor associado. No CEBTEC tem sido importante colaborador em projetos de transferência de tecnologias *in vitro* para empresas privadas.

Entre os seus vários alunos, Fernando Broetto destacou-se pelo seu grande interesse em compreender os fenômenos bioquímicos nas células vegetais. Após terminar seu curso de engenharia agronômica na Universidade Federal do Estado de Espírito Santo, na capital Vitória, em 1986 Fernando iniciou seu programa de Mestrado em Ciências no CENA. Sua Dissertação foi em parte realizada nos Laboratórios do Prof. Ulrich Zeitz, na Universidade de Tübingen, na Alemanha, em 1988-1989. Nessa oportunidade, em *1988* após participar do Simpósio da Food Agriculture Organization (FAO) sobre Biotecnologias para Países em Desenvolvimento, em Luxemburgo, Otto reuniu-se com Zeitz e Broetto em Tubingen, para discutir o uso de processos biotecnológicos para estudos de compostos secundários em plantas. Em 1991, sob orientação de Otto, Broetto obteve o título de Mestre na ESALQ e em 1995 o título de Doutor em Ciências pelo CENA, sob orientação de Eurípedes Malavolta. Nessa ocasião, Fernando já era professor de bioquímica na Faculdade de Agronomia da Universidade do Estado de São Paulo "Júlio Mesquita" (UNESP), em Botucatú, onde atualmente é Professor Associado e desenvolve intensa atividade de pesquisa em bioquímica e fisiologia do estresse em plantas, indução de resistência em vegetais superiores e trabalhos com plantas medicinais

Nessa mesma época, uma graciosa jovem, de estatura "mignon", de pele morena, cor própria dos que vivem no norte do Brasil, estava terminando seu Doutorado, sob a orientação de Otto, no CENA. Natural de Belém, capital do Estado do Pará, graduada em farmácia e bioquímica, Irenice Maria Santos Vieira iniciou seus estudos de pós-graduação em janeiro de 1980. Dona de uma invejável capacidade para trabalho em laboratório, e muito dedicada aos estudos, desenvolveu uma intensa e extensa pesquisa sobre a influência de níveis de potássio sobre os teores de açucares e a atividades das enzimas invertases em folhas de plantas jovens de cana-de-açucar cultivadas em solução nutritiva, em casa de vegetação. Os seus resultados foram apresentados e defendidos em sua Dissertação de Mestrado em 1983. Continuando a trabalhar com plantas jovens de cana-de-açucar, completou os resultados anteriores, estudando as relações entre os açucares em quatro variedades dessa gramínea em sua Tese de Doutorado defendida em 1988. Regressou então a Belém, retomando suas atividades na Universidade Federal Rural da Amazônia, aposentando-se em 2009. Em 2011 passou a fazer parte do Grupo de Biotecnolohia Vegetal, na Universidade Federal de Minas Gerais, em Belo Horizonte.

Jaime A. Cury, com formação acadêmica em odontologia, após seu curso de Mestrado na Universidade Federal do Paraná, em Curitiba, fez seu Doutorado em Bioquímica pelo Instituto de Química, da USP, *campus* de São Paulo, sob a orientação de Otto. Atualmente é professor na Faculdade de Odontologia de Piracicaba e é conhecido internacionalmente pelos seus trabalhos sobre o uso de flúor na proteção dentária. Também como orientado de Otto em seu Doutorado em Bioquímica por esse mesmo Instituto da USP, Luiz Eduardo Gutierrez desenvolveu toda a sua carreira acadêmica na ESALQ, aposentando-se como professor titular de bioquímica. Tem o reconhecimento da comunidade científica pelos seus conhecimentos sobre as bases bioquímicas da alimentação humana

Ao longo dos anos muitos alunos de graduação foram seus orientados e muitos outros destacaram-se em seus programas de Mestrado e/ou de Doutorado sob sua orientação. Entre esses últimos, Maro Ran-Ir Sondhal; Guilherme Guimarães; Maria Aparecida Schiavuzzo; Marina Yukie Murayama; Roberval de Cássia Salvador Ribeiro; Sandra Aparecida Tabai; Carlos Eduardo Corsato; Geraldo Eduardo Cuzzuol; João Chaddad Junior; Daniela de Argollo Marques; Fátima Odahara Kajiki; Antonio Barioni Gusman; Isaac Stringueta Machado; João Suzuki. Sidival Lourenço foi seu primeiro orientado de Mestrado, em *1967*.

XI
Os Seminários de Biotecnologia Agrícola – SEBIAGRI

Deixando seus laboratórios no CENA, tomou o caminho interno que une esse Centro à ESALQ. Por entre altas palmeiras aspirava o ar fresco de um inverno que teimava em não se fazer presente naquela manhã do mês de *junho de 1983*.

Um único pensamento vinha-lhe à mente: como fazer para impor sua idéia de um CEBTEC atuante no campo científico. O Simpósio de dezembro de 1981, que mostrara aos seus pares do Brasil e do Exterior que o CEBTEC fora criado, tinha sido um sucesso. Mas, isso não bastava. O mundo científico é exigente. E assim tem de ser.

Contornando os prédios do Departamento de Zootecnia da ESALQ, que estava em seu caminho, dando as voltas por entre as árvores, ele desceu a ladeira que, à sua esquerda, faz limite com os campos experimentais do Departamento de Genética, e finalmente chegou ao prédio do Instituto de Genética.

- A Dra. Aline está em sua sala? perguntou ao atendente. Ao receber a resposta positiva, adentrou o corredor à sua direita e foi recebido por Aline.

- Aline, gostaria que você me ajudasse a organizar uma série de palestras focalizando o uso de processos biotecnológicos na produtividade agrícola, disse Otto.

- Será muito bom, disse Aline, acrescentando: "estou pronta para colaborar, o que será preciso fazer para tornar realidade essa idéia?"

Aline Aparecida Pizzirani-Kleiner, docente do Departamento de Genética, era entusiasta dos trabalhos desenvolvidos pelo CEBTEC e sua colaboradora no Conselho do Centro.

- Estou pensando em convidar o Wilson Mattos e o Antonio Roque Dechen para colaborar conosco na organização das palestras, afirmou Otto.

Wilson Roberto Mattos um atuante docente do Departamento de Zootecnia, desenvolveu excelentes trabalhos em seu campo de pesquisa e ensino. A partir de *2007* passou a exercer a função administrativa de Prefeito do *campus* "Luiz de Queiroz". Antonio Roque Dechen, Professor Titular de Nutrição de Plantas, do Departamento de Química, foi um incansável colaborador na séries de palestras e simpósios do CEBTEC. No período *2007-2010* foi Diretor da ESALQ e Vice Reitor de Administração da USP (*2011-2013*).

- Como irão chamar-se essas séries de palestras? perguntou Aline

- Poderiamos designá-las "Seminários de Biotecnologia Agrícola", com a sigla SEBIAGRI, sugeriu Otto.

- Qual seria a melhor data para esse primeiro SEBIAGRI?. Obviamente, Aline estava interessada na realização desses seminários.

- A melhor data seria no próximo outubro. Antes, em setembro, deverei participar do Encontro Nacional de Cultura de Tecidos Vegetais, na Universidade Federal de Viçosa, durante a IX Reunião da Sociedade Brasileira de Fisiologia Vegetal, afirmou Otto.

Viçosa, no Estado de Minas Gerais, no centro-oeste brasileiro! Recordou-se imediatamente da primeira vez em que esteve em Viçosa: 1956, Congresso de Estudantes de Agronomia! O bosque de pinheiros! Os sonhos de um jovem estudante desejoso de iniciar-se no mundo acadêmico! E agora estava ele, em *junho de 1983*, fazendo planos para voltar para Viçosa. E ele voltou para o "Encontro sobre Cultura de Tecidos de Plantas". Nesse Encontro foi criada a Associação Brasileira de Cultura de Tecidos – ABCTP. Foi aclamado seu primeiro Presidente. A Diretoria completou-se com Walter Handro, Secretário, Gilberto B. Kerbauhy, Secretário Adjunto e Rolf Dieter Illg, Tesoureiro. Otto ocupou a presidência da Associação até 1987. Em *3 de outubro de 2003* foi homenageado pela ABCTP, agora sob a presidência de Renato Paiva. Em Viçosa, novamente!

No dia *11 de outubro de 1983*, foi realizado o SEBIAGRI I discutindo o tema "Produtividade agrícola e produtividade animal". A comissão organizadora foi formada por Otto, Aline, Dechen e Mattos. Outros Seminários se seguiram. O SEBIAGRI II, aconteceu nos dias *29 e 30 de agosto de 1984* com o tema "Biotecnologia da cana-de-açucar: da célula à planta", quando então foram apresentados e discutidos os resultados das pesquisas do grupo do CEBTEC sobre biotecnologia de cana-de-açucar, além de outras palestras. Otto, Dechen, Cristina Gonçalves e Joaquim Albenísio Silveira formaram a comissão organizadora. O SEBIAGRI III teve lugar no período de *10 a 13 de setembro de 1985*, com o tema "Bases fisiológicas da bioprodutividade vegetal e animal". Além de Otto, Mattos e Dechen, fizeram parte da comissão organizadora: Antonio Natal Gonçalves, Joaquim Albenísio Silveira e Raul Machado Neto.

Nos dias *29 e 30 de novembro de 1986*, "Biotecnologia florestal: perspectivas e aplicações" foi o tema do SEBIAGRI IV, organizado por Otto, Dechen, Natal Gonçalves e Albenísio, com a colaboração de Mario Ferreira, Luiz E. G. Barrichelo, João W. Simões, Paulo C. T. Carvalho, Angelo DiCiero Neto e N. Nicoliello, os dois últimos representantes da iniciativa privada.

Em abril desse mesmo ano de *1986*, Otto participou de uma Reunião Científica promovida pelo Ministério da Educação da Alemanha, em Bonn, juntamente com outros professores da Universidade de São Paulo, liderados pelo Reitor José Goldenberg. Nessa ocasião apresentou seus trabalhos na área de biotecnologia de plantas.

Com o tema "Carbono e nitrogênio na produtividade das plantas", o SEBIAGRI V foi realizado nos período de *28 a 30 de setembro de 1987*, organizado por Otto e Albenísio. Anos depois, em *23 de maio de 1994*, Helaine, de volta de seus estudos nos Estados Unidos, colaborou com Otto na organização do SEBIAGRI VI com o tema "Biologia molecular de plantas", na Sala "William R. Sharp" no prédio do CEBTEC, na ESALQ.

Tão logo regressou dos Estados Unidos, em *meados de 1990*, Helaine passou a fazer parte de projetos envolvendo biologia molecular de plantas. No Projeto Genoma, financiado pela FAPESP, com o auxilio de Enio e de Valentina Fátima de Martin, e vários alunos de pós-graduação, liderou no CEBTEC o sequenciamento de X*ylella_fastidiosa*, bactéria causadora de doenças em plantas economicamente importantes, como a praga do amarelinho que afeta laranjeiras. Desde então, seu talento para a pesquisa científica desenvolveu-se enormemente. Sua incansável capacidade para trabalho em seus laboratórios, orientando alunos, e seus contatos científicos internacionais, fazem com que Otto tenha certeza de que o Centro que, com Flávio Tavares, criara em *1981*, terá continuidade.

XII
1988. O novo Prédio do CEBTEC

Fevereiro de 1988. Enio e Solizete já haviam sido transferidos do CENA para o Departamento de Química da ESALQ, para prestar serviço junto ao CEBTEC. Maria Solizete Granziol como Secretária e Enio como Técnico de Nível Superior. Enio concluira seu Curso de Biologia pela Universidade Metodista de Piracicaba (UNIMEP) em *1987*.

Nesse mês começaram os preparativos para a transferência do CEBTEC para o novo prédio que estava sendo construído em uma área doada pela Diretoria da ESALQ e pela então Prefeitura do *campus* "Luiz de Queiroz". Alunos, estagiários de outras instituições de ensino e pesquisa e técnicos revezavam-se na utilização do espaço dos laboratórios, no Prédio da Química, agora exíguos para acomodá-los adequadamente. Entre os Técnicos contratados estavam o Farmacêutrico-Bioquímico Antonio Francisco Campos do Amaral, Engenheiro Agrônomo João Chadad e Romeu Aparecido Rocha.

Setembro de 1985. Joaquim José de Camargo Engler, então Diretor da ESALQ e Antonio Sanches de Oliveira, Coodenador do *campus* foram recebidos por Otto em sua sala no CEBTEC, no prédio da Química.

- Viemos aqui propor para você a construção do prédio de CEBTEC em uma área de 9000 m², aqui na ESALQ, disse Engler.

Sanches de Oliveira acrescentou: "é uma área previlegiada, próxima às instalações da FEALQ".

Com a expressão emocionada, aceitou de imediato a oferta e, acompanhado por Engler e Sanches de Oliveira, foi visitar a área que fazia limite com a Avenida Centenário. Em *6 de janeiro de 1986* foi lançada a pedra fundamental do prédio com a presença de Antonio Guerra Vieira, em final de seu mandato como Reitor da USP, Engler, Paulo Fernando Cidade de Araújo, Antonio Roque Dechen, Enio, Basso e outros professores e funcionários da ESALQ. A construção foi feita com recursos de seus projetos técnico-científicos e de seus colaboradores, administrados pela FEALQ. A planta do prédio foi elaborada por Justo Moretti Filho, professor de Engenharia Agrícola da ESALQ. Anos mais tarde, como Diretor-Presidente da FEALQ, Justo Moretti deu especial atenção aos projetos do CEBTEC. Antigo professor de Otto, compartilhou com ele todos os movimentos para a construção do novo prédio, entusiasticamente. Também é de sua concepção o designer do logotipo que encima a fachada do prédio: um plastídeo encerrado em um círculo como se uma organela fosse. Justinho, como seus amigos e colegas o chamavam, faleceu em *setembro de 2011*.

A fachada do prédio foi desenhada por Antonio Natal Gonçalves que, além de dedicar-se ao estudo da fisiologia de árvores é também renomado artista plástico. Três colunas em arco romano, simulando tubos de ensaio invertidos, recebem os visitantes no pórtico de entrada do prédio. Essa mesma metáfora arquitetônica está presente nas janelas da sala da Coordenadoria do Centro, à esquerda do prédio. À direita, um Auditório, com recursos audiovisuais, tem suas janelas circundadas por circunferências em alto relevo simulando caixas de Petri. Bioquímica e Biologia Celular simbolicamente representadas na fachada do prédio do CEBTEC (foto 4)!

Outubro de 1988. Otto estava envolvido com a transferência de aparelhos científicos, que haviam sido adquiridos com verba de seus projetos científicos, do seu laboratório no CENA para o novo prédio. Enio e Leonardo Carneiro encarregavam-se da logística do transporte. Maria Tereza e Eric Derbyshire não mais pertenciam à sua equipe no CENA, desde janeiro de 1982. Helaine estava em Rutgers, às vésperas de iniciar seu programa de Ph. D.. Ela,

Murilo Mello, Ph. D. em Fisiologia Vegetal e Luiz Antonio Gallo, M. S. em Ciências, tinham sido contratados como professores do Departamento de Química para trabalhar no CEBTEC.

Solizete adentrou a sala da chefia do CEBTEC no Prédio da Química na ESALQ e entregou uma pasta para Otto, informando: "professor, aqui está a pasta com todo o material gráfico para o Simpósio de Biotecnologia de Plantas".

- Voce fez um excelente trabalho, eu não sei o que faria sem seu auxílio na secretaria, principalmente agora que estamos às vésperas da inauguração do novo prédio e a abertura do Simpósio, disse Otto, sobraçanco a pasta. Para ele, Solizete, como sua Secretária e Enio, como seu Técnico, eram duas pessoas essenciais para o bom desenvolvimento das pesquisas do CEBTEC.

Ao examinar o conteúdo da pasta, uma vez mais retornou ao passado. Recostando-se em sua cadeira, deixou-se embalar pelas lembranças do passado recente. Sete anos haviam se passado desde a criação do CEBTEC em *junho de 1981* e a realização do Simpósio de Engenharia Genética em dezembro desse mesmo ano.

Outubro de 1984. No dia 19, em uma reunião de 1 dia, participou, juntamente com outros 6 cientistas de diferentes países, do "Tokyo Meeting of Plant Biotechnology", realizado por Plantech Research Institute, em Toquio, Japão. Em seguida, viajou para Beijing, China, onde discutiu resultados das pesquisas de seu grupo no "International Symposium on Genetic Manipulation in Crops", na Academia Sínica. Na capital chinesa hospedou-se no Friendship Hotel, visitou a Muralha da China, caminhou pela enorme Praça da Paz e foi envolvido pelo vozerio das pessoas nas casas comerciais.

- Professor, podemos levar os relatórios de suas viagens ao exterior para a sua sala no novo prédio? perguntou Enio interrompendo os devaneios de Otto.

- Por favor, estão ali sobre aquela mesa, respondeu Otto apontando para uma pilha de livros, revistas e relatórios de suas pesquisas.

Enquanto Enio e Solizete sobraçavam os relatórios, lembrou-se das muitas visitas que fizera aos laboratórios de Rod Sharp, em Ohio State University, na Campbel, em Cinnaminson, e na DNA Plant Technology, em New Jersey. Lembrou-se então que, em uma dessas visitas, durante o almoço de confraternização dos participantes do "Annual Scientific Advisory Board Conference" da DNAP, em Philadelphia, em *setembro de 1986*, Rod Sharp e David Evans fizeram-lhe uma surpresa outorgando-lhe o *"1986-DNAP Biotechnology Award"*. Vestindo somente uma camisa de mangas longas, enquanto todos os demais estavam de terno e gravata, recebeu a honraria das mãos de Norman Borloug, detentor do Prêmio Nobel da Paz.

- O senhor está pensativo, professor, disse Enio dirigindo-se à porta, para deixar a sala.

- Estou me lembrando das muitas vezes em que Rod e eu estivemos juntos em reuniões científicas aqui no Brasil e nos Estados Unidos, respondeu Otto. E continuou: "neste momento estou me lembrando quando, em *1982*, Rod convidou-me para participar de um workshop sobre prioridades na pesquisa em biotecnologia para o desenvolvimento internacional, em Washington e em Berkeley Springs, em West Virginia. Esse evento foi promovido pela Academia Nacional de Ciências dos Estados Unidos. Fiquei hospedado em um chalé no meio de uma floresta. Lembro-me de como esse lugar foi especial para minhas meditações matinais. Aliás, Enio, eu sempre gostei de meditar em meio a árvores", completou. Uma vez mais, recolheu de sua memória as experiências quase contemplativas em momentos de silêncio.

A esse workshop somou-se o simpósio "Biotecnologia nas Américas: prospectos para países em desenvolvimento", realizado em San José, Costa Rica, *em 1983*, quando então ele e Rod Sharp discutiram a importância do uso das técnicas de cultura de células e tecidos vegetais para o avanço da agricultura. E *em 1985*, participou como debatedor das mesas redondas sobre tecnologias celulares emergentes e o "scale-up" em biotecnologia, durante o Scientific Meeting of the DNAP Advisory Board, em New York.

- Mas, e o livro que o senhor e o Prof. Sharp editaram? perguntou Enio.

- Ah! É o livro "Biotechnology of Plants and Microorganism", editado em colaboração com Evans, J. E. Bravo, Flavio Tavares e E. F. Paddock. Foi lançado em *maio de 1987*, em Columbus, Ohio, durante a comemoração de 20 anos do Convênio entre a ESALQ e a The Ohio State University. Humberto de Campos, que à época era Diretor da ESALQ, também estava presente nessa solenidade.

Esse diálogo com Enio parecia-lhe uma "sessão de prestação de contas". Mas, agora estava em *outubro de 1988*. Há alguns meses, em *março*, estivera na Ilha de Trinidad, no Caribe e em Caracas, Venezuela. Em *agosto* desse mesmo ano em Toronto, no Canadá, proferiu, como convidado de A. M. Chakrabarty, uma conferência sobre os seus trabalhos, no XVI Congresso Internacional de Genética; em seguida, uma vez mais, reuniu-se com Rod Sharp nos Estados Unidos.

Finalmente despertou de suas lembranças e dirigiu-se com Enio e Solizete ao novo prédio do CEBTEC.

XIII
1988. Simpósio Internacional de Biotecnologia de Plantas

O Simpósio Internacional de Biotecnologia de Plantas teve lugar no anfiteatro do Pavilhão de Engenharia na ESALQ de *26 a 29 de outubro de 1988*, organizado por Rod Sharp, Otto e Murilo Melo. Na Cerimônia que deu início aos trabalhos, presidida por Paulo Fernando Cidade de Araújo, Diretor Presidente da FEALQ, Rod Sharp fez a conferência de abertura.

Na *noite do dia 27* deu-se a inauguração do novo prédio do CEBTEC. Ele nunca se esqueceria desse momento, impar em sua vida acadêmica. Eram vinte horas quando o acender paulatino das luzes do prédio recebeu os convidados que para ele se dirigiam. A emoção tomou conta de todos quando, ao se aproximarem do pórtico de entrada, foram saudados pelo som dos acordes iniciais da 5ª Sinfonia de Beethoven. Um conjunto de sete jovens instrumentistas da Escola de Música de Piracicaba "Maestro Ernst Mahle", sob a regência de Antonio Ribeiro, dispostos ao longo do pórtico, brindaram os convidados, embalando-os com peças de Bach e Mozart à medida que adentravam o recinto. Uma placa comemorativa da inauguração foi decerrada por Otto e Paulo Cidade de Araújo.

Entre os convidados, estavam presentes os conferencistas P. V. Ammirato (Columbia University, New York, USA), P. A. Arruda (UNICAMP), J. A. Betti (Instituto Agronômico, Campinas), J. Daussant (Laboratoire de Physiologie, Meudon, França), P. C. Debergh (State University Gent, Bélgica), D. A. Evans (DNAP, USA), J. E. Feldeberg (Cia. Suzano de Celulose), E. I. S. Floh (USP, São Paulo), E. Gander (CENARGEN, Brasilia), W. Handro (USP, São Paulo), R. D. Illg (UNICAMP), H. W. Janes (Rutgers, USA), A. Leite (UNICAMP), S. A. Malacrida (UNICAMP), C. Costa (UNESP, Botucatú), R. C. S. Ribeiro (UNESP, Ilha Solteira), J. A. Sampaio (CENARGEN, Brasília), H. E. Sommer (University of Georgia, USA), J. B. Teixeira (EMBRAPA), M. R. Söndahl (DNAP, USA), M. J. O. Zimmermann (EMBRAPA, Goiania) e A. Natal Gonçalves e José R. Postali Parra, este último, anos mais tarde, se tornaria Diretor da ESALQ.

Uma surpresa aguardava Rod Sharp. Uma placa, decerrada por ele e por Otto, no hall de entrada do Auditório, dizia para todo o sempre que aquela era a "Sala William R. Sharp". A emoção pairou no ar quando Otto e Rod repetiram o abraço que se deram ao encontrar-se pela primeira vez no Aeroporto de

Congonhas, em São Paulo, na manhã de junho de um inverno ameno, em *1971*(foto 5) !

Rod Sharp, Otto e seus familiares e convidados foram recepcionados pelo Diretor da ESALQ, Humberto de Campos e sua esposa Wanda, na bela residência dos diretores no *campus*

Livro de Biotecnologia. No palco do Auditório do Serviço Social do Comércio de Piracicaba, uma entidade de caráter privado, ao lado de Otto estavam sentados à mesa Murilo Melo, Rod Sharp e Flávio Fava de Moraes, Diretor Científico da FAPESP e ex-Reitor da USP. Eles dirigiam os trabalhos do lançamento do livro "Biotecnologia para Produção Vegetal/Biotechnology for Plant Production", co-editado por Otto, Rod e Murilo. Era uma tarde do dia *29 de novembro de 1990.*

Perante uma plateia de professores, técnicos, estudantes e convidados, Otto apresentou o livro que contém as conferências proferidas no Simpósio Internacional de Biotecnologia de Plantas, além de capítulos escritos por outros cientistas. Com Prefácio de Otto e Introdução de Rod Sharp, o livro foi dedicado por Otto para seu primeiro mentor nos meandros da química, Demosthenes Santos Corrêa, para seu orientador na ESALQ, Euripedes Malavolta, do qual fora assistente, e para o professor da ESALQ e seu grande amigo André Martin Louis Neptune, que faleceria no ano seguinte.

XIV
Os Congressos Internacionais de Cultura de Tecidos de Plantas

8 de junho de 1994. No saguão do Aeroporto Internacional de Guarulhos, na Grande São Paulo, Diva e Otto esperavam ser chamados para tomar assento na aeronave que os levaria para a Itália. Otto iria participar de mais um Congresso Internacional.

- Pai, finalmente Maria Paula chegou para despedir-se de vocês, disse Daniel dirigindo-se a Otto. Daniel foi quem levara seus pais para o Aeroporto, vindo de Piracicaba.

Diva sorriu ao observar sua filha dirigir-se a eles apressada, pois estava atrasada devido ao transito intenso na capital paulista: "que bom que você está aqui conosco", disse.

- Infelizmente o Adolfo Egídio não poderá despedir-se de vocês, disse Maria Paula. "Ele está no plantão de otorrinolaringologia, na residência médica", acrescentou.

Carla Maisa, de cabelos longos e tez morena, como a do pai, diplomada em língua inglesa, com compromissos inadiáveis em Piracicaba, também não estava presente na despedida.

Daniel, o filho caçula, de cabelos castanhos e tez clara, como a da mãe, e Maria Paula, de cabelos longos e tez clara, como a de Adolfo Egídio, juntaram-se para abraçar seus pais. Daniel cursava a Faculdade de Turismo em São Paulo e Maria Paula já era tradutora bilíngue. Eram dois dos quatro filhos que ficaram para Diva e Otto, após o falecimento do primogênito Marco Augusto, *em 16 de junho de 1991.*

A morte de "Maguto", como era chamado pelos seus irmãos, é constantemente lembrada por ele, Diva e por seus irmãos. A sua memória, como se vivo estivesse, está sempre presente, pelo menos conviveu-se com ele durante 29 anos!

Era *25 de junho de 1991* quando, no Auditório do Pavilhão de Química da ESALQ, o Diretor da ESALQ, João Lúcio de Azevedo, abriu os trabalhos da Sessão Científica de comemoração dos 10 anos de criação do CEBTEC. Paulo Cidade de Aráujo, Rod Sharp, Otto e Philip Ammirato foram os conferencistas. Nessa oportunidade Jeff, filho de Rod, entregou a Otto uma contribuição financeira homenageando a Memória de Marco Augusto. A doação foi utilizada, através da FEALQ, para construção de parte de um Laboratório, anexo ao prédio principal do CEBTEC. Uma placa, no pequeno hall de entrada do laboratório traz versos seus homenageando o seu filho Marco Augusto Lovadino Crocomo...

"voar...voar...../ para as estrelas/ e lá ficar/ contemplando o universo..."

Medalha Luiz de Queiroz. Em *1998*, Júlio Marcos e Antonio Roque Dechen, então Diretor e Vice-Diretor da ESALQ, respectivamente, apoiaram a iniciativa de Otto em conceder a Rod Sharp a "Medalha Luiz de Queiroz". Em solenidade realizada na ESALQ, em *3 de junho de 1999*, William Rod Sharp recebeu a Medalha.

Nessa mesma ocasião, ao mesmo tempo em que a alegria de conceder a Rod Sharp a Medalha Luiz de Queiroz se fazia presente, a morte do médico ortopedista Caiuby de Souza Arruda abalou a sua família. Caiuby e sua esposa, a poetisa Jahyra, eram muito próximos a ele, Diva e seus filhos. Ambos foram padrinhos de batismo de Daniel. Nessa ocasião, Rod, acompanhado por Otto, visitou Caiuby no hospital, como se fora uma despedida do grande amigo: havia um excelente relacionamento entre ambos, desde os primeiros tempos em que Rod chegara a Piracicaba.

No avião da Alitalia que os levaria a Milão, ele e Diva falavam sobre seus filhos.

- Maria Paula vai sentir-se sozinha quando os seus outros irmãos viajarem para a Suiça para nos encontrar após o Congresso em Florença, disse Otto.

- Ela há pouco obteve seu primeiro emprego após terminar seu curso de Tradutora de Inglês da Faculdade, respondeu Diva, com o seu permanente sentido prático sobre a vida. "Ela é responsável", acrescentou.

Adolfo Egídio, Carla Maisa e Daniel iriam encontrar-se com seus pais no Aeroporto de Genève, na Suiça, no dia 21 de junho, e então, de carro, viajariam por vários países da Europa.

A aeronave aterrissou no Aeroporto de Milão. Ao lado de Diva, visitou a cidade. Sentiram-se inebriados diante da grandiosidade do "Teatro alla Scala a Milano" e da Catedral, o "Duomo" de Milão, com imponente estilo gótico. De trem partiram para Florença.

A cidade do "David" de Michelangelo, de Maquiavel e Leonardo Da Vinci e Dante Alighieri, das ruelas contando histórias, das esfuziantes joalherias da Ponte Vechhio sobre o Rio Arno, esta foi a Florença onde ele participou, de *12 a 19 de junho*, do VIII Congresso Internacional de Cultura de Tecidos de Plantas. Foi Membro da Comissão Científica desse encontro científico organizado pela International Association of Plant Tissue Culture – IAPTC. No Congresso, apresentou resultados de suas pesquisas sobre produtos naturais em plantas. Foi também o Moderador do Simpósio "Juvenilization and Maturation", que organizou como parte do Programa do Congresso.

Em 1978 participou do IV Congresso da IAPTC, em Calgary, Canadá, nos últimos dias de *agosto*. Encantado com o país, para lá voltou várias vezes, em missões científicas.

Na ocasião do Congresso, os pilotos da companhia aérea do Canadá estavam em greve. Tinha o compromisso de comparecer à Reunião do Projeto sobre Proteínas de Leguminosas, que aconteceria em Munique, na Alemanha, no *início de setembro*. Finalmente, ao final do Congresso, foi autorizado um voo noturno para Toronto e em seguida para Paris, de onde partiria para Munique, após alguns dias. Os relógios marcavam 8 horas da manhã quando, deixando o Aeroporto Charles de Gaulle, entrou em seu quarto, em um hotel em Paris.

Estava demasiado cansado e sonolento para rever Paris à luz do dia. Acordou às 19 horas. Banho rápido, desceu as escadas do hotel e, tomando um metrô, chegou até o Arco do Triunfo. Deslumbrado, assistiu o acender das luzes ao longo do Champs Elisée. Naquele momento, que lhe pareceu mágico, recordou-se de ter caminhado, com Diva em *março de 1969* e com ela e todos os seus 5 filhos, desde o Arco do Triunfo até o Louvre, passando pelo Jardin de Tulleries, em *agosto de 1976*. Essa caminhada repetiria várias vezes em muitas de suas visitas a Paris, em uma delas com Diva e três de seus filhos em *julho de 1994*. E, então, uma vez mais se deslumbraria com a beleza arquitetônica do Hôtel de Ville, próximo a majestosa Catedral de Notre Dame de Paris, considerada um dos exemplos proeminentes do estilo gótico. Anos mais tarde se extasiaria lendo os relatos da construção de catedrais pelos mestres artesanais da Idade Média em romances ambientados em acontecimentos históricos, como os do escritor inglês Ken Follet.

Em *10 de julho de 1982* deixou a aeronave da Japan Airlines que aterrissara no Aeroporto de Narita, em Tóquio, no Japão. A viagem tinha sido longa: de São Paulo, no Brasil, a Los Angeles, nos Estados Unidos e depois Japão. Já acomodado na aeronave que o levaria a Tóquio, iniciou a leitura do romance "Adriano", de Marguerite Yourcenar. A trajetória do grande imperador pelo mundo das conquistas de novas terras para Roma, entusiasmou-o durante todo o trajeto aéreo de Los Angeles a Tóquio. Tendo vivido entre o primeiro e segundo séculos da nossa era, a vida de Adriano é descria com maestria e de forma romanceada pela grande escritora belga, naturalizada americana e Membro da Academia Francesa de Letras.

Desde muito cedo admirou os conquistadores pela coragem e destreza nas estratégias de suas lutas. À medida que se desenvolvia e atingia a maturidade, passou a admirar também os grandes industriais, os grandes engenheiros e arquitetos, os filósofos, os escritores, os professores, os cientistas que não se deixavam abater pelos percalços em suas pesquisas. Procurava sempre as biografias das figuras humanas que ajudam a construir a história da civilização. Quais as causas, circunstanciais ou biológicas, que fizeram com que assim o fossem? Apaixonado por música clássica escolheu, como presente de término de seu Curso Ginasial, uma biografia de Beethoven. Era *dezembro de 1949*. Tinha então 17 anos de idade

No V Congresso da IAPTC, que teve lugar em um centro de convenções aos pés do Monte Fuji, apresentou resultados de suas pesquisas com células irradiadas de cana-de-açucar cultivadas *in vitro* A troca de experiências com cientistas japoneses nas idas e vindas diárias, de ônibus, entre o hotel, onde se hospedara, e o centro de convenções, e durante as sessões de discussão, abriu-lhe novas perspectivas para sua vida de pesquisador.

A possibilidade de participação do CEBTEC em futuras reuniões científicas era promissora. E assim aconteceu. Em *agosto de 1986* ele liderou um grupo de jovens pesquisadores de empresas privadas de fabricação de papel e celulose, em sua participação no VI Congresso da IAPTC, em Minneapolis, Minesota, nos Estados Unidos.

Nesse Congresso apresentou trabalhos sobre resgate de embriões híbridos de feijão, sobre o controle da morfogênese em *Pinus* e sobre a reversão á juvenilidade durante a micropropagação de *Eucalyptus*. Após o Congresso o grupo foi recebido por Rod Sharp e David Evans na DNAP, em Cinnaminson, New Jersey, onde foi discutido o estado da arte das pesquisas do CEBTEC.

Era uma manhã do mês de *junho de 1990* quando deixou a aeronave da Air France e uma vez mais percorreu os corredores do Aeroporto Charles de Gaulle, em Paris, de onde logo partiria para a Holanda. O VII Congresso da IAPTC teria lugar entre *24 e 29 de junho*, em Amsterdam. Apresentou resultados de suas pesquisas sobre o controle da formação de embriões somáticos da palmeira macaúba (*Acrocomia aculeata*), que possui elevado teor de óleos com potencial para uso como combustíveis alternativos, de interesse estratégico nacional que, nos anos 2000 seriam denominados biocombustíveis. Como sempre fazia em suas viagens científicas, o seu interesse pela cultura universal fez com que apreciasse as obras de grandes pintores e de Van Gogh no Museu que leva seu nome, em Amsterdam, em companhia de Phillip Ammirato, então colaborador de Rod Sharp. Nessa viagem visitou vários laboratórios de biotecnologia na Itália, Alemanha, Suiça e Bélgica.

XV
Aplicação das técnicas in vitro: biotecnologia de plantas

1979 – 2001

A *década de 1980* havia sido emblemática para ele: o "boom" da biotecnologia, no exterior e também no Brasil, que se evidenciara no final da década anterior, agora mostrava toda sua potência como possibilidade de introduzir inovações tecnológicas na agricultura. Em princípios dessa década, o projeto com o PLANALSUCAR estava em plena atividade.

Em sua sala, conversando com Enio sobre as realizações do CEBTEC, ele lembrou: "esse projeto nasceu do convite que recebi para apresentar nossos resultados sobre a regeneração de plantas a partir de células de cana-de-açucar cultivadas *in vitro* no I Congresso Nacional de Técnicos Açucareiros do Brasil, em Maceió, capital do Estado de Alagoas, no nordeste brasileiro, em *janeiro de 1979!*" E lembrou: "os técnicos do PLANALSUCAR entusiasmaram-se ao perceber o potencial que essa técnica tem como auxiliar no melhoramento de plantas de cana".

Esse fora o primeiro projeto de pesquisa e desenvolvimento que realizara com empresas privadas. Com a colaboração de Enio e Neftali Ochoa-Alejo, apresentou resultados de suas pesquisas com a tolerância de células de cana-de-açucar ao herbicida ametrin, no II Congresso Nacional de Técnicos Açucareiros do Brasil, no Rio de Janeiro, em *agosto de 1981*.

Outros resultados das pesquisas com células de cana-de-açucar, também foram sucessivamente apresentados por ele em reuniões internacionais durante o *ano de 1984*. Em setembro, participou do "International Symposium on Plant Tissue and Cell Culture: Application to Crop Improvement" em Olomouc, na antiga Czechoslovakia, na época sob o regime comunista.

Aquele era um momento de recordações. Lembrou-se da sua cruzada na divulgação da aplicação das técnicas *in vitro* em agricultura. Diante de uma audiência de fruticultores, a maioria dos quais não aceitava o uso dessas técnicas, durante o "Simpósio sobre Biotecnologia em Fruticultura", realizado em Campinas, no Estado de São Paulo, *em 1986*, preconizou: "virá o tempo em que todos vocês utilizarão essas técnicas para obter melhores produtos e em maiores quantidades".

E assim aconteceu.

De *1983 a 1988* as empresas brasileiras Duratex, do Grupo do Banco Itáu, e Duraflora, financiaram o projeto "Clonagem e Seleção de Clones de Eucalipto" sob sua responsabilidade, com a colaboração de Antonio Natal Gonçalves. Pelas empresas foi inestimável a colaboração de Angelo DiCiero e Antônio S. Rensi Coelho. Durante a realização do projeto houve treinamento de pessoal e instalação de laboratório nas empresas. Os clones obtidos no CEBTEC foram mantidos em campos experimentais das empresas.

Abril de 1986. Em Bonn, na Alemanha, apresentara resultados de suas pesquisas com várias espécies vegetais, e em Bruxelas, na Bélgica, em companhia de Paulo Cidade, defendera com sucesso a continuação de um projeto patrocinado pela Comissão Científica da Comunidade Europeia, iniciado em *1983*. Após o seu regresso a Piracicaba foi procurado por investidores para a formação de uma parceria público-privada para produção de frutíferas. Após várias reuniões, a empresa Brasil Venture Capital assinou com a FEALQ o projeto "Engevit", de pesquisa e desenvolvimento para produção em larga escala de moranguinho livre de vírus, a ser conduzido no CEBTEC. Aclimatizadas em casa de vegetação, as mudas obtidas *in vitro* através da técnica de cultivo de meristemas, eram transferidas para o campo, em uma fazenda situada em Itatiba, região serrana do Estado de São Paulo, onde não se registravam incidências de vírus, e comercializadas.

Em outubro de 1989, acompanhado por Walter Traldi, um dos financiadores do projeto "Engevit", visitou vários laboratórios de biotecnologia de plantas na Itália, França, Bélgica e Espanha. Nessa ocasião apresentou resultados de suas pesquisas utilizando as técnicas de cultura de plantas *in vitro* durante o "World Food Day: Biotechnology for the Third World" no Deutsches Ubersee Institut, em Hamburgo, Alemanha.

Outros investidores também procuraram por ele para produção de moranguinho livre de vírus. De 1991 a 1992, as mudas produzidas *in vitro* e aclimatizadas no CEBTEC, foram plantadas em uma fazenda no alto da Serra de São Pedro, em altitude de 700 metros, em clima apropriado para esse tipo de cultivo. Foi o Projeto ENGEVEG, com participação de Enio e de seu orientado de Mestrado Vitor Smoller Sheffer.

Ao longo da *década de 1990*, atendendo a uma intensa demanda por mudas de banana livres de doenças, o CEBTEC instalou uma estrutura apropriada para sua produção. Várias empresas produtoras de diferentes variedades de banana firmaram convênios com o Centro, através da FEALQ, proporcionando também a formação de pessoal. Muitos alunos desenvolveram trabalhos científicos utilizando essas mudas, sob sua orientação, entre eles Humberto Zaidan. Humberto, com auxilio de Enio, estudou o metabolismo de poliaminas em microplantas de banana, a detecção dessas substâncias tendo sido feita nos laboratórios de Fernando Broetto, na UNESP, em Botucatú. Atualmente Humberto é professor associado na Universidade Federal de Campina Grande, no Estado da Paraiba, no nordeste do Brasil.

O projeto Citrus
1989-1995

Fevereiro de 1989. Ao atender uma chamada telefônica, foi surpreendido pelo convite a ele feito por um dos Diretores Administrativos do Grupo Votorantim: expor as vantagens do uso das técnicas *in vitro* para propagação de plantas.

- Onde e quando será a reunião?, perguntou, acrescentando: "nos próximos dias deverei ir a Lima, no Perú, para participar do Encuentro Peruano de Cultivo de Tejidos Vegetales, a nossa reunião poderia ser antes?"

O Encontro Científico em Lima, no qual participaria, aconteceria de *23 a 25 de fevereiro*. Seriam discutidas estratégias para incrementar as pesquisas em biotecnologia de plantas nos países andinos. Ele propôs que a reunião com o pessoal do Grupo Votorantim se realizasse no dia 21 pela manhã.

Nos escritórios que a Votorantim mantinha em São Miguel Paulista, na Grande São Paulo, foi recebido por Ricardo Ermírio de Morais, um dos responsáveis pela Empresa Citrovita. Após a apresentação de dados científicos, discutiu-se a participação do CEBTEC em um projeto de parceria universidade/empresa para produção de plantas de laranja livre do vírus da tristeza, que, quando presente, dizima os laranjais. A colaboração de Ary A. Salibe, especialista em vírus de plantas, de renome internacional, e seu amigo desde 1953, quando ambos iniciaram seus estudos de graduação na ESALQ, foi uma das exigências de Otto para a realização do Projeto. Salibe era professor titular de Fruticultura na UNESP, em Botucatú.

O Projeto "Microenxertia de citros: obtenção de laranjas livres do vírus da tristeza e de viroides" desenvolveu-se de *1989 a 1995*. Nos laboratórios do CEBTEC foram produzidos microenxertos de laranja pera, a partir de meristemas obtidos das melhores árvores cítricas selecionadas por Salibe, que percorreu fazendas nos Estados de São Paulo e da Bahia. As mudas de laranja obtidas dos microenxertos foram mantidas em ambiente anti-afídico, em casa de vegetação no CEBTEC e, em seguida, enviadas para os laboratórios da UNESP, Botucatú, onde foram indexados por Salibe para o vírus da tristeza, em casa de vegetação. As plantas selecionadas foram transferidas para casas de vegetação da fazenda da Citrovita, na cidade de Itapetininga, ao sul do Estado de São Paulo e, após serem aclimatizadas, mantidas a campo. Todos os testes comprovaram a não presença do vírus da tristeza nas árvores de laranja pera, as quais apresentam elevada produção. No CEBTEC, o projeto teve a colaboração de Enio e Gallo; na UNESP, a colaboração de Antonio Tubelis.

XVI
A memória viva do tempo não perdido

2011

Naquela manhã de uma primavera sub-tropical, abriu a porta da sala de estar e lentamente iniciou a descida pela escada que o levaria a área de lazer de sua casa. Dos largos patamares que separam os degraus, ao som dos cantos de pássaros, apreciou as plantas dos vasos que ladeiam a escada e, através do muro que limita o terreno onde a casa se situa, lançou olhares de admiração pela beleza das flores e arbustos no exterior das casas do outro lado da rua. Desde o alto de uma esquina, a casa oferece a ele a ampla visão das duas ruas. Muitas vezes, caminhando pelo condomínio surpreendeu-se admirando as árvores nas calçadas, algumas de alto porte e outras mais baixas, as altas palmeiras, e as flores enfeitando a entrada das casas parecendo dar as boas vindas aos visitantes. A elas juntam-se as altas tuias, os ciprestes, os arbustos, as flores, que rodeiam o exterior de sua casa. Um verdadeiro jardim aos seus olhos.

Deixou sua casa onde vive há 40 anos e, a passos lentos, caminhou em direção à parte da rua que termina em um beco sem saída. As árvores, com cachos de flores amarelas que olham para o chão de asfalto, em ambos os lados da rua, formam um túnel vivo encobrindo, do alto, os arbustos verdes e as flores azuis, vermelhas e rosas que circundam as paredes das casas. De repente, viu-se nesse mesmo espaço, a brincar de bola com Rod Sharp e seus filhos Marco Augusto e Adolfo Egídio, ainda crianças, nos *inícios dos anos 1970*. As casas e as árvores ainda não existiam naquele beco, naquele tempo. Daniel, o seu caçula, dava os primeiros passos. As meninas, Maria Paula e Carla Maisa brincavam com bonecas. Tudo era diferente, suspirou. Era um tempo em que as primícias do vir-a-ser estavam burilando no seu coração e na sua mente, preparando o terreno para as conquistas científicas que se sucederiam ao longo dos anos vindouros.

A admiração pela beleza, naquele momento uma vez mais, mágico, exuberante no verde e nas flores multicoloridas, e nas suas formas, estava na raiz de *sua inquietude sobre a causa das coisas serem como são*. Ah! o controle bioquímico da formas das coisas: a função das enzimas alostéricas na regulação das reações bioquímicas! Como foi crucial o desenvolvimento das pesquisas de Jean-Pierre Changeux nesse campo, em princípios de 1960, lembrou-se. Essa e outras perguntas sempre estiveram presentes na sua mente ao observar a natureza.

Essa inquietação era fruto de seu profundo interesse pelo mundo das idéias filosóficas que, desde sua juventude, aprendera a garimpar, como que a procura de ouro e diamantes, na tentativa de explicar a vida. A companhia dos filósofos gregos aos contemporâneos, passando pelos filósofos medievais, renascentistas e iluministas, foi importante para que ele procurasse, na interpretação dos dados de suas pesquisas e no transmitir conhecimentos a seus alunos, investir no pensamento científico. Muitos anos haviam se passado desde que penetrara no domínio das células vegetais e procurara nele as reações bioquímicas que estão na base da vida. Essas andanças pelo "mundo do invisível" permitiu-lhe depois que vivenciasse o casamento entre a biologia celular e a biologia molecular, em sua vida de professor e pesquisador, adentrando no terreno espetacular da biotecnologia de plantas. Da célula vegetal à planta produzindo no campo! No início dos anos 80 a transferência da tecnologia celular para o setor privado estava a um passo e esse, e outros muitos, foram dados por ele e sua equipe ao longo dos últimos trinta anos.

Deixou o beco dirigindo-se para a avenida, a uma quadra de sua casa, e iniciou a descida da ladeira que o conduziria a uma outra avenida, larga e movimentada. Ao seu lado direito estava o Clube de Campo, uma ampla área, outrora uma chácara pertencente a Rodolfo Lara Campos, o Conde Lara, elegante no seu trajar e nas suas maneiras, que costumava passear pela cidade em seu coche puxado por cavalos negros, assim Otto dele se recordava em sua infância. No início dos anos 1950 um grupo de piracicabanos comprou a área e, em parte dela, foi instalado o Clube; a outra parte foi dividida em lotes, em um deles Otto construiu sua casa. As piscinas do Clube e a piscina da sua casa! Caminhando ao longo do muro que circunda o Clube, lembrava-se bem: em uma de suas visitas a Piracicaba, no deque, ao sair da piscina da casa, Rod Sharp escorregou e no tombo machucou o cotovelo. Foi uma correria. Rod foi socorrido, conduzido ao hospital e medicado. Fora somente um susto!

Atravessou a Avenida Armando de Salles Oliveira e em uma longa caminhada dirigiu-se à margem esquerda do Rio Piracicaba, decantado em prosa e versos pelos piracicabanos, dos pescadores aos intelectuais. Descendo pelas curvas da Avenida Beira Rio, apreciava a beleza do espetáculo das águas escorregando pelas enormes pedras de basalto como que dispostas em degraus irregulares, parecendo imitar uma cachoeira: o "Salto do Rio Piracicaba". Mais abaixo, as águas, há pouco caudalosas, formavam agora um lençol majestoso que calmamente seguia o destino de todo rio que caminha para o mar.

Pelas subidas e descidas da avenida, chegou até uma construção datada de meados do século XIX: a chamada "casa do povoador", em homenagem ao capitão Antonio Correa Barbosa quem, *em 1767*, fundou o povoado que posteriormente se transformaria na cidade de Piracicaba, com a sua atual pujança econômica. Apesar de não ter sido construída pelo capitão Barbosa, a casa é considerada como se o fosse, um símbolo da memória histórica. É ponto turístico da cidade, com frequentes eventos culturais e exposições de artes plásticas. Feitos de madeira e pano por Elias Rocha, pescador piracicabano, descendente de índios e morador das margens do rio e que, com ele, durante sua vida, manteve uma convivência de respeito, os "bonecos de Elias" espalham-se pelo entorno da casa e pelas margens do rio, representando pescadores, como guardiões de reminiscências históricas.

No seu caminhar apreciando a natureza, deteve-se no Largo dos Pescadores. Rodeado de pequenas casas que vendem produtos turísticos, o Largo é palco de reuniões musicais, as "noites de serestas". Cantores, os seresteiros, entoam músicas sentimentais, com letras que decantam as maravilhas do amor e, muitas vezes, as dores de amores não correspondidos. Olhando à sua volta, deparou-se com um grupo de seresteiros preparando-se para cantar o Hino de Piracicaba, de autoria de Newton Almeida Mello. *"Piracicaba que eu adoro tanto, cheia de flores, cheia de encantos..."* é a estrofe cantada por todos os que vivem em Piracicaba. Nesse momento, viu-se, ainda menino, *nos idos da década de 1940*, em companhia de seu pai João, após a morte de sua mãe Tereza, muitas vezes adentrando na espaçosa casa dos DiGiaimo, italianos de origem napolitana, não mais existente nesse Largo. Uma das filhas dos DiGiaimo, Ada, casou-se com Francisco, seu irmão. A lembrança dos sons da música italiana, que em coro cantavam nessas visitas, brincaram em sua mente e em seu coração, misturando-se com os acordes das serestas piracicabanas, que, agora, ouvia.

Ao som do murmurar das águas do rio, buscava pelo tempo passado. Alguns sonhos e desejos não conseguira realizar, mas agora, chegando aos 80 anos de idade, não tinham mais importância. Estavam no mais íntimo de seu ser e lá continuariam. Não precisaria rebuscá-los, expondo-os. Os altos e baixos dos contatos humanos durante seu caminhar pela vida acadêmica lhe haviam ensinado a lutar por algo que lhe parecia ser importante. As conquistas no mundo técnico-científico que, para ele, não lhe parecem ser tão grandes quanto o consideram ser os seus pares e os seus amigos, ocupam agora um lugar de destaque em suas realizações, apesar de tudo e de todos. Parafraseando Marcel Proust, o seu tempo não fora perdido!

Continuando a caminhar, viu-se entrando no Parque da Rua do Porto. Vinte mil metros quadrados de área verde, de preservação histórica. Muitas e muitas vezes caminhou pela sua pista para exercícios físicos, ladeada de frondosas árvores, serpenteando o extenso lago onde são realizados campeonatos de canoagem. À frente do lago, no outro lado do calçadão e da avenida, estão bares, quiosques de côco gelado e restaurantes típicos de frutos do mar e de água doce.

Em uma das curvas do caminho que vagarosamente percorria sentiu em sua pele o ar que envolvia as árvores sob as quais, agora, descansava em um banco de madeira. Sorveu o oxigênio que exalava das folhas, resultado do "milagre" da fotossíntese. E respirou. Como é grandiosa a natureza e sábio quem a fez! As reações bioquímicas que acontecem silenciosamente por traz daquilo que os nossos sentidos experimentam, e que não vemos: sempre presentes em sua mente!

Há alguns quilômetros dali está o CEBTEC. O seu coração bateu mais forte. Desde 2001 não era mais seu Diretor Científico, mas a sua sala ainda está lá, com todas as suas recordações e presença frequente. Quantos alunos discutiram com ele os resultados de suas pesquisas, naquela sala!

O contacto não só com Enio, Helaine, Gallo e tantos alunos que transitam pelos laboratórios do CEBTEC, mas também com os seus colegas da Associação dos Professores Aposentados da ESALQ, a sua tão querida ADAE, da qual atualmente é o presidente, dão-lhe contínuo alento para manter acesa a sua vontade de continuar estudando a natureza que está ao seu redor. Sempre há tempo para aprender!

Os galhos das árvores balançaram. Sentiu o toque da brisa que o envolveu, suavizando o calor que se anunciava no quase final daquela manhã de recordações. A brisa ...o vento....*"às vezes ouço passar o vento....e só de ouvir o vento passar vale a pena ter nascido"*: versos do poeta português Fernando Pessoa brincando em seus pensamentos! Não simplesmente viver, mas degustar cada momento da vida... assim Otto percebia a sua vida! E sempre percebendo-a, ali, sentado naquele banco de madeira, os seus pensamentos, como em revoada, fizeram-no recordar: em *março de 1998*, nasceu Pedro Augusto, o seu único neto, filho do caçula Daniel e de Fernanda. Alto e esguio, com vivaz inteligência, é a alegria de toda a família, confiante em seu futuro.

"Muitos anos se passaram, estou em 2011!", admirou-se e então alegrou-se: "o CEBTEC está completando 30 anos! há poucas semanas, em princípio de outubro, a ESALQ e a Associação dos Ex-Alunos da ESALQ (ADEALQ) entregaram-me uma placa em homenagem a essa data, para mim repleta de significados. Recebi-a das mãos de Helaine e de Roque Dechen, durante a solenidade da 54ª Semana Luiz de Queiroz." Esse reconhecimento da sua Escola jamais seria esquecido, tinha certeza.

Durante essa solenidade ele e seus colegas do Curso de Graduação da ESALQ foram homenageados pelos seus 55 anos de formatura. Uma outra placa lhe foi oferecida celebrando a sua indicação para receber a "Medalha Fernando Costa" que a ele será entregue pela Associação dos Engenheiros Agrônomos do Estado de São Paulo (AEASP), em cerimonia a ser realizada no dia 15 de junho de 2012, na cidade de São Paulo (fotos 6 e 7).

Caminhou até a Avenida Beira Rio e sentou-se em um dos bancos sorvendo água de côco gelada. Em um dos bares, à sua frente, o relógio lembrou-lhe que já eram 12 horas. Levantou-se. "Helaine e Enio estão esperando por mim no CEBTEC", falou para si mesmo, apressando seus passos. "Há 10 anos, em junho, Helaine e Raul Machado foram à minha sala para comunicar que a Rutgers University iria me homenagear", de repente recordou-se.

Rutgers, 2001. Como um clarão, viu-se ao lado de Diva e da filha Carla Maisa, que viera de San Francisco, na Califórnia, onde cursava Mestrado, para New York, encontrar-se com seus pais, abraçando Rod Sharp no John F. Kennedy International Airport. Eram 7 horas da manhã do dia *4 de agosto.*

No dia 8, com Rod, partiram para New Brunswick. Na tarde desse mesmo dia, na presença do Presidente e de professores da Rutgers University e de Helaine, Raul Machado e Carlos Alberto Labate, e alunos de pós-graduação da ESALQ, ele foi homenageado pelo Cook College & The New Jersey Agricultural Experiment Station, Rutgers, The State University of New Jersey.

Absorto pelas suas recordações, caminhando pelo calçadão do Parque, não percebeu que um carro estava estacionado mais adiante, no meio fio da Avenida Beira Rio. Ao aproximar-se dele, foi surpreendido por uma voz conhecida que o chamava: "pai, estamos aqui". Era Daniel que deixava o carro para encontrá-lo. Daniel, dono de voz potente, com agradável timbre, herdado de sua mãe Diva.

Um jovem, ainda menino quase adolescente, saiu do carro e correu em sua direção. "Eles estão esperando por nós", disse Pedro Augusto (fotos 8 e 9).

No carro, que percorreu o caminho inverso daquele que, naquela manhã, fizera quando saíra de sua casa em direção ao rio, o seu olhar, refletindo uma saudade futura daquelas curvas, subidas e descidas, procurava pelas águas deslizantes, perceptíveis entre as copas das árvores que, circundando as bordas do rio, lançavam uma acolhedora sombra deixada para trás pelo carro dirigido por Daniel.

Finalmente chegaram ao CEBTEC. Os seus olhos encontraram os olhares de seus amigos. Ao lado de Diva, seus filhos e neto, ouviu a notícia transmitida por Helaine, secundada por Enio: José Vicente Caixeta, Diretor da ESALQ, informara que a Congregação da Escola havia aprovado a solicitação do Departamento de Ciências Biológicas, feita através do seu Chefe, Ricardo Ferraz Oliveira. O Prédio do CEBTEC passaria a ter o seu nome!

Palmas foram ouvidas, porém não lhes prestava atenção. Sua mente estava povoada pelas lembranças e emoções de sua infância na Rua Rangel Pestana, em Piracicaba. Recordou-se uma vez mais. Estava dormindo. Viu-se brincando em uma casa enorme, rodeada de flores e árvores. Acordara. Era apenas um sonho: estava na casa de seus pais! No decorrer dos anos, com o tempo passando, aquela casa enorme seria o lugar onde estudaria e trabalharia, mas ainda assim não era sua casa (foto 10).

Abraçado por Enio, Otto sussurrou-lhe aos ouvidos: "Agora sim estou também em minha outra casa!" Lembrou-se de Shakespeare:

"somos feitos da mesma matéria dos nossos sonhos"

Agradecimentos

Agradeço os comentários recebidos de Cecílo Elias Neto, Plínio Montagner, Celso Rossi, Enio Tiago de Oliveira, Leonardo Alves Carneiro, Maria Helena Aguiar Corazza e o apoio de meus filhos Daniel, Maria Paula e Adolfo Egídio e da minha esposa Diva. Agradeço especialmente à minha filha Carla Maisa pelo auxilio inestimável na incansável edição do texto em inglês. E a todos aqueles que compartilharam comigo as aventuras nessa caminhada pelo mundo do pensamento científico.

PHOTO 1:
OTTO E SUA ESPOSA DIVA (INICIO DA DÉCADA DE 1980)

PHOTO 2
DIVA, MARY BEATRICE, AVÓ DE ROD E ROD. JARDIM
DA CASA DE OTTO. PIRACICABA. 1972

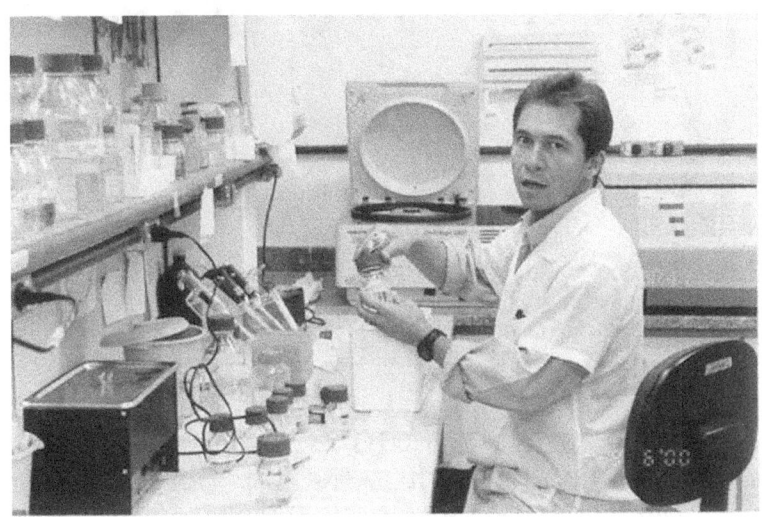

PHOTO 3
ENIO TIAGO DE OLIVEIRA NOS
LABORATÓRIOS DO CEBTEC (2001)

PHOTO 4
VISTA AÉREA DO CEBTEC, ESALQ, PIRACICABA, SP, BRASIL

PHOTO 5
NAUGUIRAÇÃO DA SALA "WILLIAM R. SHARP" NO PRÉDIO DO
CEBTEC. DA ESQUERDA PARA A DIREITA: ROD SHARP, OTTO,
PAULO F. CIDADE DE ARAÚJO (27 DE OUTUBRO DE 1988).

PHOTO 6
OTTO AGRADECENDO A OUTORGA DA MEDALHA
FERNADO COSTA (15 DE JUNHO DE 2012)

PHOTO 7
OTTO E SUA ESPOSA DIVA E A MEDALHA
FERNANDO COSTA (JUNHO, 2012)

PHOTO 8
REUNIÃO FESTIVA DA FAMILIA DURANTE A CELEBRAÇÃO
DE 50 ANOS DE CASAMENTO DE OTTO E DIVA, HOTEL
SOFITEL, EM GUARUJÁ, BRASIL. DA ESQUERDA PARA
A DIREITA: DIVA, ADOLFO EGIDIO, DANIEL, MARIA
PAULA, CARLA MAISA, PEDRO AUGUSTO E OTTO

PHOTO 9
CELEBRAÇÃO DOS ANIVERSÁRIOS DE ROD SHARP E OTTO
EM PIRACICANA, SP, BRASIL (23 DE SETEMBRO DE 2013

PHOTO 10
TOTEM EM FRENTE AO EDIFICIO DO CEBTEC,
ESALQ, PIRACICABA, SP, BRASIL

CHAPTER 7.

ODE TO OUR FATHER, OTTO

Adolfo Egidio Lovadino Crocomo, Carla Maisa Lovadino Crocomo, Daniel Lovadino Crocomo, Maria Paula Lovadino Crocomo, Pedro Augusto de Toledo Almeida Crocomo

1963. The year I was given the opportunity to be born in this family, thanks to this union that truly embodies intelligence, kindness, persistence, and love: OTTO JESU CROCOMO and DIVA LOVADINO CROCOMO. The man and the doctor that I have become I owe to the role models provided by these two good souls, to whom I am eternally grateful. I grew up experiencing their dedication to their studies and the education of our family. "Knowledge is an asset that can never be lost," Dad would say as I was growing up, and this valuable piece of advice has been with me ever since.

Dad, with a proud heart, I congratulate you for this remarkable book.

Adolfo Egídio Lovadino Crocomo, 50.

******

You need iron to keep you strong and protein to help you think! Dad would say to remind us, youngsters, of why we rolled out of bed to a breakfast meal with seared ox liver every morning before we embarked on our studies. Most people usually expect the eggs, but I don't suppose many kids around had ox liver in their 6-am-schoolday meals. We did, and that's how Dad introduced me very early to the world of chemical properties and their effects on our body and brain functions. Later I'd be introduced to the secret of well-combined elements: it's all in the mixing! Squeezed navel oranges combined with a hint of lime and ice cubes would make for a perfect drink at the lunch table. But not after Dad ceaselessly stirred and stirred and stirred the mixture were you served that tall glass of healthy, tangy, fresh orange juice mixed up to perfection to quench your thirst during those typical warm days in our hometown, Piracicaba. What a delight! We all confirmed. This is pure chemistry—it's all in the mixing! Dad would teach us. That's what chemistry is about! He'd reiterate—hence establishing my second lesson about chemistry, which I apply to this day to my cooking. It's all in the mixing.

I've gained a lot from my father over the years. Growing up with a scientist father very early opened up the world of knowledge before my eyes. From childhood to my adolescent years, I saw my father coming and going from trips to the airport that would take him to the scientific world overseas. Listening to his stories from the foreign lands cultivated in me the curiosity of the unknown. Witnessing Dad's accomplishments throughout his successful career taught me that knowledge is power. Following his encouragement to pursue a career taught me that dedication pays off. Having long conversations with Dad about the chemical properties in foods has helped me to further understand the human body and maintain a lifestyle geared toward optimum balance. Having Dad as a father has inspired me to strive to keep up with today's fast-changing world where chemistry continues to be in all the corners you dare to look.

For all the things that my father has taught and shared with me, I have feelings of gratitude and hope—hope that this book will inspire others, and they too will dive into the waters of knowledge. Congratulations, Baba!

A Proud Daughter,

Carla Maisa Lovadino Crocomo, 46.

******

I was invited to write a testimonial about my living with this extremely vital and important person to me, my father. With great honor, I will try to put my perception into words and add a few lines to this wonderful biographical work by the distinguished scientist Otto J. Crocomo, who is also the father of three sons and two daughters, and a grandfather.

I'm the youngest and I'm sure my siblings, including the oldest, Marco Augusto "Maguto," who has been among us as light for 23 years, will agree with every single word I say.

My father, my hero. Some say that as we grow up the figure of the father as a hero will naturally fade to be more realistic, closer to what we, the offspring, turn out to be: human beings in the eternal process of maturation in search for understanding, in the unconscious search for emotional intelligence and harmony, in love and for love. For such abundant love and admiration, I cannot say that this deconstruction happened to me. My father and my mother are still my heroes, and that is so because they have never given up on us.

While coexisting with my father since childhood, I've seen in my scientist father a dedicated, vivacious and optimistic being ahead of his time, a devourer of books, and a person with a sui generis ability to merge expertise with giving, strictness with generosity. I've witnessed a person who laughs, listens, and counsels intellectually and emotionally, interested in human relations, in being critical, empathetic, loving, poetic, and a dreamer who follows with his light spirit the development of future generations.

With so many personal and professional accomplishments as we find in this book and being recognized both nationally and internationally, this man has in his veins the science of life in a poetic, authentic, and admirable way. "Never put off that which you can do today" is one of the advice phrases that he customarily repeats to me, and which sums up one the pillars of his success.

I will confide one last impression to you. The union of science and faith is part of a whole, and that has made up the education my parents imparted to us. In this regard, I emphasize here that this force is, has been and always will be in the love cultivated in this union: My mother, faith and my father, science.

This work is a merited celebration of his life and a gift to readers. Dad, I am grateful for everything.

With love from your son,

Daniel Lovadino Crocomo, 40.

******

How is it to be the daughter of a scientist? It's to feel constantly inspired by his vivacity and wit. It is like having a tutor at your fingertips, 24 hours, in eternal learning about the world, the human being, history, science, arts, music, and philosophy. It's to admire his daily achievements while watching their development in a real network of scientific romance. It's to pray every night for his safe return from all the many trips he has made. It is to be part of a team led by an endlessly generous family. I thank you, Dad, for this work of art!

Maria Paula Lovadino Crocomo, 48.

******

I feel privileged to have the best grandpa in the world that is also my buddy. His name is Otto Jesu Crocomo, a scientist renowned for his competence and intelligence. He is a person with strong features, well known for being studious and determined. Although a serious and intellectual being, he always carries a smile on his face making me laugh several times a day. With his infinite kindness, he has taught me to never give up on my dreams saying that it is never too late to start again. A good man, he never forgets of his only grandson bringing him, at the end of the day, the simplest treats such as a bar of chocolate—simple actions that I will always remember. There have been countless times when he drops everything to help me with my homework or to quiz me before an exam. I don't have enough words to describe what my grandfather means to me—without him life would have no meaning. Thanks, Grandpa, for all your advice and for the wonderful being that you are. A big hug from your only grandson,

Pedro Augusto de Toledo Almeida Crocomo, 15.

CHAPTER 8.

ODE AO NOSSO PAI, OTTO

Adolfo Egidio Lovadino Crocomo, Carla Maisa Lovadino Crocomo, Daniel Lovadino Crocomo, Maria Paula Lovadino Crocomo, Pedro Augusto de Toledo Almeida Crocomo

1963. O ano em que foi me dado a oportunidade de nascer nessa família como fruto dessa união que personifica verdadeiramente a inteligência, bondade, persistência e o AMOR: OTTO JESU CROCOMO e DIVA LOVADINO CROCOMO. O homem e médico que me tornei devo ao exemplos dessas duas boas almas, pelas quais sou eternamente agradecido. Cresci vivenciando a dedicação deles aos estudos e à educação da família. "O conhecimento é a única coisa que não se perde na vida", dizia meu pai ao me ver crescer—e essa máxima ainda caminha em meus pensamentos.

Pai, com o coração orgulhoso lhe parabenizo por esse livro memorável.

Adolfo Egidio Lovadino Crocomo, 50.

******

Voces precisam de ferro pra ficarem fortes! E proteína para pensarem bem! Papai dizia a nós, pequeninos, para nos lembrar do porque saíamos da cama para uma refeição que incluía fígado de boi todas as manhãs antes de embarcarmos em nossos estudos. A maioria das pessoas é acostumada com ovos nas refeições da manhã, mas eu não acho que muitas crianças ao redor tinham fígado como parte da refeição das 6 horas da manhá em dias de escola. Nós tínhamos. Foi assim que meu pai me apresentou muito cedo para o mundo das propriedades químicas e seus efeitos em nossas funções fisicas e cerebrais. Mais tarde eu seria apresentada ao segredo dos elementos bem combinados: é tudo uma questão de misturar bem! Laranjas baianas espremidas combinadas com um toque de limão e cubos de gelo fariam as vezes para uma bebida perfeita na mesa do almoço. Mas somente depois que meu pai incessantemente mexia e mexia e mexia a mistura que você seria servida em um copo grande de suco natural, saudável e refrescante de laranja misturado à perfeição para saciar a sua sede durante aqueles dias quentes típicos da nossa cidade natal, Piracicaba. Que delícia! todos nós confirmávamos. Isso é pura química—é tudo uma questão de misturar! Papai nos ensinava. Isso é química! ele reiterava—estabelecendo-se assim a minha segunda lição de química, lição essa que atualmente aplico ao meu modo de cozinhar. É tudo uma questão de misturar bem!

Eu aprendi muito com meu pai ao longo dos anos. Crescer com um pai cientista muito cedo abriu o mundo do conhecimento diante dos meus olhos. Dos meus primeiros anos de infância à minha adolescência, eu vi meu pai indo e vindo de viagens para o aeroporto que o levariam para o mundo científico no exterior. Suas histórias sobre as terras estrangeiras cultivaram em mim a curiosidade do desconhecido. Ao testemunhar as muitas realizações do meu pai ao longo de sua carreira bem-sucedida, aprendi que conhecimento é poder. Seu encorajamento para que eu seguisse uma carreira me ensinou que dedicação vale a pena. As nossas longas conversas sobre as propriedades químicas encontradas em alimentos continuam a me ajudar a melhor entender o corpo humano e manter um estilo de vida voltado para o equilíbrio ideal. Ter um pai cientista me motiva a me esforçar para me manter atualizada com o mundo acelerado de hoje, onde química continua a ser encontrada em todos os cantos que voce se atrever a olhar.

Por todas as coisas que meu pai tem me ensinado e compartilhado comigo, eu tenho sentimentos de gratidão e esperança— esperança que os contos deste livro irão inspirar outros e esses também mergulharão nas águas do conhecimento. Parabéns, Baba!

Uma filha com orgulho,
Carla Maisa Lovadino Crocomo, 46.

******

Fui convidado a escrever um pequeno depoimento sobre a minha convivência com essa pessoa extremamente importante e vital para mim, meu pai. Com muita honra tentarei traduzir em palavras o meu olhar, e assim somar algumas linhas a esta maravilhosa obra biográfica do ilustre cientista, Otto J. Crocomo, pai de tres filhos e duas filhas e avô.

Sou o caçula, e tenho certeza que meus quarto irmãos farão deles as minhas palavras, incluindo o mais velho, Marco Augusto "Maguto", que está entre nós em forma de luz há 23 anos.

Meu pai, meu herói. Alguns dizem que ao crescermos a figura do "pai herói" naturalmente vai se esvaindo para uma figura mais real, mais parecido ao que nós, filhos, nos tornamos: seres humanos no processo eterno do amadurecimento, na busca por entendimento, indagações, na busca inconsciente pela inteligência emocional, por conquistas, harmonia, no amor e por amor. Por tamanho amor e admiração, não posso afirmar que tal desconstrução ocorrera comigo. Meu pai e minha mãe são ainda meus heróis e só o são porque nunca desistiram de ser.

Na convivência e sob meus olhares, desde criança vejo em meu pai, e neste cientista, um ser estudioso, dedicado, vivaz e semeador do otimismo, à frente de seu tempo, devorador de livros, com uma capacidade *sui generis* de mesclar competência com doação, de mesclar firmeza com generosidade. Uma pessoa que ri, que escuta, que aconselha, que sabe o que faz, inteligente também emocionalmente interessado nas relações humanas, crítico, compreensivo, amoroso, poético e sonhador, e que segue com seu espírito leve a evolução das gerações.

Com tantas conquistas pessoais e profissionais como constatamos neste livro e reconhecido tanto em seu país como internacionalmente, este homem tem correndo em suas veias a ciência da vida numa forma poética, autêntica e admirável. *"Nunca deixe para amanhã o que se pode fazer hoje"* é uma das frases que costumeiramente repete a mim, e que soma-se aos pilares do seu sucesso. Gostaria de confidenciar uma última impressão. A união da ciência e da fé é parte de um todo, e compõe a educação transmitida à nós, seus filhos, e neste sentido sublinho que esta brava força está, esteve e sempre estará no amor cultivado desta união: minha Mãe a fé, meu Pai a ciência.

Esta obra é um marco histórico não somente de sua vida, mas um presente registrado aos leitores.

Pai, sou grato por tudo, com amor, seu filho,
Daniel Lovadino Crocomo, 40.

******

Como é ser filha de um cientista? É sentir-se inspirada constantemente pela vivacidade e sagacidade dele. É como se ter um tutor ao alcance das mãos, 24 horas, num eterno aprender sobre o mundo, o ser humano, a história, as ciências, as artes, a música e a filosofia. É admirar suas conquistas diárias acompanhando os acontecimentos numa real trama de romance científico. É rezar, todas as noites, pelo retorno seguro de todas as inúmeras viagens que ele faz. É fazer parte de uma família liderada por uma generosidade sem fim. Obrigada, Pai, por mais essa obra!

Maria Paula Lovadino Crocomo, 48.

******

Tenho o privilégio de ter o melhor avô do mundo em minha companhia. O seu nome é Otto Jesu Crocomo, um cientista renomado e conhecido internacionalmente pela sua competência e inteligência sem fim. Ele é e sempre foi uma pessoa com características muito fortes, conhecido por ser muito estudioso e determinado em tudo o que faz. Mas ao mesmo tempo que é uma pessoa séria e intelectual nunca deixou de andar sempre com um belo sorriso estampado em seu rosto me fazendo dar altas gargalhadas várias vezes ao dia. Com uma bondade sem fim, me ensinou a nunca desistir dos meus sonhos me dizendo que nunca é tarde para começar. Um homem boa praça, ele nunca se esquece de seu único neto, sempre lhe fazendo agrados mesmo com as coisas mais simples que possa existir, como um chocolate, ao final da tarde quando chega de seus afazeres—coisas simples que ficarão sempre marcadas em nossas mentes. Quantas vezes já deixou de fazer coisas importantes para me ajudar a estudar para uma prova ou uma tarefa onde tinha dificuldade. Não tenho palavras suficientes para descrever o quanto o meu avô significa para mim, pois sem ele a vida perderia o sentido. Muito obrigado por todos os seus conselhos e por ser esta pessoa maravilhosa e a cada dia mais. Um grande abraço do seu único neto,

Pedro Augusto de Toledo Almeida Crocomo, 15.

CHAPTER 9.

A Life in Plants

David Lee

Many people, activities and institutions, particularly while growing up, strongly influenced my development as a plant biologist and university professor. At different stages of my career I have reflected on those influences, but now I have now finished my formal work as a scientist and professor and especially welcome the opportunity to reflect on my life. Writing this chapter is an opportunity to do just that, contemplating with more than the usual focus and discipline, easy to skip in the normal day to day activities, even in retirement. Writing it also gives me another opportunity to honor those individuals who helped me along the way, and adds to my feelings of humility and gratitude.

CHILDHOOD

I was born and grew up on the Columbia Plateau of Washington State. My mother and father, John and Mary Lee, moved to the small town of Ephrata, which was the seat of Grant County, recently married and starting a new life together. My mother's family was originally from Seattle, but moved frequently in the state following the work of my grandfather as a civil engineer. She came from English and German background, from ancestors who lived in Pennsylvania in the early 19th century. My father's family moved to the town of Chelan, from Michigan and Illinois, in 1920. I was told that we came from the Lees of Virginia, but genealogical research by a distant relative indicates that our ancestor, William Lee, arrived in New York around 1675. A look at my family tree shows the good and bad sides typical of a family; a family name is Webster (my middle name) after Noah Webster, another family name is Hawkins (my dad's middle name) after the notorious pirate. My father's family moved to the town of Chelan, from Michigan and Illinois, in 1920. My grandfather practiced dentistry and was unsuccessful in starting an orchard.

Both families moved to Grand Coulee to find work associated with the dam construction in 1933, and my mom and dad met there. My father came to know Jim O'Sullivan, who was an early booster of the dam and associated irrigation project (for which Ephrata became the headquarters). Dad ran a small luncheonette near the movie theater in Grand Coulee, and he noticed that people could find the money to escape to the fantasy of movies, even in the depths of the great depression. So they moved to Ephrata, purchased and renovated the local movie house, the "Kam" renamed the "Capital", and started the "sagebrush circuit", showing movies in grange halls in the small dry land farming communities in the area. The business prospered, and my parents eventually owned and ran a chain of some twenty theaters and drive inns in that part of the state, the Columbia Basin Theaters.

I was born in Wenatchee, the largest nearby city, in December of 1942; there was no hospital in Ephrata until three years later (and my father was instrumental in its establishment).

I grew up in this little town, in a small house, later enlarged, on the edge of town, Ephrata was an isolated but progressive place in the middle of the sagebrush steppe (or cold desert) of the Columbia Plateau. It served outlying agricultural activity and consisted of retail business, support of the Great Northern

Railway (including a row of grain elevators), county government employees, and the skilled professionals who designed the
Columbia Basin Project, one of the largest irrigation projects in
American history. Ephrata was known for its excellent school system, and the town was like an oasis in a harsh treeless landscape.
As a child, I had a bicycle and the means to travel anywhere in
town, as well as into the dry washes (alive with salamanders and
frogs in the spring time) in the countryside. During the summer
months, my friends and I would spend entire days together away
from our homes and families. The landscape, with the Grand
Coulee just north of town, was a powerful presence: blue cloud-
less summer skies, hot days and cold nights, dark basalt canyon
walls with flecks of lime green and orange lichens, and mineral-
rich lakes on the canyon floors. As a child, I did not have any real
mentors leading me towards a career in biology, but was strongly
influenced by my personal experiences in nature with my friends.

There were some strong themes in my childhood, associated
with the institutions that were in the town and prepared children for adulthood. These institutions were the schools and the
churches, and organizations for children and youth: cub and boy
scouts, girl scouts, and Future Farmers of America.

I was the middle child, with an older brother and younger sister. There were fights among us kids, but we got along pretty well
and stay in contact with each other today. It was a stable family
situation. My mother took care of the home and family, and also
helped my dad with financial aspects of his business. My father
worked hard to make his businesses successful and that made him
a little remote from us. He also performed a lot of volunteering in
the community. He was the first president of the hospital association and a leader of the Lions Club campaign to raise funds for
the first local hospital. He was active in the Lions, serving as chapter president when they built the first park and swimming pool.
My mother and father also were active in the concert association,

which brought performing artists to Ephrata, with concerts in the Lee Theater. Eventually, my dad served on the city council, and then served as mayor. Neither of my parents completed college, my mother went to a business school for a few months, and my father attended the University of Washington for two years before his money ran out. At 21, he was running a small mercantile and dry cleaning business, with two stores. By 23, after the beginning of the Depression, he was bankrupt. Despite his business background, he was a lifelong Roosevelt Democrat. My parents often joked that their votes cancelled each other. Education was important in the family, and each of the three kids graduated from college.

I attended the public schools in Ephrata, first Parkway Elementary, just a half block from our house (and I was frequently observed still polishing off the remains of my breakfast while walking into my classroom!). Then I attended junior and senior high school at the north end of the town. Early in the third grade, our teacher, Miss Storm, took the children to the town library. We were allowed to check out a single book. Mine was about a Dutch child who stuck his finger in the dike and held back the flood waters to save his town. It sounds pretty stale now, but it was a revelation, that a book could open up a new world to me. From that time on, I read voraciously. In the fourth grade, with crowded schools, my classroom was the Parkway School Library. I remember reading Raymond Ditmars *Snakes of the World*, and *Reptiles of the World*, which I discovered on the library shelves. By the end of the fourth grade I read books at an adult level. I read widely, but particularly enjoyed mystery novels for kids and science fiction. A notable aspect of my childhood, even fairly unusual in our town, was the absence of television in our household, until my senior year of high school. TV came to Spokane around 1950 and could be picked up in Ephrata with a high antenna, but my father refused to own a set for many years; TV was the single biggest threat to his movie business. However, in those days we had two movie houses owned by my dad, the Marjo (short for Mary and Johnny) and the premier Lee Theater, with two changes of double bills per week plus the Saturday afternoon kids' matinee—so I saw 4-5 movies every week.

I was not a great student, and was even a pest in junior high school. When I began to take studies more seriously, in high school, I did not excel. In my graduating class, and I am still friends with many of my classmates, I was a "B" student; some of my friends were quite bright and later attended some prestigious universities, including MIT, Yale, The Air Force Academy, and Cal Tech. I took the normal pre-college curriculum of Biology, Chemistry, Physics, a language (French), pre-college English, and pre-calculus mathematics. I had some notable teachers during that time. Mr. Bob McIntee was a well-trained and excellent chemistry instructor, and very enthusiastic about science. Mr. Bob Atkinson, who was also my basketball coach, taught world affairs and very carefully included a unit on Marxism (unusual during those times of McCarthyism). I had excellent English language instruction. My senior instructor, Miss Gerhardstein, was a literature Ph.D. candidate at the University of Washington. My writing was good enough for me to win the writing award for my class during graduation, but that was partly the result of the awards being spread around. Otherwise, Larry Reeker who finished Yale in three years and became a respected expert in artificial intelligence, would have one all of the awards.

An important activity for kids growing up in Ephrata (as in all towns in our region) was sports. I had some older neighbors who had excelled in high school sports, whom I idolized. My brother, three years older than I, was active in football, basketball and track, and I followed his footsteps in those sports. I was a bit overweight in junior high school, and didn't begin to have much success until the 9th grade, when I grew several inches. At 6' 2 1/2" and 185 pounds, I was the tallest of the athletes at Ephrata High School—the Ephrata Tigers. I played end on offense and defense in football, center on the basketball team, and ran hurdles, 220 yard dash, and even competed in high jump for the track team. I played those sports for three years, and lettered my junior and senior years—was the captain on the football and basketball teams. I was very serious about sports, and from that participation I learned much about effort, focus, patience, endurance, and teamwork. These were qualities that later became important in graduate school and my career in scientific research. During that career I saw many colleagues, who I gauged as much brighter than I, fall by the wayside for lack of those qualities.

As all small towns in the region, Ephrata was a religious town, with far more churches than taverns, and lots of protestant denominations. The one family with Jewish cultural roots, the Agronoffs, eventually became Episcopalians. My family was not particularly religious, and we did not belong to any denomination when I was a young child. However, we shared a backyard boundary with a young Lutheran Minister, Ray Pfleuger ("Pastor Ray"), and he frequently was invited for dinner. He established a congregation, and the church was constructed in our neighborhood, so our family became Lutherans (of the more progressive ALC variety). I was baptized in this church when I was about 10 years old, and was confirmed in the church at 13. I had many deep questions about death and the immensity of the universe (which was easy to imagine from the star-filled skies above our town), and got a little comfort from Christian devotion....for a while.

I had many experiences in nature, both near the town and further afield, growing up. As a child, I was active in the Scouts. In Boy Scouts, we had a great troop leader, who organized camping trips on the Columbia Plateau and in the Cascades. I have vivid memories of the places we visited. However, a new troop leader took over, insisted on treating us like little soldiers, and

I quickly left. During the summers, many from my town went to the Lake Wenatchee YMCA Camp, deep in the Cascades and along the shore of Lake Wenatchee. I first was a camper for many years, and then helped out in the kitchen and as an assistant counselor as a young adolescent. My experiences at the camp, hiking and studying nature, deeply affected me. The YMCA movement had a religious/ethical program installed at these camps that combined Christianity and the native American reverence for nature. The programs involved accepting a challenge of behaving nobly and receiving a bandana (in colors denoting different stages) in a sacred ceremony. It had a profound effect on me, much more than learning the tenants of Lutheranism. My father was a very busy man, but he did take my older brother Jack and me on fishing trips, as up the Sanpoil River, and on a long trip up the Alaskan Highway to Fairbanks. We drove up on an old Dodge truck with a heavy canvas cover on the bed and side stakes—and three bunk beds. We came back with a booth of movie projection equipment he'd bought from the Air Force on auction. We stopped to sit in hot springs and fished for lake trout and grayling. I grew up a lot during that trip. We also took some family vacations, as to the Oregon coast and Glacier and Mt. Rainier National Parks, which exposed me to the beauty of those sublime landscapes.

I was part of a close group of friends, first as a child and then as an adolescent during the high school years. We talked about all sorts of topics, including the existence of evil, free will, time in the universe, utopias, and on. As we began to drive (a few owned old cars) we went on back packing trips into the High Cascades Mountains, to the west. Often, we'd hike up over a mountain pass to a high lake on Friday evenings, using carbide miners' lamps to light the way, and then return on Sunday afternoon. Occasionally the trips were a bit longer. The scenery was majestic, and inspired our conversations about the nature of the universe.

I worked summers in an aluminum fabricating business my father had started, manufacturing and installing commercial and storm windows and doors. I learned how to use many tools, including welding aluminum. This experience was invaluable in giving me some manual skills that were important to me later on in scientific research.

COLLEGE YEARS

I graduated from Ephrata High School in spring of 1961. All of my friends were bent on attending college, and so was I. I had written a letter to the Chairperson of the Botany Department at Washington State University, asking questions about carnivorous plants. He had sent a long typewritten letter in reply, very impressive to me—and it was natural for me to be interested in attending Washington State. However, I had attended a senior open house weekend, and all of the students were drunk. It didn't seem like the place for me. In my junior year, I had been selected as a delegate to attend Boys State, which was held on the campus of Pacific Lutheran University, just south of Tacoma. I was leaning towards studying biology, and PLU had a good reputation in the sciences, so I decided to attend there. It was a good place for me, with small classes and a couple of inspiring teachers.

My freshman year I took Zoology I and II, taught by a parasitologist, Prof. Keith Strunk. This rigorous course (and I found the parasite life cycles fascinating) was the portal for the selection of pre-med students, and Prof. Strunk was their mentor. I received a high "A" and was invited to an interview with this iconic figure. When I told him that I was thinking about studying plants and certainly not medicine, a look of incredulity and disgust took over his face and the interview was cut short. I had excellent chemical background from high school and was invited to take an experimental general course, with Dr. William Giddings, a recent Harvard Ph.D., in which I spent much of my time devising experiments; it was a great entry into science and the scientific method. Most impressive among the biologists was Jens Knudsen, who was a zoologist with a love for nature and teaching, and a research program in marine invertebrates. The summer of my sophomore year, I attended a field course taught by Dr. Knudsen and his colleague Dr. Harold Leraas, a very kind elderly man, at Holden Village, high in the Cascades off of Lake Chelan. Holden Village was an old mining town converted into a Lutheran retreat site, on the edge of the Glacier Peak Wilderness and accessed by road from a remote landing on the lake. We had lectures, but spent much of our time hiking and camping in the high cascades, learning the plants and ecology as we travelled.

PLU was a liberal arts institution, and I benefited much from a variety of courses: British Empirical Philosophy with Curtis Huber and George Arbaugh, Asian History by Walter Schnackenberg, and American literature by Martin Hillger, among others. I retain an interest in these subjects to this day.

I did not return to school the fall of 1963. I was having some emotional problems; I was quite immature really, and decided to give myself some time for work and travel. This was also a time of religious yearning, not met by any experiences in Christianity, and I was reading theology: Bergson, Teilhard de Chardin, The Varieties of Religious Experience, and more. I found a job at the Weyerhaeuser Paper Mill in Longview, along the Columbia River and downstream from Portland. I found a room in a boarding house and hitch-hiked to my job at the mill. It was shift work, and I could make pretty good money during holidays with overtime. I saved most of my earnings, preparing for a trip to the South Pacific, mainly New Zealand. I wanted to visit places with great natural beauty, so this trip was definitely more attractive to me than studying in Europe, as several of my college friends did. I quit the mill in January, and boarded the S.S. Oronsay in Vancouver for an 18-day voyage to Auckland. We had layovers in San Francisco, Long Beach,

Honolulu, and Suva (Fiji Islands). Several of us took advantage of these stops to visit areas away from the port. Arriving in Auckland, New Zealand, I soon got on the road, hitch-hiking up and down the North Island, down the South Island, even to Stewart Island. I saw much of the great scenic beauty of this island nation, and also studied botany and zoology for one term at Victoria University of Wellington. I left in July, flying to the Fiji Islands and Society Islands, then Hawaii, on my way back to the U.S. In Fiji, I travelled up to a hill station on the largest island, Vita Levu, and ascended the highest peak, looking for the famous *Degeneria* tree (thought to be the most primitive angiosperm at the time). I didn't find it, but travelled through a magical cloud forest to the summit and enjoyed a Kava ceremony with some elders in a nearby village. After my return, a young British couple who were volunteering as foresters, invited me for tea. They said to me, "you can do what we're doing." They made a great impression on me, and I decided to become a botanist and study plants in different parts of the world, helping people along the way. I wasn't sure how one did this, but knew I had to study plants in earnest.

I had many opportunities to travel in mountain wilderness areas as a college student, both in New Zealand and in Washington State, and experiences during some of those sojourns were transformative. In July of 1962, I hiked up the Surprise Creek Trail in the Alpine Lakes Wilderness Area, and then joined the Pacific Crest trail to switch back up onto the ridge above Trap Lake. Just north of the trail on the ridge crest was a boggy area with low mounds of sphagnum moss amidst the subalpine setting and wildflowers. Time stopped, the vegetation began to glow with its own light, Glacier Peak beckoned to the north, and I was transformed. In April of 1964, I hitchhiked to the end of the road on the west coast of the South Island. Then I followed trails set up by surveyors (a highway now traverses the area) towards the road terminus to the south). It was late in the day. There were several trails and I got lost. Then it began to rain. And I became ecstatic; time stopped and the setting became luminous. Eventually, I found my way to the terminus and stayed the night in an abandoned house, and the memory of that experience stays with me now, just over a half a century later. These powerful experiences added to my motivation to study plants and do research in nature, a big part of the mix of motivations that led me to become a professional botanist.

So, I returned to study at PLU, finishing my science requirements and took the plant courses that were offered. I wanted to do a research project, and the physiology professor, Earl Gerheim, suggested that I start a project studying evolution by comparing the proteins of plants using immunological techniques, techniques he knew something about. I had been struck by the beauty of members of the heather family (Ericaceae) in my studies in the Cascades the previous year, and then saw other taxa in the family in the high mountains of New Zealand. I read about the geography of this widespread family and decided that I would study the evolution in this family using immunology of plant proteins. My approach was pretty crude, even by the standards at that time, but I worked entirely on my own (well, I got some help in injecting and bleeding rabbits!), conducted experiments and wrote up a report. I was given a key to the science building, so that I could use the equipment at any time of the day. In doing this research, I learned that it was conducted in a few other places, mainly in Europe, but notably in the laboratory of Prof. David Fairbrothers at Rutgers University, in New Jersey. I wrote to him, describing what I had been doing and asking some questions about methods. He quickly replied, answering the questions and suggesting that I look into the possibility of continuing this research as a graduate assistant in his laboratory.

GRADUATE STUDIES AT RUTGERS UNIVERSITY

In January of 1966, David Fairbrothers met me at the plane in a driving snow storm in Newark, and took me home to spend my first days with him. Then I quickly settled in graduate student housing and started my graduate studies as his student. I was a teaching assistant in biology during my first semester, but was supported as a research assistant for most of my five years at Rutgers. I began to continue research on the seed proteins of members of the Ericaceae, but found it impossible to extract proteins from their tiny seeds, and I abandoned that project after eight months of hard work. I took advanced courses in immunology and biochemistry, plus plant systematics from David Fairbrothers. I became friends with several very human and supportive professors, whether I had a course from them or not. Jim Gunckel instructed me in anatomy and microtechnique. Carl Price instructed in Physiology and Plant Biochemistry, Barbara Palser (who was a member of my graduate committee) in plant morphology and embryology, Charlotte Avers in cytogenetics and Ovid Shifriss (also a committee member) in the evolution of domesticated plants. I liked all of these courses, particularly because the instructors were active in research and expressed their enthusiasm in the material. I didn't take courses in ecology, but was surrounded by a very strong graduate ecology program. I knew Murray Buell, Jim Quinn, Paul Pearson, and hung out with a number of ecology graduate students. I attended the excellent ecology seminar series every week, and learned quite a bit about ecology through this exposure. Richard Forman, today acclaimed for his work in establishing landscape ecology as a legitimate field of study, arrived my second year of studies. I took bryology and lichenology from him, which involved a long field trip to West Virginia, and I helped with some of his field research on Mount Washington, in New Hampshire. A decade later we renewed our friendship when we both pursued our research interests in Montpellier, France.

Although the coursework was useful, I learned most from talking (and arguing) with my fellow graduate students, quite a diverse crew. Some new friends from my graduate immunology course took me under their wings to teach me something about urban life, classical music, and Jewish culture (of which I was totally ignorant). Much of our conversations were in labs and corridors rather late at night, In Nelson Laboratories. I met and became friends with Rod Sharp at this time. I also became friends with the Lutheran Chaplain and his wife, Warren and Joan Strickler, organizing a tutoring program for disadvantaged kids and attending organ and choral concerts of sacred music. It was a very interesting and stimulating period for me as a graduate student. Although my research involved local collecting and lots of laboratory work, I was able to take some time to travel to the tropics (Puerto Rico and Venezuela) and walked in tropical rainforests. These experiments helped sustain my interest in studying plants, counterbalancing the grind of long hours of laboratory research. My period of graduate studies was a time of great social upheaval in the United States, catalyzed by protests over the war in Indochina and civil rights issues in the U.S. Race issues had not been important to me while growing up, because I hardly knew any black students, other than a couple of athletes from the nearby town of Moses Lake during my high school years.

My graduate study years were important in my emotional maturation, and also in developing some social confidence in meeting and working with a variety of people, from diverse cultural backgrounds. I became aware of the gay and lesbian community, and my tolerance of different people and lifestyles was greatly expanded. I was active in the Graduate Student Association, playing in their basketball league, attending the weekly receptions (free beer) and editing the Graduate Student Guide.

I switched my research from the phylogeny of the Ericaceae to a masters project on hybridization and introgression in cat-tails

(*Typha*), which expanded more broadly in the monocots, and to the study of the inheritance of isoenzymes for my Ph.D. In addition to the immunological techniques, I acquired other techniques of protein identification and purification, particularly polyacrylamide gel electrophoresis (PAGE as we now know it) and isoelectric focusing. I expanded my research to include the inheritance of seed proteins in castor beans, using varieties provided by Prof. Shifriss. The use of electrophoresis and the study of multiple forms of enzymes were very recent developments in biology, and our lab was on the forefront of this research. My dissertation committee consisted of David Fairbrothers, Barbara Palser, Oved Shifriss, and Nicolas Palczyuk (an immunologist).

My advisor, David Fairbrothers, was an excellent mentor. He was an accomplished scientist, a leader in the University, and led an active research program with several graduate students, undergraduates, and occasionally a senior faculty member visiting on sabbatical leave. Despite his busy schedule, he met regularly with us, and was available for personal advice when it was really needed. When I was finishing my dissertation, the University shut down over protests against the invasion of Cambodia. There was such turmoil that the completion of my Ph.D. seemed meaningless. When I talked to David about this he related to me an event in an all-university faculty meeting just held about the shutdown. A young faculty member stood up and declared to all present that he was fed up with the attitude of the administration and the stance of the University, and declared that he was resigning. Another even younger faculty member near him also declared that he, too, would resign. Later on, David learned that the first speaker had been denied tenure and was leaving anyway, but not the second speaker! The story was shared to encourage me to keep a balance and a vision for the longer run about things. He had suffered grave danger and privation during his service in Europe during World War II, particularly during the Battle of the Bulge. He didn't share any battle stories with us, and I only began hearing about his experiences during the last eight years, but those experiences gave him perspective and patience.

I received my Ph.D. in June of 1970, with my parents in attendance. I continued research that summer and then drove my old black '57 Chevy station wagon full of all of my possessions (mostly books and records) to Columbus, Ohio, to begin a post-doctoral fellowship at The Ohio State University.

THE OHIO STATE UNIVERSITY

I had received an offer of an Ohio State University Post-doctoral Fellowship, to work under the supervision of Don Dougall, who was an Associate Dean in the newly formed College of Biological Sciences. Within days of my arrival I helped the lab move into a brand new life sciences building, devoted to research in cell biology and microbiology. My academic home was actually the Department of Microbiology. Don did research in plant tissue culture, looking at the biochemical clues to cell and tissue differentiation in the wild carrot system. He wanted someone with some background in protein chemistry to help him identify enzymes, particularly glutamine synthetase (which he believed existed in two different forms that responded to the growth regulator treatment that led to the formation of embryos). Rod Sharp had mentioned me, and my training (my "tool kit"), and he was instrumental in my selection.

In biological research there has always been a little tension between "questions"—and hypothesis testing—and "techniques" (the tool kit part). The danger of focusing on the techniques was that I'd lose touch with the scientific reasons for conducting the research, and I was, at that time, a little critical of focusing on techniques. Yet, I was the beneficiary of practical learning of all sorts that boosted my scientific research, including carpentry and metal work. I didn't think about it too much at the time, but it was clear that the techniques I'd mastered gave me the post-doctoral opportunity I accepted, and not any basic research results I'd achieved.

My fellowship lasted one year, and Don moved on as Director of the Alton Jones Cell Science Center in Lake Placid, New York. I continued doing research a second year, and obtained support from various academic units, even the Institute of Polar Studies. I did research on the wild carrot system, looking at differences in isoenzymes, and helping Don with his glutamine synthetase research, but I did not enjoy the research and found the long hours in a windowless building to be oppressive. I could not envision myself spending the rest of my life doing such work. I did get involved in many side projects that were rewarding, such as creating multi-media programs for teaching basic biology, and I had a wide circle of friends. One of my childhood friends, Larry Reeker (the artificial intelligence guy) was a faculty member in the Computer Science Department. Through Larry, I made friends with several grad students and young faculty members. Tom Defanti, who later became an acknowledged expert for his work on internet-2 and virtual reality environments, shared an apartment during my second year. Also, my time in Ohio helped me do some additional research beyond my dissertation, and provided the time to write manuscripts that led to the three articles that originated from my Ph.D. research.

The most important event during my stay at Ohio State was meeting and ultimately marrying Carol Rotsinger. Carol was an art student, recently returned from a long trip to Europe and Asia and finishing her degree in Art Education. We met through mutual acquaintances in the Computer Science Department. We met and talked one evening in late March of 1972 at Bob Jones' house, and that brought us together; we married the following August in a remnant hardwood forest, The Gahanna Woods, on the edge of Columbus. I write this chapter after four decades of marriage.

From the beginning, Carol and I shared a formless spiritual yearning, not Christian or denominational in any sense. For Carol it was supported by her experiences from travelling in the east, for me by my experiences in nature. We became interested in the teachings of G.I. Gurdjieff, who had come to the west to help people have an experience of their Self. We heard about these teachings from two childhood friends, Allan Lindh and Curtis Amo. We began a correspondence with a leader based in Warwick, NY, where Curtis and his wife Laile had moved to participate in an intentional community, The Chardavogne Barn, under the leadership of Dr. Willem Nyland, a student of Gurdjieff's.

THE UNIVERSITY OF MALAYA

As a graduate student, I had corresponded with Benjamin Stone who was an American plant systematist working at the University of Malaya, in Kuala Lumpur. Ben was quite a remarkable man and was the world's expert on the classification and evolution of the Pandanaceae, a family of ecological and economic importance in Asia. I had needed seeds of members of this family for my immunological work, and he had helped me. He had also wondered if I might be interested in a faculty position in experimental taxonomy at the University of Malaya. This idea stayed with me and I begin to pursue this possibility as a post-doc. In 1972 I received an offer of a lectureship and tentatively accepted it. There was a long process of visa application that took time, and I was anxious about the reality of this actually happening as the wait stretched into months. I was attracted by the possibility of studying plants in the rainforests of tropical Asia, and also by the cultural and spiritual traditions of the east. The position at the University of Malaya was not only a work opportunity for me, but also a life for a newly married couple. We arrived in Kuala Lumpur in February of 1973.

When I went to the campus to begin my work I found a mixed faculty of European/ American expatriates, along with the Chinese, Malay and Indian Malaysian faculty. The Chinese were shocked and a little dismayed to see me, David Lee, of European descent; they were certain that a Lee would be Chinese-American. The Malays were quite pleased. It was a good faculty, the students were bright and hard-working and, at that time, represented the cultural diversity of the country. My teaching load was quite modest; I had a nice laboratory, and a young Chinese-Malaysian woman assigned to me as a laboratory assistant. We lived in university-subsidized housing a short distance from campus. My knowledge of the natural history of Malaysia was primarily derived from reading Alfred Russel Wallace's *The Malay Archipelago*, but I learned quickly from my colleagues and frequent field trips.

It was an interesting time in Malaysia. It was a new country, having split from Singapore a few years earlier. There were also racial tensions, primarily between Chinese and Malays, that had led to riots four years earlier. In the aftermath, there were sincere efforts of reaching out, particularly manifested during the cultural celebrations of the separate communities. We saw the seeds of Islamic fundamentalism taking root, particularly in the more rural east coast states, but the Islam of Malaysia in the 1970s was tolerant. There was a strong British influence in local institutions, befitting its colonial past. I understood the phrase "red-tape" (bureaucratic inefficiency) when I visited government offices and saw the shelves over-loaded with bundles paper.....all bound in pink or red fabric tape. I joined the Hash-House Harriers, an old running association. We met at regular intervals to follow a paper trace (with numerous false leads, and there's another phrase rooted in British tradition) through secondary forest, oil and rubber plantations, and scrub (belukar) eventually returning to our starting point, where a truck full of iced beer was awaiting. We inherited a dog, Pooch, who was an institution on the "Hash." Pooch was with us until shortly before we left Malaysia.

The shock of being in such a new place, foreign and yet quite westernized, far from family and friends, was hard on our new marriage, but we weathered the storm and had a really wonderful four years in Malaysia, with occasional trips to India, Indonesia (Java, Bali and Sumatra) and Thailand. Carol found enjoyable work in teaching printmaking to fine arts students at the Mara Institute of Technology (the MIT of Malaysia!). My mentors in learning about the biodiversity in Malaysian rainforests were Benjamin Stone (for plant diversity), Brian Lowry (phytochemical diversity) and Peter Ashton (of the University of Aberdeen and a frequent visitor, plant diversity and ecology). With frequent trips to rainforests in the mountains east of KL, I gradually became more familiar with the diversity and more comfortable in exploring the forests. Each year, faculty and the 20 or so B.S. Honors students would organize a research trip to a rainforest area to teach learn botany and ecology, and collect plants. These were also a valuable mechanism for me to learn about plants and natural history, by listening carefully to my colleague's discussions with students, as Engkik Soepadmo on plant diversity and Ratnasabapathy ("Ratna") on limnology.

Although I had been selected for the post because of my research expertise in cutting-edge techniques to address plant systematic problems, I became more and more interested in the general phenomena that I saw in the forest: (1) the physiognomy of understory plants adapted to very low light conditions; (2) the presence of iridescent blue plants in the understory; (3) the production of brilliant red young leaves; (4) the frequent presence of red undersurfaces of understory plants, (5) the presence of lectins in seeds of legumes, and much more. Gradually my interests turned from systematic botany (although I published a couple of systematic papers from work there) and moved in the direction of plant functional ecology. That meant learning more ecology (my experience at Rutgers among ecologists gave me a good start), and learning new techniques. Basically, my experiences in Malaysian rainforests gave me the research questions that I pursued for the rest of my scientific career.

After a year of living in KL, Carol and I decided to move to a rural valley south of the city, Ulu Langat. A couple of faculty friends lived there, and we learned about an empty schoolmaster's house (the "rumah guru besar") in Kampong Sungei Serai (the village of the lemon grass stream). I pursued its rental within the bureaucracy of the Ministry of Education and eventually persuaded them to rent the small house to us. We occupied it for about 6 months before the electricity was turned on. It was a great move. There was secondary rainforest behind our house and the protected forest of a drinking water catchment within walking distance. At the head of the valley there was an old British hydro-electric project that provided walking trails up to the points of water intake and into the hill forests and the highest mountain, Gunong Nuang, with moss forest at its summit. Near the head of the valley there were a couple of villages of orang asli, or aboriginal people. These were the Temoin, and they became important sources of information, as well as guides for more extensive trips into the forest. I set a goal of learning all of the walks to places of scenic beauty and of natural historic value in the valley, and our dog Pooch was an invaluable companion on those trips.

We became involved in various affairs in KL and the university. Carol's work in the arts led to contacts with local artists. I became a board member of the Malayan Nature Society, which is now the *Malaysian* Nature Society and the most important nature conservation and education organization in the country. I went on, and eventually led, nature walks in different areas near the city. Ben, who became a great friend and mentor, was in the process of establishing a new botanical garden, Rimba Ilmu, on the edge of the University of Malaya campus. I helped, particularly with the symposium held at its dedication. I was able to invite David and Marge Fairbrothers, and David gave a plenary address at the symposium; they had a wonderful visit.

At the beginning of my stay, I enrolled in an intensive Malay course offered by the same instructors who trained Peace Corps volunteers. There was a large contingent of volunteers, several became friends. Jack Putz, now a Professor at the University of Florida, was a volunteer working at the Forest Research Institute, and became a friend. My skill in Malay was adequate for me to travel comfortably in rural areas throughout Malaysia and Indonesia (it had been the market language for the region) to meet and greet local people and ask them about the plants they collected and used. We also learned phrases in Mandarin (spoken by the educated Malaysian Chinese) and Tamil (most of the local Indian residents had originated from South India); this helped us to function socially, although the students and educated older population spoke excellent British style English.

During our sojourn in Malaysia, the magnitude of the pressures of development and the environmental problems became clear to me, and my personal response was to help develop programs in environmental education. I wrote the first book on environmental problems in the region, *The Sinking Ark* (published by Heineman Books in 1980), and I wrote articles for general publications, as business magazines, etc. University-bound Malaysian students sat for a British style comprehensive exam, the scores for which determined whether they could gain entrance into a local university. Traditionally, the exam had included questions and tested on materials more appropriate for a British student than a Malaysian one (the exam was for the *Cambridge* Higher School Certificate). My colleague and neighbor, Norman Williams, and I wrote the first biology exam review to fully use Malaysian examples and thereby influenced the direction of curriculum development. I also received a grant from the Ford Foundation to produce filmstrips and slide sets on various topics to put more local examples in the science curriculum. On days away from the university in Ulu Langat, I began to explore the valley for specially attractive village settings (and the villages were settled by Malays from different parts of Indonesia, with different cuisines and arthitectures), attractive forest and streams, waterfalls, and routes towards Gunong Nuang. I had made contact with R.E. Holttum, a great British botanist who had worked in Malaysia, and I collected thelypterid ferns from various locations for which he was producing a taxonomic monograph. The results of my travels in the valley was the production of the first hiking guide written in Malaysia, and published in the Malaysian Naturalist in 1976: *Trips and Tracks in Ulu Langat.*

There was a rhythm of life in Malaysia that was initially frustrating to us, but eventually was very attractive. Things needed for research arrived with much delay, just like the visas that had frustrated our arrival. However, once those obstacles were accepted I was quite amazed by my productivity in completing projects and publishing research. My first letter in *Nature* was published in 1975 from research on an iridescent plant (and a second Nature letter was co-authored in 1977). We were blessed with some good friends, and I remain grateful for the friendship and knowledge of Ben Stone. Carol and I were also fascinated by the different sacred traditions in Malaysia, especially the Hindu and Buddhist ceremonies and the Sufi teachers, and they reinforced the sacredness of the nature I was privileged to study.

Ben Stone and David Fairbrothers shared a trait that eventually influenced me. I never heard either Ben or David speak badly of another person, no bad-mouthing or speaking behind the back. If they thought ill of someone, they remained silent. That silence, then, would speak volumes.

A number of eminent scientists visited us in Ulu Langat. Paul Richards, the author of the iconic *Tropical Rain forest,* was one. I had read this classic as a college student, and actually went up to Seattle to hear him lecture at the Forestry School of the University of Washington. I was gratified to see my work on understory plants covered and cited in his 2nd edition of 1996. Egbert Leigh also visited, as did Max von Balgooy, the Dutch plant anatomist. Francis Hallé, known for his work on tree architecture and the canopy research platform, *Le Radeau des Cimes*, stayed with us in Ulu Langat, and we became good friends.

My initial contract was for three years, and I asked for, and received, a one year extension. One important purpose of the expatriate faculty was to identify talented local students, train them, and help them obtain graduate training overseas—all so they could return and take over our jobs. In my last visit to Malaysia in 2005, to study the physiological function of blue iridescence in understory plants, I visited my old university (now the University Malaya) to meet an old student as the Dean of the Faculty of Science, and another as the Chair of the Institute of Biological Sciences. Former graduate students whom I knew had senior research positions at the Forest Research Institute of Malaysia, where I later did research on shade responses of tree seedlings with NSF support.

Thus, we made plans to leave at the beginning of September of 1976, and began a long odyssey back to the United States.

All told, I have lived and worked in tropical environments for some seven years, including living in Malaysia, the dozen or so 1-2 month trips for research throughout the tropics, along with two years of living in India. Our return to the U.S. took about eight months, all of this travel by someone who had lived in an isolated and rural town in eastern Washington State, and did not leave the northwest until that trip to the South Pacific at the age of 21 years. Well, that first trip certainly was a stimulus, and as a child I read voraciously about other parts of the world, devouring every issue of National Geographic. I also collected stamps, and enjoyed learning about the countries from where the stamps came, so that I could understand the importance of the images on the stamps. I mention this here, because the experiences were an inspiration for later research and writing. An important part of these travels has been meeting with scientists and visiting educational institutions, almost always giving seminars. I also became interested in the use of plants by local people, always visiting farms and local markets and learning of various products raised and collected by them. These observations (including color transparency photographs) were important for later teaching in introductory and economic botany, as well as tropical botany and ecology.

On our return voyage, we spent ten weeks in India, starting in the south and residing in the high foothills of the Himalayas, in Himachel Pradesh, for one month. We also spent some time in Delhi and in Rajasthan. We left India in early December on a Syrian Arab Airlines flight to Damascus (the cheapest ticket) and spent 10 days in Syria, before flying on to Greece. We expected to spend some time in Athens and then move along the southern coast of Europe. However, we did not anticipate the winter coldness in the Mediterranean region, so we decided to stay some time on Crete, in the middle of the Mediterranean—but it was quite chilly there, as well. From Crete we flew back to Athens, and then decided to travel across North Africa. We flew to Cairo and visited for a week before travelling south to Luxor, and then on to Aswan. We visited this venerable city during the street riots of 1977, the last to precede the revolution of 2011. We travelled south to avoid the winter temperatures, and Luxor was about right in January, with warm days to ride bikes to the Valley of Kings, and chilly evenings with star-filled skies. We travelled across North Africa through Tunisia and Algeria, with a longer stay in Fez, Morocco. We then ferried across the Straits of Gibralter to Spain, then France and Switzerland, to England, and then home on April 20, 1977. While in France, we visited Francis Hallé, and in the United Kingdom I visited Eric Holltum at Kew and gave a seminar there, and we travelled north to stay with Peter and Mary Ashton in Aberdeen.

We found the U.S. a different country from the one we left, more at peace and more tolerant. We had decided to accept an invitation by Francis Hallé to work and teach in his Tropical Botany Laboratory at the University of Montpellier II. In the interim, I practiced my high school and college French, and we travelled around the country for almost five months. We did this with care, as Carol was pregnant with our son Sylvan, but we had been gone well over four years and had family, relatives and friends with whom to renew ties. We also visited some local groups involved in Gurdjieff work as we continued those studies.

THE UNIVERSITY OF MONTPELLIER

I had been appointed a Maître de Conférènce Associé (Visiting Professor) at the Institute Botanique, Université Montpellier II, for a year with a possible extension for a second year. My duties would be to assist Francis with the running of a post-graduate diploma program in tropical botany that attracted students from all over the world. This meant giving lectures in tropical botany, including illustrated talks on tropical plant families. We arrived in Montpellier on 24 September. Carol was now over 7 months pregnant, but our settling in was smoothed by the support of Francis and Odile Hallé and their four interesting and loving children. With their help, we bought an old Simca automobile (a baignoule!) and rented an apartment at the Chateau de la Mogère, just to the south of the city. It was a beautiful 18th century country estate (** in the Michelin Guide), once owned by the great French paleobotanist Gaston Saporta in the 19th century and with a formal garden. We actually lived in one of the outbuildings of the farm, a massive two story structure with ~2' thick stone walls (which made the place super cold during the winter). Francis quickly learned how bad my French was, so we both enrolled in intensive courses, Carol for survival and me to support more professional lecturing and interactions. I'm reminded of the extent of my learning from a visit to the rustic countryside outside of Montpellier where a local farmer asked if I were from Belgium; I later learned that the French found Belgian French to be crude and unacceptable! After I finished the French course, Francis told me that I could teach graduate and senior undergraduate students, but that the first and second year students would laugh at me. So, I limited myself to that group of students.

My old friends Richard and Barbara Forman from Rutgers days took their children to Montpellier for a year where Richard worked with Michel Godron of the CNRS Ecology Laboratory to write the first book on landscape ecology, thereby establishing a new field of investigation in ecology. They were particularly supportive after the birth of Sylvan on November 9th, as were Francis and Odile.

In late January 1977 Francis and I flew with the diploma students, and some other scientists, including Patrick Blanc, to Cayenne, French Guyana, for a month of tropical rainforest field research. During that time, we stayed at locations adjacent to, or within, tropical rainforest. A highlight was a flight to Sa l, where we made a three day trip to Mt. Galbao, high enough to support montane forest. This was my first experience of forest of the Amazon region, and having Francis and other botanists well-acquainted with the vegetation enabled me to learn a lot of botany and take lots of photographs. I was also able to make observations on the physiognomy of plants, compared to my experience in Southeast Asia.

After our return from French Guyana, I was busy lecturing and helping students work on their short graduate theses. I became quite interested in the history of science, botany in particular, in this old university city. French professors took 2-3 hour afternoon breaks for lunch and a snooze. Since I usually rode a bike from La Mogère to the University, it was inconvenient for me to commute during lunch. So I'd have a quick lunch with the diploma students, and then explored the city and libraries for a couple of hours, looking for documentation of the importance of Montpellier (as locating the old homes—"hotels"—of Pierre Magnol, and other luminaries) and visiting libraries, particularly that of the University Medical School. Since the science of botany rose out of medicine in 15th century Europe, the Medical School library, adjacent to the institute and botanical garden, and near the cathedral, was a rich source of old botanical books and illuminated manuscripts. Montpellier had been an important center for the study of medicine and natural philosophy since the 12th century. Many great figures in botanical history (the Bauhin brothers, Clusius, Magnol, Rondelet, and others) had studied there as medical students. I photographed hundreds of illustrations from old botanical works and learned about them. At that time, I could request an old illuminated parchment manuscript of Hippocrates or Galen, have it brought to me, and could photograph it as I wished. I photographed hundreds of illustrations from these works and learned about them. This documentation assisted me in injecting history into my lectures on botany and biology in the next 30 years of university teaching.

I didn't do very much research in Montpellier. I did establish a collaboration with Charles Hébant for electron microscopic research, which we published in 1984, and wrote several manuscripts from late Malaysian research that were published in 1980 and after. However, it was a stimulating year for thinking and talking about tropical botany, with Francis, and many eminent figures in tropical botany. I first met Barry Tomlinson in Francis' lab, and proof-read the galleys for *Tropical Trees and Forests. An Architectural Analysis,* one of the seminal works in tropical botany in the past half-century.

By early June, the university term ended, and students and faculty dispersed. Francis and family went to their summer cottage on the Île de Groix, off the coast of Brittany. We visited them for a time, but mainly relaxed in Montpellier, staying in their vacant home. I made many bicycle trips to small villages with old Roman monuments and Romanesque churches. After a brief rest on the Mediterranean coast, we returned to The United States 22 September, 1978.

WARWICK, NEW YORK

We moved to Warwick, New York, to join the Chardavogne Barn community to more intensively involve ourselves with the Work, as outlined by G.I. Gurdjieff. We had corresponded with a member of the community and had read extensively, and I had also written my old friend Curtis Amo. Curtis and Laile were there for us when we arrived, and we rented the house they left to move into another home they had purchased. Activities at the Barn included group meetings, service/work days, and movement classes. There were special days, as Gurdjieff's birthday, when many students came from New York City to the "wilds" of upstate Orange County, Warwick was a township, and included a fair amount of countryside (fields and forests) as well as several villages. We lived in an early 19th century farmhouse in the village of Amity. Shortly after our arrival, Carol became pregnant, and our daughter Katherine was born in July of 1979.

There was the question of making a living in this new place, and we were naively optimistic. I had decided to write articles for magazines and produce educational materials as a major supplement to our income. I also prepared a proposal for a trade book about tropical plants, and met some editors in New York City. There was some interest by Viking Press, but their market analysis discouraged any further discussion. Thus, I pursued a number of jobs, involving skilled manual labor, to make a living. Members of the "Barn" ran businesses (often cooperatively and with the intention of making that labor an extension of the Work). In Warwick Village there was a bookstore, craft gallery, and auto repair shop, all run by members of the Barn. The people at the Barn came from very diverse backgrounds, but generally were highly educated and with interests in the creative arts; we made several friendships. We helped in the community day care program, and I was a volunteer teacher in the community school, which taught children from kindergarten to grade six. So, I performed the following work: landscaping, carpentry and home building (my principal occupation), apple picking, and piano re-building and regulation. I looked around for some part-time teaching. I was hired to teach biology to nurses at Orange County Community College in Middletown for a semester. Then I was hired to teach general biology and environmental science courses at a new private community college campus run by Upsala College, in Sussex, New Jersey. Upsala was a Lutheran liberal arts college, much like Pacific Lutheran where I had studied as an undergraduate, and this new campus was hoped to breathe some life into a college that was shut down a few years later. Also, through the support of Iaian Prance, who was then the Vice-President for Research at the New York botanical Garden, I became an Honorary Research Associate, to help me keep my academic life alive.

We continued the spiritual work at the Barn, but it was clear that no one was truly knowledgeable of the teachings of Gurdjieff, as his student and the founder of the Barn, Dr. Nyland, had passed away two years earlier. On the 26th of August, 1979, just a few weeks after the birth of our daughter, the Lee family drove an hour north to South Fallsburg, on the edge of the Catskills, to meet the Indian meditation teacher, Swami Muktananda. We had seen posters about his visit in Warwick. My first meeting with Baba, as we called him, changed my life. I began to meditate regularly, and had deep experiences of the truth that exists within each of us. It was revolutionary. Other members of the Barn were also visiting the Sri Muktananda Ashram, and having similar experiences. Eventually, we gradually pulled away from many of the activities of the Barn, and took on the discipline of meditation and the ancillary activities of chanting (to still the mind) and the study of various teachings in the Hindu spiritual tradition, such as the Upanishads and the Bhagavad Gita.

At about the same time we first encountered Baba Muktananda, Barry Tomlinson, an eminent Harvard Professor of tropical botany (who was a friend of Francis and a co-author of the book *Tropical Trees and Forests: An Architectural Analysis* and who I had met in Montpellier), told me about a job at a new university in Miami: Florida International University. These were the days before the internet, and it was difficult for me to learn about this place, other than it was a state university which had begun to instruct students in 1972. At any rate, with little expectation (and I had been applying to other schools, with little interest—mostly no response at all), I sent in an application. Sometime later, I received a response and, eventually, an invitation to come for an interview. Then I was offered a position. We accepted. I needed a job badly to properly support my family, and it seemed that I would be able to pursue the research interests that had been ignited in Malaysia in this strange place on the edge of the tropics. Thus, we continued our life in Warwick, with friends from the Barn, teaching, building houses, and visiting the Ashram—until it was time to move.

It is useful to review what I had learned in that I would take with me to this new life and professional position in Miami, at the age of thirty-eight years. I had developed a "spiritual" curiosity and yearning, deeply informed by experiences in nature and in eastern philosophy. I had accumulated a strong classical background in botany with familiarity of the physiology and biochemistry of plants (at FIU I would teach Plant Physiology and Tropical Botany, and help out in General Biology). I had a rich practical experience of the tropics, particularly the Asian tropics but with some field work in the new world tropics, as well. I had learned much by observing traditional cultures and their uses of plants (I would later use this in the courses of Introductory Botany and Economic Botany). I had moved my research from experimental systematics into functional ecology, although I still had much to learn about new techniques being developed to study the function of plants. I had also matured as a person, and that was a slow process. Coming from a good family, with the support of my parents, made a good start. I had learned about persistence and hard work through sports, and was fortunate to have some really good teachers in high school. I was aided by the support of mentors, as David Fairbrothers and other Rutgers faculty, by Benjamin Stone and Francis Halle. Mostly, I learned through my marriage to Carol and acknowledging the responsibility I had for nurturing the development of our two children. Although I had accepted the offer of a position at the Assistant Professor level, my experience was far beyond that.

I had been a member of one professional society throughout that time, and even today: the Botanical Society of America. David Fairbrothers strongly recommended such a commitment, and I joined in 1966—now 47 years! I also learned an invaluable lesson from an old neighbor and family friend, Paul Hamilton, from a hospital bed as he was dying of cancer. Paul had risen from running a hamburger restaurant in my home town of Ephrata to become the Secretary of the Department of Ecology for the State of Washington. He shared with me the secret of his success: work hard to make the institution a more effective and successful, but never take credit for any of the fruits of your labors—leaving that to your superiors, as they will take care of you, encourage and support you in your career.

MIAMI AND FLORIDA INTERNATIONAL UNIVERSITY

We arrived, moved into a temporary home, and I arrived for work at a new university. I found Florida International University (FIU) a new, highly immature in good and bad ways, and very idealistic place. It had opened eight years previously with about 6000 students, the largest initial enrollment of any university in U.S. history, and was at about 11,000 students in 1980. The statistics were a bit misleading, because the number was a head count, and did not reveal that 80 % of the students were older and with jobs, and part-time enrollment. They were a lot of fun to teach; they were mostly older (mean student age of ~28) and serious about their studies. FIU was in the process of becoming a more mainstream university, with a four-year program (in 1980 it was taking community college transfers and had a few graduate programs in education and business) and a breadth of graduate programs. It had two campuses. The south, and principal, campus where I taught was on the site of an old municipal airport, and a smaller campus was located in the northeastern corner of the county, on Biscayne Bay—the site of a botched trade exposition. As a young university, there was an urgent need for faculty involvement in institution building. It was important for the university, students, and the ultimate success of faculty and faculty scholarship. There was also a need for institutional involvement in the community. Despite its glitz, Miami has always been a poor city, seen in low family income and the percentage of school-age children eligible for free meals. FIU as a public low-cost university was a needed addition, to the few private institutions, particularly the University of Miami.

The Department of Biological Sciences had an interesting mix of faculty members. Many had unusual records, with international experience similar to my own, and were mostly relatively early in their careers. However, not much research was happening, and there were no real opportunities for graduate study. The department was located in an interesting setting, on the edge of a very cosmopolitan and Latin city and surrounded by tropical marine and terrestrial ecosystems. So, the opportunities to study tropical plants were excellent. Furthermore, Miami was well-connected to countries throughout the neotropics, and there were lots of opportunities for research there, supported by the university's strongest academic program, The Latin and Caribbean Center (LACC).

In Miami, I renewed my friendship with Rod Sharp. He arranged my giving an invited talk at a national food science meeting (on exotic tropical fruits and vegetables), and enlisted me as a horticultural consultant in providing plant materials for the newly formed DNA Plant Technology, Inc. (DNAP).

My efforts at FIU were well divided into teaching, service, and scholarship. In retrospect, I believe that my most inspiring college and graduate school mentors led the way in showing that it wasn't just about research (and certainly the most efficient way to success and salary increases was to minimize teaching and service and focus on the success of research—visibility and outside funding). The two individuals who were the most influential were Jens Knudsen at PLU and David Fairbrothers at Rutgers. In my previous academic position at the University of Malaya, I had little impact on the institution, partly because I was an expat; my efforts went into research and teaching.

Teaching. I was hired to teach two courses that had been regularly taught by my single predecessor and were part of the regular curriculum: plant physiology and tropical botany. Plant Physiology was taught with laboratory, and most of the biology majors took it. It fulfilled the graduation requirement for (1) a plant course; (2) a physiology course; and (3) a lab. I quickly developed the reputation of being organized, fair and interesting (I was able to integrate my extensive experience into the course).

This was also true for Tropical Botany, a course taken as an elective by biology majors and Environmental Studies majors. That course included lecture and laboratory, and the latter was primarily visiting local plant communities and learning about plants: their functional ecology, development, systematic relationships, and uses. In 1988, we hired a plant physiologist, Steve Oberbauer, and I stopped teaching that course. I had taught other courses, as plant morphology and economic botany, to later drop them as we hired faculty in those specialties. I taught a portion of the general biology course, for a special program for gifted high school students. In 1984, FIU was granted the ability to establish four year programs, and I began to teach in the general biology course for our own freshmen students. As the lower division program expanded, a need to add our offerings in first year non-major science courses was partly fulfilled by my offering of a non-major introductory botany course. I developed a style of teaching that was most effective for upper division students, but over time (particularly in the last 10 years of my work) I became more frustrated by my ability to reach the first-year general biology students; as the university grew, those students became more traditional, i.e. 18-19 years in age, fewer with jobs and supported by their parents. Our teaching loads were somewhat intermediate between a teaching college and a high level research university. It was more of a challenge to do research, but definitely doable.

The aftermath of hurricane Andrew in 1992 was a traumatic time for Miami and our family. Our home was seriously damaged during the storm, and it took four months of repairs before we could move back into our home. I was particularly struck by the damage to our urban landscapes and how emotionally affected we were by it. It was difficult to drive around with familiar tree landmarks lost. I contemplated the impact of these changes and came up with an idea for a totally new and inter-disciplinary course: "The Meaning of the Garden". The course was offered by the Liberal Studies Program, and I received students with all sorts of interests. The course met once a week, on Friday, and began with about two hours of physical work on various garden projects, including weeding and planting. The students then washed up, had a cold drink, and we re-assembled for a lecture/discussion section. The topics often were delivered by outside experts in landscape architecture, ethnobotany, horticultural therapy, art history, poetry, and more. I found that the physical work by the students together created friendships and broke down barriers that inhibited free discussion. I assigned writing projects, including an autobiographical essay on "my first garden experience" and a Haiku. The greatest student effort went into a project, and I worked with each student on defining what that project would be and how the student was connected to it. At our final class meeting, we created a reception like environment with excellent snacks (some of which were products of their projects) and the students gave presentations on their projects to everyone present. This is the only course I'd taught where several students told me that it had changed their lives. However, even today I occasionally run into former students who mentioned that my lectures on plants had made them interested in the environment or in gardening. In our non-major botany course, each laboratory section produced and maintained a garden plot on campus, and students could consume its products at the end of the spring term (gardening is a winter activity in Miami).

I reached out beyond the university by giving talks and making friends with groups interested in plants. I made a particular effort to connect to the large nursery and landscape nursery, mainly centered in the Redlands, an agricultural area south of Miami. I gave talks at their meetings and was able to obtain a book fund from them that added plant titles to our library collections. Early in my teaching career, I was involved in the training of science teachers in botany (many of them had learned human physiology and kinesiology, and were coaches) during the summer. This also helped to enhance my nine month salary a bit. I also took up carpentry jobs during summers for several years.

Service. On one hand, tenure-earning faculty members like me were not encouraged to take on too much service, but circumstances in the department demanded it. My first service assignment was as head of the Curriculum Committee. In that I also was a member of the College Committee. It was interesting to me to see how new courses were developed and approved, and then became part of the course descriptions of the State University System, and assigned a course number. Issues came up for courses of an interdisciplinary nature that required consultation with cognate departments. In time, I also served as member of the Faculty Senate.

When I arrived at FIU, there was no advanced degree program in the department. The amount of research improved steadily after my arrival (not just because of me, as a majority of the faculty was motivated to pursue research). We could produce M.S. students through a master's program run by the School of Technology, in "Environmental and Urban Systems". Then, we established a cooperative program with Florida Atlantic University. This was a sister member of the State University System, slightly older than us, and some 60 miles north, in Boca Raton. They had an M.S. program and allowed us to produce master's theses through them. This was onerous; theses were defended in Boca Raton, and also demeaning because our faculty was more energetic and research-minded than theirs. I became chair of our small graduate committee, and created and guided our proposal for an M.S. degree through the university. By 1984 we had received approval for our own independent M.S. degree. We had serious problems with a disruptive and abusive senior faculty member, which blew up over the treatment of a graduate student. I had to take care of this issue, which resulted in protection of the student and the transfer of the professor to the north campus, and he soon retired. Some of the elements of the M.S. program, as the core beginning course, Introduction to Biological Research, also became an important feature of our Ph.D. program. The size of our M.S. program, and the growing quality of students (as well as the need for graduate teaching assistants to serve our expanding undergraduate population) led to our development of a Ph.D. program. At first, it was to be a cooperative program between us, FAU and the University of Central Florida (UCF) in Orlando. However, a very strong external review mandated by the Board of Regents gave us the opportunity to push through an independent Ph.D. program in 1989; I also served on the Graduate Committee at that time.

I had been interested in environmental issues since my days as a graduate student, increased by what I learned in living and travelling overseas, and by my brief teaching position at Upsala College. FIU had offered B.A. and B.S. degrees in Environmental Studies since opening in 1972. Ironically, it was one of the oldest such programs in the country in one of its youngest universities. Its director and cheerleader was Jack Parker, and he was aided by the Environmental Council, a group of individuals with environmental interests in various departments, as Philosophy, Economics, Political Science and International Relations. I became a member and enjoyed the interactions with those people.

In addition to dealing with program issues, the Council had also supported the maintenance of the Environmental Preserve, since its establishment in 1978. I became particularly concerned with the Preserve, along with Jack, and had to defend/protect it from development on numerous occasions. I became good friends with Charlie Hennington, the Grounds Superintendent, from the beginning of my work at FIU. Charlie understood the value of the university landscape for teaching, and was always looking for opportunities to plant unusual natives and interesting tropical plants. I was asked by the Campus Architect to serve as the chairperson of a new committee, the Landscape Advisory Committee, which advised on the master planning process and on landscape plans for new building projects. That committee then became the first Environmental Committee, later banished by a vice-president, and then re-established. I had to stick my neck out on many an occasion to uphold environmental values on campus, and received a University Service Award in 1991 for my environmental work. In the late 1990s the Preserve was threatened again, and I organized a charrette for the proper development of the area, the first such design activity on campus involving faculty, staff, students and outside professionals. The results of that chrarette are still being used to guide development of that part of the campus.

I was asked by the Dean of the College of Arts and Sciences to lead a review of the Environmental studies program around 1993. The committee recommended that the program become a free-standing department, and the Dean concurred. In 1995, I became the first Chairperson of the new department, helping it to add faculty and eventually finding a physical home for it. I am reminded of the ground-breaking work David Fairbrothers was doing at that time, in preserving natural areas in New Jersey, and helping with the establishment of the Pinelands National Preserve. My service was tiny compared to that. I also remember that David was supportive of athletics, being a good personal friend of the football coach. I also became active in athletics, being appointed to the Athletic Council and serving for some time. Later, I became a vocal critic of the campaign to start a football team, and was a faculty leader in trying to reign in some abuses of the administration with regard to sports.

I also served as Chairperson of the Department of Biological Sciences for a year, not long before my retirement. That term was cut short when I took up another service responsibility as Director of the Kampong, of the National Tropical Botanical Garden,

mainly based in Hawaii but with a garden in Miami, the historically important Kampong--the old home of the great plant explorer David Fairchild. I pushed for a close relationship with FIU,
to give the garden more of an academic/scientific mission. As I
write this, those plans are being realized.

I feel some satisfaction in serving a university that is so vital
to the well-being of this area, with its problems and opportunities. Now, FIU is a university approaching 50,000 students, with
professional schools of Law and Architecture, a full Engineering
College and a new Medical School (which I helped as chairperson
by serving on organizing committees).

Research. After leaving Malaysia until arriving in Miami I
was not able to engage in research. I did write up the rest of the
papers from my four years in Malaysia, and they were published
during my first year at FIU. I had a lot of motivation to get that
research going, but there were obstacles. It was months before my
lab was made available to me, and there was no startup funds for
my position. Furthermore, it was clear to me that I would have to
do the research myself; there was not a graduate program and
students available. That became the case for most of my research
in Miami, although some excellent students came through my lab
towards the end. My mechanical skills were helpful as I renovated
my laboratory by building my own furniture, and renovated an
old spectrophotometer to make optical measurements of leaves.
I received my first outside funding at FIU from the Whitehall
Foundation in 1984. My research agenda was to study the functional ecology of tropical rainforest plants. I soon discovered that
the tropical forests of south Florida were seasonally too dry for my
research. I travelled to Costa Rica my first year at FIU and found
ideal conditions at the La Selva Research Station, run by OTS.
Later, I also visited the station at Barro Colorado Island, run by
the Smithsonian Institution. These sites had good infrastructure,
and it was possible to quickly learn about plants of interest to me,

and the stations were a short flight from Miami. I wanted to document the light environments of these forests, to help me in my study of the adaptations of the plants, and I was able to purchase a new Li-Cor spectroradiometer through the support of FIU's Division of Sponsored Research.

An underlying theme in my research was beauty; I was attracted to phenomena and structures that were esthetically beautiful and emotionally attractive to me. Two phenomena I returned to again and again were the subject of blue iridescence (structural color) in tropical plants and the function of anthocyanic coloration in leaves. I collected leaf samples of iridescent plants from La Selva and analyzed them in Miami, and I found iridescent fruits from exotic trees in Miami to study. For studying anthocyanic leaves, I initially chose the production of anthocyanins during the development of leaves of mango and cacao. I also got interested in the leaf surfaces of understory plants, particularly the lens functions of convexly curved epidermal cells. The papers from this research began to appear in 1985.

When I first arrived at FIU, I was the only person in the department who studied plants. I made friends with scientists at Fairchild Tropical Botanic Garden, particularly Jack Fisher, but there was not much research collaboration. However, with the expansion of the department, and the university's recognition that tropical plants were a natural for research development, we added colleagues. The first was Jennifer Richards, a plant development specialist. Then we hired a plant physiologist, Steve Oberbauer, who was a physiological ecologist and was primarily studying in tropical rainforests at that time. We hired an ecologist, Suzanne Koptur, who worked on the interactions between insects and plants. Then we hired a plant biochemist, Kelsey Downum. These were colleagues and friends with whom I published papers. We hired several other plant specialists and formed a strong area of botanical research within the department. My approach in this was not to push to aggressively and selfishly for plant people, but partly to encourage candidates to apply for positions that were generally very broadly described. Kelsey's hiring was the result of plans to hire a biochemist. Probably, I promoted the hiring of more plant faculty through my own example.

My research on tropical rainforest light climates stimulated me to find means to duplicate the differences in the spectral quality between extreme shade and light flecks (with the spectral quality of sunlight). I was interested in the plasticity of plants, as well as the advantages of plants producing different kinds of leaves at different life history stages, what we call heteroblasty. In 1983, I developed a spray paint that could be applied to any transparent medium, as commercial greenhouse plastic, that would produce the spectral quality of rainforest shade. That could be contrasted with a normal black spray to distinguish between the two light signals of quantity and quality. I published those methods in 1985 and used them on several research projects. However, applying the spray was pretty cumbersome, requiring multiple light coats of spray paint. I eventually figured out that one pigment most crucial for altering spectral quality was purchased by companies that manufactured commercial energy films, i.e. the plastic films applied to automobile windows and windows of commercial buildings to reduce energy costs. Then I settled on a couple of film products from 3M and got them to supply me rolls of film for building walk-in shadehouses.

Back to Tropical Asia. Our meditation teacher, Swami Muktananda, died in 1982, and he announced a brother and sister as his successors, and eventually that became the sister, Swami Chidvilasananda. We were able to visit them in the summers a bit in South Fallsburg, near our former home, but were attracted to the idea of living in the home ashram in India. Gurudev Siddha Peeth is a large ashram in the small village of Ganeshpuri, northeast of Mumbai. I applied for an Indo-American Fellowship, which would provide funds for travel and subsistence in India for a year. I received the fellowship, and made connections with two educational institutions, one with K.R. Patel at Bhavan's College in Mumbai, and with Kailash Paliwal at Madurai Kamaraj University, a new and very progressive research university in South India. We would be able to live in the ashram, and it had a small school for our children, and I would be able to conduct research in nearby forest and at an experimental farm run by a family that was connected to the ashram. Then I could occasionally travel to Mumbai, and less frequently but for a longer time, visit Kailash Paliwal in Madurai. We arrived in September of 1984 and departed at the end of July, 1985. I set up a forest plot for assessment of tree phenology and light measurements, an hour's walk north of the ashram and at the foot of a sacred mountain. I constructed shadehouses at the ASPEE farm, near the ashram, and studied the developmental plasticity of three vines. Those vines responded well to the treatments and resulted in a paper in the Journal of Tropical Ecology, as did the results of my forest measurements. It was an excellent year for my research, and also for our family staying in the ashram.

We travelled to Ganeshpuri for another year's stay in 1988-89. This was a year of sabbatical leave (one year at half salary). My service work in the ashram included frequent travel to collect sacred and medicinal plants to add to its gardens. So, I travelled to the northeast (Kalimpong), the south (Cochin and Madurai), and the west (southwest, Uttarkhand) looking for plants. I also received some funds from the American Philosophical Society to travel to Malaysia to establish some collaborators for research that would require submission to NSF for funding. During my trip to Malaysia, I formed a collaboration with the Seed Technology Laboratory at the Forest Research Institute of Malaysia (FRIM). I hadn't done much with them other than visit occasionally when I worked at the University of Malaya, but some former UM graduate students were on staff there, and S.K. Yap as head of the seed lab became my grant collaborator. Later on Marzalina Mansor took over. This was a logical connection, because I proposed to study the effects of light on the development of seedlings of tropical rainforest trees. We had a good year in India, in the ashram, but Carol had come down with the symptoms of what came to be called chronic fatigue and immune deficiency syndrome (CFIDS), and it took many years after our visit for the symptoms to largely disappear. We returned to Miami in early August of 1989.

We obtained funding from NSF to start the research at FRIM, in Malaysia, and I began to visit Malaysia most summers for approximately a month, starting in 1991 and lasting through 1998. This was a successful project, resulting in some solid publications. Its success was due, in part, to the skill of our research technician, Haris Mohamed., who was adept at constructions and electronics (such as installing and running the dataloggers and sensors). Its success led to the establishment of another NSF-funded project with Kailash Paliwal in Madurai; this project was not successful, but I continued to travel to India until 2000. In all of these research trips there were opportunities to visit natural areas of great interest and beauty, partly through visiting scientific colleagues and offering seminars. These tropical trips were typically 2-4 weeks, limited by me need to be home and help my wife recover her health, and help care for our children.

In 1999, I had the opportunity to join Francis Hallé on a canopy raft research expedition to a tropical rainforest site in Gabon. That was particularly valuable to me because it provided the opportunity to experience the third major tropical rainforest region and compare it to Central America and Asia. I have truly been blessed to be able to visit such beautiful places and study such remarkable plants. Kelsey Downum and I, with Francis, tested his hypothesis that secondary and biological active compounds will be more numerous and at higher concentrations in the leaves of target species in the forest canopy than in individuals in the understory. We completed that research and published the results in the Journal of Tropical Ecology

My last research trip to tropical Asia was in 2005, with support of a National Geographic Society travel grant. The primary purpose of this trip was to test a hypothesis on the function of blue leaf iridescent in these deep shade plants. This was a two month trip that included a short visit to Chiang Mai and a 10 day trip to Xishuangbanna and Lijiang, in southwest China. From this trip I discovered a novel mechanism of structural color production involving silica nanoparticles in an understory plant.

Red Leaves and Autumn Color. From my initial interest in the function of anthocyanins in red developing leaves, and also in the undersurface of shade plants, I still continue a little research. The classical explanation had been protection against UV, but I showed that not to be the case for mango and cacao. However, the discovery of photoinhibition in photosynthesis suggested that these pigments could be photoprotective by reducing photoinhibition. Kevin Gould, then at the University of Auckland, proposed that he work with me on this problem during a sabbatical leave, in 1993. We published a short article in Nature in 1995 that generated some interest—but not to the level I'd hoped. It came to me that the phenomenon that could be studied and would create some interest in anthocyanin function in leaves is the autumn coloration in temperate deciduous forests, particularly in New England. I approached Missy Holbrook, now a professor at Harvard, about working on this problem. I received a Bullard Fellowship from Harvard to work at the Harvard Forest on autumn color. My initial stay was late summer through early winter in 1998, and I returned for a second stay in 2004. Our hypothesis was physiological, that anthocyanins are produced during senescence (producing the red color) to provide photoprotection against the destructive activity of light-activated reactive oxygen species, thereby reducing the amount of nitrogen resorbed by the plant for the next growing season. At the same time another group proposed a co-evolutionary hypothesis: that the color could warn potential herbivores and reduce the load of insect (aphid, in their case) eggs that could attack the tree the following year. We published a cover article in *Plant Physiology* with evidence in support of photoprotection, and also a data-rich general paper about color change in these forests. These ideas got a lot of exposure, including an article in the *New York Times* and other papers. The controversy over these two hypotheses continues today. Kevin and I also organized a symposium on "anthocyanin function in leaves" at the 2001 Botanical Society of America meetings in Albuquerque, and those papers were published as an issue ("Anthocyanins and Leaves: the function of anthocyanins") of *Advances in Botanical Research* in 2002

The Everglades. I began to work actively in the Everglades around 1996. I became part of a team of workers conducting a project mandated by the legal settlement between the National Park Service and various federal and state agencies. My role was to assess plant (as they say, "macrophyte") responses to dosages of phosphorus. I soon discovered a link between the responses of these plants to phosphorus, and the responses of my rainforest plants to shade: developmental plasticity. My final research project, in collaboration with Jenny Richards and still underway, is a study in the environmental plasticity of a common sedge, the spike rush.

Research in Perspective. There are a number of themes that my research interests followed over the decades. First, I became interested in studying plants. I remember in high school that I took a vocational aptitude test. There were dozens of questions about likes and dislikes, and then you were matched up with a vocation or profession, based on those answers. An answer that came up: botanist. It seemed rather unlikely to me at that time, and I didn't really know how a botanist made a living. My interests in botany solidified in college, particularly after my trip to the South Pacific. None of my mentors in high school or college pointed me towards plants in particular. Certainly, my experiences in nature, surrounded by plants, influenced my choice, and my sacred personal experiences as well. As I moved into botanical research, my work was almost entirely laboratory- based, although I did collect plant specimens in nature. I struggled with the dichotomy between my working environment and my love for the natural world, and the resolution of this conflict was our moving to tropical Asia, where exploring for plants and studying them in natural environments became an important part of my research. From then on, studying plants in tropical forests required occasional travel to very beautiful places, and the plants themselves were of great beauty. Even the subjects that I studied were esthetically beautiful, iridescent colors in fruits and leaves, and red-purple leaves in development and senescence. I was certainly drawn to that.

I was also drawn to research projects that were beautiful and interesting, and not in any way because of their material importance, or value for human health. The closest I came to an application of economic value was in my study of blue leaves. In some species, the structural basis for iridescence was a periodic layering of the cellulose microfibrils in the cell wall, to produce the layers that were responsible for constructive interference in the blue wavelengths. Thus, controlling the layering of the cellulose could produce iridescence......artificially. I wrote a popular article on blue leaves for American Scientist in the 1990s. Afterwards, I received an email message from a scientist working at the National Institute of Standards and Technology (NIST), who mentioned to me that "mother nature has beaten us to the punch." I deduced that he was a paper technologist working on producing iridescence patterns that could make it very difficult to counterfeit paper for currency. I never heard anything more about that, and don't think that the technology was actually applied in that way. My graduate friend, Rod Sharp, worked in an area of high commercial potential, as I did for a bit in Ohio. It was interesting basic research, and it also had potential applications. Having such a research agenda, interesting and with some potential applications, would've been beneficial in several ways. It would be possible to produce results that could directly benefit society, it could be interesting, and it might be easier to get financial support. However, it didn't work out that way for me.

I also chose research topics so that I could learn more about an area of knowledge. I started out studying plant systematic and evolution. Then I moved to the functional ecology of rainforest plants. I also began to do research on the responses of those plants to light, more in the area of photobiology, including the documentation of light climates and the effects on plants. I studied phytochemistry (made easy by my strong graduate background in plant biochemistry) and researched plant pigments and their function in leaves. Finally, I began to study the functional ecology of plants of the Everglades, particularly their response to light and water depth. This research was partly motivated by our shared concern for the future of this important ecosystem, but allowed me to transfer some knowledge and interests in the developmental ecology of rainforest plants to this new environment. These different projects allowed me to learn new things, not only in the experiments and observations, but also in the extensive literature research that is required to move into any new subject.

Studying plants in relatively undisturbed ecosystems in the tropics helped me become aware, fairly early in my career, of their fragility in the face of the increasing human population and growing impacts on these systems due to our exploitation of cheap energy resources. Much of my career, I have lived with an almost daily sense of grieving over the accelerating loss of nature, beautiful and sacred. That knowledge motivated my teaching, always including exposure to environmental issues in my botany courses, and solutions to those problems. This awareness was a strong motivation for me to write for a more broad public audience. This started in Malaysia with *The Sinking Ark,* along with articles in magazines. This continued in the United States. My last two books *Nature's Palette* and *Wayside Trees of Tropical Florida* both "preach" a bit about the intrinsic value of nature and the importance of conservation. My present book, about leaves, will have a similar message.

That environmental stewardship was taught to me by Jens Knudsen and David Fairbrothers. David had a great love for the natural areas of his native state of New Jersey. His later work on natural area preservation was an inspiration to me and, I am sure, to many other students. My writing has been influenced by their examples.

I never considered myself to be a brilliant person. I probably was better at mathematics than I thought. However, as a teenager I took math courses with some friends who did higher math at Cal Tech, Yale, and the Air Force Academy; I suffered in comparison with them. My high school friend, Jim Arnold, later commiserated with me that he was discouraged about taking higher math in college because of the same high school experience. He ended up as a philosophy major when he really wanted to pursue electrical engineering to fulfill his passion for radio communication. He later studied Indian philosophy and music, living in India for many years. There he began to work for the State Department, and gradually moved back into his love for radio communications, helping to set up communications networks for the distribution of foods in drought stricken regions, finally working for FAO. I certainly could've benefited from a stronger math background. My strengths in graduate school and as a working scientist were a strong work ethic (past summer work and sports), persistence and endurance, and an ability to focus on subjects or work for long periods of time. Those traits contributed to my relative success as an adult scientist. Although I've made a bit of a name for myself, as for work on iridescence in plants and in autumn leaf color, none of my research led to any "paradigm shifts." I was one of many people "mopping up after Darwin", as a colleague Doug Schemske once remarked.

I spent most of my career at a growing public university in a rather poor large city, to which the university contributed greatly. Considerable demands were made for teaching and serving to establish new programs. In such a setting, with no graduate student tradition, research productivity was bound to suffer. I learned to function without students, especially in comparison with more research-productive faculty who came along later. I published an average of about two articles per year (peer reviewed research publications or book chapters) during my career, my first paper published in 1969. By today's standards at FIU, and at larger more established universities, this is not a strong record. Yet, I enjoyed doing the research, it helped inform my teaching and my service, and a place like FIU probably was the best career choice for me. If I had been hired at a larger university with greater demands on the levels of external funding and in publishing in high impact journals, I might have been much less happy. At FIU, I also had greater opportunities to meet interesting colleagues with a range of interests, from science, history, philosophy, art and music. I learned much from those friendships.

I did not have a lot of graduate students during my career at FIU, six M.S. students and five Ph.D. students. I tried to treat them well, being available whenever I was needed but also trying to empower them to develop and test their own ideas. I was definitely influenced by the mentorship of David Fairbrothers as an example. I encouraged them to publish their results as quickly as possible, and generally did not co-author articles with them unless I was substantially involved in the research. In all cases, I spent substantial time helping each student to develop skills in science writing: concise and direct.

WINDING DOWN

I write this short autobiography moving towards the age of 71, over three years past my retirement now. I completely stopped teaching, but visit the university a couple of times a week. I attend most of the departmental seminars, and other lectures as well. I am completing a couple of research projects, and writing papers with collaborators on research already completed: perhaps another four articles to finish my research career. I travel quite a bit, primarily to see family and friends, and plan make some longer trips now that my wife recently retired from her work at the USDA Agricultural Research Service—tropical Asia beckons. I am writing a book about leaves, intended for a general audience, to be published by the University of Chicago Press. This book is the culmination of decades of research on leaves, just as the successful book on plant color, *Nature's Palette* (Chicago, 2007), was. It is enjoyable to pull ideas together from history, philosophy and biology to weave a coherent story; I'm thinking the title might be "Nature's Fabric". Partly based on my work at The Kampong, I am compiling an anthology of David Fairchild's writings, with commentaries. There may be another book or two, depending upon how I last. I'm enjoying gardening more, and spending time with our two children and grandson. We are thinking seriously about moving out of this large city to a smaller place, probably in the West. However, we are not in a hurry to move.

I've received two awards that recognize the achievements of this career, not a prolific one in a scientific sense. I received the Botanical Society of America's Bessey Award for contributions in education, in 2005, and I received The FIU Alumni Association Golden Torch Award as outstanding University Professor in 2007.

I've greatly appreciated the invitation by Rod Sharp to write this autobiographical chapter, and I look forward to seeing what others have written. I have seen Rod's story, and also that of Julius Kreier, a person I remember affectionately from my days at Ohio State. I remember Julius once telling me: "You know, a disease is nothing but an ecological problem." Thinking about that, every disease manifests in an environmental context. Writing this chapter has evoked in me a deep sense of gratitude for the life I've lived (with a few regrets), blessed by family and friends, influenced by some important mentors, and privileged to immerse myself in the sacred nature of the world's diverse ecosystems. Contemplating this long life of pursuit of truth in nature and in my own life, I am reminded of what Albert Einstein wrote about his pursuit of truth:

I have never imputed to Nature a purpose or a goal, or anything that could be understood as anthropomorphic. What I see in Nature is a magnificent structure that can be comprehended only very imperfectly, and that must fill the thinking person with a feeling of humility.

Figure 1. David Lee, at the age of 8 (1950) on the shores of Blue Lake, about 25 miles north of Ephrata in the Grand Coulee. My sister Mary Ann is on the right, and my mother Mary is on the left. I learned to swim at this beach.

Figure 2. Close high school friends, taking a break on the high ridge above Deep Lake in the Alpine Lakes Wilderness of the North Cascades, 1961. Left to right: Curt Amo, Alan Lindh and John Freer

Figure 3. The Fairbrothers lab group taking a roadside rest during a field trip in upstate New York, 1968. From the left: David Fairbrothers, Jerry Pickering, Gary Hildebrand, Roy Clarkson (a professor on sabbatical leave from the University of West Virginia), Steve Osborne (sitting), and Art Tucker.

Figure 4. Rod Sharp, taken during a free-wheeling night discussion at Nelson Labs, Rutgers University, around 1967.

Figure 5. My Malaysian mentors on the Genting Ridge, in the mountains outside of Kuala Lumpur, 1973. We were looking for specimens of the rare and newly discovered Malaysian citrus, *Citrus.* Left to right: Peter Ashton, Benjamin Stone, and Brian Lowry. Our dog Pooch (whom we received from Brian) is in front.

Figure 6. David Fairbrothers and myself, showing off our new hand-printed batik shirts at the banquet for the inauguration of Rimba Ilmu, in 1974.

Figure 7. Carol and myself on the steps of our
home at Kampong Sungai Serai, in Ulu Langat,
Malaysia, in 1975. Pooch is with us.

Figure 8. Carol and I with our newly born son Sylvan,
at the Mogère near Montpellier, spring of 1978.

Figure 9. The Lee family at our home in Miami, winter of 2012. Top, left to right: David, son Sylvan, Sylvan's former girlfriend Liz. Bottom, left to right: Carol, daughter Katy, grandson Shaun.

CHAPTER 10.

Our Journey

Sally Miller and Chip Styer

We were happy to be asked to contribute to "Reflections and Connections", and decided early on to write it together as a couple. It would be hard to reflect on our lives separately, as we have been together more than 40 years. We met while still teenagers, married young, and were blessed with three wonderful children. Our professional interests at first ran in parallel, then diverged as we grew into careers that suited us. Sally remained in biological science and Chip moved on to his true calling in computer science and data management. Our professional stories are not yet complete – we don't anticipate retirement for another 6-8 years. However, this is a good time for reflection. When our son entered OSU in 2012, we did not have a child at home with us for the first time in 32 years. Our elder daughter will be married next year, and we made the weighty decision to downsize and will soon move from the 1892 Victorian that has been our home for the past 13 years to a brand new downtown condominium. So we are in a period of personal transition, which inevitably leads to thinking about where we have been, where we are and where we are going.

Coming of Age – Sally's Family Influences

I was born in Canton, Ohio into what would eventually become a rowdy family of seven children. I was third in line, for most of my childhood the official middle child, and the middle daughter. My parents, Stanley and Eileen Larke Miller, were very hard working people who had high expectations of all of us. Neither had attended college; my father couldn't wait to get out of high school, and joined the army just after World War II ended. He was posted in US-occupied Germany and re-capped tires in the Ober-Ramstadt Ordinance Tire Repair Shop, rising to the rank of sergeant. My mother was an excellent student and had been offered a partial scholarship to attend a university in Ohio. However, her family could not afford the remaining tuition and housing costs, and she accepted work in a Hercules engine factory in Canton. She started dating my father after he returned from the service, and they were married in 1950. Although attending, and much less graduating, from a university was uncommon for women in the 1940s, my mother regretted the lost opportunity, and later made sure that all of her children would go to college. My father found work in a steel mill in Canton, rapidly rising to foreman, and my mother started having babies. But my father was restless in the mill and wanted something more for his growing family. As a teenager he had helped a family friend with carpentry work, but had no formal training. He bought a book entitled "Your Dream House for $2,000" and proceeded to build the modified Cape Cod home where, through numerous expansions over the years, my siblings and I grew up and my parents still live today. Based on their success with their own home, my father left the steel mill and my parents started a construction business. My father built things and my mother handled all of the bookkeeping. They started very small but eventually built a business that continues today in the capable hands of two of my brothers and my younger sister. That construction business provided the hard-won economic security that allowed my siblings and me to grow up with the expectation of being able to do whatever we wanted to do in life. My parents were both extraordinary people but in particular my mother set an example of strength in the face of many challenges. My father was known for his physical strength and upright character. My mother took care of things – her husband, her children, and her own parents, and was generous to others. She was an excellent and efficient cook, providing us all with a hot breakfast and dinner and a packed lunch every school day. But her life as a mother was not easy. Two of my brothers developed pyloric stenosis as infants and had to undergo major surgery. My youngest brother was born about 2 months prematurely and my mother's recovery from the Cesarean section was very difficult. Most difficult of all, my older brother contracted a particularly virulent form of rheumatic fever as a small child, and the damage to his heart valves eventually led to his death at the age of 23. Through it all, my mother persevered. She was in some ways the quintessential 1950's woman, family-oriented and (mostly) deferential to my father. On the other hand, she managed not only the financial aspects of the construction business, but also handled our family finances. She was an active member of several organizations, including the Federated Women's Clubs and the National Association of Women in Construction, for which she served as chapter president from 1977-79. I am certain that the teenage me did not grasp that my mother was my first and best role model.

Coming of Age - Chip's Family Influences

I was born in El Paso, Texas, the youngest of the five children of Donald and Carol Opal Hancock (Kit) Styer. My father was a dentist in the US Army Dental Corps, and my mother was a homemaker. My dad was born and raised on a prosperous dairy farm in Dunn County, Wisconsin. My grandparents recognized early on that he was exceptionally intelligent, and saved enough money to pay his tuition at Stout College in Menomonie (now University of Wisconsin-Stout). Having decided early in his life that milking cows would not be a part of his future, he was happy to go to college. After 2 years, in 1939, he transferred to the University of Minnesota where, in addition to his studies, he ran a boarding house, served drinks at "The Coliseum" and set bowling pins to pay his expenses. After graduating in 1941, he entered dental school at the University of Minnesota under the auspices of the US Army, and graduated in 1944. In his military career he served in Japan, Korea, Germany and many posts in the US. He opened many dental clinics, trained clinic staff, and interestingly, pioneered computer technology for dental analysis in the 1960s. He met my mother at the University of Minnesota, where she was a student. My mother was born and raised in the small town of Appleton, Minnesota. Her father, General Lester Hancock, was the commander of the Minnesota National Guard as well as an officer in the US Army during World War II. She and my father were married in 1944, and after his brief stint in private dental practice, became an Army wife for good. Although my mother did not graduate from the University of Minnesota, she and my father both placed a high value on education. I haven't mentioned yet that I was the only son in our family. Growing up with four older sisters could be considered a challenge to some, but my sisters mostly doted on me and I learned at an early age how to get along with women. Like other military families, we moved often. After El Paso, I lived in Neosho, Missouri, and then in Albuquerque, New Mexico. After first grade, my father was transferred to Heidelberg, Germany. That meant all seven of us drove across country in a station wagon from New Mexico to New Jersey to board a ship for the Atlantic crossing. After three years in Germany, we moved back to the states and settled in Annandale, Virginia, while my father had several assignments in the Washington DC area. Thanks to his willingness to commute, we lived in the same house for 8 years, and I was spared any more moves between 5th and 12th grade. I was pretty much left to my own devices academically in high school, but my sisters were all exceptionally good students and I followed suit. I graduated at the top of my class of 700 from W. T. Woodson High School in Fairfax, Virginia.

Chip's Early Interest in Science and Technology

My friends and I spent almost every summer day playing outside in northern Virginia. We dammed the small stream running through a nearby field, or played combat in a dense pine forest. While we were learning important lessons about nature, none of the learning was formal. The only structured learning took place while I helped my mother with her flower gardens. It was not until my eldest sister, Linda Styer Caldas, installed a dendrometer around a pine tree in our backyard that the idea that people actually studied nature occurred to me. And it was several years later, while visiting Linda in her lab at The Ohio State University, that I realized that "being a biologist" was a real profession. Linda was working on her doctoral degree and showed me dozens of small carrot plants growing in sterile gel medium in glass jars. That was my first formal exposure to what I considered real science.

When it came time to pick a college, my choices were OSU, Indiana University, where my sister Sandy had graduated, and the University of Alabama in Huntsville. I applied to UAH since it had an undergraduate major in aerospace engineering. This was an interest I developed during the summer after my junior year in high school while volunteering as a tour guide at the Smithsonian Institution's Air and Space Museum in Washington D. C. I built and launched model rockets during my senior year, but at the time I did not appreciate the differences between a career in engineering and one in biology. When it came time to decide on college, I stayed in my comfort zone; I had spent a few days at

OSU, a few hours at IU, and had not set foot in Alabama, so I chose OSU.

Sally's Mushrooming Interest in Science

My favorite childhood memories were of exploring the woods, grasslands, hills and streams near my home. While I didn't know the names of more than the most common plants, I loved being outdoors among them. I loved the spring violets, May apples and dogwoods, the summer milkweed and asters and the truly spectacular autumn colors of the maple trees that I had climbed all summer long. Every spring I stood on the back of my father's old Farm-All tractor as he and I bumped along the uneven path bordering the power lines to the woods where we searched for morel mushrooms. We often came home empty-handed from our morel expeditions, but when we hit the jackpot the whole family enjoyed the tasty morsels that my mother dredged in flour and fried in butter. I truly believe that those carefree expeditions into the woods and fields formed the foundation for my life-long interest in, and excitement about, plant science. I didn't know at the time that hunting morels was my first introduction to mycology.

My parents believed that each of their children should have a solid Catholic education, and we attended Catholic elementary and high schools in Canton. I could diagram sentences with the best of them and won the class spelling bee every other year, including 4th place in the Youngstown Catholic Diocese spelling bee in the 8th grade. However, with the exception of Sister Mary Florence's class in the 7th grade, I had very little science education in elementary school. That changed immediately upon entering Central Catholic High School, where I had the great privilege of being taught honors freshman and senior biology by Sister Mary Dolores Staudt. Sister Dolores was an outstanding teacher who introduced us to Charles Darwin and the theory of evolution, photosynthesis and respiration, Louis Pasteur's germ theory, and much more. She also taught experimental design and encouraged research and independent study. I was taught by other excellent nuns and lay teachers in math, chemistry, physics, French, English, history and literature, so while I felt somewhat socially awkward in high school, the quality of education CCHS afforded me set a firm foundation for my career in science. I was also introduced to discrimination in my high school. I had written an editorial for the school newspaper about unequal treatment of women in the Catholic Church, and was summarily called to the (male) principal's office and roundly berated for my views. My article was never published in the school newspaper but my commitment to women's rights was strengthened from that day onward.

I knew very little about colleges and universities and applied only to OSU, where my older sister was enrolled. I didn't know until later that my father was staunchly opposed to my sister and me going to Columbus – mainly a result of anti-war demonstrations on the OSU campus and many others in the late 1960s and early 1970s. The Kent State University shootings on May 4, 1970, practically at our back door, had deeply affected many parents, and my father was very concerned about sending his daughters to what he worried was a potentially dangerous and possibly subversive environment. Couldn't we just stay home, find husbands and start families? My mother, however, weighed the benefits of advanced education over very unlikely dangers, and insisted that we go. So we did, and happily. I met Chip in an Honors Classics course my very first quarter in Columbus. It was through Chip that I met Rod Sharp, a connection that has influenced almost every phase of my professional career. Chip knew Rod through Linda Styer Caldas, although Linda had already earned a Ph.D. and had left for Brasil by the time we arrived on campus. Rod was then a professor of Microbiology at OSU, and agreed to mentor us both in biology, including our Honors thesis studies. My project in Rod's lab involved an analysis of the effects of asbestos on hydroponically-produced plants. That small project introduced me to the nuts and bolts of scientific investigation, and to the rigor required to conduct valid studies. My coursework focused on plant science and microbiology, which I found very interesting. When the time came to think about graduation and what to do next, Rod suggested graduate school. At that time in the 1970s, the US economy was struggling and prospects for jobs in biology with only a Bachelor's degree did not appear bright. Rod was aware of some good labs involved with plant tissue culture research in the Department of Plant Pathology at the University of Wisconsin-Madison and encouraged me to apply. I had never heard of plant pathology; at OSU plant pathology was in the College of Agriculture, while biology, microbiology, botany, etc. were in the College of Biological Sciences. "Ag" was "across the river" and I knew nothing about it. But I learned that plant pathology combined the disciplines of plant science and microbiology that I found so interesting. I was elated to be accepted in a M.S./Ph.D. program in Professor Doug Maxwell's lab in the Department of Plant Pathology at UW-M. I started out as a buckeye and became a badger, only later to become a buckeye again.

Chip's Take on Plant Science

With an undergraduate degree in microbiology, my most likely job was working in a hospital laboratory. Having taken immunology at OSU, I had already figured out that medicine was not for me, so working in a hospital was not very attractive. So I also applied to graduate school at the University of Wisconsin-Madison. During my interviews in the biochemistry department, I realized that I probably wasn't interested in or prepared for a career as a biochemist. Having finished in the mid-afternoon, I waited at the Memorial Union for Sally. She spent another two hours in the Plant Pathology department, and, in the days before cell phones, I just had to wait. It was very cold and snowing like crazy and I wondered if coming to Wisconsin would really be a good idea. But we compared notes at the end of the day, and Sally suggested that I look into plant pathology. After all, plant pathologists are microbiologists who work on plants. This was a field that I thought could interest me, and I subsequently accepted a graduate appointment with Dr. Rick Durbin.

Graduate School at the University of Wisconsin-Madison
Sally Makes Plant Pathology her Career

Chip and I arrived at UW-M on July 1, 1976, 2 weeks after we were married in Canton, Ohio. Professor Maxwell was in Germany on a Humboldt Fellowship for the year, so his postdoc and former student, Dr. Vivien Armentrout, took charge of my introduction to plant pathology. My introductory plant pathology lab was taught by Professor Steven Slack, who is now the Associate Vice President for Research in the OSU College of Food, Agricultural and Environmental Sciences and Director of the Ohio Agricultural Research and Development Center (OARDC). Chip and I both work under Steve Slack now, giving credence to the old adages that 1) the world is a smaller place than you think, and 2) building bridges is much smarter than burning them. Fortunately we must have done the right things, as we have good working relationships with Steve at OSU-OARDC.

The UW Department of Plant Pathology was highly ranked within the discipline, and in the 1970s included two members of the National Academy of Sciences, Professors Luis Sequeira and Arthur Kelman. The atmosphere was vibrant and collegial, with many leading-edge research labs, active Extension programs, and a graduate student cohort of 50, including for the first time a substantial number of women. Professor Maxwell's lab was just transitioning away from fungal physiology and into work on *Phytophthora* diseases of field crops. My research focused on "host" and "non-host" interactions of alfalfa with *Phytophthora megasperma* formae speciales from alfalfa and soybean, respectively. These formae speciales have now been elevated to species and renamed *P. medicaginis* and *P. sojae*. I spent countless hours at the light microscope, microtome and transmission electron microscope examining the process of colonization or attempted colonization of alfalfa seedlings by these pathogens (*Can. J. Bot.* 62:109-116). I was also able to document specific physiological changes associated with these interactions at the cellular level (*Can. J. Bot.* 62:117-128). We were interested in the comparative biochemical changes occurring in alfalfa challenged with *P. medicaginis* and *P. sojae*, and focused on analysis of phytoalexin production, using an alfalfa tissue culture system (*Phytopathology* 74:345-348). Based on this work, Rod Sharp asked me to contribute a chapter "Evaluation of Disease Resistance" to Volume 1 of his *Handbook of Plant Tissue Culture* series. Rod had kept in touch with Chip and me during our graduate studies and was aware of our research interests. As a mere graduate student I was apprehensive about taking on a task I believed was the purview of much more senior "experts", but Rod, in his usual encouraging and supportive way, convinced me that I could do it.

Our first daughter, Allison, was born in 1981 while we were graduate students at UW. Doug Maxwell was supportive and allowed me the extra time I needed to complete my research while I shared childcare responsibilities with Chip. We organized our schedules so that one of us left very early for his or her lab, and the other came back late, and we only needed a babysitter for about 4 hours a day. The first day I left Allison with the babysitter was one of the hardest of my young life. I was very busy and sometimes rather stressed between research work, writing my thesis and taking care of an infant, including breastfeeding. Fortunately, having helped care for siblings from 7 to 16 years younger than myself, I was confident and comfortable with our little miracle. Chip did not have this advantage but was a quick learner and soon was a first-rate example of modern fatherhood, taking on responsibility for her care, with the obvious exception of breastfeeding, at a level of effort at least equivalent to my own. I finished writing and defended my Ph.D. thesis in the late spring of 1982, when Allison was about a year old, and prepared to start the next phase of my career.

Chip's Interests Extend Beyond Plant Pathology

I joined the lab of Dr. Rick Durbin, one of several USDA Agricultural Research Service (ARS) scientists embedded in Plant Pathology Department. Dr. Durbin worked in the area of microbial toxin characterization and mode of action, and I chose a project on *Pseudomonas syringae* pv. *tagetis*, a pathogen of marigold, sunflower, ragweed and other plants. This pathogen produced tagetitoxin, a toxin that caused a striking apical chlorosis of infected plants. Most of my memories from graduate school are not science-related, but more of the brat and beer celebrations after a student passed an exam or earned a degree, and the volleyball and softball games we played. The intensive summer field course, where we learned about crops and diseases also was memorable. So was walking to or from the lab in below-freezing weather – our first winter in Wisconsin in 1977 was one of the coldest on record. We really wondered what we had gotten ourselves into so far north. But, we were fortunate to work and socialize with so many smart students who were dedicated to science. Many of our friends later became faculty members at Land Grant Universities, while others contributed to plant science by working at seed and crop protection companies, the USDA, the US Agency for International Development (USAID), and others. My office mate and softball teammate Cleo D'Arcy joined the Department of Plant Pathology at the University of Illinois just after graduating from UW, and along with a stellar career in research and teaching, including co-authoring an introductory plant pathology textbook, served as president of the American Phytopathological Society (APS). We played volleyball with Jan Leach, a student of Professor Sequeira and now an internationally known researcher in the area of host pathogen interactions, and John Sherwood, now the chair of the Plant Pathology Department and Associate Dean at the University of Georgia, who were also APS presidents. Sally has just been elected APS vice-president, and will serve as president in 2015/2016. That is a lot of leadership from one cohort of students.

DNAP Calls Us – What Sally Remembers about DNAP

During our last year of graduate school, we started talking with Rod and Doug about careers. Doug had suggested postdoctoral positions at Cornell University in Ithaca, where I would work on *Phytophthora* and Chip in plant bacteriology. By then Rod had left OSU to found DNA Plant Technology Corp., and as usual was looking out for us. Rod was able to pull together two positions, one in plant pathology and one in plant tissue culture. Rod asked us to choose; we talked it over and as Chip had not found plant pathology to be as compelling as I had, the choice was easy – and Rod was very persuasive. So I continued to pursue my interests in plants and microorganisms, and Chip was able to satisfy his knack for engineering. We picked the burgeoning biotech industry over academia – while we loved the academic environment of UW-M, we both felt that we had the opportunity to experience something new and exciting with DNAP. So we packed up our meager grad student belongings and headed to New Jersey, with Allison in tow. We began working at DNAP on July 1, 1982.

For my first year or so at DNAP I worked on tissue culture projects to improve disease resistance in tomato. However, Rod had something else in mind, and I soon began working with him on his idea to start a spin-off company that would focus on developing and marketing diagnostic tools for plant diseases and other uses. The technology for producing monoclonal antibodies had only recently been introduced into the mainstream, and made rapid detection of relatively complex microorganisms like fungi and bacteria feasible. Until then, use of serological tests in plant-based agriculture had been almost exclusively limited to virus detection. Rod chose to start with products to detect fungal pathogens of turfgrass, focusing on the golf course industry. The idea was to tie fungicide use to accurate diagnostics, removing the guesswork for the golf course superintendent and at the same time eliminating incorrect or unnecessary applications of these pesticides. Rod secured the funding and pulled together a technical team that included Dr. Dave Grothaus, an OSU grad and microbiologist, Dr. Frank Petersen, an expert in monoclonal antibody/hybridoma development, and Dr. Jim Rittenburg, who was very clever with the technology side of serological diagnostics. My role was on the plant pathogen side, and I was asked to participate in many stages of product development, from marketing to applications testing, quality control and tech sheet preparation. I also enjoyed the more basic biology, working with pathogens and cropping systems. The company was called Agri-Diagnostics Associates (ADA), and during its time it was at the cutting edge of plant disease diagnostics. We developed the forerunner of today's rather ubiquitous Lateral Flow Device (LFD). Our product, called "Reveal", was a "flow-through" assay that for the first time allowed pathogen diagnostics in the field in a short time (less than 10 minutes). The test could be done in a superintendent's office or even out on the course. We soon also branched out into other pathogens and crops, and in collaboration with Ciba-Geigy (now Syngenta Crop Protection) developed the "Alert" product line that included both the flow through technology and lab-based 96-well enzyme-linked immunoassay (ELISA) formats.

Ciba-Geigy commissioned these assays to promote the proper use of its fungicides against *Phytophthora* on soybean and other crops, *Septoria* diseases (now *Stagonospora nodorum* and *Septoria tritici*) of wheat, and *Mycosphaerella fijiensis* (black Sigatoka) of banana. This new technology caused quite a stir in the world of plant pathology, and I was invited to present seminars at various universities and make presentations at annual meetings. I was also invited to write book chapters and articles about monoclonal antibody applications in plant-based agriculture, including and Annual Review of Phytopathology paper "Molecular diagnosis of plant diseases" published in 1988 (Annu. Rev. Phytopathol. 26:409-432). Beginning in 1988, I organized a National Workshop on Rapid Diagnostic Assays for Plant Pathogens held during the APS annual meeting. Its purpose was to acquaint professional plant pathologists and other scientists with modern techniques for pathogen detection and disease diagnosis. I met hundreds of plant pathologists during the 10 years that I conducted these workshops, providing untold networking opportunities that have contributed to my entire career.

My US contact at Ciba-Geigy during the 1980s was Dr. Allison Tally, a plant pathologist of the same age and family situation as myself, and our early interactions led to a friendship that continues today. As the only female Ph.D. scientist at ADA, and one of very few in DNAP, I appreciated the opportunity to work with someone who understood the challenges of family and career for working women. Chip and I expanded our family by the addition of our daughter Carly in late 1985. I found motherhood very challenging away from the nurturing environment of academia, with little support among my colleagues and superiors (with the exception of Chip and Rod Sharp) for my choice to have another baby. I worked hard to maintain my workload as if I had not even had a baby – in fact I found myself on an airplane for a business trip a mere 3 weeks after giving birth. Now I can only think "how crazy was that", but at the time I felt that my career would stall without that type of effort. Fortunately, like her sister, Carly was a happy, healthy baby and grew into a beautiful, sweet child. Allison and Carly were wonderfully close as children, and I can't remember ever hearing a serious disagreement between them, even as adolescents and now adults. While life was hectic, home was a peaceful, calm respite from the pace I kept at the lab.

I learned a great deal about career and life during the 9 years I worked at DNAP and ADA and despite some difficult times, I am grateful for the opportunities the companies afforded me. I was able to pursue my interests in plant pathology while working on cutting edge science. My lab at ADA was staffed with smart and dedicated researchers and support staff, particularly Drs. Mike Klopmyer, Tony Joaquim and Karen Plumley. Karen was an excellent technician who had the knack of knowing what needed to be done at any given time and the almost boundless energy to do it. A real Jersey girl, Karen left ADA to pursue a Ph.D. in plant pathology at Rutgers University, and has made a successful career in the crop protection industry. I learned how to work, despite a few bumps in the road, sometimes of my own making, with a multidisciplinary group of scientists who did not always share the same vision but in the end came up with some truly exciting products. Developing applications for the diagnostic tests allowed me to travel widely in the US, as well as in Europe and even Costa Rica. I met and worked with excellent plant pathologists like Professors John Menge and Jim MacDonald at the University of California, Pete Timmer at the University of Florida, Bruce Clarke at Rutgers University, and Alex Csinos at the University of Georgia. In Europe, I worked with Ciba-Geigy scientists Drs. Ludwig Mittermeier and Wilhelm Dercks. Their enthusiasm for this technology was infectious – they were all excited about the chance to apply these diagnostics to real-life problems in agriculture. Professors Fritz Schmitthenner and Mike Ellis at OSU did some very interesting work with the kits. Both were located at the OARDC in Wooster, and I made a habit of visiting the department twice a year when Chip and I came back to Canton, only 30 miles away, to visit our extended family. While my intention was to check in on their research progress with our kits, I also had the opportunity to meet the plant pathology faculty at OARDC. I was very impressed by their collegiality and dedication to education, outreach and research. Fritz was a leading authority on *Phytophthora* and *Pythium* species in soybeans, while Mike was an applied researcher and Extension specialist for fruit crops whose advice was highly sought after. Other members of the faculty were equally impressive. Professor Randy Rowe had excellent research and extension programs on vegetable disease management, and Professor Harry was known worldwide for his work on composting and biological control of plant diseases. Professor Larry Madden was a highly regarded plant disease epidemiologist. Professor Pat Lipps was a very knowledgeable applied researcher and extension pathologist working on grain crops. I was often invited to coffee break during my visits for informal chats and I enjoyed those opportunities to become immersed in plant pathology, something I truly missed at ADA.

Return to the Buckeye State

I became aware of a faculty position in the Department of Plant Pathology at OSU in Wooster in the spring of 1991. Chip and I talked it over at length and we both agreed that I should apply. While the work at ADA was going well, we felt that our daughters would benefit from being closer to their grandparents and aunts and uncles. And I was beginning to feel the urge to move on to something different. So I applied, interviewed and received an offer to join the OSU faculty. What I remember about the interview was wearing one of my favorite dresses – a black shirtwaist with white collar and cuffs - and a particular question after my seminar. Someone asked me how the faculty could judge my work since I had very few peer-reviewed publications during my time in industry. I thought about it for a few seconds then replied that every time someone uses an ADA diagnostic kit he or she has the opportunity to judge my contributions to science. I think that is how we should always consider scientists working in industry, where peer-reviewed publications are not usually encouraged. New, innovative, effective products are the result of the contributions of many hands and minds, and these contributions should not be underestimated.

Chip's Experience at DNAP

I first had to learn somaclonal plant tissue culture, the basic technology for plant improvement at DNAP. We worked on tomato and carrots, and later coffee and other crops. My engineering interests finally had a chance to be expressed, during the development of plant bioreactor systems for somatic embryos. While I found the engineering part of the projects interesting, I still did not recognize it as my true professional destination. In 1986, DNAP purchased my first personal computer, a Macintosh Plus. Soon, I was doing simple data analysis with a spreadsheet, and making, what are by today's standards, very crude illustrations for DNAP presentations. But at the time they were innovative, and relatively easy to produce on demand. So that part of my job continued to grow. The next big step, and one that proved to be irreversible, was when I designed and installed a data acquisition system from Hewlett Packard that captured and analyzed the raw digital data from lab instruments, in this case primarily gas chromatographs. This automation increased the number of samples we could analyze by ten-fold, and flipped the analytical lab from a cost center to a revenue source. At this point, I understood the transformative power of these new machines. My plans to expand the network of PCs into other research groups at DNAP were not appreciated, so I set off on my own as a computer consultant. So when Sally was offered a faculty position at OSU, I was both ready and able to move back to Ohio. Unlike many dual-career couples, Sally and I never faced the difficult collision of competing careers.

Chip's Ohio Experiences

After we moved, I continued consulting for several years until Sally became pregnant with our third child. The attraction of steady income led me to take a job in a computer sales company, working with major corporate clients. During the seven years of selling computers, I witnessed first-hand the shrinking profit margins on hardware that would inevitably follow the wide scale adoption of personal computers. My initial typical sales commission was based on a gross margin of 20% on a $5000 computer, which collapsed to 3% on an $1800 computer. At that point, I was looking for another opportunity.

That opportunity first appeared in a phone call from Dr. Skip Nault, then the OARDC Director, asking if I would come in for an interview. The job description was for a support position, specifically for Macintosh computers for the directors, a not-too-appealing job. However, during the interview with Dr. Bill Ravlin, Assistant Director at the time, it became clear that the position could be much more interesting with a broader range of responsibilities including more data management. Judging from my previous professional history, I accepted the job figuring I would stay at OARDC for five or six years. Now in my fourteenth year, I guess I have found my niche. I did spend some of my time in Mac support, but my more interesting projects involved making sense of reams of financial, operational and human resource data. In a time of almost continuous budget cuts from state and federal sources, it became clear that universities had to be able to document their impact to stakeholders. Dr. Ravlin recognized the need to collect faculty activity information in a digital format, and assigned the task to me. Only four months after purchasing the AICS application from Washington State University, and after making some OSU-specific customizations, OARDC and our college launched our version of the application, the Unified Reporting System in January 2001. Many of the faculty, especially from departments that had not taken annual reporting seriously in the past, did not like the new system, to say the least. I had to draw on most of my patience (and I have a lot of patience) while cajoling the departments into acceptance and adoption of this system. Over the next couple of years we made many design and infrastructure improvements to improve performance and address user requirements. It was then my job to analyze the data submitted by the College faculty and present it in a form that was usable by the administration. For the first time we had in place a centralized data source that documented the depth and breadth of faculty activities and their impacts. One of the biggest advantages of the URS was that administrators could obtain information on faculty research and impacts, without the direct involvement of the faculty member or the department chair. I was often asked to find out which faculty members were working in a particular research area and what funding was being generated, for example, and this system, the first of its kind at OSU, made it possible to mine the data efficiently. While the College of Food, Agricultural and Environmental Sciences (CFAES) was using the URS, the OSU Medical School started an independent development project of a similar reporting system, called OSU: Pro, geared towards the needs of the medical faculty. OSU: Pro had many advantages over the URS, including direct links to other internal OSU databases that streamlined some data entry. However, the new and improved user interface caused headaches for many faculty as they were required to switch reporting systems when the university adapted OSU: Pro university-wide in 2009. The one component that OSU:

Pro was missing was a module to track activities of the college's Extension faculty and staff. We had built an Extension module for the URS, so I collaborated with the OSU: Pro development team to duplicate this module in OSU: Pro. Eventually most faculty became proficient at using OSU: Pro, only to see it replaced by the Thomson Reuters "Research In View" application a couple of years later. This remains a work in progress, and I continue to assist OARDC and CFAES in reaching acceptable middle ground between the data needs of administration and reporting burdens on the faculty.

While there are many examples of the importance of education on a person's life, there is none better than Sally's experience. I often tell the story that Sally, who grew up in Ohio, had never flown in an airplane when we met at 18, and aside from Canada, had never been in another country, indeed, she had never even been west of the Mississippi River when we were married at 22. And now Sally has literally flown around the world dozens of times, and is a "Million Miler" on Delta. Graduating from OSU and UW-M opened the entire world to her. Obtaining a Ph.D. most directly built on her undergraduate degree, and opened opportunities in industry. Being part of a commercial venture expanded Sally's worldview, and provided her with new contacts both here in Ohio and abroad. Moving into the academic world propelled an exponential growth in her international activities, and Ohio and the world are better for it. It is education that was the pivot point, one that turned a small town girl in to a plant disease expert and an international traveler who can land on any continent knowing there will be welcoming hearts to greet her. It was only many years after my migration from biology to computers, especially when Sally became immersed in international development projects at OSU, that I realized my role in plant pathology was never as a scientist, but to serve as a scout to help her find her calling. If I had gone to the University of Alabama to study engineering, we would never have met, Sally would never have crossed paths with Rod Sharp, and she would likely not have become a plant pathologist.

Sally's OSU Experience

I was hired as an Assistant Professor of Plant Pathology, and as a State Extension Specialist (35%) and researcher (65%) with a focus on vegetable crops. I was the first female member of the faculty of the Department of Plant Pathology, although not the first woman to work in plant pathology at OSU. The first was Dr. Frederica Detmers, who in 1891 was also the first person to receive an M.S. degree in botany from OSU, which was also the first advanced degree in plant pathology in Ohio. The intervening 100 years were a long period of male domination. After my experience working primarily with men at DNAP and ADA, I was not particularly apprehensive about joining an all-male faculty. And I am happy to say that the OSU faculty, not only in my department but in our College, were very welcoming and eager to collaborate.

Chip and I expanded our family unit one more time in 1993 with the birth of our son Mike. OSU, like other major universities, had made progress by then in helping assistant professors who became new parents to navigate the promotion and tenure process successfully. I was still an assistant professor and might have had some concerns, but my worries were soon eased. Dr. Rowe, my fellow faculty members and even the vegetable growers with whom I interacted so frequently were supportive. OSU had instituted a policy whereby new parents could add a year to their "tenure clock". While in the end I did not need to make use of this policy, it was comforting to know that it existed. Chip was by this time a father *par excellence* and supported our growing family not only emotionally but also by taking more than his share of responsibility for child care and household duties. My third pregnancy was by far my easiest from a professional and personal point of view, although having a baby at nearly 40 is a lot harder physically than having one at 28. I never felt the need to prove to my colleagues at OSU that pregnancy and motherhood would not make me less of a scientist or contributor.

My job description at OSU specifically required me to conduct applied research on the management of bacterial and other diseases of vegetable crops. As I had never worked with bacterial phytopathogens, or with vegetables, I experienced a pretty steep learning curve the first few years at OSU. However, I had a lot of help from the other faculty members at OSU, as well as an amazingly collegiate network of vegetable pathologists nationwide. One of my first and strongest mentors was Professor Steve Johnston of Rutgers University. He was a leading expert on Phytophthora blight of peppers and cucurbits, bacterial diseases of fruiting vegetables (e.g. tomatoes and peppers) and many others. In fact, he was a walking encyclopedia of disease management knowledge, who was always willing to help out. Unfortunately, Steve passed away far too early in a tragic accident. Randy Rowe was very generous with his time in helping me learn the ropes in vegetable pathology, and personally took charge of me my first year, introducing me to growers and in general providing invaluable guidance. Randy later became the chair of our department, and continued to serve as a valued mentor until his retirement a few years ago. Although I had spent the first part of my career working on plant disease diagnostics, I was not a diagnostician and I had to learn to diagnose diseases of the great variety of Ohio's vegetable crops. I did not consider myself a skilled diagnostician for a number of years after joining OSU, and still am stumped by unusual specimens and situations I encounter after more than 20 years in the field. My lab serves as part of the Ohio Plant Diagnostic Network, an umbrella group that includes OSU's C. Wayne Ellett Plant and Pest Diagnostic Clinic, the Ohio Department of Agriculture Diagnostic Laboratory, and diagnostic labs run by faculty in our department. We routinely diagnose hundreds of vegetable samples, mostly from commercial growers in Ohio, other states and other countries. In addition to these physical samples, we also respond to many (I have not actually counted them) digital samples sent by email and text. We recognize that accurate diagnosis is an absolutely necessary first step in plant disease management, and make every effort to support growers struggling to manage yield-robbing diseases. So Rod Sharp's approach to making diagnostics easier and more fool-proof in order to take the appropriate steps to control disease was remarkably on-target.

In addition to learning how to be a "Plant Doctor", I continued to conduct research on plant disease diagnostics and sustainable approaches to vegetable disease management. Without the backup of a lab dedicated to monoclonal antibody discovery and someone like Frank Petersen to manage it, I found it impractical to continue to do research on monoclonal antibodies against plant pathogens. However, we have returned to serological tests when needed. For example, a collaboration with Drs. Doug Luster and Reid Frederick of USDA ARS, Jill Czarnecki of the Naval Medical Research Center and Professors Anne and Mike Boehm of my department resulted in the development of an immunofluorescence assay to detect airborne spores of the soybean rust pathogen, *Phakopsora pachyrhizi* (*Plant Dis.* 92:1387-1393). This disease has caused severe economic damage since its recent introduction into Brasil, and it continues to be monitored in the US to predict its potential entry into the soybean belt. Monoclonal antibodies were also developed (*Appl. Env. Microbiol.* 78:3890-3895) and have been licensed to a diagnostics company for field tests.

I was pleased to learn several years ago that a monoclonal antibody against *Phytophthora* that we developed at ADA in the 1980s was incorporated into a lateral flow device to screen woody plants imported into the UK for *Phytophthora ramorum*. This destructive oomycete causes sudden death of oaks and other trees and woody ornamental plants, and cannot be treated effectively with fungicides; therefore exclusion is the most important preventative measure. In the UK, the *Phytophthtora* LFD test was carried out at import inspection stations, and positive-testing samples were sent to specialized labs for confirmation. This proved to be highly effective in preventing the pathogen from entering the UK. While I was disappointed that ADA did not remain in business after the mid-1990s, at least some of our products are still being used today.

By the early 1990s, polymerase chain reaction (PCR) assay technology was being adopted quite rapidly in plant pathology, and proved to be an excellent tool for diagnostics. The ability to generate genome sequence data took much of the guesswork out of assay development. PCR primers could be generated easily based at first on a number of tricks used to identify unique sequences, and later on sequencing targets within the genome. Mega-databases such as GenBank could be used with BLAST software to compare sequence data among many more organisms than could be screened physically. While we continued to use serological assays in laboratory and field diagnostics, PCR became our go-to technology for detection and classification of plant-associated microorganisms. For example, Dr. Pervaiz Abbasi, a postdoctoral associate in my lab, utilized Random Amplified Polymorphic DNA (RAPD) analysis to identify unique amplicons from an isolate of *Trichoderma hamatum* (Th382), an effective biocontrol agent. These were then sequenced and PCR primers developed. The subsequent PCR assay (*Appl. Env. Microbiol.* 65:5421-5426) has been used to confirm the identity of Th382 in our lab and others during product development and testing for more than 15 years.

We utilized PCR technology to identify new pathogens on vegetable crops in Ohio and to better understand the taxonomic relationships between groups of phytopathogens. While we have worked with a number of genera of bacterial phytopathogens, our early efforts in the 1990s involved characterization of strains of *Xanthomonas campestris* causing very serious and economically important diseases of vegetables in Ohio and much of the eastern half of the US. My first Ph.D. student, Fikrettin Sahin, came to OSU on a scholarship from the Turkish government and dove into this problem with a great deal of energy. He characterized hundreds of strains of *Xanthomonas* and helped establish our lab in the field of phytobacteriology (*Plant Dis.* 80:773-778; *Plant Dis.* 81: 1443-1446). Fikrettin completed his Ph.D. degree and returned to Turkey, where he has had a stellar career and currently leads a very large, diverse group of scientists and students as Chair of the Department of Genetics and Bioengineering at Yeditepe University in Istanbul. We have continued to interact in a number of areas, and I have truly enjoyed watching his career progress so successfully.

While I mentored a number of students in similar areas, the bread and butter of my program continued to be applied research to address specific disease problems in vegetables. For bacterial diseases of vegetables, this has been frustrating since we have tried many approaches, but as with researchers that came before us, have not come up with a silver bullet. There are no bactericides or antibiotics that are highly effective against bacterial diseases of vegetables. The treatment of choice is currently to apply a copper-based product multiple times throughout the growing season. Unfortunately, rapid evolution of phytobacteria towards insensitivity to copper rapidly diminishes the utility of this approach, and overuse of copper also causes environmental problems. My students, staff and I have worked hard to develop or adapt intermediate, integrated measures that when taken as a whole will reduce the negative impact of bacterial diseases of vegetables, including clean seed protocols, biological control, and use of systemic resistance inducers and cultural practices. The bully pulpit afforded to me as a state extension specialist has been an effective conduit for delivery of these messages.

The crop protection industry has supplied a pretty steady stream of fungicides to treat problems caused by fungi and oomycetes, and we have conducted hundreds of trials to assess the efficacy of these products and provide results and non-biased interpretations to farmers. For example, several years ago some Ohio farmers came to us with a disease never previously observed in parsley, which is widely grown on our muck soils. We identified the pathogen as *Septoria petroselini*, immediately set up on-farm trials with various fungicides, and quickly identified an appropriate treatment. This saved the farmers hundreds of thousands of dollars in crop losses in a single season. I am always grateful for the assistance of Jhony Mera, my field research manager, for our ability to run our large and active field research program. Jhony came to the US from Ecuador about 13 years ago. When he started working in my lab his English skills were not perfect and he had no experience in agriculture or research. However, his intelligence and work ethic soon became evident and every year he took on more responsibility in my program, all the while taking classes at OSU Agricultural Technical Institute and earning an Associate's degree. Another critical team member is Dr. Fulya Baysal-Gurel, who came to us as a research scholar through Fikrettin Sahin about eight years ago. Fulya conducts research and is critical to the efficient operation of our vegetable disease diagnostic lab. Many of her research outputs have been picked up by the vegetable industry. For example, Fulya recently tested biocontrol products for the greenhouse tomato industry in response to increasing problems with postharvest tomato diseases. Her results led to recommendations that within a year were widely adopted in greenhouses in the US and Canada.

We have continued to work with the greenhouse vegetable industry nationally and internationally. One of the most important diseases of tomatoes in temperate climates in the field and greenhouse anywhere is bacterial canker, caused by *Clavibacter michiganensis* subsp. *michiganensis* (Cmm). This pathogen is systemic, seedborne and easily transmissible via plant sap, water and contaminated surfaces. We were interested in factors affecting the colonization of tomato plants from infested seed to seedlings and older plants. My Ph.D. student Xiulan Xu inserted Lux genes into the Cmm chromosome, resulting in marked strains that emit light visible with a special charge couple device camera (*Appl. Env. Microbiol.* 76:3978-3988). Xiulan showed using real time bioluminescent imaging that relative humidity influences endophytic colonization of tomato plants by Cmm, and that the pathogen moves relatively rapidly from the point of entry up and down the plants, also colonizing roots (*Phytopathology* 102:177-84). This work would not have been possible without the collaboration of Professor Gireesh Rajashekara, a young molecular biologist in OSU's Food Animal Health Research Program (FAHRP) in Wooster. FAHRP is part of OSU's College of Veterinary Medicine, and although Dr. Rajashekara works primarily in the area of bacterial diseases of food animals, he is keenly interested in plant-pathogen interactions and lent his expertise to these studies. This is one of many examples of cross- disciplinary collaboration at OSU that have been instrumental in moving science forward. Another good example that pre-dates the Cmm work is our entry into the area of food safety. After several serious outbreaks of zoonotic human pathogens such as *Salmonella* and *E. coli* O157: H7 on fresh produce, my lab began to collaborate with Professor Jeff LeJeune, also in FAHRP, an international expert on food safety based on his work with cattle and birds. We have worked on several

USDA-funded grants exploring the interactions of plant and human pathogens on plants, means of reducing the risk of human pathogen contamination of plants pre-and post-harvest, and approaches to optimize the delivery of food safety information to farmers. My long-time Research Associate Melanie Lewis Ivey chose one of these projects for her Ph.D. research, focusing on both the information delivery aspects (*Food Control* 26: 453-465) and co-management of plant and human pathogens in surface irrigation water (*Water Research* 47:4639-4651). Dr. Lewis Ivey is the first to earn a Ph.D. while a full time staff member in our department. She has been an outstanding team player and contributor to every area of our vegetable pathology program. It was inevitable that with her skills and experience she would strike out on her own, and after 15 years with my lab has departed from OSU to take a faculty position in the Department of Plant Pathology at Louisiana State University. We all recognize that our job in academia is to train the best of the best and send them out to make their own contributions to science, and I will enjoy following Dr. Lewis Ivey's progress at LSU.

Sally's International Development Work

In 1994 I received a call from Dr. Tom Payne, then the Director of OSU-OARDC, asking if I would join the Integrated Pest Management Collaborative Research Support Program (IPM CRSP). Dr. Payne chaired the advisory board of the IPM CRSP, which was managed out of Virginia Tech and had several US Land Grant university partners including OSU, Penn State, UC-Davis and Purdue. This large, multidisciplinary project was funded by USAID and had projects all over the world. The project in the Philippines was in need of a plant pathologist, but I had concerns about starting up something like that, not having been promoted yet and having two young daughters and a toddler son at home. However, Dr. Payne was persuasive, Chip was amenable, even encouraging, and I started on a road that has led me all over the world and helped me build relationships I never would have imagined. The IPM CRSP is one of several CRSP projects funded by USAID, most of which cover commodities such as bean/cowpea, aquaculture, horticultural crops, and sorghum/millet. The IPM CRSP focuses on integrated management of horticultural crops and takes a participatory approach that encompasses pathology, entomology, weed science, horticulture, economics and gender studies. We work closely with institutions in host countries to solve technical and social problems in agriculture identified through a participatory process that includes comprehensive stakeholder input. The CRSPs are designed to build capacity in host countries through short- and long-term training and research support.

I joined the Philippines site of the IPM CRSP in 1995 to provide advice on plant pathology research. My partners were Professors Ed Rajotte, an entomologist from Penn State University and George Norton, an agricultural economist from Virginia Tech. We worked on this site for about 10 years, traveling to the Philippines together up to three times a year. I can easily say that I learned most of what I know about entomology and all of what I know of economics from these two people. We also became good friends and easy traveling companions. In general in developing countries, as a result of congestion, lack of driver discipline, frequent accidents and poor road maintenance, it takes a long time to get from point A to B. Therefore easy chatter is crucial for those long periods shared in uncomfortable vehicles. In the Philippines, we interacted with scientists at the International Rice Research Institute (IRRI), the Philippines Rice Research Institute (PhilRice), the Asian Vegetable Research and Development Center (AVRDC), the University of the Philippines at Los Baños (UPLB) and other institutions, to develop IPM methods for diseases of vegetables grown in rotation with rice. I started out as the US pathologist and in 1998 became site chair. The project involved 40 scientists, technicians and students. Among its many research accomplishments, the IPM CRSP project was first to characterize the effects of the root-knot nematode on onion yield and quality in the Philippines, and developed cultural and biological strategies to manage this destructive disease. Project scientists also combined host resistance and grafting technology to manage bacterial wilt disease of eggplant at the small farm level, and evaluated methods of pink root control in onion. Our UPLB collaborator Professor Aurora Baltazar developed strategies to reduce the amount of hand weeding, often done by children, in onions and other vegetables, and eventually became our site coordinator. Ed Rajotte brought pheromones from the US in his suitcase for years for research on pest monitoring, which eventually resulted in major changes in the pest management regulatory framework in the country. Dr. Sally Hamilton, then a gender specialist at VT, found that women in farm families often held the purse strings and made decisions about spending for farm inputs, including pesticides, but rarely received any training in IPM. I particularly enjoyed working with Mrs. Dulce Gozon, the leader of a large onion growers' cooperative in the northern part of the main island of Luzon. Our Philippine co-operators conducted field trials on the farms of organization members and met with them often to provide results and recommendations. There was always a good give-and-take between the farmers and researchers that influenced the direction of the research.

The Philippines project was expanded in 1998 to include Bangladesh, which I joined in 2000. My first visit to the Philippines in 1995 profoundly influenced the way I saw the world. I had never seen poverty in my previous travels to the degree I had seen in the Philippines, with so many people lacking proper housing and clean water. However, my experiences in the Philippines scarcely prepared me for what I saw in Bangladesh. A dangerously overcrowded country, it is one of the poorest in the world. Ed Rajotte likes to refer to the capital city Dhaka as something straight out of the Ridley Scott science fiction film *Blade Runner.* Our project continues in Bangladesh today, and we have made much progress in developing IPM solutions for the benefit of smallholder vegetable farmers, including tomato and eggplant grafting to manage bacterial wilt and root knot nematode, a range of biological controls for insect pests and improved vegetable varieties. During one visit, a young farmer approached and asked us to see the small patch of land where he was growing cabbage. He proudly pointed out his new house nearby, a small but sturdy structure, which he said was built because he had made money using practices developed by the IPM CRSP. This is what we are striving for – to use research to solve problems that hold poor farmers back and prevent them from obtaining the basics like nutritious food, adequate shelter, clean water and education for their children.

During this time I experienced odd physical symptoms, primarily severe tiredness, that I tried to ignore and attributed to work, travel and being the mother of three children. I learned in 2003 during a medical exam that my thyroid gland was enlarged. My grandmother had a very visible goiter so I was surprised, as I never noticed the swelling. An ultrasound test showed multiple tumors, and I underwent a complete thyroidectomy. A subsequent biopsy revealed papillary carcinoma and the next step was to burn out any remaining thyroid tissue with radioactive iodine. The survival rate for thyroid cancer is very high, and I have been cancer free for more than 10 years. I don't even think about it. I am most grateful for the fact that shortly after the treatment ended my energy returned, my other symptoms disappeared and I felt like a much younger version of myself. I am ever thankful for

Chip's support and that of my family, friends and colleagues, both before and during my treatment.

During the second phase of the IPM CRSP, which began in 2004, all of the sites were regionalized and additional countries were added. We had finished our work in the Philippines and added Nepal and India (Tamil Nadu) to the project in Bangladesh. I also began working with sites in East and West Africa. There is much in the IPM CRSP to recommend it, but I believe that its management as a system of interconnected programs, rather than isolated projects, is critical to its success. This was a requirement of the original principal investigator of the project, Dr. S. K. DeDatta, now retired. Since then well-known IPM proponents including Dr. Short Heinrichs and Dr. R. Muniappan have served as project directors and encouraged communication and technology sharing between regional programs. The second phase also ushered in cross-cutting global theme projects in areas such as insect transmitted viruses, impact assessment, gender and diagnostics. I became the PI of the diagnostics effort – the International Plant Diagnostic Network (IPDN). This project was predicated on the deep conviction that diagnostics are important but neglected in the developing world, and that something ought to and can be done about it. The IPDN operates in four regions and 12 countries in collaboration with the IPM CRSP regional programs. These are Guatemala, Ecuador, Honduras, Uganda, Tanzania, Kenya, Senegal, Mali, Ghana, Nepal, India and Bangladesh. The goal is to connect pathologists and entomologists within regions to each other and to experts in and outside their regions. We have documented the needs of scientists and technicians, standardized diagnostic protocols, developed diagnostic assays and built human resource capacity through intensive regional training programs. In many if not most developing countries, scientists have neither training nor infrastructure to identify invasive pathogens. When the aggressive bacterial wilt pathogen *Xanthomonas musacearum* "jumped" from ensete in Ethiopia to banana in Uganda about a decade ago, samples were shipped to the UK for identification. The IPDN hopes to remedy this situation in all of our member countries. We have trained hundreds of scientists in intensive regional programs focusing on diagnosis of diseases and pests important in their regions, utilizing technology appropriate to their infrastructure capacity. US colleagues Drs. Sue Tolin (VT), Bob Gilbertson (UC-Davis) and Carrie LaPaire Harmon (UF), have joined me and regional experts including Drs. Fen Beed, Ranajit Bandyopadhyay and Lava Kumar (IITA), Marco Arevalo (Agroexpertos, Guatemala), Zachary Kinyua and Monicah Waiganjo (Kenya Agricultural Research Institute), Delphina Mamiro and M. Mwatawala (Sokoine University of Agriculture, Tanzania), Mildred Ssemakula and Jeninah Karunji (Makerere University, Uganda), and Eric Cornelius and Rodney Oswu-Darko (University of Ghana), among others, in providing these training opportunities. Interestingly, in some regions, particularly in Africa, our training programs have provided opportunities for pathologists and entomologists to network with one another, as other opportunities such as annual meetings and workshops that we take for granted are rare. The value of diagnostic networks becomes more and more clear as factors such as climate change and increased movement of people and goods across borders enhance the risk of invasive species arrival and establishment, especially as funding for applied research and development continues to decline (*Annu. Rev. Phytopathol.* 47:15-38).

While the IPM CRSP occupied most of my time spent on international development, I have also been privileged to be involved in other USAID-funded projects including the Pest and Pesticide Management Program in D'nepropetrovsk, Ukraine between 1997 and 1999. I worked with my OSU colleague Dr. Pat Lipps and

Ukrainian scientists to upgrade laboratory facilities and develop field and laboratory research projects on wheat and tomato disease management. I was honored to be the first American woman given the title of Honorary Professor of the D'nepropetrovsk State Agricultural University to recognize the accomplishments of the project. Currently we are beginning a new research project in Tanzania through the iAGRI program, a Feed the Future initiative managed by OSU. It seems clear to me that there are more opportunities for scientists in the US and other developed countries to participate in projects that will benefit developing country agriculture than are people willing or available to do so. Applied scientists with direct field experience are particularly in demand, as our numbers shrink with each round of retirements followed by shortsighted neglect to refill such positions in many land grant universities. However, I am encouraged by the increasing number of students interested in international development and willing to commit their time and effort to help reduce poverty in the world.

Synthesis

Chip's sister Linda Styer Caldas observed that biologists get better with age, and pointed out the biology is a "synthetic" science, in that a great deal of synthesis of many observations and experiences are required to truly be a great biologist. We can both agree with her statement and would extend it to other disciplines as well. We both feel that we are smarter than we were at 25 and hope to continue to become smarter still. We both look forward to the next 8 years or so before retirement, and will do our best to continue to make useful contributions. We are proud of our daughters Allison and Carly who have careers of their own and of our son Mike who is just starting out. We have often been asked how we managed two careers and three children, a difficult question to answer since there are so many small things done along the way and long forgotten. However, the main reasons were shared responsibilities, trust in and support of each other, and focus on the children and their needs. So we never had time for hobbies but we didn't miss them. We made an effort to travel with our kids within the US and internationally, and consequently they all have a sense of adventure. Sally took each of them on one of her extended work trips when they were teenagers – Allison to Europe, Carly to China and Mike to Tanzania – that helped them develop a worldview that still influences their daily lives.

We thank Rod Sharp who was such a positive force in our personal and professional lives. We can't know what our lives and careers would have been like if our paths hadn't crossed, but we are happy and grateful that they did.

Fig. 1 Sally & Chip
OSU, 1973

Fig. 2 Sally
Bangladesh, 2009

Fig. 3 Styer Family
APS, 2010

CHAPTER 11.

John Peters' Story

John Peters

Early years

I can't remember a time while growing up that I didn't want to be a biologist. I don't remember wanting to be a specific type of biologist, but I was certain that I would have a career in some study in biology. My father used the GI Bill following World War II to go to college to become a school teacher. He attended Mount Union College in Alliance, Ohio and completed certification requirements to teach and when I was two years old, he accepted a teaching position in Dover, Ohio, a typical small Midwestern town of about 10,000 residents. Dover was my home until I went to college. Dad taught a variety of courses between grade school and high school for a few years and earned an administrator certificate. He was promoted to a grade school principal before becoming the supervisor of principals for the town's four grade schools and eventually the assistant superintendent. Although I never had my father as my principal, he was always very aware of my academic performance, especially after he became an administrator. My mother was a "housewife" which has become a lost profession. Her mastery of her career field was highlighted by raising my brother, sister and me and being the president of the Parent Teacher Association and a Girl Scout leader. More impressively, she made all of our beds for us every day, did all our laundry and ironing, and had dinner on the table every evening at 5:30PM. Dad would leave work a little after 4:00PM, go to the Elks Club for a beer, and then come home for a sit down dinner with the entire family. I forgot to include that he came home every day for lunch which my mother had ready. If she would have worn pearls and a dress, our life would have been a prototype for "Leave it to Beaver".

My grade school was a part of the same building that housed our junior and senior high schools. Early in my grade school education, my Dad took me to the other end of the building to meet Mr. Shough, the high school biology teacher. I believe that I had a question to ask him about biology and Dad took me to the expert. What an inspiring individual he was! I could not wait to get to ninth grade to take his biology course. In the interim, I contented myself with frequent visits to his classroom with complete disregard to the probability that I could be interrupting a class. I distinctly remember going to see him with a fossil that I had found in our garden. It had some form of tubular structure in it and I was sure that Mr. Shough could identify it. I don't remember what he said it was, but I had no doubt that he could identify it. When I made it to ninth grade, I got to take biology with him which did not disappoint. I did a science fair project as a part of the course. The science fair was always a big event at our high school and I remember visiting it every year as a grade school student. My favorite exhibit was the metal orb in physics that you could touch and make your hair stand on end. When my turn came to do an experiment for the science fair, I decided to study the effect of smoking on hamsters. I am unsympathetic to people today who claim they had no idea that smoking was dangerous and blame tobacco companies for not making them aware of the risks. Today, my experiment is entirely unacceptable for scientific and moral reasons, but in 1963 no one questioned my experiment. My first problem was to make a smoking machine. I got some business in town to cut the bottom off a gallon glass jar. I then covered the end with a piece of latex. This was then attached to a hand-made pump using an Erector set motor as a piston to flex the latex in and out. I used the other end to insert the hamster and plug with a rubber stopper that had a hole in it large enough to hold a cigarette. The smoking machine was remarkably efficient at inhaling, but poor at exhaling. I am surprised that none of the hamsters died during their smoking. I kept daily records of the hamster's weight until it was time to "sacrifice" them for necropsy. Mr. Shough opened the lab on a Saturday morning so students could work on their project and I used this time to euthanize the hamsters. Each was placed in a jar with some chemical that would kill them. My classmates had gathered around to watch the process and as the first hamster died, it convulsed. We made juvenile comments and laughed as it died. Following its death, Mr. Shough pulled me aside and asked me why we treated the hamster's death like a performance at the Roman Coliseum. What a teaching moment... I never forgot that lesson. I went on to gain a blue ribbon, but my girlfriend got the prize for the best project. She had done a newfangled project in plant tissue culture where she cultured some carrot explants

Undergraduate Education

In high school, I was very active with the Thespian Club and the production of two plays and a musical every year. I was especially interested in technical theater, but also did some acting. When I graduated from high school, I was convinced that I would become a high school biology teacher and be involved with the school's theater program. I received a significant theater scholarship from Otterbein College in Westerville, Ohio and I entered college in 1967 to obtain a double degree in theater and biology.

When I registered for my first semester of college, my father helped me complete my course choices. He told me to register for R.O.T.C. . . . . I didn't know what this was, but being respectful of my father, I did what he said. R.O.T.C. stands for Reserve Officer Training Corps and my college had an Air Force unit on campus. This meant that I had to attend a military course every semester related to Air Force training and wear a uniform on Tuesdays. During my freshman year, my grade point was high enough to be invited to join a scholastic honorary organization and at the end of the year I was offered an R.O.T.C scholarship which paid full tuition and gave me a $50 stipend each month. I would be required to serve four years as an Air Force officer upon completion.

During my second year of college, I realized that my goal of a double major in theater and biology was unattainable, especially with my R.O.T.C. obligation. Although I gave up on theater in college, my former high school theater teacher and I produced and directed a summer musical in the high school auditorium every summer. We did popular musicals like "Annie Get Your Gun" and "Cabaret". Many members of the community participated in acting, technical production and the orchestra. Every year we made a few hundred dollars profit to give to the high school foreign student exchange program.

Although I was now committed to a career in biology, I hadn't made much of an impression on the biology faculty at my school. As a small college, we only had four full-time professors. The school was small enough to have annual interviews between the biology professors and the students. When I had my meeting near the end of my sophomore year one of the professors asked me where I had been hiding. They indicated that they were impressed with my performance in their recent classes, but hadn't been aware of me before. They offered me a job as a lab assistant which was mostly just washing dishes and cleaning animal cages, eventually moving up to making microbiology media and prepping materials for classes. From these professors I learned that a true mentor goes beyond the classroom in opening doors for their advisees. My close working relationship with these professors led to being invited to do a senior research project and graduation with distinction in biology. I studied the large salivary chromosomes of fruit flies and active regions called "puffs" where genes were turned on. I didn't make any significant contribution to science with this project, but I learned the discipline to conduct research on my own, to review the literature about research in this area, valuable laboratory technique especially in microscopy, and scientific photography techniques including film development and printing. This opportunity was the foundation for my success as a researcher.

Graduate School

Prior to graduation from Otterbein in 1971, I asked the Air Force if I could delay entry onto active duty for two years to obtain my Master's degree. Activity in the Vietnam War was slowing and the Air Force gave me permission to go to graduate school. Otterbein College was only about 15 miles from The Ohio State University and I never gave any consideration to going to another school.

My interests were in an emerging biology study called Developmental Biology. Ohio State University did not have a specific developmental biology program, but the departments of Botany, Genetics, Zoology, and Genetics had assembled a new interdisciplinary program in developmental biology. I thought this was a good fit with my interests and interviewed with a few professors connected to the program. One of those professors was Rod Sharp. Rod was a new professor in a new Biological Sciences building and lab. He described his lab's research in plant tissue culture and asked me if this interested me. I remember telling him that I thought that I would rather work with animals than plants. I then saw Rod's skill as a salesman for the first time. He pointed out that he could clone plants which could not be done with animals. He also could give plants cancer and study this with controlled experiments. I was sold and began my studies in the summer of 1971. Almost immediately, Rod left for a summer in Brazil, but before leaving he got me a summer job in the OSU botany greenhouses and set me up with a teaching assistant's position for the fall quarter. Two graduate students in Rod's lab introduced me to plant tissue culture technique.

Harry Sommer was a Ph.D. candidate who was a few years older than me and had served as an Army officer prior to returning to school. Before computers, literature was reviewed through using a publication called <u>Current Contents</u> which was published every other week. Harry was extremely disciplined and had established a routine where he would visit various university libraries including the Botany/Zoology, medical, pharmacy and agriculture libraries on a recurring basis. I would visit libraries, but not nearly on the schedule that Harry did. Every few days I would come to the lab to find index cards with references that Harry found that he thought I might be interested in. He shared an enthusiasm for plant tissue culture that resulted in both of us collecting flower buds from around campus and putting them into tissue culture bottles to see if we could establish cultures of their cells. The other graduate student was Rosa Raskin who had only been in the lab a few months prior to my arrival. She was quick to show the new guy how to prepare media and use sterile technique to establish cultures. We worked well together and each of us was aware of the other's research and made constructive contributions. Rod asked me to use tobacco tissue culture to see if I could generate a nicotine-free tobacco line. As a part of these studies, I decided to examine the effects of adding nicotine to the culture media of undifferentiated tobacco cells. To my surprise, I found that I could reproducibly generate roots by adding nicotine to the growth medium. Nicotine was believed to be a waste product of tobacco, but my data suggested that it might play a role in differentiation. Rod taught me the importance of publishing results and attending scientific meetings to present papers. I don't know where he got the funding, but he sent me to meetings at Argonne National Laboratories near Chicago, to national meetings of the American Botanical Society at the University of Massachusetts, and to international meetings at Dalhousie University in Halifax, Nova Scotia. As a new person to plant tissue culture, I didn't realize the significance of having drinks in a university guest house with giants of the plant tissue culture field such as Herbert E. Street of England. At the time, it all seemed so casual and just an exchange of ideas. If I could revisit this with what I know today, it would be a moment of awe and inspiration. As an educator today, I use historical advancements as a part of my lectures to present the foundation of great discoveries.

Working with Rod, I always knew there was a possibility that I could accompany him to Brazil in one of his annual research excursions. I completed all the necessary work for my Master's degree and during the first week of the summer quarter of 1973, I turned in my thesis and my wife and I departed for Brazil. I knew that I would be entering the Air Force upon completion of my degree and this gave us approximately three months to do research. We were permitted two checked bags each and I filled one bag with chemicals to use in our research. It seems so naive now, but when we arrived, the one bag was confiscated by customs as questionable. Why would anybody confiscate a bottle of white powder labeled as "500 grams of caffeine"? Then, I got to see how government agencies in developing countries work. The laboratory we worked in was government funded and within a few days the suitcase of chemicals was delivered. I never knew how Rod managed the funding to take me to Brazil, but I had a small stipend and we lived in a university owned house with several other visiting scientists and their wives. The real difference in doing research in Brazil and doing research at Ohio State was that we had support in the form of multiple technicians. Our work was to design experiments and we had technicians who could prepare media and wash glassware. In approximately three months, we set up 3,000 tissue cultures to study tissue culture of the common bean (<u>Phaseolus</u> <u>vulgaris</u>).

This bean was a food staple in Brazil and tissue culture could lead to improved varieties. Although the bean had been grown in tissue culture, no one was able to get the tissue to differentiate into plants. Instead, the cells would divide in culture as callus with no specialized structures. We were attempting to use different plant hormones and chemicals to induce differentiation. In all of our cultures, we only produced two plants and both came from a trial where we added an extract from ground beans to the media.

We had defined the results of varying different hormones on growth of bean cultures and upon entering the Air Force, I sent Rod several manuscripts that he reviewed, edited, and published. This short summer provided the most productive period of research that I have ever experienced because I had Rod's support in experimental design and the assistance of other scientists and technicians in making media and inoculating cultures. Brazil's funding of research in the 1970's has led to their emergence from a developing to a developed country.

When Rod brought me into his laboratory in 1971, I was a 22 year-old new graduate student and he was "Dr. Sharp". He immediately informed me that we were now colleagues and he was to be called "Rod". At that time, this was a difficult transition. I had never known a professor on a first name basis. I will always be proud to be a colleague of Rod's, but that word does not say enough for my admiration and respect for him as a mentor and as a friend.

Military Experience

In November of 1973, I entered into the Air Force to meet my R.O.T.C. obligation. I was disappointed to be placed in a field using radar to guide fighter planes to complete interception of unidentified aircraft entering the United States air space. After a little over 3 years in the Air Force, I had contacted Rod about returning to Ohio State University to pursue my Ph.D. He graciously welcomed me back to the lab, but then the Air Force offered me the opportunity to cross-train into the biomedical field.

This opportunity was due to one individual who I have lost contact with and who probably doesn't have any idea of the impact that he had on my professional career. Lieutenant Colonel Fullerton was my immediate supervisor in late 1977 when I was a Captain in the Air Force in a radar position as a part of the North American Aerospace Defense Command (NORAD). I was stationed in Syracuse, New York. I don't remember the context of the conversation with Lieutenant Colonel Fullerton, but I had expressed my disappointment at not being able to cross-train into the Biomedical Laboratory Officer career field. These individuals supervised military hospital laboratories. I had read manuals specifying how to apply to the Laboratory Officer Career field and thought that I was following procedures, but I never received any follow-up to my inquiries. My four year commitment was coming to an end and I was planning to complete my time in the Air Force and return to Ohio State to enter a Ph.D. program with Rod as my advising professor. Lieutenant Colonel Fullerton asked me if I had ever talked to anyone in the Laboratory Officer Career field about my inquiries. I told him that I didn't know anyone in the field and really didn't know where to start. He sat down at my work station and said that he wasn't leaving until I was talking to the senior Laboratory Officer in the Air Force. He was ordering me to start calling. It only took two phone calls. First I called the nearest Air Force hospital and asked to speak to the Laboratory

Officer. I then asked him for the phone number of the senior Laboratory Officer in the Air Force. He gave me the number of Colonel Billy Joe Robertson at Lackland Air Force Base in San Antonio, Texas. I called Colonel Robertson and told him my story. He said that he remembered my "application" when he was on the selection committee for new entries into the career field, but that my "application" was not complete. I told him that I didn't realize that I was applying for the field and that I had just sent a letter of inquiry. He said based on his memory of my "application", I should be very qualified for selection and recommended that I send a complete application to the next meeting of the committee. I did this and was selected for entry into a training program beginning in the fall of 1978. As a supervisor and in a position of authority as a senior officer in the Air Force, Lieutenant Colonel Fullerton ordered me to complete a task which ultimately resulted in a 21 year career in the Air Force. My career included supervising a small hospital laboratory including hematology, chemistry, microbiology, and blood banking, being a member of the biology faculty at the Air Force Academy for five years, and supervising the largest medical microbiology laboratory in the Air Force which was the second largest clinical microbiology laboratory in the state of Texas for four years. Lieutenant Colonel Fullerton taught me about the responsibilities that I would face as a supervisor in the Air Force and as a college professor. I had an obligation to help my subordinates and my students achieve their goals. In the Air Force this meant giving them career guidance and insuring that they got the training, recognition, and awards that they deserved so that they could achieve promotion. As a teacher, I am responsible for the competent delivery of instruction and the design of coursework that enables students to retain and utilize the skills learned in my classroom.

My decision to stay in the Air Force exemplifies a big difference between Rod Sharp and me. Rod is a risk-taker. He was able to leave the security of his position in academia to build a successful career in commercial enterprises. I don't think I could have done the same. For me to return to Rod's lab would have been to leave the security of a good job in the Air Force with my wife and daughter. It would have taken at least three years to complete a Ph.D. followed by at least one post-doctoral experience before seeking employment. Ultimately, the decision to stay in the Air Force was a good one for me.

My work in Rod's lab made my internship as a medical technologist relatively easy. I successfully managed a small hospital laboratory at Vandenberg Air Force Base for almost three years before being invited to join the faculty at the Air Force Academy. This was a rewarding experience where I taught a variety of courses including introductory biology, microbiology, developmental biology and molecular biology. I also directed student research projects testing media variations for the tissue culture of guayule, a desert plant that produces rubber. I encouraged the cadets to present their research at scientific meetings and publish their results. Air Force Academy cadets were exceptionally bright and almost all biology majors at the Air Force Academy were accepted into medical school. In 1986, I received an Air Force award for innovative research and development and then was selected to return to The Ohio State University to do my Ph.D.

Graduate School Again

Another significant contributor to my professional career was my Ph.D. advisor at The Ohio State University, Dr. Darrell Galloway. While teaching at the Air Force Academy, I was aware of an opportunity to have the Air Force sponsor me to obtain my Ph.D. Once selected, I chose to return to The Ohio State University to pursue my Ph.D. in Microbiology/Immunology. I knew that Rod continued to have close association with The Ohio State University Microbiology Department and I asked him to make some of the initial contacts to begin the search for an advisor. He put me in touch with Dr. Robert Pfister. Dr. Pfister's lab was next to Rod's when I was a student with Rod and I was happy to learn that Dr. Pfister still remembered me after 14 years. Dr. Pfister recommended that I contact Dr. Darrell Galloway. This was an incredible fit. Darrell was only a year or two older than me. He was a Reserve Navy officer (at the time, I held a higher military rank than he did). As a military member, he was aware of the military requirements, and that I would return to my assigned career field after three years whether I had completed my degree or not. Darrell and I had an unwritten contract that we would both work toward successful completion of my degree. Darrell asked me to use monoclonal antibodies to study a bacterial enzyme called elastase. This enzyme is associated with "flesh-eating" bacteria and dissolves connective tissue. For my project, I learned how to make monoclonal antibodies and began to assay elastase activity. Darrell had some commercially obtained elastase in the lab and when I began to set up assays, I found the activity to be extremely low. I went to Darrell and asked him to purchase some fresh elastase so I could perform the assays. He said that he thought that it would be a good exercise for me to purify my own elastase. I remember being frustrated with his decision. In my mind, this was not research. The procedure to purify elastase had been described in the literature over twenty years earlier and I knew that I would have to suspend my research for at least two weeks as I went through the busy work of purifying the enzyme. But, my military discipline came through and I said, "Yes, sir" and pressed on. Darrell's request for me to purify my own enzyme was directly responsible for my dissertation, the discovery of a new enzyme, a new understanding of how elastase mediates tissue destruction, and grants to further this research. After purifying my own elastase, I found that it had the same low activity as the commercial elastase Darrell had in the laboratory.

For reasons that I still don't understand, I had kept the various fractions of proteins obtained during column chromatography to purify the elastase. I began to add various purified fractions back to the elastase and very quickly found one that greatly enhanced elastase activity. The purified new protein had negligible activity by itself, but when combined with elastase, the combination was explosive. Our conclusion was that the new protein "nicked" the elastin protein in connective tissue. Following this "nicking", the elastase enzyme could quickly dissolve the elastin protein. We sequenced the amino acids in a portion of the new protein and found them to match a sequence of the genetic code of a protein that was known to participate in some unknown way in elastolytic activity. The gene had been labeled the "lasA" gene and we named the new enzyme the "LasA" protein. We also recognized that the "LasA" protein was larger than the published sequence, so we were able to resequence the gene and publish the corrected sequence. The newly discovered enzyme became the core of my dissertation and led to a significant grant and continued study in Darrell's lab following my new assignment in the Air Force. I will always be grateful for the opportunity to be a part of Darrell's lab and his guidance as my advisor.

College teaching upon completion of 21 years in the Air Force, I retired from the Air Force to become what I always wanted to be - a teacher. Although I had enjoyed success in research, I knew that my first love was teaching and I sought a teaching position at the small college or junior college level. In 1994, I obtained my current position as a biology instructor at McHenry County College in Illinois. I was hired for my expertise in microbiology. Our primary mission is to prepare students for entry in the health professions career field, especially nursing. After 18 years of teaching at this community college, I have taught over 3500 students' introductory biology and microbiology. Many of these students do a laboratory exercise where they investigate the effects of plant hormones on tobacco and carrot plant tissue cultures. Using the training that I gained in Rod Sharp's lab, the transfer of knowledge continues.

Most of the students who have been in my classes entered nursing and other allied health professions. There are fewer resources at a community college than at a major university, but I pride myself in presenting courses at a comparable level to those that I have experienced in my education and in my prior teaching experience. My philosophy has been that the course content needs to be directly related to their educational progression in the allied health fields. I have used my personal experiences as a clinical microbiologist and researcher to make the course objectives relevant to the students' future careers. After 18 years of teaching at MCC, the job is as rewarding as it was when I began teaching. I continue to hear from successful graduates returning to praise the contribution that I made to their careers. Former students are now a part of the allied health workforce and one of the greatest rewards as a teacher is to meet my former students working in their chosen career field as a part of the community. Many of these former students were single mothers who came to my class intimidated about returning to school several years after a largely unsuccessful high school or early college experience. They gained confidence with their success in my classes and then completed their degree in nursing. I share in the pride they feel as a professional. This is why I always wanted to be a teacher and I look back on my career with the same sense of fulfillment and pride as my students.

Figure 1. Rod Sharp designed and built growth chamber used in his Ohio State laboratory in 1972.

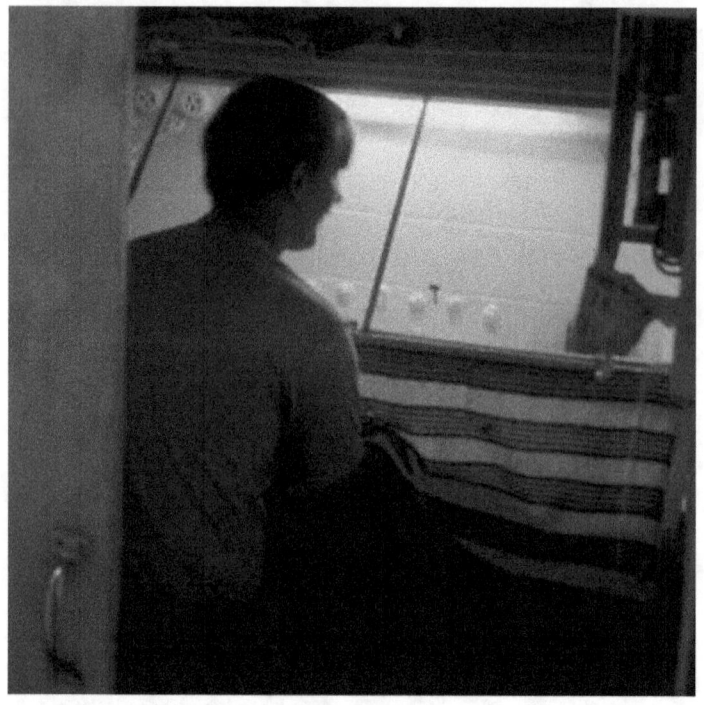

Figure 2. John Peters in homemade tissue culture room designed and built by Rod Sharp to limit airborne contamination of cultures.

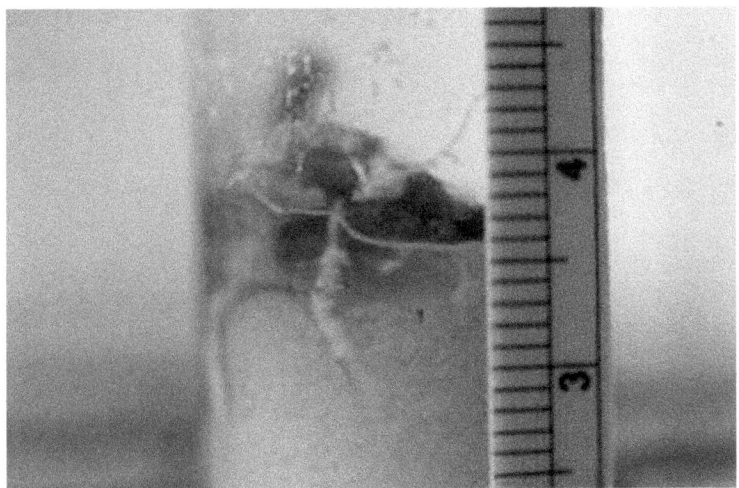

Figure 3. One of the two bean callus cultures (out of 3000 cultures) to differentiate into plantlets.

CHAPTER 12.

From an Oklahoma Farm to Ecology Professor at Rutgers

James A. Quinn

In 1946, Dr. Murray Buell arrived at Rutgers and immediately began to develop a nationally-recognized program in plant ecology. At that time I was seven years old, moving to a farm in the plains of Oklahoma. This chapter will focus on my serendipitous path to plant ecology and Rutgers, my challenging climb of the academic ladder and suggestions to those contemplating an academic career, and a brief review of my contributions to post-Buell terrestrial plant ecology at the Rutgers Nelson Biological Laboratories.

INITIAL INTERESTS AND BACKGROUND

My early years on the farm emphasized chores and field work, school studies and activities (my parents were strong supporters of both), and 4-H activities and projects. I was blessed by the strong, loving support of my parents, and in the knowledge that they would proudly celebrate my achievements. In the process of finishing Guymon High School as the Valedictorian of my class, it was often suggested that I should become a teacher. However, my dad, who had been a successful vocational agriculture teacher (many accolades for his Future Farmers of America Chapter's accomplishments) in Sayre, Oklahoma, before he left in 1946 to take over the family farm from his ailing parents, was strongly opposed to a teaching career in the public schools. This was due to his prior experience with school politics, parents, and school boards. As our family was struggling financially, he was also opposed to me "staying on the farm." In my high school years, I was exposed to the 4-H grass identification contests sponsored by the Soil Conservation Service (SCS) and became familiar with the SCS personnel and agency activities. In 1957-58, I served as State 4-H President, graduated from high school, was inducted into the Oklahoma 4-H Hall of Fame, and enrolled at Panhandle A. & M. College (now Oklahoma Panhandle State University) to study agronomy (crops and soils), with the goal of becoming a Soil Conservationist with the SCS. In the summer of 1958, I jumped at the chance of working with the SCS in Liberal, Kansas, as an Aide. One of our projects involved monitoring grass seedling establishment in the Great Plains program for grassland restoration. This enhanced my desire to work with grasses and grasslands and led to a Student Trainee position in Range Conservation in Hattiesburg, Mississippi, during the next two summers. Upon graduation from Panhandle A. & M. College (PAMC), I returned to Mississippi as a Soil Conservationist (it was discovered that I lacked the necessary courses to satisfy the Civil Service requirements for Range Conservationist). However, the Mississippi SCS had already groomed me (with visits from the D.C. national office personnel) to be their first State Range Conservationist and suggested that I go back to school to pick up the necessary courses in plant identification, physiology, and ecology to qualify as a State Range Conservationist. However, I had utilized my 4-H and academic scholarships and a 30-hr work week to pay for my undergraduate education (and my parents were still struggling financially), so my only alternative was to apply for a graduate assistantship at the three rangeland schools suggested by my D.C. mentor, Robert Williams. Because of strong letters of support and undergraduate honors such as Valedictorian and Student Body President, I received offers from all three. However, the offer from Colorado State University came so quickly that I accepted it before receiving the other two offers (but CSU was my first choice in any case).

SEQUENCE OF EVENTS LEADING TO A CHOICE OF A CAREER IN PLANT ECOLOGY

At Colorado State University (CSU), I joined the Department of Range Science as a Graduate Research Assistant, completing my thesis research ("Forage Losses due to Trampling by Cattle on Sandhills Range") at the Eastern Colorado Range Station and receiving my M.S. degree in May 1963 (Dr. Donald F. Hervey, Thesis Director). My M.S. studies included many basic courses in botany and ecology that had not been available at PAMC and were required by the CSU Range Science Department and the Civil Service. I performed well in those courses, leading to suggestions that I apply for a National Science Foundation (NSF) Graduate Fellowship. I took my first plant ecology course (taught by Dr. Richard T. Ward), which I enjoyed immensely. I was especially intrigued by the studies of variation within and among populations of a grass species, as I had noticed such variation when preparing for and training teams for grass identification contests. Dr. Ward was involved at that time in an NSF-funded project on ecotypic variation, and after several stimulating conversations, I decided to apply for an NSF Ph.D. Fellowship, with Dr. Ward as my Thesis Director and ecotypic variation as my proposed research. Dr. Ward was incredibly helpful and supportive, and an ideal role model, during my Ph.D. studies.

Upon receipt of the NSF Fellowship, I immediately began developing my proposal and collecting plant materials in the summer of 1963. As I had been scheduled to rejoin the SCS following the receipt of my M.S., I sent them a letter resigning my position and letting them know of my intentions to pursue a career in plant ecology. This was in spite of the fact that the other graduate students in botany made it clear that they considered ecology to be a new fad and a "less rigorous" science. However, Dr. Ward's students have accomplished far more professionally and scientifically than those students as a group.

THESIS COMPLETION, JOB SEARCH, AND RUTGERS INTERVIEW

An early start (summer of 1963), the NSF Fellowship, and strong support from Dr. Ward allowed me to complete my research and Ph.D. requirements by August 1966. My dissertation was entitled "Ecotypic Variation in Switchgrass (*Panicum virgatum*) and Sand Dropseed (*Sporobolus cryptandrus*)." In anticipation of a summer completion I began applying for academic positions in the prior winter and spring. Due to the breadth of my coursework and expertise in crops, soils, range science, botany, and genetics (Ph.D. minor), I received eight offers (some with no interview, others with an interview to be scheduled). Robert Funsch, my apartment roommate, was a Rutgers alumnus and New Jersey native and urged me to apply for the Rutgers position in plant ecology. Robert said "you haven't applied to any schools in the East"; I replied that I had applied for a University of Wisconsin position. Robert was perplexed at my answer, as he did not share my opinion that Wisconsin was an eastern school (even though it sits in the middle of the eastern half of the United States). My advisor, Dr. Ward, also strongly urged me to apply to Rutgers, pointing out that Dr. Murray Buell had developed an outstanding ecology program and produced many students who were also making major impacts. He also pointed out that the Department of Botany at Rutgers overall had a national reputation. About the same time, I received a letter from the SCS offering me an Area Range Conservationist position in scenic southwest Colorado. I was amazed that they knew that I would be finishing at CSU but found out later that Robert Williams (my D.C. mentor) had visited my hometown in regards to some range meetings and asked a local barber about me, hearing that I was about to finish my degree at CSU. The Colorado offer and several of the academic offers were quite attractive, but I decided to wait for a Rutgers response.

Rutgers then called and suggested that I fly out for an interview and seminar presentation in 7-10 days. I was not ready for a quality slide presentation (only had hand-drawn figures and some glass-mounted slides), so I rushed color photos and dissertation figures to the CSU Photo Lab. The lab promised that the slides would be ready prior to my departure. I began to check with them daily and wound up being promised that they would be available in the early morning of my departure. That morning I waited as long as I could before I gave up and tried to make it to Denver to catch my flight. Fortunately, I had just purchased a new Chevrolet Nova; I drove at speeds up to 90 mph on the interstate, and arrived just as the plane was ready to leave its Gate. I stopped at the ticket counter, and they called ahead and asked that the plane wait for me. As I boarded the plane, the doors were closed, and the passengers were glaring at me. I often wonder if the whole trajectory of my life would have changed if I had missed that flight.

I arrived at the Newark Airport and was greeted by Drs. Murray and Helen Buell and Dr. James Gunckel (Chair, Department of Botany). When they saw me, they feigned relief that I was a young guy because PAMC had sent them my Dad's transcript. I stayed with the Buells and spent the next few days interviewing, going on field trips with the Buells, and giving my seminar at the Thursday evening Ecology Seminar. The Buells took me to Hutcheson Memorial Forest (Murray had played a leading role in its preservation and was its first Director). I was fortunate that I had worked with the SCS in Mississippi, so I actually knew some of the tree and shrub species (remember that I grew up in the grasslands of Oklahoma and took my Ph.D. working on Colorado grasses). As they were telling me about the long-term successional study that they had started, we encountered some dead herbaceous plants from the prior summer, and they asked me if I could identify the species. I did not have a clue, but I crushed some dry capsules, looked at the seeds, and exclaimed excitedly "that's Velvetleaf." (I had never seen live plants of Velvetleaf (*Abutilon theophrasti* Medikus) but knew the seeds from our Crops Commercial Grading exercises in which you have to recognize seeds of weedy species in the crop sample.) The Buells were visibly impressed and told others about it. I consider that event and my seminar to be turning points in my evaluation. I was quite impressed by Murray Buell, his ecology students, the people at the Ecology Seminar, and the members of the Department of Botany (especially the Chair, Jim Gunckel, and the systematist, David Fairbrothers, who took his Ph.D. in grass systematics and was ecologically-oriented).

Rutgers offered me the Assistant Professor position, and I accepted, much to the delight of Dr. Ward, Robert Funsch, and my dad, who was very proud and excited that I would be teaching at Rutgers. My dad was never opposed to teaching and/or research at a university, only to public school teaching. I immediately declined the other offers and interview requests, although some of them had been increasing their salary offers when I delayed my decision, so that my Rutgers salary ($8,500) was more than $2,000 less than some of their offers. I believed that the faculty, academic environment, departmental reputation, and personal opportunities more than compensated for the lower salary.

EARLY YEARS AT RUTGERS

The next day after August graduation ceremonies, I loaded up my car and a U-Haul trailer and drove to New Jersey. I arrived only a week or so before the beginning of Fall 1966 classes and stayed with the Buells prior to finding a basement apartment in Highland Park. This was to be Murray's final year, and I was privileged to sit in, give lectures, and participate in his courses. I was also assigned additional responsibilities, e.g., recitations and labs in General Biology. Certain of the General Biology recitations and labs were extremely difficult for me, because although I had a diverse training in agriculture and plant science, I had never taken a course in vertebrate zoology or human anatomy and physiology, and never had the delightful experience (ha, ha) of dissecting a frog, fetal pig, etc. Dr. Richard T. T. Forman, a community and ecosystem ecologist, also came to Rutgers in the fall of 1966. We joined a nationally-ranked Department of Botany consisting of Murray F. Buell, David E. Fairbrothers, James E. Gunckel, Edwin T. Moul, Barbara F. Palser, and H. Bruce Reid. All of this small but diverse group were committed to scholarship and teaching. This collegial department was chaired by Jim Gunckel, who had weekly faculty meetings in which all business was conducted as "a committee of the whole." Jim Gunckel was a self-sacrificing team player who was incredibly helpful and supportive in my first years as an Assistant Professor. He and his wife, Jean, even provided my first Ph.D. student (Larry Lindauer) and his family a place to stay until they located an apartment.

As an ecologist, I was delighted by the ecological and field interests of the department. That first fall and spring, Drs. Gunckel and Buell organized several field trips consisting of faculty and interested graduate students. Also, many of the botany and ecology courses had a strong field experience component that Dick Forman and I were delighted to continue.

As a boy from the Oklahoma plains with a Ph.D. in grasses from Colorado State University, I learned much about New Jersey plants and plant communities (and had some amusing experiences) in my first few years. In the fall of 1966, I joined Dr. Buell and Dr. Paul Pearson (zoologist who worked closely with Buell to develop the ecology program at Rutgers) in the General Ecology course. When Dr. Pearson introduced me to the class as an "expert on grass," the students cheered and applauded. When I gave my first lecture, I asked for feedback, and Dr. Pearson replied that he liked the content and organization, adding with a smile that he was amazed to hear an Oklahoman talk so fast. This led me to put SLOW DOWN at the top of each page of my future lecture outlines. In the spring semester with Dr. Buell on a field trip to Hutcheson Memorial Forest in the Plant Ecology course, he asked me to go help a couple of students that were slow in completing their plot sampling. However, he forgot about us, and the bus went back to New Brunswick. The students and I with our threatening sharp pointed surveyor pins and quadrat slats then had to hitchhike a ride back to campus. Fortunately, a Budweiser salesman stopped for us.

With the retirement of Dr. Murray Buell at the end of the 1966-67 academic year, Dick Forman and I were left to continue his terrestrial plant ecology program. This was extremely difficult for us because at that time, Assistant Professors were expected to "do their share" of the teaching, advising, and department/college committee responsibilities. We were also not given start-up funds. (Beginning in the mid-1980s, Rutgers modified its policies to provide start-up funds and minimize teaching, advising, and committee responsibilities for the first few years.) Needless to say, we both made many personal sacrifices in order to meet our responsibilities. I cannot speak for Dick and his experiences, but I had no time for a "social life." In fact, I was referred to as the "oldest graduate student" by the current graduate students because I was still at the Nelson Biological Labs building after they (except for Rod Sharp) had gone home for the night. Several years later, one of my most promising M.S. students, Robert J. Cartica, responded to my suggestion that he continue for a Ph.D. by saying that "I don't want to work as hard as you do for the rest of my life." (However, Bob is now Director, Office of Natural Lands Management, NJ Department of Environmental Protection, and Executive Director, NJ Natural Lands Trust, working as hard as I have ever worked.)

Despite the 60-70 hours per week, I was excited and pleased with my Rutgers position, enjoying my research, teaching, and student interactions. Several interesting (some amusing) events occurred in my second year at Rutgers. For example, on an all-day General Ecology trip to North Jersey, Paul Pearson was in charge of one bus and I was in the second bus. While in Stokes State Forest, the students began asking for a "rest stop." Since there were no facilities available (in the bus or at our stops), Paul and I had decided prior to the trip that, if necessary, "boys" would be in the forest on one side of the road and "girls" on the other side. Unfortunately, our directions in the two buses were opposite to each other, creating chaos. Also in General Ecology, on a beautiful fall day, we led a field trip to study old-field succession that culminated in a stop at Hutcheson Memorial Forest. I was feeling very good about my performance and the trip in general, and asked one of the students boarding the bus for the trip back to campus if he enjoyed it. He replied "It was OK, but I'm puzzled as to how this will help me in Med School." In the summer session that year I was teaching Ecology and Systematics and spent a lot of time in one lecture on the human population problem and the need to control population growth. At break, I walked up to several of the guys and the one nun in the course, and one fellow then said loudly to the nun "what do you think about Dr. Quinn's comments on controlling population growth?" She smiled and said "I'm doing my part!"

Many of our faculty and graduate students regularly attended meetings, lectures, and field trips of the Torrey Botanical Club, and our faculty frequently served as officers and council members of the Club. (From 1970 to 2000, I served as a member of the Council for 15 years, also serving as President of the Club and as an Associate Editor of the <u>Bulletin of the Torrey Botanical Club</u>.) In my first full year at Rutgers, the Club celebrated its Centennial Year (1967) with meetings and symposia at Rockefeller University, Brooklyn Botanical Garden, Fordham University, New York Botanical Garden, Long Island University (Brooklyn Center), C. W. Post College, and Columbia University. Jim Gunckel made certain that any interested faculty member or graduate student had a ride to these events, which for me, recently from the West, served as an opportunity to meet botanists and see the facilities at these local institutions.

Another aspect of our closeness as departmental faculty was our block of seats at the Rutgers men's basketball games (Jim Gunckel, Barbara Palser, Bruce Reid, and I plus new faculty, Richard Triemer and Aurea Vasconcelos).

Rutgers already had an undergraduate botany major when Dick Forman and I joined the department in the fall of 1966. I believe that Dick Forman was the botany advisor from 1966 to 1973. At that time I became the advisor for botany undergraduates until 1980, when I left for a sabbatical year in Australia. We had some excellent Botany majors in the 1970s, e.g., Dr. Tom Zanoni (New York Botanical Garden) and Dr. Keith Clay (Indiana University). I also served on the Honors Committee of several outstanding Cook College students, e.g., Dr. Joseph Colosi (DeSales University) and Dr. Howard Neufeld (Appalachian State University).

REFLECTIONS ON CLIMBING THE ACADEMIC LADDER

Rutgers faculty have always been asked to make significant contributions in teaching, research, and service to the university, community, and society. From the 1960s through the early

1980s, these standards made the attainment of tenure extremely difficult as Rutgers began to morph from an "ivy league" or liberal arts college persona to that of a major research university and emphasize the early development of a grant-supported research program and de-emphasize teaching contributions. Accordingly, Rutgers in the mid-1980s began to limit teaching and committee services of new faculty, providing fellowships and start-up monies. This was a fair and wise move, but, unfortunately, limited the early access of our undergraduate students to these young, cutting-edge, and enthusiastic faculty. This has been followed more recently by the hiring of "superstars" expected only to do research and obtain mega-grants, and the hiring of other faculty (coadjutants) to do their teaching. It could be argued that both the researchers and the coadjutant teachers benefit from doing what they do best, but the undergraduates miss the exposure to the researchers' new techniques, recent research results, and theoretical models, and have a greater chance of having teachers who may have limited research experience or may not be up-to-date on the latest advances in their field.

Teaching.—In my early years, faculty, regardless of rank, were generally expected to teach two courses with labs per semester, sometimes supplemented by another course or by recitations or lab sections in General Biology if they were co-teaching a course. Allowances were made for running a huge multi-section course and for time-consuming administrative positions (Chair, Graduate Director, etc.). In my later years, as mentioned earlier, teaching expectations were reduced. My undergraduate teaching at Rutgers included General Ecology (21 semesters), General Ecology Lab (21), Ecology and Systematics (1), Pollution and Ecosystems (3), Populations and Evolution (10), Plant Ecology (11), Principles of Botany (2), General Biology (6), and TV tapes that were used in General Biology 101 (1974-83). Graduate courses were Population Ecology (23), Reproductive Ecology of Grasses (3), and Seminar in Botany and Plant Physiology (1).

A positive influence on me early by Drs. Buell and Pearson, and later by Dr. David Fairbrothers in our course on Populations and Evolution, was the importance of keeping lectures up-to-date and relevant to the students. I made a point to keep up-to-date files of newspaper and journal articles on each topic included in my lectures. This led me to cover such topics as global warming long before they showed up in textbooks; in 1973 I began spending considerable lecture time documenting the causes and effects of global warming.

Growing up in the West led to some interesting exchanges with the students: 1) I asked a student if he had ever traveled in the West, and he replied "Yes, to Pennsylvania"; 2) I forced myself to accept that "Midwest" in New Jersey referred to those states in the middle of the eastern half of the United States; and 3) It has been fascinating how colloquial language and word usage changes over time: A) In the late 1960s, when referring to the spatial arrangement of creosote bush individuals in Nevada, I said that they were "spaced out," and the students immediately laughed loudly (at least they were listening!); B) When covering mathematical indices of species diversity, I mentioned the Simpson Index, saying "it's something Bart worked up in his spare time"--this elicited an instant response that I'm not certain would occur today!

When sitting in (observing) Dr. Buell in the Plant Ecology course my first year, Dr. Buell (perhaps *only* that year) often forgot to warn students about what they might encounter on the field trips or to tell them that the trip activities would take place regardless of conditions. On one trip with heavy rain the students *en masse* refused to get out of the bus, and on a trip to the Pine Barrens were not dressed properly or did not have proper footwear for walks through a bog and a recently burned Pine Plains forest of stump sprouts, complaining loudly. This was a valuable take-home lesson for me. As a result, several lectures before a field trip, I always made a point to emphasize "Dress for the weather and/or conditions, whatever they might be" and then give suggestions relevant to a specific field trip, e.g., footwear/change of shoes for a bog, old clothes for the Pine Plains due to the burnt stems and tearing of clothes. I also tended to overemphasize potential hazards or difficulties, so the students were often pleasantly relieved, and seemed to enjoy the trip even more when they were fully prepared.

A key responsibility in teaching at the graduate level is the advising of individual students, either as the Thesis Director or as a member of the student's graduate committee. During my years at Rutgers, I served as the Thesis Director for 12 M.S. and 13 Ph.D. recipients. Many of these students pursued successful academic careers at colleges and universities, but others have had equally productive careers in high school teaching, state and federal ecology and environmental agencies, environmental consulting companies, nature centers, and private industries such as landscaping and biotechnology. It always gives me a feeling of personal satisfaction and pride when I hear of their accomplishments. I also served on 176 graduate committees of students advised by other faculty. Many of them were taxonomy and systematics students of Dr. David Fairbrothers, and others were students of faculty in plant and animal ecology, botany, forestry, entomology, and landscape architecture. I also served as Graduate Director of the Botany and Plant Physiology Graduate Program during a transition period (1983-85) in which it became more inclusive and diverse.

Research and Scholarly Activity.—My Ph.D. research dealt with adaptive genetic differences among populations of two grass species (*Panicum virgatum* L. and *Sporobolus cryptandrus* (Torr.) A. Gray) in relation to environmental factors, species interactions, and ecological history. Since this research area contains elements of both ecology and systematics and Dr. David Fairbrothers' Ph.D. was in grass systematics, we submitted an NSF proposal entitled "Ecological Differentiation within the *Danthonia sericea* Complex." This proposal was funded in 1968 and led to six papers in referred journals and seven abstracts of papers presented at professional meetings. I benefitted greatly from this collaborative research with David and from our continuing interactions over the years at Rutgers. These papers led to sabbatical research with Dr. Ken Hodgkinson (Commonwealth Scientific and Industrial Research Organization, Australia) on population variability among populations of *Danthonia caespitosa* Gaud., a widely distributed and important forage species in Australia.

My investigations of adaptive variability among populations of a species next progressed into cutting-edge areas at that time, e.g., *r*-K selection, density-dependent *vs.* density-independent selection, and resource allocation strategies. Over the years, my students and I were also involved in studies of demography, germination ecology, succession and species interactions, heavy metal tolerance, pollination ecology, breeding systems, and life history strategies and sex ratios. Beginning in the 1980s, my research focused mostly on the reproductive ecology and sex expression of grasses and the delineation of fixed genetic differences and genetically-based, adaptive phenotypic plasticity. Buffalo grass (*Buchloe dactyloides* (Nutt.) Engelm.) was an excellent study organism for investigating the life history strategies, sex ratios, and evolution of dioecy. Dr. Tongjia Yin developed a mechanistic model of one hormone regulating both sexes, validating it with tests utilizing buffalo grass and cucumber (*Cucumis sativus* L.).

My vitae lists 62 articles in referred journals and 105 additional publications (e.g., book reviews, abstracts of papers presented at meetings). I am proud of my research contributions and those of my students. However, the volume and impact were reduced due to my teaching responsibilities (mentioned earlier) and my willingness to serve on department, college, graduate program, graduate school, university, and professional society committees. Taking the committees listed in my vitae and multiplying each by the number of years on that committee gives a total of 218 "committee years". In addition to chairing many of these committees, I was elected or appointed to 23 offices (department, college, graduate school, and university levels). I strongly suggest that an Assistant Professor should actively limit committee responsibilities, if it's not a firm policy of his/her department. I accepted far too many administrative and committee responsibilities as a young faculty member. When you do these jobs well, it snowballs, i.e., the old expression--"if you want a job done, give it to a busy person." Administrative duties and committee overloading can greatly reduce research productivity and chances for tenure. In my case, I was scooped at least three times on the cutting-edges of reproductive ecology ideas and/or research approaches because I was unable to find the time to get my completed research submitted to a journal. One specific example was a project that analyzed the variance components in the plasticity of density stress responses among populations in relation to ecological history. It would have been a "first." However, a similar experiment that had been initiated *after* I had completed my research was published before my research appeared in print.

In summary, it is important for a young professor to be involved and have input in department and college policies and decisions. It also fosters collegiality and a sense of being a part of the academic process. However, I urge moderation and caution young faculty against administrative duties and committee overloading, or getting overly involved in environmental issues and community service, even if strongly aligned with their interests and expertise.

Upon receipt of tenure, the faculty member then owes it to his discipline and department to be involved in decision-making committees and environmental issues. My service on college, graduate school, and university committees was often quite helpful to botany and ecology, e.g., my influence on the Graduate School Biological Sciences Area Committee in support of Plant Biology at Cook College.

The Tenure Process.—I have no complaints of a personal nature about the tenure and promotion process at Rutgers. I joined Rutgers as an Assistant Professor in the fall of 1966, was promoted to Associate Professor (with tenure) in 1971, and promoted to Professor in 1977.

I am fully acquainted with the tenure and promotion procedures. In the first year (1981-82) of the 55-member Department of Biological Sciences, I served as the Associate Chair for Personnel, preparing eight promotion packets. In this case, the discipline (former department) provided me with documentation and references (sometimes initially incomplete) to submit to the tenured faculty for the departmental evaluation. Rutgers has an elaborate procedure with all sorts of guidelines at each step or level (department, college, and university) to make tenure and/or promotion decisions as fair and unbiased as possible. However, humans are part of the process, and a young faculty member, unfortunately, should be careful not only to meet the criteria but also to stay on good terms with senior faculty members and to provide them with a complete accounting of professional accomplishments and departmental contributions.

Certainly, a critical and careful review of the tenure applicant's past and potential research, teaching, and university service is necessary, because a mistake ("false positive") is costly to the department and university. However, I remember voting in favor of applicants when the overall vote was not sufficient for tenure. I vividly recall two cases which I consider "false negative" mistakes. In both cases, an undue emphasis was placed on the lack of NSF grants (very important to the university because of the indirect costs returns and the contribution to public stature) despite the fact that the candidates had secured multiple state, foundation, or private industry grants, supporting graduate students and an active research program. They had published quality research in referred journals. They were also excellent teachers and valuable mentors to both undergraduate and graduate students.

In a case with a happy ending, I was a replacement appointee to the Rutgers College Appointments and Promotions (Tenure) Committee. As they were bringing me up-to-date, they informed me that they had tentatively rejected the tenure application of Dr. Edmund (Ted) Stiles. I was horrified because I had co-taught General Ecology and Population Ecology with him and knew that his lectures were dynamic and up-to-date. He also interacted extremely well with the students after class and on our field trips, receiving great student evaluations. When I mentioned this, the response was "his research just deals with birds and bees, and who has heard of the journals anyway?" I responded that <u>Evolution</u> and the <u>American</u> <u>Naturalist</u> were two of our most prestigious journals, and that his papers are widely cited. After about 30 minutes of discussion, the committee took a final vote, and he was unanimously approved. There were two major initial problems: 1) Prior to my joining the committee, there was no one who understood the significance and quality of his research, and 2) the Zoology Department had done an incomplete job of preparing his packet and had picked references who knew very little about his professional contributions. Thankfully, Ted continued his outstanding contributions to teaching, research, and Rutgers, was promoted to Professor, and served several years as the Director of the Hutcheson Memorial Forest Center until his death in 2007. As Director, he more than tripled the acreage in the Center, and was also quite successful in his land preservation efforts elsewhere in New Jersey.

Value of Sabbaticals and International Meetings.—When a faculty member is overwhelmed for several years with teaching, administrative, and committee responsibilities, his/her research creativity can suffer from a lack of time for an evaluation of the literature to suggest future research needs. He/she may also lack the time to finish data analyses and submit the results for publication. A sabbatical offers a chance for a faculty member to restart or retool his/her research program, maintaining his/her scholarly and research activity. Having the chance to become totally immersed in thinking and designing new projects restores the excitement of discovery that sometimes gets lost in the day-to-day myriad of responsibilities. It becomes especially stimulating and invigorating if a person has the opportunity to pursue research and interact with colleagues in a foreign country; the opportunities for insight and exposure to new ideas are enormous. I was given this opportunity in 1972-73 when my Chair, Jim Gunckel, strongly supported me in the attainment of a Rutgers Research Council Faculty Fellowship (full salary for 12 months!) for a year in Australia. I joined a group of rangeland ecologists with the Commonwealth Scientific and Industrial Research Organisation (CSIRO) Division of Plant Industry in Deniliquin, New South Wales (NSW). Not only was I exposed to new ideas and new landscapes, but I was able to work closely with a young scientist, Dr. Ken Hodgkinson, on population variability among populations of *Danthonia caespitosa* Gaud, a widely distributed and important forage species in Australia. This research allowed me extensive exposure to the flora and landscapes of Australia during our plant collections ranging from northern NSW to Tasmania. We then brought these plants back to Deniliquin and established both a transplant garden and seed production plots. Throughout the year I was constantly impressed by CSIRO's technical and financial support. The collections from Tasmania are an example. We drove our station wagon to the Melbourne airport and boarded a plane with our shovels and collection materials. When we arrived in Hobart, Tasmania, we were met by CSIRO personnel who provided us with a station wagon and, according to our specifications, began building a large box for the airplane transfer of our plants. We then spent several days locating and collecting plants from several sites throughout Tasmania, driving back to Hobart, where we packed our plants (each in a plastic bag) into the box that CSIRO had constructed, drove to the Hobart airport, oversaw the loading of the box onto the plane, and boarded the plane for the flight back to Melbourne. At Melbourne, we picked up the vehicle we had left at the airport, inserted the box into the vehicle, and drove to Deniliquin. The plants were quickly transplanted, each to its own population block, along with other population blocks from our prior collections. The blocks were separated by an appropriate distance to eliminate any significant pollen contamination in this predominately inbreeding species. Subsequently, we were able to collect seeds from each population block which were then used in a series of controlled environment experiments in the CSIRO Canberra Phytotron, comparing the populations in growth rates, reproductive allocation, phenology, responses to various combinations of temperature and photoperiod, etc. We had the assistance of several technicians in the establishment of these experiments, which wound up utilizing numerous controlled environment rooms and growth chambers. The phytotron supervisor threatened to rename the facility the Danthoniatron! Ken and I moved to Canberra for the duration of these experiments. I was housed in a hostel for federal employees (Canberra is the capitol of Australia). The interactions with Australians and their viewpoints at Deniliquin, at the phytron at "tea breaks" with the technicians and other scientists, and at the hostel where our dining table consisted of federal employees from multiple agencies and regions of Australia are something that I will always remember and cherish.

Because of my awesome research and cultural experiences of 1972-73, my next sabbatical in 1980-81 was a return to Australia to finish data analyses and preparation of publications from our 1972-73 research (two papers had already been published in the <u>Australian Journal of Botany</u>). I again joined the CSIRO lab at Deniliquin to work with Ken Hodgkinson. Dr. Allan Wilson was the Officer-in-Charge of the lab, and although I had gotten to know Allan in 1972-73, I had a chance to get to know him even better and experience his hospitality. Ken and I worked on data analyses and a symposium invited paper that we would present at the International Botanical Congress in Sydney at the end of August at the end of my sabbatical year. For March through August, I moved to the University of New England at Armidale, NSW, to work with Dr. R. D. B. Whalley, who shared my interests in adaptive differences among grass populations and in their breeding systems. This was a fulfilling and thoroughly enjoyable interaction, not only with "Wal" but also with the other members of the Botany Department. Leaving Armidale, I loaded my car and drove to Sydney for the Botanical Congress. At the conclusion of the Congress, I sold my car to an American scientist who was beginning his sabbatical year at the Congress. Ken drove me to the Sydney airport and returned the car to its new owner. This concluded my fantastic two sabbatical years in Australia, and Drs. Ken Hodgkinson, Wal Whalley, and Allan Wilson not only have my profound appreciation but also my enormous respect for them as scientists and individuals

The preceding paragraphs should make it apparent why I strongly endorse sabbaticals away from your home institution. However, the need for free time for research, data analyses, and publishing, plus a need to provide support to graduate students finishing their research and writing their theses, also led me to take 6-month sabbaticals in 1986, 1990, and 1994.

A faculty member should attend as many international meetings as possible. They offer new cultural experiences and acquaintances in foreign countries and international networking and exchange of ideas. For a botanist/ecologist, the associated field trips provide introductions to new landscapes and plant communities that you could never hope to duplicate on your own. Because of my varied interests in botany, ecology, grasslands, and grasses, I usually had several attractive meetings each year from which to choose. My criteria were a chance for exposure to a new region (haven't been there) and whether or not I had been invited to present a paper. Over the years, I attended meetings in Australia (3 meetings), Canada (2), Denmark, France, Japan (2), Switzerland, USSR (1975), and West Germany (1987) and had meeting-associated field trips or arranged stops in Azerbaijan, East Germany (1987), Fiji, Finland, Georgia, Great Britain, Netherlands, Philippines, Poland, Portugal, Spain, and Sweden.

REFLECTIONS AND CAREER CONTRIBUTIONS/ ACHIEVEMENTS

As I reflect on my career at Rutgers, there are certain contributions and recognitions that are most meaningful to me.

Research.—A primary focal point of my research at Rutgers was the genetic diversity within and among populations of a species, specifically the diversity of physiological, morphological, and life-history adaptations for establishment, reproduction, and survival in different habitats. This led to studies of population genetic differentiation along latitudinal, altitudinal, moisture, soil pollution, and grazing pressure gradients, and to studies of the relative amounts of genetically-based phenotypic plasticity among populations of differing ecological histories (resource levels, environmental predictability, disturbance history). In my later years, my students and I were involved in studies of the evolution of plant reproductive strategies and breeding systems among populations of the same species in different habitats and geographic regions.

My early studies on population genetic differentiation led me to question the use of the term "ecotype" for genetically-based variation that is habitat-correlated. It was originally used to apply to a grouping of populations by habitat or latitude, but soon became a buzz word to indicate almost any degree of genetic difference below the level of species. In 1978 I published a paper that emphasized that the evolutionary unit is the locally adapted population and that if "ecotype" was used, it should only refer to a single locally adapted population. In 1986 I was invited by the Organizing Committee of the International Organization of Plant Biosystematists to justify and update this interpretation. One of my most cherished memories came at the conclusion of my presentation; Dr. Anthony Bradshaw (an icon widely known for his research on genetic divergence over short distances, especially in regard to heavy metals) asked why it wasn't OK to use "ecotype" for convenience. I replied that he and his students, in their multitude of papers, had never used it instead of genetic differentiation, genetic differences, locally adapted populations, etc. His response was a smiling "you got me," and the audience gave me loud applause at the end of questions. In 1999, in the 3rd edition of <u>Terrestrial</u> <u>Plant</u> <u>Ecology</u> (Addison Wesley Longman, Inc.), it was stated (page 45) that "The term *ecotype* is now used in Quinn's sense."

Earlier, it was mentioned how stimulating and rewarding my sabbatical years in Australia had been. My 1½ years of research with Dr. Ken Hodgkinson on adaptive population differences in *Danthonia caespitoa* Gaud had resulted in four papers in major journals which have been frequently cited. I was especially proud of Ken's decision to include them among his best papers when documenting his qualifications for an honorary D.Sc. from Massey University in New Zealand.

Our Rutgers research on the evolution of plant reproductive strategies and breeding systems, and the importance of genetically-based plasticity in reproduction and reproductive systems in grasses, led to invited papers at national and international meetings, book chapters, and a review paper in the Annual Review of Ecology and Systematics. The Annual Review paper on "Cleistogamy in Grasses" was a collaboration with Dr. Christopher Campbell and my Ph.D. students Gregory Cheplick and Timothy Bell.

In my later years at Rutgers, I focused mostly on the life-history strategies, sex expression, and sex ratios of dioecious plants, and the evolution of dioecy. I spent portions of almost 20 years on buffalo grass (*Buchloe dactyloides* (Nutt.) Engelm.), and my last Ph.D. recipient, Dr. Margot Bram, published a couple of excellent papers on water hemp (*Amaranthus cannabinus* (L.) Sauer). Dr. Scott Meiners and I also published on the dioecious *Juniperus virginiana* L. My research on buffalo grass led to three invited speaker presentations in international symposia, and one of my most outstanding Ph.D. students, Dr. Tongjia Yin, working with buffalograss and cucumber, developed a one-hormone model of sex determination in plants.

Professional Recognition/Contributions.--*Professional societies and organizations.*—A scientist should actively support those societies and organizations that promote his/her discipline and research interests. They provide valuable opportunities to interact, network, and exchange ideas at meetings and to publish in their peer-reviewed journals.

The two societies in which I was most involved were the Torrey Botanical Club (now Society) and the Botanical Society of America. As mentioned earlier, the Torrey Botanical Club provided lectures and field trips within 1-2 hours of New Brunswick. In addition, my students and I published many papers in the Bulletin of the Torrey Botanical Club (became the Journal of the Torrey Botanical Society in 1997), and along with my fellow botanists at Rutgers, I made an effort to "give back" by serving on the

Council for 15 years, while also serving on several committees, and as President (1982-83). I also served as an Associate Editor of the Bulletin. In 2008, the Council of the Torrey Botanical Society elected me to the category of Life Member, citing my "long and dedicated service to the Society."

I served on the following committees of the Botanical Society of America--Conservation, Elections, Guidelines for Sectional Officers, and Karling Graduate Student Awards. My most important contribution was to assist in the establishment of an Ecological Section within the Society, and served as Vice-Chair and Chair of that section in its formative years. I also served as the elected Representative of the Ecological Section to the Editorial Board of the American Journal of Botany, one of the most prestigious journals in the biological sciences.

The New Jersey Academy of Science has a prominent role in enhancing communication among NJ scientists and promoting science and science education. Its annual meeting provided our graduate students a chance to present their research in contributed paper sessions identical in format to those at the national and international meetings. Continuing a tradition of active participation by Rutgers botanists, I served as Member of the Council for 9 years and Treasurer for 5 years.

A major honor for our graduate students and new faculty was admission to Sigma Xi, the honorary scientific research society. I served the Rutgers chapter of Sigma Xi as Vice-President, President-Elect, and President.

Finally, lumping all of the societal meetings in which I was involved, I served as the Presiding Chairperson of Ecology Sessions (Contributed Papers) at 23 state, national, and international meetings.

National and International Symposia.—In later years, I was frequently an invited speaker at national (4) and international (4) meetings. In four cases, I was also asked to organize that symposium. I am most proud of my contributions to the breeding systems and reproductive ecology of grasses portions of the International Grass Evolution and Systematics meetings in 1986, 1998, and 2003. In both 1998 and 2003, I was the organizer and chairperson of the featured symposium on reproductive ecology.

External Examiner/Consultant.—I was selected as the External Examiner for Ph.D. dissertations in Canada (2) and Australia (4) and as an external examiner/consultant for graduate programs at the University of Western Ontario, Queens College of CUNY (The City University of New York), and the University of Ohio. I also served as an "expert" on *Buchloe dactyloides* (Nutt.) Engelm. for the Canadian Species-at-Risk Program, and was certified as a Senior Ecologist by the Board of Professional Certification of the Ecological Society of America (1990-2011).

Graduate **Students**.—Rutgers has always been able to attract high-quality graduate students in botany and ecology, beginning in the mid-1940s with Drs. David Hammond, Werner Baum, and Calvin Heusser (Murray Buell's first Ph.D. recipient). Subsequent students of Murray Buell, such as John Cantlon, Gily Bard, and William Niering, and many more in the 1950s and 1960s, had such an impact on North American ecology and the Ecological Society of America (ESA) that when I attended my first ESA gatherings, I often heard the phrase "the Rutgers Mafia." Over the years, our Rutgers graduate students were supported by fellowships, research assistantships, and teaching assistantships (TAs). Because of the finite number of TAs, the Department of Botany, and later the Department of Biological Sciences, insisted on a conscientious, high-quality teaching performance and a time limit on semesters of support. I was consistently informed that my students performed well for other faculty, and they definitely performed superbly for me in Plant Ecology and General Ecology. These TAs provided necessary financial support and invaluable experience for future high school and university teaching positions. Although others could be mentioned, I will focus only on my last two Ph.D. students, Drs. Margot Bram and Tongjia Yin. Margot joined us after receiving an M.S. degree at Pennsylvania State University, with letters indicating her interests and abilities in teaching. She constantly received very high student evaluations, ultimately receiving the 1996 Award of Distinguished Contributions to Undergraduate Education in Rutgers-New Brunswick. After a post-doctoral position at the Philadelphia Academy of Sciences and beginning a family, she returned to teaching by accepting a couple of adjunct professor positions and a high school biology position near the family home. Tongjia Yin came to Rutgers after receiving an M.S. from Sichuan University, People's Republic of China. Upon arrival in New Brunswick, Tongjia had some minor problems in understanding and speaking English. He worked very hard and conscientiously on improving his English, and we would meet every Friday afternoon in my office to discuss NJ culture, slang, idioms, and world politics. His first responsibilities in General Biology were mostly lab preparation rather than presentations to students. Over the years, he improved so dramatically in student evaluations that he became the "head TA" with responsibilities for the coordination, training, and lab prep for 30+ TAs. Upon graduation, Tongjia held a couple of academic positions, but pursued his overriding interests in biotechnology and is now a Senior Scientist at Gen-Probe in San Diego.

The insistence on time limits for TA support caused much stress for both faculty and graduate students. My faculty colleagues and I were constantly discussing how much we hated badgering our students to finish their research and thesis writing. However, we knew that if the students did not finish their research and thesis before they lost TA support, they might not be able to finish their degree program and get the type of job they desired. Those students on fellowships or research assistantships (without the time demands of a TA) usually had fewer problems finishing their degree work on schedule. I was incredibly fortunate to have my first Ph.D. student on an Ecology Training Grant Fellowship. He was Dr. Larry Lindauer, a biology teacher on sabbatical from the Denver Public Schools. Working lengthy days in an efficient, diligent manner, he was able to complete his coursework and thesis research in two years, returning to his position in Denver. He culminated his career teaching college prep courses and as an administrator in the Denver Public Schools.

Another early student, Dr. Richard Frye, not only did some outstanding research on floodplain forest succession (M.S.) and old-field succession (Ph.D.) but also served as Caretaker of Hutcheson Memorial Forest for four years, giving me some invaluable assistance during the period I served as Acting Director. He recently retired from the Department of Ecology, State of Washington.

Speaking at national meetings with colleagues who have hired our plant ecology graduates, I have frequently heard positive comments about their teaching performance, and likewise, our graduates were known for their "unusual" knowledge of grasses (apparently due to my lecture and field coverage in Population Ecology and Dr. Fairbrothers' excellent coverage of the grass family in Biosystematics and Advanced Taxonomy).

Seven of my Ph.D. students have had successful academic careers at colleges and universities. In regards to research, my two most productive Ph.D. students both at Rutgers and at their respective institutions have been Gregory Cheplick and Brian McCarthy. Dr. Cheplick is a full professor at the College of Staten Island, CUNY, and is known particularly for his research in reproductive ecology and cleistogamy, sibling competition, and the grass-endophyte symbiosis, and has served as Program Chair and Corresponding Secretary for the Torrey Botanical Society. He has served on the editorial boards of <u>Plant Ecology</u> and <u>Evolutionary Ecology</u>, edited a book on the population biology of grasses, co-authored a book on the grass-endophyte symbiosis, and is currently writing a text on "Plant Evolutionary Ecology." Dr. Brian McCarthy is a full professor at Ohio University and has worked on a wide range of projects including the reproductive ecology of hickories, restoration of the American Chestnut, invasive species, forest ecology, fire and oak regeneration, and disturbance and succession. In addition to his many grants, research projects, and graduate students, he has served his department both as Chair and Graduate Director, and is currently the Associate Dean for Faculty, Research, and Graduate Studies. In addition to serving as Associate Editor for four other journals, he has served as the Editor-in-Chief of the <u>Journal</u> <u>of</u> <u>the</u> <u>Torrey</u> <u>Botanical</u> <u>Society</u> (the oldest botanical journal in the Western Hemisphere) for the last 10 years.

I was known as a "hard-nosed, detail-oriented" editor of student reports, research proposals, theses, and publications, and often for this reason, I suspect, other faculty suggested that their students add me as a graduate committee member (hence, the 176 graduate student committees during my Rutgers career). Upon retirement, a couple of letters from former students thanked me for helping to develop their writing skills, and spoke of my red pen. There may be a connection between the writing and editing abilities of Drs. Cheplick and McCarthy and my efforts to provide constructive feedback in editing their research proposals, theses, and manuscripts. In addition, the Editor-in-Chief (Dr. Beverly Collins, Western Carolina University) for the <u>Journal</u> <u>of</u> <u>the</u> <u>Torrey</u> <u>Botanical</u> <u>Society</u> for the five years prior to Brian McCarthy's tenure was one of my best M.S. students (although she took a Ph.D. with Dr. Steward Pickett, an extremely accomplished writer and editor). Also, another M.S. student (Dr. Diane Byers, Illinois State University) has served as the Editor of <u>Plant</u> <u>Biology</u>, although again it must be mentioned that she took her Ph.D. with Dr. Tom Meagher, also an excellent writer and editor.

Diane and Beverly also have been active in the Ecological Section of the Botanical Society of America; Diane has served as Chair, and Beverly has served as Secretary.

Several of my students have contributed significantly to ecology and environmental issues through federal and state agencies, but I will mention only Robert Cartica, who is Director of the Office of Natural Lands Management for the NJ Department of Environmental Protection, and Executive Director of the NJ Natural Lands Trust.

Finally, several of my former students will almost certainly equal or surpass my professional achievements and contributions, and for me this is the most meaningful contribution of all.

Establishment of the Ecological Section of the Botanical Society of America.—At the joint meeting of the Botanical Society of America (BSA) and the Canadian Botanical Society (CBA) in Edmonton, Alberta, in 1971, I was forced to present my research paper in a BSA General Session (mixed bag, miscellaneous). Dr. Paul Cavers, an ecologist and Secretary of the CBA in 1971 (President in 1973), made a point to attend even though he had little interest in the other papers. He asked me "why doesn't the BSA have an Ecological Section, we have had one for years?" I mentioned his comments to my Rutgers colleague Dr. Barbara Palser (Secretary of the BSA in 1970-74, President-Elect in 1975, President in 1976), who had always been supportive of ecology. She said that she would bring the suggestion to the BSA Council. Apparently, her comments, directly or indirectly, led Dr. Peter Raven, BSA President in 1975, to push for such a section. He wrote Dr. Dwight Billings in this regard, and Dwight wrote me in 1976, asking me to serve on the Committee on Formation of an Ecological Section, and urging me to call or write other plant ecologists. Our goal was to have our founding meeting at the BSA meetings in New Orleans in 1976 and ask for approval by the BSA Council (chaired by Barbara Palser) at that same BSA meeting.

After a Tulane campus housing mix-up that put someone in Dr. Billings' room who then put Dr. Billings' luggage outside the door, forcing Dr. Billings back to the long lines at housing, he asked me to make and post signs for the meeting around campus. In spite of these logistical problems, our meeting was attended by 22 enthusiastic botanists, a preliminary set of By-Laws was distributed, and the officers named in those By-Laws were elected. The 48 botanists who had earlier responded positively to the idea of an Ecological Section but were unable to attend the meeting were added to those attending the organizational meeting to give 70 charter members of the Ecological Section. Dr. Billings was elected the first Chair, and I was elected as Vice-Chair. I served as Chair the following year. Both of us sent out Section Newsletters and organized symposia for subsequent meetings. As of August 15, 1977, we had 335 members and sponsored or co-sponsored four symposia at the 1978 BSA meetings. The Section is now one of the strongest and most active sections in the BSA.

EPILOGUE

I accepted the Rutgers position based on the strong ranking of the Department of Botany in general and on the national reputation in plant ecology produced by Murray Buell and his students. Suddenly, terrestrial plant ecology at the Nelson Biological Laboratories was left in the hands of Dick Forman and myself, and we had to develop our research programs from scratch with no start-up funds or released time from teaching. Dr. Roger Willemsen joined us in 1971 and helped immensely, teaching plant physiological ecology and doing research in germination ecology and allelopathy. He left us in 1976 for a research position in agricultural industry. In 1977, Dr. Steward Pickett (physiological plant ecology, plant populations and community organization, vegetation dynamics) was added to our Department of Botany faculty, and Dick, Steward, and I began to make considerable progress in spite of our ongoing teaching loads. Losses of Dick to Harvard University (1984) and Steward to the Cary Institute of Ecosystem Studies (1986) then led to the additions of Dr. Steven Handel (population ecology and genetics, plant-animal interactions, restoration ecology) in 1985 and Dr. Tom Meagher (plant evolutionary biology, quantitative genetics, sexual dimorphism, linkages between molecular variation and phenotypic evolution) in 1987, both making outstanding contributions to the prominence of terrestrial plant ecology at Nelson Biological Laboratories. It should be emphasized that although Steward Pickett had left in 1986 for the Cary Institute, he continued to contribute in a major way to the botany and ecology graduate programs, serving as either Thesis Director or a graduate committee member for many graduate students.

In the late 1990s, Biological Sciences in New Brunswick underwent a major reorganization that led to a breakup of the mega department in Nelson (formed in 1981) and the necessary transfer of certain individuals and groups to new locations. Dr. William R. (Rod) Sharp, Dean of Research at Cook College, provided enthusiastic support and leadership in relocating most of the botanists and ecologists to the Cook campus in 1996-97. This resulted in many Nelson ecologists (including Handel, Meagher, and Quinn) joining the Cook College Natural Resources faculty to form a strong and diverse Department of Ecology, Evolution, and Natural Resources, and some of the botanists joining an equally high-quality Department of Plant Biology and Pathology. The initial moves and new facilities for the ecologists at the Cook campus were organized by the tireless, awesome efforts of Dr. Steven Handel, the first Chair of the Department of Ecology, Evolution, and Natural Resources.

In 1966, I made the decision to choose Rutgers from my range of offers and would make the same decision today if I had to do it all over again. I retired at the end of 2000. Rutgers provided the environment for an exciting and fulfilling 34-year career, followed by an extra five years on the Cook campus finishing research projects and student committee responsibilities. In my retirement, I continue to be professionally involved and supportive of Rutgers and botanical, ecological, and environmental organizations.

James A. Quinn Photograph

CHAPTER 13.

Dr. William "Rod" Sharp's Graduate Student #1

Rosa Shine Raskin

One day while researching the *Columbus Dispatch* online in regard to an article for one of our hospital physician patrons, I momentarily saw the name William R. Sharp scroll quickly across my computer screen. My heart seemed to stop beating and I held my breath, as I paused to notice his name cited in the obituary section. I stopped to read the article as regrets floated through my mind. Dr. Sharp had attended the funeral of his former wife in Columbus. The article referred to Dr. Sharp as William R. Sharp of New York City. I mentioned the article to my co-worker boss, who suggested I contact Dr. Sharp as it had been many years since I last spoke to him. I searched through ten William Sharp's in New York City for his address and phone number. I found my lifelong adviser, Rod, once again in 1999 and planned never to let him slip away.

As far back as I can remember I loved finding answers to questions. I devoured my mother's home health and cookbooks written in German. At age 16 years, I walked to the Coventry branch of the Cleveland Heights-University Heights Public Library to apply to be a library page. While others were joining clubs, I was a student volunteer grading chemistry exams for a chemistry teacher at Cleveland Heights High School.

I trace my interest in information science to my immigrant parents who valued education and researched answers in print sources. My father prided himself in asking "smartness" questions posing a mathematical or logical problem which we were to solve in our head or look up in library books. My mother purchased the *World Book Encyclopedia* for our growing family and while working fulltime, took courses at the Cleveland State University (CSU). My mother continued her college studies on retirement, taking interesting classes in philosophy and women's issues that I wanted to take, but never seemed to have the time. My mother made the time.

In her late 70's while a student in the Project 60 program at CSU, my mother was offered a position to tutor students. She enjoyed reading until her last days when she noticed that one of my articles was the subject of the cover of a trade magazine and replied, "My children will be famous." My writings herein are dedicated to my mother, Louise V. Shine, who died at the age of 94 on September 14, 2011, while this chapter was in its first draft.

In memory of my parents, Louise and Herman Shine, I recently published the book, **Walk Forward** at http://www.amazon. com/Walk-Forward-ebook/dp/B009H6Y7AC, about the story of our immigrant family and my father's four years in captivity during the Holocaust. In **Walk Forward** I reveal the hidden secrets which I uncovered in my fifty-year search for my sister, Eugenia Chimowicz, who was lost in the Holocaust. My half-sister was caught, along with my father and his first wife, in a hellish spider's web as members of the "last 500." As pawns in the chess game between Hitler's assistants, Heinrich Himmler and Albert Speer, my father survived the Lodz Ghetto, Auschwitz, Stutthof, the burning city of Dresden, Germany, and released from concentration camp Theresienstadt on May 9, 1945. My sister, however, was lost in concentration camp Stutthof. She was last documented alive on her arrival in Stutthof from concentration camp Auschwitz on September 3, 1944. Her three male first cousins were transported, along with their young mother, back to Auschwitz on September 10, 1944, one week after they had arrived in Stutthof from Auschwitz. To date, Eugenia and her mother are not listed on this return transport, nor have their names appeared on any list. This is most unusual as mother and child were members of the 500 metal workers charged to move to the East to make munitions. The whereabouts of each member of the 500 was most closely monitored by the Nazis in charge. Since mother and child were replaced, their fate would have had to have been documented before the replacements would have been sought. Whether the record does not exist or is illegible, is the question which spurs my continued research on the fate of my older sister, Eugenia Chimowicz, who was born in November, 1935, in Schwersenz, Poland, to Herman and Sylvia Fabian Chimowicz.

While writing my book for over 20 years, I thought much about how my parents influenced my life and that of my younger sisters. My father seemed to know much about medicine, was an avid inventor, taught me how to plant a vegetable garden, and made covers for books we purchased. My father died the summer he was building a new driveway as the Romans did. Our extended family continues to comment on my father's delicious tomato crop, which we consumed for weeks after his funeral in early August 1974. I ate Rutgers tomatoes twice in my life, after the death of my father in 1974, and after the death of my mother in 2011. Dr. William Rod Sharp had sent the highly valued hybrid Rutgers tomato seeds, which my mother and I lovingly nursed from seed to seedling on my dining room table and bow window during the unusually long Cleveland winter of 2010-2011. My mother closed her eyes *to Brahms Lullaby* and died peacefully on September 14, 2011. I find comfort in the memory of my mother eating the wonderful tomatoes; one lonely tomato lingered on the vine through November of that difficult year.

Around the age of three, I remember my maternal grandfather taking me to local gardens in the city of my birth, Karlsruhe, Germany. My maternal aunt accompanied me on nature walks to pick wild berries and observe native plants of the Black Forest; she would periodically stop and pick delicious treasures for us to share. These wonderful memories are before I entered the displaced persons' camp Vegesack (a concentration camp until 1944 near the port of Bremen), from which we departed for the United States of America on the troop carrier, the USS General William C. Langfitt.

Immigrating to the U.S. was my mother's goal as she willing entered the displaced persons' camp. My youngest sister and I compare our mother to Ruth in the Old Testament. Like Ruth, my mother embraced the religion of my father, "Your people shall be my people, your G-d, my G-d." My father became ill almost immediately on entering the camp, was taken to the camp hospital, and my mother was alone to care for her two young children. I remember waving to my father and being happy that he waved back, while my mother and I patiently waited for our food. We were handed bottles of formula for my two year old sister, stood in lines for my mother's adult meals, and mine. We slept in a long room of bunk beds, waited with strangers to use the shared bathroom facilities, little boys being with their mothers. I realized first hand; there is little privacy in a displaced person's camp for a young girl.

While spending hours in the many lines for each of our meals and for the daily shots that children were given, being so close to the floor, I noticed the dirt under my feet; there were no beautiful clean floors like the ones I was used to. I had been a somewhat spoiled child up to this point, being a member of the first generation, the generation born in spite of the horrors of WWII. The camp was surrounded by a high fence as if once in, one could not get out. During our stay at the camp, my mother taught me to walk forward, carefully observe, be kind to everyone and never give up, important qualities for a future researcher.

My mother left her parents and worldly possessions on entering the camp, waiting to be called to the ship. At 3 ½ years old, I remember clinging to my mother's skirt while my teddy bear was securely anchored in my rucksack. My mother was not only a model student in Germany and in the U.S., she demonstrated the spirit of a true pioneer, never looking back to what she had lost, but embracing the future. It was her medical books that I devoured as a child, her trips to the Black Forest that spurred my interest in nature, and her green thumb which I was lucky to inherit! She was not only my mother, but a best friend.

My interests, biology and information science, became intertwined when I met a student at Cleveland State University (CSU) who told me she was obtaining a degree in biology and transferring to library school for a graduate degree to become a medical librarian. Cleveland had two library schools at nearby universities, Case Western Reserve University and Kent State University. I planned my educational path to obtain an undergraduate degree in biology, followed by a Master of Library Science degree at Kent State University, to qualify to work in a "special" library in industry or medicine.

While a senior at Cleveland Heights High School, I talked a girlfriend into considering CSU as it was much cheaper than attending Western Reserve University. My girlfriend invited me to a picnic at Cleveland Metropark's Squire's Castle sponsored by a club at the new state university. The college senior who volunteered to drive a group of girls and me to the picnic is my husband, Jules.

Jules was the only business student enrolled in biology courses. When he picked me up for a date or drove me to my job at the library, he had something from his comparative anatomy class wrapped in the backseat of the car. His Mom would not allow animal parts in the house. I quickly realized that I preferred green plants to frozen cat limbs.

At CSU I was hired (there were no graduate students in 1966) as a student assistant to teach freshman biology laboratory, my specialties being the life cycle of the fern and the dissection of the clam and frog. In addition to teaching and grading biology exams, I created CSU's first herbarium.

Before his graduation, Jules and I had one quarter together at CSU. After two years, Jules asked me to marry him by giving me an engagement ring stored for the day in the glove compartment of the same car that formerly housed his cat limbs. Our wedding was shortly after my 20th birthday on March 21, 1968, the first day of spring. Jules had been working for Dun & Bradstreet in Cleveland, but accepted a position with Industrial Risk Insurers as a fire protection engineer in Columbus, Ohio, hoping to one day transfer to his dream, underwriting.

We were married on a rainy Thursday after I finished my organic chemistry final at CSU. Jules' father drove us to the airport as our car was already in Columbus. We took the short airplane flight to Columbus with our wedding dinner, Aurora's spaghetti sauce, in my carry-on luggage which everyone on the small airplane must have smelled. A stranger bought us small bottles of alcohol, but I was under the legal drinking age and the flight was much too bumpy to eat or drink. We unpacked boxes and moved into our first apartment in Columbus, while our parents had what we later heard was a wonderful wedding party without us! I entered The Ohio State University the following Monday to complete my undergraduate degree. My father had made Jules promise that my education was not to be interrupted by our marriage.

Teachers were much in demand at the time, so I enrolled in the College of Education that, as luck would have it, sent me back to Cleveland to work in the public schools for most of my credits in education. During my next to the last quarter at OSU, I decided to run across the street to Battelle Memorial Institute to ask if Battelle would hire me with a degree in science education. Battelle responded, "No", and suggested I transfer to the College of Arts & Sciences with a major in biology.

It was to my advantage that CSU had stricter requirements than OSU. I had the necessary language and other credits to transfer my major at the last moment. After graduation, I was offered a position with the Ohio Department of Health testing well water and milk, but Jules suggested I continue in graduate school as he loved going to OSU football games when home from his heavy traveling schedule. Our little family included Goldie Tiger, a red tabby kitten, found in the parking lot of an industrial plant Jules was inspecting in Mansfield, Ohio.

Shortly after moving to Columbus, I was disappointed to learn that the university did not have a School of Library Science or an evening law school program for Jules, who had completed his first year in the night program at Cleveland Marshall Law School. Looking back, his hectic travel schedule would have made it impossible for him to attend night classes.

We settled into married life and quickly found OSU's University Hospitals as I had appendicitis in May 1968, necessitating an appendectomy six weeks after starting spring quarter at my new university. Since I applied to OSU under my maiden name and was under 21, my parents were sent letters about my poor attendance. Luckily, my parents knew I had an appendectomy. I told them that I would be returning to class as soon as OSU's University Hospital allowed me to leave, the delay being a persistent, low grade fever.

Transferring credits from Cleveland State, the new state college in Ohio, to Ohio State was beyond a nightmare in 1968-69. I had to find equivalent courses to ones I took at CSU and provide class notes, and/or laboratory workbooks to the Ohio State University professor in charge of a similar course. I met many biology professors while seeking credit for classes completed at CSU. I visited every laboratory in the Botany & Zoology Building (B&Z) and spent much time in the B&Z's Library, where I was tear-gassed near an open window in the spring of 1970. The OSU riots had started on Neal Avenue, directly in front of B&Z Building.

Kent State University

The spring of 1970 was a time of great turmoil in Ohio colleges. We had to show identification to the Ohio National Guard when we walked on the OSU campus. Since we lived in a nearby apartment complex, University Arms Apartments, I went to and from campus on the apartment bus showing my identification card several times a day. Ohio State had fires on campus, but was nothing like Kent State. It was beyond belief that Kent State, with its beautiful campus and famous black squirrels, just 45 minutes from our home in Cleveland, would harbor such tragedy. The killing of students at Kent State was beyond our comprehension. My younger sister was a student at Kent State and near the cafeteria at the time of the shootings. Her future husband had his clothes and possessions burned in the fire at the ROTC Building. Our family remained in shock in regard to the events at Kent State.

Per the recent riots, my plan to attend Kent State after receiving my undergraduate degree in biology appeared to be impossible. However, ten years later I commuted to Kent State University from my home in Cincinnati, Ohio, in pursuit of my original plan.

Choices

Despite the turmoil and early dismissal of classes, I graduated from OSU in June 1970 with a degree in Zoology, and decided to continue graduate studies in microbiology. I was heart-broken, having taught freshman biology laboratory at CSU as a sopho-more, that I did not receive a teaching assistantship in the OSU Department of Microbiology. To my surprise, I received a University Fellowship in the summer of 1970. I began to think about the professors I met while transferring credits and contin-ued my search for the right person to guide me to a career com-bining biology with my love of information. It amazes me that I was able to pick a wonderful husband and an incredible adviser at such a young age in a time when society, including the role of women in the home and workplace, was changing.

I was raised with an incredible working mother and intended to follow my chosen profession, a combination of science and in-formation. When I realized that Ohio State did not have a School of Library Science, I decided to pursue science from every angle. I continued to spend much time in the library, greenhouse, and microbiology laboratory. I often thought "there must be a better way to review the literature than read every important and non-relevant citation," this was before the emergence of online data-bases. I was lucky to have found an adviser that not only taught me plant cell culture, but how to review the scientific literature, how to prepare an article for publication, and introduce me to the value of intellectual property.

Dr. William "Rod" Sharp

While meeting with biology professors in 1969, I noticed a new, blond-haired researcher working with an older professor in biochemistry. I had seen the young professor before, in Cleveland near my house on Coventry and/or at Case Western Reserve University, where I took a year of chemistry the summer Case Institute of Technology merged with Western Reserve University. I introduced myself to the professor, Dr. William (Rod) Sharp, in what seemed to be a busy plant laboratory shared with Dr. Dougall and several Ph.D. candidates in the basement of B&Z. The professor was going to move into the Biological Sciences Building under construction on the other side of the OSU Greenhouses near the B&Z Building parking lot. Dr. Dougall subsequently accepted a position in Lake Placid with the American Tissue Culture Association leaving Dr. Sharp alone in the upcoming new laboratory in the soon to be completed Biological Sciences Building.

Once Dr. Dougall left, Dr. Sharp seemed more confident. Sometime later, I had a formal meeting with Dr. William Rodney Sharp who told me that he was accepting graduate students and had a grant from the American Cancer Society to lower the nicotine in tobacco. It was my understanding that the Nicotiana (tobacco) plants in the OSU greenhouse belonged to Dr. Dougall. Dr. Sharp mentioned that I could chose the plant species I might like to grow in cell culture including but not limited to geranium, lily, or tomato. Dr. Sharp was excited, energetic, and shared his visions of the future. In addition, to being the most convincing professor in the building, he was an incredible salesman.

I was blessed with a good memory for faces and as I spoke with Dr. Sharp, I remembered seeing him near our family home in Cleveland Heights, Ohio, near the streets Avondale, Hillcrest, or Coventry with a young wife and a golden-haired little boy, the perfect family. When Dr. Sharp mentioned doing research on finding a tomato to grow in the desert, he echoed my father's wish --- to grow a desert tomato. My father had secretly and successfully grown tomatoes in the cold winter of Poland as a slave laborer. I was impressed by Dr. Sharp's knowledge and surmised that being his first student he would have more time than a seasoned microbiology professor, a conclusion based on my inexperience in academia.

Acid mine drainage and clinical virology were of great interest, but I decided to devote my efforts to plant cell culture. I concluded that a new professor would have more time to brainstorm, share his knowledge (I had no clue as to the work required of a junior professor), and was impressed when asked for input in redoing "our" new laboratory. I had much to learn about being a graduate student, the field of cell culture later termed "biotechnology," and was intrigued at the prospect of creating potentially useful plants and/or isolate desired substances from a cell culture.

Plant cell culture was a relatively new field, what could be more important than trying to feed the world?

The objective of plant cell culture in 1960-70 was to discover and define external and internal factors which control or influence the development of cells (White, 1963). A cell culture requires less space than an entire plant and can be more easily subjected to controlled conditions. However, in order to be successfully cultured, the chosen cells must grow and divide in a nutrient medium containing essential substances for continued cell division. We limited ourselves to the culture of haploid cells (at that moment in time) as haploids with desired characteristics could be doubled to form plants with the diploid (normal) number of chromosomes, thus, reducing the traditional time, effort, and growing seasons needed to produce generations of new and improved plants.

I was eager to learn sterile technique, watch pollen grains divide, stain chromosomes, prepare media, pot and water hundreds of plants in the greenhouse, wash and autoclave laboratory glassware, and clean refrigerators. Dr. Sharp was not yet given teaching assignments in the Department of Microbiology and taught me how to place anthers in culture using what he termed "nurse" culture. I was in my element, backed up by a forward-looking adviser, and the vast resources of the Ohio State University Libraries.

The journey that was about to begin has continued for more than 40 years, with advice from Dr. William "Rod" Sharp along the way, to this day.

The Journey

Moving into a new science building is work, work, and more work. New construction is never clean. We had building failures such as elevators getting randomly stuck, contamination of the building's water supply ruining months of research, and it was rumored that the new building was sinking.

Dr. Sharp proved most decisive and resourceful in designing his laboratory. After visiting our new lab, Dr. Sharp remarked; "Rosa, don't you think these counters are a bit too long and run too close to the wall, I will cut them."

The counters were indeed long, being close to the wall, but solid with stainless steel sinks. I could not imagine that we were permitted to cut the new counters and suggested there must be someone in charge of adjusting them for the size of the room. To my relief, the counters were resized by appropriate workman. I realized that Dr. Sharp was not aware of or did not consider turf issues. He appeared to follow the principle, "Just do it."

At 3:00 A.M. one morning, Dr. Sharp called to say, "It was raining". Not quite awake, I wondered if he was inside or outside the building when he asked me to bring a broom, bucket, whatever I might have to wipe up water. Pipes above our Room 417 burst and the water was several inches deep. We had four new refrigerator-like growth chambers to grow plants in cell culture, autoclaves, and a sterile culture room at risk.

On entering the laboratory with my broom Dr. Sharp said, "Rosa, write your name on the broom!" Being more than a bit surprised, I decided not to write my name on my broom, but wrote his name on the broom, "Dr. Sharp, Room 417." I treasured that broom for many years after leaving OSU as it made me smile as I remembered Dr. Sharp.

Colleagues often came into "my" lab, as it was essentially unsupervised. Everyone seemed to be interested in refrigerator-like growth chambers full of anthers, green callus, and plantlets. Trying to grow plants in sterile conditions where microbiologists culture bacteria presents a challenge on two fronts; we share the same potentially polluted air, and those growing micro-organisms may accept animals in the study of immunology, for example, but rarely work with green plants. I was not going to add fuel to the fire by putting my name on a broom to be teased by colleagues in Dr. Sharp's absence. Several months later I had the courage to ask Dr. Sharp why he wanted me to write my name on the broom and he responded, "I was trying to teach you about intellectual property." I learned my lesson well as I wrote my name on every tree when the bulldozers came to remove trees on our lot twenty years later and eventually followed my original path to become a technical information specialist searching intellectual property, prior art, patents, and trademarks. I eventually formed my company, Rosa S. Raskin & Associates, LLC.

A few months after I entered Dr. Sharp's lab, he told me he received a Fulbright-Hays Fellowship and was going to Brazil for the summer of 1971 to work on coffee, more specifically to save the coffee plant species from a rapidly spreading coffee rust fungus. I had not heard of the Fulbright-Hays Program, was not yet a coffee drinker, but saving something appealed to me and my father looked forward to my participation. Unfortunately, my husband, Jules, feared letting me go to Brazil. It had taken me almost one year to recover from the complicated appendectomy and Jules imagined I might get sick overseas. I reluctantly remained in Columbus and Jules helped me maintain my research plant collection in the OSU Greenhouse Complex after work, including potting tomato plants on our third wedding anniversary. I did not go to Brazil, a great disappointment to my father who had told many about my research and built two greenhouses. My father turned his Cleveland backyard into one big tomato patch should I run out of pollen grains. My youngest sister came to live with us in Columbus; Brazil was no longer an option. I wondered how I might complete my research on haploid plantlets.

The preparations for Brazil are as vivid in my mind today as they were 40 years ago. I was more than a bit jealous of the travel plans, scientific meetings, dinner engagements, and laboratory supplies which were discussed as Dr. Sharp made arrangements to leave. One of Dr. Dougall's graduate students married a scientist from Brazil, left for her new country a few months before Dr. Sharp, and made plans to meet with Dr. Sharp in Brazil. Dr. Sharp talked much of the upcoming work in Brazil while I wondered how I would manage, but I remained confident that I would figure it out.

One of Dr. Dougall's biochemistry students remained in the laboratory and I could ask him a question or find answers in my favorite place, the library. I ran many experiments from published articles until Harry told me that data included in an article may not necessarily be correct as published. The idea that a scientist purposely published incorrect information in a peer-reviewed journal was foreign to me. Unfortunately, we continue to hear about such incidents on occasion, in the most prestigious scientific journals.

Harry mentioned Dr. Sharp's family life as he thought I should be aware of the stress Dr. Sharp was facing in divorce. Harry wanted to tell me details, such as Dr. Sharp had better offers than OSU, but came to Columbus to please his spouse. I did not wish to hear that my "model" family was falling apart. I feared the fate of Dr. Sharp and worse, his losing daily contact with his beautiful young son. On hearing Harry's words, I decided that I would go with my husband wherever he wanted to go rather than have Jules harbor resentment, little knowing that such a situation would soon be my reality.

While separated from his wife, Dr. Sharp lived with his grandmother. I wondered why his shirts looked much better than the ones I was ironing for Jules. One day I asked Dr. Sharp who did his laundry? A bit flushed and surprised by my question, he responded that he washed and dried his shirts, but never ironed them. He did not realize that he had discovered a secret, by immediately removing them from the dryer, the shirts did not wrinkle. Many young wives' lives were made easier as I told this trick to my family and friends. Once an iron touches the shirt, it was doomed to wrinkle. If the shirt comes out wrinkled, rewash and re-dry rather than iron!

In regard to my research, I placed pollen grains of the tomato, lily, and geranium in culture and hoped they would divide. I captured cell division in timed experiments, sacrificing pollen grains for staining.

In 1970-71, our laboratory wished to document the origin of callus cells from various plant species. Callus, an undifferentiated mass of cell tissue, is formed in cell culture. The plants of interest to us were tomato, geranium and lily. Our goal was to produce haploid cells in the form of callus and/or plantlets originating from pollen. We placed pollen grains in culture (with or without the plant's anther, techniques that continue to this day) in appropriate media with the hope that callus and/or haploid plantlets would eventually emerge. Pollen grains contain two nuclei and one or both might prove to be the source of haploid callus. We observed that tomato callus formed in two weeks. Which nucleus was significant remained a mystery. Geranium anthers in culture developed callus in three weeks, but the nuclei did not stain at all. Lily, however, served as excellent cytological material; lily chromosomes being identified as the Drosophila of the plant kingdom. The large pollen grains of lily were ten times larger than that of tomato. The two nuclei within the lily pollen grain could be differentiated in the mature grain, but proved more difficult to differentiate in culture. The problem with culturing lily was that it took two months for multi-celled pollen grains to appear, multi-celled grains being presumed to be the beginning of callus cell formation. Once lily callus was obtained, we tried to determine what size bud and anther was best for callus production.

Using staining techniques previously identified, the first task was to correlate the bud and anther size with the state of pollen development to identify the contents of the anther, the stage of development of the pollen grains placed into culture. The anther may include five categories of cell types. In contrast to the synchronous populations of cell types reported in the literature, I observed all five categories of cells present in all but the youngest of the lily anthers. The staining techniques differed per the populations of cell types. In some cells, the stain (Feulgen) disrupted the cell's coat. Luckily for my time course study, each lily bud had six anthers allowing me to culture five and sacrifice the sixth for cell analysis. Both nuclei within the pollen grain were documented as dividing after one month in culture and new cells emerged. The green callus appeared, but all too soon growth stopped. I suspected that growth ceased, because the original anther serving in nurse culture died. Refreshing media did not appear to promote growth, an unknown substance from the anther was needed for my cultures to continue, but what?

I wondered what would happen if I encountered a breakthrough? How could I best document my research as a solo student while Dr. Sharp was gone?

I tried adding plant hormones, auxins, for cell enlargement and cytokinins for cell division, but had no luck in getting my cultures to grow past the death of the anther. I determined that the role of the anther in culture needs to be analyzed. The presence of a healthy anther appeared to be a prerequisite for continued cell divisions in the mass of cells emerging from my pollen grains.

Luckily, Harry came to the rescue when I observed something new or unusual in my cell cultures. Harry had attached his Minolta camera to a microscope and told me that I was most welcome to photograph anything that might be of interest. Without Harry's camera, I would not have been able to document for the long term, the exciting cell division of the pollen grains in culture. My slides long since disintegrated, but the photographs remain to this day.

My objective was to produce haploid tissue in culture and prove its origin from pollen grains. The problem to date had been that when a plantlet or callus formed in culture, it was difficult to determine the origin of the tissue.

In order to ensure that callus derived from cultured anthers did not result from the anther sac itself, Dr. Sharp taught me to culture pollen grains in "nurse culture". I learned how to manipulate single pollen grains in micro-drops and place them on filter paper discs above or next to the anthers. The small green callus obtained from the pollen grain was easily observed on the white filter paper background and the discs could be periodically removed. Through microsurgery, single cells could be isolated and used in the production of cloned cell populations and for genetic analysis as to chromosome number.

I documented that when starch was present in the pollen grain, neither a multi-celled pollen grain nor callus developed. I thought that perhaps starch physically hindered mitosis? I started to wonder about the role of sucrose in callus formation as Dr. Sharp had obtained plantlets, differentiated cells, from high sucrose while other researchers reported differing results. My nutritional study involving sucrose was on the horizon, perhaps sucrose might shorten the time needed to produce callus in the lily? My efforts were rewarded as I observed callus formation as early as 17 days at 60 grams/liter sucrose, but at the highest sucrose concentrations callus was temporarily inhibited!

Dr. Sharp impressed upon me that many discoveries are made by accident. I waited patiently for serendipity as my pollen grains began to divide in culture.

We incubated our cultures under a 12 hour photoperiod. Reading the methods sections of articles on haploid tissue culture today, I noticed that newly incubated cultures are initially placed in the dark. I can't help but wonder if the electricity in the new biological sciences building had remained off for a longer period during the "raining" incident described earlier, would our cultures have developed faster?

I tried to hold back my excitement and double check any new observation. I tried to stay calm on those occasions when my cultures displayed contamination and I had to identify the culprit. Dr. Sharp prepared me that research takes time. He knew more than I did; I thought he knew "everything."

Alas, in varying sucrose concentration, I accidentally found a way to store pollen grains within an intact anther. Freezing was a problem as when the grains became wet the pollen coat was disrupted, while a dry anther was a dead anther and of no help in promoting haploid cell culture.

Plant cell culture was time consuming and labor intensive, but the prospect of contributing to the identification of new and improved plant species motivated our laboratory.

Correlations were made relating bud size to stage of nuclear development. At maturity both tomato and lily pollen possess two nuclei, while anthers used in successful culture contained mononucleated microspores. We suspected that in tomato, a plant in which only callus tissue was formed, the generative nucleus divided while the vegetative disintegrated, or neither formed in the "normal in vivo" manner while in culture. Thus, it would be advantageous to observe and identify nuclear interactions.

I observed that in lily, I could use the size of the anther up to 2.4 cm to determine the developmental stage of the pollen grains (gametic cells) but if the anther was larger than 2.4 cm, I needed to consider bud size in addition to anther size; larger anthers included mixed populations of immature and mature pollen grains. These observations were important as the success in haploid cell culture centered on the precise stage of the development of the pollen grain when placed in culture.

After one year in culture, our lily callus tissue organized and formed plantlets! Geranium callus also formed plantlets. Only tomato callus did not form plantlets in our laboratory in the period 1970-71.

I will never forget Dr. Sharp's face and where he stood in his office the day he told me that OSU's intellectual property team told him that nothing we did was patentable. I was walking into his office when he told me the news. He was visibly upset and neither of us was convinced that what he was told was correct. He stored his issues of *Current Contents* on top of a filing cabinet on my right as I entered the office. After telling me the news, I thought he was about to break down when he gave me a huge stack of *Current Contents* issues to review. He rarely got upset, but was most visibly disturbed by what he had been told. He could barely say one word clearly as he handed me the stack of journals.

Dr. Sharp left for Brazil for the summer and on returning told me that he was planning additional trips. He was always aware of potential opportunities. I never figured out his source, but remained impressed as he nominated me to represent OSU at a seminar on aging in Chicago. It was my second flight on an airplane as one of three students selected to represent OSU. It was one of the worst airplane flights I have ever taken. One of the students complained about the side to side motion, while the up and down movements reminded me of a bad ride on a roller coaster. Being Dr. Sharp's only student was an advantage after all, I attended the seminar at Argonne National Laboratory, a place that I would frequent in the future as a federal employee.

Life in the Laboratory

In Dr. Sharp's absence I was queen of the laboratory and greenhouse. I went to classes and spent hours each week watering my Rutgers tomato plants, my source of pollen grains, and other plants such as tobacco which were growing in the greenhouse. Dr. Sharp sent me to pick young buds of Easter Lily plants at a commercial nursery a bit south of the university. I was amazed how much Dr. Sharp seemed to know about Ohio State, the Department of Biology, and the city of Columbus having moved to Columbus shortly after I did.

I conducted studies on the geranium and raided the gardens of friends for specimens. Dr. Sharp patiently taught me the basics, how to wrap glassware for autoclaving, the concept of sterile technique. We spent time in a small sterile room in which Dr. Sharp seemed to stand as far away from me as possible. I told Jules that I either scared Dr. Sharp or made him nervous. I did not realize that Dr. Sharp's hearing had been seriously impaired by an explosion on the deck of a US Navy Destroyer during a naval at sea exercise.

Following Dr. Sharp's lead I tried my best to place the anthers in the media, tried not to breathe on my cultures as feared contaminating my work. Dr. Sharp was practical, often sticking his nose in a bottle of bleach to determine if the bleach was fresh. My research progressed slowly (I later worked in animal cell culture and was amazed how quickly animal cells grow in culture compared to those of plants).

Our first published article (Sharp 1971) mentions the conflict in the literature pertaining to the precursor cells of haploids; Sunderland and Wicks (1969) reporting that embryoids in tobacco originated from the vegetative cell while Bernard (1971) observed that both the vegetative and generative cells contribute to embryoid formation. I attempted to go beyond chemical methods to influence the environment and tried what was then called "micromanipulation" with the goal to remove one of the nuclei in a pollen grain and place the pollen grain without one of its two nuclei in culture. I also attempted to germinate a pollen tube, sever the portion of the tube possessing the desired nucleus, but could not separate the pollen tube from the agar medium substrate. I identified media to germinate tomato pollen, but did not find a suitable medium for lily and the lily pollen burst in the tomato pollen germination medium (I later discovered that this was due to the osmotic concentration of this medium).

One memorable professor in the medical school taught me what was termed "micro-manipulation." Using very thin pipettes I learned to move cell nuclei from one cell to another without destroying the mother or target cell. This was state-of-the-art; I was fascinated with the potential of such technology to benefit humanity.

In the micro-manipulation laboratory under the direction of Dr. Milton A. Lessler, I inserted micro-needles into the pollen grains. Tomato pollen grains were the smallest, varying from 5-88 microns depending on the state of maturation. The two nuclei of the tomato pollen were easily distinguished after subjecting the grains to the Feulgen stain. The vegetative nucleus is long and thin, while the generative nucleus is round. In distilled water the grains take the form of spheres, while on agar or filter paper the grains appear football shaped and possess three pores. The pollen coat is very tough and micro-needles thin enough to properly penetrate the pollen coat were often broken in the process. Penetration into the pore itself was a possibility, but the nuclei were only apparent after sacrificing the specimen on staining. The tomato pollen grains were so small that in a liquid solution, the slightest movement of the micro-needles caused the grains to move violently through the medium and not easily caught by the holding needle. Tomato pollen grains were more easily manipulated on a solid agar surface.

Lily pollen, five to ten times larger than tomato pollen, lacked the protrusions displayed by tomato pollen and did not possess the jerky movements of tomato pollen grains in liquid solution. Three-fourths of the lily pollen grain displayed plate-like thickenings of the pollen coat, while the remaining side lacked plates. The side lacking plats seemed the most likely area into which to insert a needle. However, on inserting the needle into the lily pollen grain the entire pollen grain exploded. Later, I noticed that many of the lily pollen grains were bursting before I touched them with the needle. Several repetitions of the process revealed that the pollen taken from buds 3.5 cm and smaller burst, while the more mature pollen did not readily explode. Pollen burst in water, oil, and in sucrose solution at a concentration of 80 grams/liter. However, the grains did not burst in a Murashige and Skoog (1962) lily culture medium.

Becoming even more creative, I decided to destroy one of the nuclei of a pollen grain with an ultraviolet microbeam and a quartz needle. It worked for lily and geranium, but not in tomato pollen grains as the two nuclei in tomato pollen seemed not to be separated by a membrane. I realized that with the use of sterile technique, microsurgery had much potential in separating cell nuclei. Under the direction of Dr. Milton A. Lessler (Lessler, 1988), a medical school professor who spent much time teaching me techniques of micromanipulation, I became comfortable working with cells, isolating clinical viruses, and growing both plant and animal cells in culture.

I concluded that although techniques of haploid tissue culture permit creativity and innovation, the advantages of haploid cell culture for plant growth studies were interwoven with numerous inherent disadvantages! Much work remained in regard to identifying optimum culture conditions for timely callus and plantlet development in lily, geranium, and tomato.

I discovered the tenants of being a professor's "first" graduate student after the fact and hope this helps graduate students, when selecting a new member of a department's academic faculty:

1. Plan with your adviser, but learn to work solo.
2. Understand that it takes time for a new professor to gain grants, become established in the field, and acquire the respect of colleagues.
3. It may appear that your adviser knows everything, but as the newest member of the academic department, you and your adviser are both learning.
4. Find out who your adviser's mentors are as they may help you in your adviser's absence.
5. Make your adviser look as good as you can no matter what you may hear or read.
6. Find colleagues to share the normal ups and downs of laboratory research.
7. Never take comments to heart such as "You will never find a job doing that!" The "that" is the vision you share with your mentor. It may be emerging state-of-the-art, a yet nameless new field of science, business, or a combination of fields. In our case, the new field was biotechnology.
8. If you lose contact with your graduate school advisor, it is never too late to rekindle the relationship!

Dr. Sharp wanted me to experience fields other than plant cell culture in what would later be termed biotechnology. A technician that worked in the basement of B&Z accepted a position at an emerging genetics company, Metagen. On visiting her laboratory, she typed and photographed our own human chromosomes.

I did not realize while a student of Dr. Sharp that he attended OSU as an undergraduate student for three years before entering the Navy. Dr. Sharp knew the niche expertise of biomedical faculty members. He knew who embraced emerging technology and sent me to learn from them in one-on-one teaching sessions.

In regard to my interest in traditional publication, Dr. Sharp taught me the courtesy and protocol used to publish technical articles. He believed that our work must be documented in the scientific literature long before formal written or oral presentation to faculty. To my knowledge none of my graduate student peers in the department were publishing, I was the first. I learned about sending reprint requests to other scientists being most proud of the 100 reprints I had of our article to share with those that might contact me. I remember Dr. Sharp asking me if I thought 100 reprints would suffice. I could not imagine the number of researchers who might cite our work in subsequent years, but became skilled in tracing who was citing us in the literature. Dr. Sharp impressed upon me that everyone in the lab should be included in the author field of a publication, as each member contributed to the lab's operation.

As I look at an early draft of our first published article, "Haploidy in Lilium" (Sharp 1971), I reflect that Dr. Sharp left the length of the lily floral buds blank in the first draft of our article waiting for my data, encouraging me to publish my results in the scientific community. Little did I realize that the article would continue to be cited in the literature for more than 40 years!

Dr. Sharp was the first to teach me the importance of selecting a concise, appropriate title. He spent much time discussing the pros and cons of each potential journal for publication of our research manuscripts. It was the first time that I compared journals in regard to suitability for publication of our research. Dr. Sharp selected those he thought most influential in our field, what one would today term the journal's "impact factor." Dr. Sharp stressed the importance of publishing one's research for peer-review in the global plant cell culture community, thus, identifying ourselves as active researchers.

I taught Dr. Sharp's principles of how to read an article to hundreds of physicians, chemists, engineers, and students through the years. As an information specialist, I continue to find answers in the methods section of technical articles, seek best evidence, and review intellectual property including patents. Dr. Sharp encouraged me to do primary research and communicate with the inventor or scientist.

New Colleagues

Two male students eventually joined me in Dr. Sharp's lab. The first was a graduate student in microbiology who was interested in the bean and worked with a botany professor in Dr. Sharp's absence. The student, recently married, could not accompany Dr. Sharp to Brazil. A year or so later, an undergraduate student in genetics started working at the bench in our laboratory.

Dr. Sharp's office door was in the middle of the lab on the wall away from the laboratory benches. He rarely came out of his office when I was working solo in the lab, thus, I had no clue that he was watching. I remember one incident in particular. The new graduate student was making culture media like a cup of instant coffee. I thought that this was wrong, but just returned from class and wondered if perhaps Dr. Sharp was testing a new technique, "instant media?" Just as I was about to leave my burner and help the new member of our lab, Dr. Sharp came running out of his office to aid the student. I never figured out how Dr. Sharp could see what was going on at the distance his office was from the bench. Dr. Sharp may have had some hearing difficulties, but more than made up for it with his apparent far-sightedness. Dr. Sharp had the vision of an eagle and soared to the rescue of his student.

The laboratory team was in its infancy, but had potential. Our most interesting team consisted of four members: one new graduate student, one undergraduate student, Dr. Sharp, and me. Harry continued to work quietly in the lab. On occasion his wife, Peggy, who was working on her doctoral thesis in the Department of History on the Shaker seed industry, would stop by as did Jules on occasion if in town. Since I had to attend classes and take care of the greenhouse, I was glad Harry was ever present in Dr. Sharp's absence should something in the laboratory unexpectedly fail.

My research was making progress, the laboratory was under control, and plants were thriving in the old greenhouse. Jules and I put new aluminum siding on the trim of our home to cover the new white paint which began peeling off in long sheets after I sanded the trim too smooth and/or most probably painted the trim with the wrong paint (I do much research for paint companies today). My classes were going well, my dining room set was on order, and I planned to invite our laboratory to my house for dinner when I was approached with the unexpected.

Dr. Sharp just returned from a trip. I was working late in the laboratory and Jules picked me up at the Biological Sciences Building parking lot. To my surprise Jules had unbelievable news; he told me in the car that he had been offered a lateral position in underwriting. My heart seemed to stop beating for an instant as I heard Jules tell me that he would regret it the rest of his life if he did not take the underwriting position in "Cincinnati, Ohio." I could not believe my ears. Jules had been waiting for a position in underwriting in Columbus, but the opening offered was in Cincinnati. I wanted Jules to be an underwriter rather than continue to climb the high water towers as a fire protection engineer inspecting sprinkler systems in highly protected risks. Jules continued the conversation as I listened. I was in shock as he proceeded with details including that the company was not going to help sell our house of 2 1/2 years or cover the moving expenses. If he wanted the position, he would have to be in Cincinnati shortly. One underwriter had left, and Jules would be one of only two underwriters in the Cincinnati office. I would stay behind until I sold the house. My mind began racing, would I end up like Dr. Sharp and his wife, and eventually divorce if I insist that we remain in Columbus? Did Jules ever realize that he stopped me from going to Brazil and now he was asking me to end my four year University Fellowship? We had only been married for 3 years, we could not afford two residences; one of us would have to give in.

I dreaded leaving my research in early 1972, our beloved neighbors on Marland Drive, OSU, and Dr. Sharp, but this was not the first time I felt like this. The sadness was not new to me. I experienced it once before when leaving my grandmother, aunt, and cozy home to go to a displaced persons' camp at the age of three years. My family had to go to the camp before entering the ship which took us to the United States.

I knew things would somehow work out as my father had said, "When one door closes, another will open." My mother had taught me to walk forward, not back, but I felt literally sick. This was not going to be easy for Jules, another new city for someone who never wanted to leave Cleveland. In Cincinnati, Jules would no longer have a heavy travel schedule, he would be home more as an underwriter. I left a full scholarship and grant at Cleveland State University to come with Jules to Columbus. It was unbelievable that we were going to leave Columbus for Cincinnati, a city we had driven through the year before and commented how lucky we were to be living in a college town like Columbus!

I remember the day I told Dr. Sharp that we were moving. He tried to be positive, but was noticeably shaken. He told me that I might continue my plant cell work in Cincinnati, but I could not find anyone doing plant cell culture. I had become attached to Dr. Sharp's five year old son who would come to the lab on Saturdays. I had few regrets, but wished I had spent more time with little Jeff. I hated the thought of leaving my laboratory family, our friends in Columbus, and the dearest neighbors one could imagine.

I considered every option, but Jules had to move quickly. I had to at least try to play the role of wife, attend dinners with Jules' new boss, and ride to Cincinnati with company directors. Wives followed husbands in those days and companies had no compassion for families, certainly not for the graduate studies of a spouse. Maternity coverage was not always included in basic family health care insurance in those days; potential employers asked the most personal questions on job interviews. Dr. Sharp treated me as a professional; I was not prepared for what I would encounter.

One of my best friends at OSU was born and raised in Cincinnati. Her father had a greenhouse and her parents became our adopted family. I eventually found much opportunity for employment and life-long friends in Cincinnati, but missed my life at OSU more than I realized.

The move had been difficult, filled with mixed emotions. Our house sold in a few days to a buyer who said I was the most reluctant seller he had ever met. He offered us more for our house than the price we were asking. The sale of our house moved quickly as the buyer was Vice President of a local Columbus bank. We had not found a home in Cincinnati and our kitty, Goldie Tiger, was missing. We placed an advertisement in the *Columbus Dispatch* about our missing kitty and moved to Cincinnati not knowing what happened to our beloved red tabby.

After a few weeks, Goldie Tiger was found by our dear neighbors at Cat Welfare in Columbus having recovered from being hit by a car. After nearly giving up on Goldie, we drove back to Columbus to pick him up. Goldie and I both began the slow recovery from the move having left the familiar behind. Neighbors in our Cincinnati apartment complex began knocking on our sliding glass door to invite me for coffee; we were now in P&G country. To hasten my recovery and since we did not have a house to take care of, Jules suggested we fly overseas to visit my remaining family in Germany. In September of 1972, we found ourselves in Munich, Germany, the day after the Israeli Olympic team was brutally murdered. It was a nightmare for us and especially for my mother who begged me on the phone to cut my vacation short and get out of Germany immediately. The murder of the Israeli Olympic team had torn open unhealed wounds and my mother feared for her children once again. We left the city dumfounded by what had happened during the Olympics and were relieved to return to the United States.

Dr. Sharp's travels to Brazil turned out to be blessings in disguise. In the absence of an adviser or colleagues, one becomes self-sufficient in solving problems. If I could work alone in the plant cell laboratory I could work in any lab, but acknowledged that I would never find another Dr. Sharp.

Having enlisted in the military after completing three years as an undergraduate student at OSU, Dr. Sharp would light up if anyone asked him about military service. He would smile whenever he mentioned the Navy. I remember one day when Dr. Kreier came in and asked Dr. Sharp where he served and if he was ever on a ship, to which Dr. Sharp replied, "Yes, the USS Compton." Dr. Kreier responded that he knew many that served, but no one that actually made it to a ship. I observed that Dr. Sharp was wise to surround himself with those from whom he could learn. He selected excellent mentors, was not afraid to ask a question, or call anyone in the world. Dr. Sharp was the ultimate "information specialist."

Practical advice was a specialty of Dr. Sharp who advised me on much beyond what one might expect of an academic adviser. He encouraged me in regard to my troubled, youngest teenage sister who today is a Professor of English. Dr. Sharp seemed omnipotent. He owned property and by coincidence rented a suite to our friends, a young married couple returning to OSU after their tour of duty in Vietnam.

After moving to Cincinnati, I periodically visited Columbus on my way to Cleveland hoping to see Dr. Sharp in Room 417. The

last time I visited, the lab looked unfamiliar. Someone in the hallway told me, "The famous plant lab moved." I was surprised that
the individual referred to the lab as "famous" and was not certain
if the lab moved to another location in the university or had Dr.
Sharp left Ohio State? This may have been the time the laboratory moved to larger facilities on the roof, if so, I was unaware of
its new home.

Dr. Sharp visited me once in Cincinnati, but I was depressed
and do not remember the visit to this day. I knew he had been
there as my research notes and some of my photographs were
gone. Dr. Sharp wanted me to write up my research and complete
the M.S Degree Program. The date was set for the defense of my
research. I took the bus to Columbus the day before, but was up
most of the night in a noisy motel on High Street where the staff
continued to give my room key out to whomever was registering.
We had come home from Germany a few months earlier and neither Jules nor I had completely recovered from the shock of the
murder of the entire Israeli Olympic team in Munich. I was tired,
upset, and decided to return home on the next bus to Cincinnati,
but stopped to call Jules from the bus station. Jules in no uncertain terms said to me, "Call Rod before you leave the bus station."
It must have been around 7:00 A.M. when I called Dr. Sharp to
tell him I was leaving. He told me to wait at the bus station and he
would pick me up. I wanted to go home to Cincinnati, but waited
for Dr. Sharp.

The ride in his car calmed me down a bit. Dr. Sharp had my
notes and many of my photographs, but would I remember anything after a sleepless night? Would I be the laughing stock of the
microbiology department, trying to justify working on plant cells?
Dr. Sharp assumed I was nervous about the presentation and told
me that his adviser had told him to "Imagine the members of
the committee in their pajamas." I rarely cry, but on reaching
Room 417, the first time I had seen the laboratory since moving

to Cincinnati, I started to cry. I could not seem to stop the endless flow of tears, flooding not unfamiliar to the lab once deep in water (As I write these words, my eyes are tearing up as they did so long ago.)

Dr. Sharp's last phone call was for help with a project while I worked for the U.S. Environmental Protection Agency (EPA). I lost contact with him for many years after his last call. I heard about his success from the Dean of the Department of Microbiology, Dr. Patrick Dugan, who told me that Dr. Sharp's company, DNA Plant Technology, was doing very well on the stock market.

In Cincinnati, I spent a short time working in clinical virology in an academic medical center and soon accepted a position as an environmental scientist for the U.S. EPA. OSU won the contract to train new EPA project officers. I was sent to OSU in Columbus for in depth training in grants, contracts, and inter-agency agreements as a project officer in EPA's Solid and Hazardous Waste Research Laboratory (SHWRL).

Times were changing, less than one percent of females had traveling jobs like mine. Jules was doing what he wanted and listened to me complain for 15 years that I needed to know the secrets to obtaining the best information. In addition to monitoring grants, I was responsible (along with a most knowledgeable pharmacologist who had left academia for government service) for creating a series of criteria documents on selected pollutants. I attended meetings of the National Academy of Sciences and regularly visited federal, academic, and industrial research laboratories.

In 1978, I decided to follow the path I had chosen many years earlier and commuted to the School of Library Science at Kent State University. I drove from Cincinnati in southern Ohio to the city of Kent in northern Ohio, passing through Columbus and Ohio State each week. In 1980 I received my Master of Library Science and accepted a position in the Educational Resources

Center at Miami Valley Hospital in Dayton, Ohio, carpooling to Dayton from Cincinnati each day. I later worked in the Fortune 100 and private companies in Cincinnati as the information specialist in strategic business development, finance, and research and development.

Jules and I lived in Columbus for four years, Cincinnati for almost 15 years, and moved back to Cleveland when Jules' company closed the Cincinnati and Columbus offices. At this point, I was a medical librarian working in the hospital known for the first true surrogate birth and visits by the rich and famous for weight loss programs like Optifast, invented at our hospital. The incredible team of physicians and researchers were devoted to the best in patient care. The hospital library responded to knowledge-based research needs from companies local to Cleveland, the physicians of the Mt. Sinai Medical Center being sought by medical programs and national news. The center was the first to have a system where physicians could retrieve needed knowledge-based information by dialing into our platform of medical databases.

Reconnecting

The Mt. Sinai Medical Center and associated hospitals closed in the year 2000. Although the physical facility was no longer available to our educational resources center, the physician medical society decided to fund us for three years beyond the closing of the hospital. A hospital library surviving after the closing of the parent institution was a first. Our library team of three was to set up a new nonprofit 501(c) (3) and find a board of directors.

My boss suggested I ask my personal physician and another physician, both heavy users of hospital library services, if they might be interested in participating in our required board. To my surprise, my boss suggested I also ask Dr. Sharp. I had not seen or talked to Dr. Sharp in years, but proceeded to ask if he might join the board. He graciously accepted and knew more about details about forming a board and nonprofits than we did at the time. I picked Rod up at Cleveland Hopkins International Airport and he was the familiar Dr. Sharp. He had served as Dean of Research at Rutgers, but I could not imagine the "Just do it" professor as a dean. When I watched him read the *New York Times* a day later before departure, he seemed a bit more dean-like, but entrepreneurial endeavors seemed a better fit.

After participating on our board, Dr. Sharp referred an entrepreneurial recent medical school graduate to me. Rod was involved in a dot-com project termed, MedTower. It was a virtual medical building, but re-designed as an information resource, modeled after the work we were doing in Cleveland.

MedTower failed as originally planned, but served an even more important purpose. As the terror of September 11, 2001, was in progress, I used the Mt. Sinai-Cleveland web site to post resources and phone numbers to help those in New York City. The medical student graduate inventor of MedTower happened to be on my web site while he watched from his office in New Jersey as the second airplane crashed into the World Trade Center Towers. Being on my web site by coincidence, the student immediately contacted me via the Internet, his phone lines being down, and forwarded critical New Jersey phone numbers for me to include on our website, along with those already on the site from local New York City and federal government resources. We were one of the first to place needed information on the web, transferring information from the television to the Internet, while two other members of our Mt. Sinai team called companies throughout the U.S. to secure appropriate respiratory masks for those at Ground Zero. The Library of Congress archived the many web sites that broadcast information about 9/11 as the country prayed for those at Ground Zero, the Pentagon, and the heroic members of American Airlines Flight 93 bound for Cleveland, Ohio, that crashed in Shanksville, Pennsylvania.

A few years later, Dr. Sharp asked me to participate in one of his pharmaceutical ventures. The invention, after securing the patent from the appropriate university, would be mass produced by the start-up pharmaceutical for those suffering from acetaminophen poisoning by accident or intent. The invention made the horrible tasting antidote in pill form (the antidote also tastes terrible in liquid form) more palatable and easier to swallow. As fate would have it, I saw the need for the coating first hand as a clinical instructor, "informationist in context," providing information as a member of the patient-care team on clinical bedside rounds in the intensive care units.

Dr. Sharp was awarded many honors through the years in the U.S. and Brazil. In 2007, he was honored with the Distinguished Service Award by OSU President Gordon Gee. The award was followed by a reception in the Biological Sciences Building. I knew that Dr. Sharp would be in Columbus and wondered if I might see him sometime over the weekend. Dr. Sharp said he was "in the hands" of the University and I was a bit disappointed to lose the opportunity to see him. Shortly thereafter Jules and I were surprised with an invitation to attend the evening reception in Dr. Sharp's honor in the Biological Sciences Building. We drove to Columbus after work. At the evening reception, the Dean of Biological Sciences called on me to relate a memory as Dr. Sharp's first graduate student. I hate impromptu public speaking, but since the Dean requested, I could not refuse. I quickly thought of the broom story as an illustration of Dr. Sharp's teaching methods on intellectual property.

I never attended one of Dr. Sharp's classes, but imagine if his teaching methods were similar to those that he used with me, he had to be a most memorable professor. The event honoring Dr. Sharp allowed me to reconnect with Dr. Kreier, an important mentor to Dr. Sharp, and meet members of Dr. Sharp's team in Brazil whom I had wanted to meet for more than four decades.

Dr. Sharp's son, Jeff, who I last saw at age five years, was now in his 40's and reviewed his father's life in pictures. I finally met Sally, Dr. Sharp's sister, who he talked about often through the years. Sally related the importance of her brother, Rod, in her own life.

The following spring, my nephew played violin with his school orchestra at Lincoln Center. I called Rod (Dr. Sharp) to tell him that we would be in Manhattan. Rod invited my sister, who at age 12 years lived with us in Columbus, and visited the OSU Laboratory and Greenhouse, and me for lunch at the Tavern on the Green in Central Park. Rod looked taller than ever in his navy blue suit and wrinkle free shirt. We talked about family and days gone by in the beautiful setting of lilies and spring flowers. As I spoke, I noticed a tooth on my luncheon plate. I had no clue that a dental crown had fallen from my mouth. Sitting in the most formal restaurant, I insisted my sister check to see if anything was missing from my mouth. Rod took it all in stride saying he always loses dental work on trips!

While we were walking outside the restaurant in Central Park, Rod cautioned my sister and me to step away from the curb; distance ourselves from the passing cars. Several years earlier Rod had been hit by a car and refused to go with emergency personnel as feared the person who hit him might be an illegal alien and sent back to his home country. Just as Rod helped a dot-com team that was sleeping on the floor while developing MedTower, Rod sympathized with those less fortunate. Rod's young son Jeff, the blond-haired little boy that woke his father up early to spend Saturday in the lab, is successful in his chosen profession and has a family of his own, making Rod a loving grandfather.

After more than 40 years, Rod and I continue to write traditional letters to communicate scientific events and family news. We recently moved to email. The special seed packet Rod sent me, a limited edition of a hybrid Rutgers tomato, having the most delicate yellow blooms and perfectly shaped tomatoes, was the first step in planting a memorial garden outside of what was my mother's bedroom window at my house. I pass Rod's old street, Hillcrest, in Cleveland Heights, Ohio, each time I pick flowers from my garden for my mother's grave at the Mayfield Cemetery.

I look forward to one day catching up on the volumes of articles and books that Dr. William "Rod" Sharp published during his most active career in academia and industry.

To Success

I continue to use the principles of the scientific method which I learned in grade school. Whether I am providing knowledge-based research, writing articles for trade journals, creating an innovative product for a client, revising my company web site, experimenting with tomato plant varieties in my garden, developing recipes for my family, or blogging per Information Specialist Secrets, Precious Cooking, and Most Precious Memories, I consciously and unconsciously use and remain devoted to the principles of science.

Have I been successful?

If you met me or my family, you would not think the word "success" applies and wonder why I am included in this book? Success in the sense of fame or prosperity certainly does not apply to me. However, I believe that success defined as an achievement of something desired, planned or attempted, truly does. I probably have made more "attempts" than most, continue to set goals and desire to explore more things that I can list. As a human being, I have made more than my share of mistakes, but I have tried to rectify them to the best of my ability.

My family taught me by example that success is getting up after being knocked down harder than anyone can imagine. I consider my father to have been more than successful, having built another life with three more daughters after his first young wife and daughter disappeared in the Holocaust because of their religion and being in the wrong place at the wrong time.

My personal success includes going back to work after having Brachytherapy (internal radiation) during my lunch hour, encouraging a spouse back to life after a major heart attack, coma, and paralysis, caring for elderly family members like my mother and my aunt during their last days, and never easily giving up on anyone or anything. I make new friends, but never seem to let the old ones go. My husband, Jules, says overdoing everything is one of my best attributes. He is correct, and as not knowing when to give up, is something I have yet to learn.

Dr. Sharp continues as my adviser for life. I chose him for that role more than 40 years ago and he continues to answer my email, now on his iPhone! On the Internet or in person, I remain his most dedicated student. As I posted on the net years ago:

My hero is Dr. William Rod Sharp, an entrepreneur, dedicated to making the world a better place for all humanity . . . Our long-term relationship can best be expressed in the words of the Alma mater of The Ohio State University, Carmen Ohio:

"Time and change will surely show, how firm thy friendship . . . O HI O."

Rosa Shine Raskin, M.S., M.L.S.
rosa@raskinfo.com
References:

White, Philip R. (1963). The Cultivation of Animal and Plant Cells, 2nd Ed. New York: Ronald Press.

Sunderland, N. and Wicks, F. M. (1969). Cultivation of Haploid Plants from Tobacco Pollen. *Nature,* 224, 1227-1229.

Bernard, S. (1971). Developpement d'embrons haploides a partir d'antheres cultivees in vitro. *Revue de cytologie et de biologie vegetales,* 34, 165-188.

Murashige, Toshiba, Skoog, and Folke. (1962). *Physiologia Plantarum,* 15, 473-497.

Sharp, W.R., Raskin R.S., and Sommer, H.E. (1971). Haploidy in Lilium. *Phytomorphology,* 21, 334-337.

Revised: September 20, 2012

CHAPTER 14.

Sharing the Deepest Secrets

William R. Sharp

As I grow older and reflect on 77 years of life, the kaleidoscope colors blur into a magnificent fireworks exhibition in honor of the unique individuals contributing to my coming of age, goals, sense of values and more recently the passing of the baton to another generation. These unique individuals who influenced me for so many years are my heroes and include colleagues, family, friends, professors, teachers, students and resulting connections. I would like to borrow the following words of Ruth Mueller, a friend and author of the book entitled: *One Minute Before Midnight, A Memoir*:

"My life has been shaped and touched by so many wonderful, brave, strong, inspirational and loving people whose memory might be lost if I did not put pen to paper at this time. I have watched the world around me change in ways that no other century in human history has. I have done as I pleased with my life, and I am very happy with the results. I am the product of my colleagues, family and friends, and it is to them that I dedicate the book."

Serendipity was a major factor, in leading me into a career that placed me among the pioneers who broadened the horizons of the plant sciences. This career included participation in international research initiatives, the launch of the plant biotechnology era, tenured professorships at the Ohio State University and Rutgers University, a co-founder and executive vice president of DNA Plant Technology Corporation, a leading agricultural biotechnology company, founding of two subsequent biotechnology companies, the Dean of Research at Cook College, Rutgers University and continued involvement in university committees and start-up technology companies.

The book entitled *Outliers* says it all: "There is logic behind success, and it has more to do with legacy and opportunity than high IQ." Malcolm Gladwell, the author, casts his inquisitive eye on those who have risen meteorically to the top of their fields, analyzing developmental patterns and searching for a common thread and the author asserts "that there is no such thing as a self-made person." Instead "the true origins of high achievement" lie in the circumstances and influences of one's upbringing, combined with timing. One's venues and social ecosystem determines what we become." The social ecosystem includes parents, grandparents, extended family, siblings, teachers, professors, fellow students, friends, and serendipitous acquaintances.

The Beginning

It all began in Akron, Ohio, on September 13, 1936, followed by moves to other cities and towns tracking my father, William John Sharp's early career in the retail division of the Goodyear Tire and Rubber Company. Our city name, Akron, is derived from the Greek word "Acropolis, meaning city at high point. The fifth largest city in the state of Ohio and the county seat of Summit County, it is located in the Great Lakes region approximately 39 miles south of Lake Erie along the Little Cuyahoga River. Akron is today part of the larger Cleveland-Akron-Elyria combined statistical area with a population of 780,440 according to Ohio Central History, Retrieved August 4, 2013 from *http:// www.ohiohistorycentral.org/w/Akron,_Ohio?rec=650*

Our family returned to Akron following my father's promotion to a corporate position at Goodyear headquarters during the summer of 1947. We purchased a beautiful home on Kenilworth Drive adjacent to the homes of the Rubber Barons – The Harvey Firestone Estate and the Charles Seiberling Estate. I enjoyed our home and the relationship with my siblings and visits to the home of my aunt and uncle who lived on Castle Boulevard, a beautiful tree lined boulevard, adjacent to Kenilworth Drive. My time was occupied with school and an obsession with reading newspapers and magazines page to page. I invested my allowance and monetary gifts in subscriptions to dozens of magazines including *Life Magazine, National Geographic, Popular Mechanics, Popular Science, Readers Digest, Saturday Evening Post, Time Magazine,* and many others. My hobbies were collecting coins, stamps and building model airplanes. I was a fan of the radio serials: *Batman, Captain Midnight, Jack Armstrong, the Shadow, The Thin Man, The Fat Man, Sam Spade, Tom Mix,* and subscribed to all the radio program special offers which required remittance of cereal box tops and small change. I was a big time fan of the Cleveland Indians and in pursuit of my great dream of becoming a batboy for the team, participated in the batboy lottery for a number of years.

My dad ended his career with Goodyear during 1948 after he and his brother, John William Sharp received substantial funds from a trust account established by their deceased father. Some of the funds were used by my father for founding the Sharp Sporting Goods Company and in tandem with my uncle launched the Highland Square Hardware Store. Sharp Sporting Goods was housed in about 5,000 sq. ft. of retail space in a new shopping center located in an affluent section of Akron. The company, which retailed fishing and hunting gear, HO scale model trains, model airplanes, u-control airplanes and accessories and sports equipment and toys. Sharp Sporting Goods Company was a frequent sponsor of county-wide competitive u-control airplane meets with divisions for different age groups in which my siblings and I often participated. Winners were awarded trophies and merchandise prizes. Several times my siblings and I were subjects of articles appearing in the *Akron Beacon Journal* Newspaper along with our father. My parents encouraged my participation in all facets of retailing at the Sharp Sporting Goods Company beginning at age 7, which allowed me to evolve into an experienced salesperson. I was permitted to accept payment from customers for construction of model airplanes and HO scale railroad cars and for maintenance of model airplane gasoline engines.

I was a member of the Cub Scouts and my father was a troop leader for several years. The Cub Scouts were organized into Dens with weekly meetings at the homes of members, with parents supervising meetings and monitoring projects. Monthly troop meetings were organized at a neighborhood church hall where we displayed projects and received awards. But because of parental political issues that I never completely understood, I never joined the Boy Scouts.

My parents were important in molding me into the person that I ultimately became. I was the older child with three siblings, including a sister, two years younger, a brother, five years younger and a sister, twelve years younger. My older sister, Patricia, was interested in art, dance, fashion, and music, my brother, Bill, was an accomplished athlete, and my younger sister, Sally, was multi-talented. During her early years, her interests included the nursing of sick animals back to health, art, athletic activities, music and writing. Sally was athletic and took up bowling and was soon at the top of her game. Overnight, my parents' home was flooded with championship medals, plaques and trophies. Patricia attended Ohio State University, married two times and had a family of four children. Bill attended Ohio State University, married with three children, joined the Goodyear Tire and Rubber Company and retired as Vice President of Goodyear Tire and Rubber Company

International. Sally married and divorced with one daughter. She became an accomplished educator, a co-founder of an intercity creative arts high school, author and supporter of human rights and recently retired to return to her writing. Sally and I have always have been on the same wave length and have shared our family life which continues today.

I attended the Akron schools of my parents including my dad's King Elementary School and both parent's John R. Buchtel High School. My academic progress varied according to my interest in subjects and focused on science and the social sciences with no clear career goal. My siblings were much better students and more career-oriented. My academic performance was slightly better than average and I bonded minimally with my teachers. Elementary school was somewhat of a calamity in grades one through four. In kindergarten in Springfield, Ohio and first grade in Huntington, West Virginia, I didn't have a clue of what was happening. The confusion continued in grades two through four, when the same teacher ruled our classroom for all three years. The teacher was considered to be forward thinking at the time. She segregated our class of thirty students into six units of five students each based primarily on academic performance. The student groups had elected leaders and were organized in semi-circles based on academic achievement and appropriately named the airplane group, train group, car group, motorcycle group and the bicycle group. Sadly to say, I was assigned to the bicycle group for all three years. My classmate, Renee, a member of the bicycle group, and I bonded because we were both left handed and slow readers. We were frequently punished for attempting to use our left hand during writing exercises. Our punishment was to wear dunce hats and be seated on high stools placed in opposite corners at the front of the classroom for extended periods of time. The teacher was a believer in corporal punishment and frequently punished Renee and me by slapping our hands with a ruler in front of the students for using our left hands. Our fellow students always showed empathy and treated us as rock stars.

I consistently received failing grades in penmanship and reading but somehow was always promoted to the next grade level. My parents, in collaboration with the teacher, would assign reading and penmanship exercises, which included the copying books, such as the telephone book, but my penmanship and reading abilities seemed to never improve. Renee and I gradually matured and became better students during the fifth grade. Renee developed into a beautiful, popular and smart young woman during her high school years. Reading was never really a problem for either one of us. We were both probably immature for our grade level and somewhat impeded by our slow development. I was, in fact, an avid reader of the many magazines to which I subscribed. The stigma of being assigned to the bicycle group never really fazed me in the least. I was always happy at school and at home and had a good rapport with my fellow students. The remainder of elementary school and high school were not complicated with the exception of Latin classes. The Latin teacher's modus operandi was to whisper test questions to force students to be more attentive. This sotto voce testing method was somewhat challenging for me because of a slight hearing disability at the time resulting from childhood pneumonia and elevated fever that caused minor neural damage. One afternoon, the teacher gave a snap quiz and whispered an infinitive for the students to conjugate. The infinitive that I heard, seemed unfamiliar which prompted me to ask the Latin teacher to repeat the infinitive, but she merely ridiculed me for not paying close attention. After, straining my hearing abilities, I correctly conjugated the unfamiliar verb. The teacher gave me a grade of 40 out of 100 for the conjugation of an incorrect infinitive. After class, I went to the principal, threatened to quit school, and abruptly exited the building. My parents immediately intervened and I was assigned a new Latin teacher.

The hearing problem did create a few difficulties for me. My parents thought that I was not attentive and friends thought I was arrogant. Nevertheless I managed. The situation worsened during my years in the U.S. Navy. Two fellow shipmen and I suffered from broken eardrums following an explosion at the aft gun turret during an at sea operation. The accident further worsened the hearing disability. The problem was sometimes inconvenient but not a deterrent to my pursuit of achievement and happiness.

The only other unpleasant incident in high school was being bullied by a student with the name of Tom, who had the physique of a football player. Tom sat across from me in Geometry class. He frequently stalked me and pushed me. One day, I decided that enough was enough and developed a strategy to deal with Tom's unpleasant behavior. I dealt with his frequent encounters based on boxing lessons taught to me by my father during the first grade. That is when a similar obnoxious fellow student persisted in banging her metal lunch box on my head on our walk home from school. So when I caught a glimpse of Tom running toward me for the big push, I abruptly turned around and socked him in the right eye with all my might. He fell to the ground with a giant howl. I left other students to attend to him and proceeded home to relieve my mother at the family-owned hobby shop business.

After dinner that evening, Tom's mother, a single mother and a nurse by profession, appeared at our front door in a highly emotional state to talk to my parents. She threatened my parents with having the police arrest me on an assault and battery charge. My parents calmly explained the situation and suggested a meeting with the high school principal to resolve the matter. She stormed away and we never heard a word from her again. The following week, following his return to school, Tom threatened that he and his buddies would kill me after school. I requested a change in my seating and nothing else ever happened to me again. The episode taught me a lesson that I had to fight my own battles. In the following year, our junior year, Tom asked that we forget the incident and we became friends.

Among my early learning activities was the construction of a soapbox racing car and the entry of it into the All-American Soap Box Derby in my hometown of Akron. I did this under my dad's supervision. This was a big deal to me because Akron was the headquarters and home to the National Soap Box Derby. The Soap Box Derby building project and entry in the race car competition occurred over a five year period beginning at age eight. The Soap Box Derby provided me with experience in project management and the thrill of participating in a competitive event. I was fortunate to have had the experience of participating in the run-off competition during the final year and almost winning which could have gained me entry into the national competition. My brother followed in my footsteps, but my sisters were unfortunately not eligible because of the rules at the time excluded for girls from participating in the race.

The weekly bonding and conversations with my father in our basement workshop were enjoyable and important during the Soap Box Derby years and doubtless to say, this also sparked my interest in science and technology. My father was committed to the father and son bonding opportunity, probably because of the lack of a father figure during his early childhood. He and his brother were raised by their mother and shuttled off to the Great Lakes Summer Camp for Boys every summer. They were hired as camp counselors during their high school years. Their father, William Walter, died of pancreatic cancer when my dad was seven years old. However, he remembered some of their important times together and a few pearls of wisdom bestowed upon him by his father during the brief time they shared together. Some of these pearls of wisdom were passed down to me.

One of those lessons was imparted when my dad and his father were flying a kite together on a weekend afternoon and the kite and string became tangled in a tree. His dad removed the kite and tangled string from the tree and requested that my father untangle the string. My dad said, "Untangling the string is impossible."

"Nothing is impossible in this world," his father replied to him. "You must untangle the string before you return to the house." My dad spent considerable time untangling the string before returning to the house to acknowledge that all things were indeed possible.

My dad's father's family was of English background and immigrated to Philadelphia, Pennsylvania from Manchester, England and later to New Philadelphia, Ohio where the family owned a foundry. His father, William Walter Sharp, moved to Akron, Ohio, where he founded the Mill Mine and Supply Company which is situated on the tracks of the former Eire Railroad adjacent to the University of Akron. The Mill, Mine and Supply Company was an important provider of industrial products, materials and services and maintained a relationship with the family foundry located in New Philadelphia, Ohio. The Company's former headquarters have been incorporated into the University of Akron campus. Interesting to note that my son's film company is named Story Mining and Supply Company.

My mother, Mary Louise Richey Sharp, also supported the bonding experience of the Soap Box Derby. My mother, too, lost her father in her childhood. Her father, Earl Franklin Richey, was of Irish heritage. Her father died unexpectedly, a few months after her birth of gas asphyxiation from a living room fireplace. She and her mother, Mary Beatrice Hale Richey at the time, were attending religious services with her grandparents. After the death of her father, her mother assumed a position at a financial company in Ashland, Ohio while my mother spent the week with her grandparents on the family farm in Pavonia, Ohio. The farm was located on the outskirts of Mansfield, Ohio. There Rodney and

Elizabeth Hale raised their four children who were Martha Ruth, my grandmother, Mary Beatrice, a son, Mervin K, who died at the age of 21, and Elizabeth June.

My mother's grandfather, Rodney D. Hale was president of the Richmond County Agricultural Society at the time and served as secretary at the time of his death. He always retained an active interest in farming since he had lived on a 160 acre farm prior to moving to Mansfield, Ohio. He was elected to the position of county auditor and subsequently purchased a large home and moved to Mansfield, Ohio with his wife, Elizabeth Mary Kagey Hale. The county books were audited by the State Auditor for whom a man by the name of Walter Garrison worked at the time.

During one of their audit meetings, Walter Garrison disclosed that his wife had died while giving birth to his daughter and he was a single parent. Rodney Hale suggested that his daughter, Mary Hale Richey, a single parent with a daughter and Walter Garrison meet. After a brief courtship they married and my mother acquired a stepfather and stepsister, Mary Emmaline. The family of four took up residence in Akron, Ohio.

Family Life

Together my dad and I purchased construction materials for building the soap box derby racecar, learned about the use of shop tools, explored the principles of tear drop race car design for reduction of wind resistance, selected paint color for maximum absorbance of solar heat, developed expertise for conditioning the ball bearing based wheel mounts with graphite and various lubricants, plus, and selected the wheels for rubber hardness. A rubber gauge was used to measure the hardness of rubber and selection of appropriate racecar wheels. My dad and I would scout retail outlets and examine hundreds of race car wheels to inspect the ball bearing wheel mounts and hardness of the rubber tires prior to the purchase of the multiple wheel sets. My dad's university education was business oriented; however, he definitely possessed superior intellect and the mindset of a scientist and/or engineer. Our conversations in the basement shop between us while constructing the race car dealt with science and engineering topics and beyond. I'm certain, if my grandfather's had lived, that he would have encouraged my father to enroll in engineering school. Topics of our discussions with my dad, ranged from the origin of life to the peaceful applications of nuclear energy. In addition to science, and engineering, both my dad and mom, were interested in the political and social science aspects of current events with a special concern for human rights issues.

My maternal grandparents were likewise great influencers in the education for me and my siblings. My grandmother was of English and Dutch origin and her first husband and my mother's father was of Irish origin. Walter Garrison, my grandmother's second husband was of English ancestry. They were both avid readers with large collections of books, especially in history, literature and political science, and were loyal subscribers to the "Book of the Month Club." Walter Garrison, my step-grandfather, served as the deputy auditor for the state of Ohio, a political position, attached to the Governor's office. At the time, the gubernatorial elections were held every two years which meant that my grandparents were continuously engaged in political campaigns. Walter was fortunate to have survived both Democratic and Republican gubernatorial elections and enjoyed a full career of public service in the governor's office prior to his retirement at age 65. My siblings and I spent summer vacations and holidays with our grandparents in Columbus, Ohio. Our grandparents were special people to us and provided a window of learning on the cultural aspects of Columbus and beyond including museums, theaters, and the university, and sporting events at the Ohio State University Stadium, where we sat in the block of seats reserved for the Governor's office. Our grandparents were our heroes – they believed in us, invested in us and cheered for us. I can never thank them enough for their continuous love and support.

My life was totally consumed by school, church and work at the Sharp Sporting Goods Company. Church membership and regular attendance were encouraged by industry and schools in Akron as were restrictions on the sale of alcohol and the operations of bars and taverns on Sunday. Akron was an industrial and factory city where families were expected to relax on Sunday and be prepared for school and work on Monday mornings. Our parents did not attend religious services themselves but insisted that we do so. We were constantly told that religion was part of our overall education and that one day we could make our own decision regarding to our religious preferences.

So Sundays in our family included mandatory religious services at the Holy Trinity Lutheran Church, a magnificent miniature gothic inspired architecture cathedral that was founded by Akron notables John F. Seiberling, Charles Miller and J.H. Hower in 1868. Holy Trinity Lutheran Church has been home to many of Akron's prominent business and civic leaders. The church's present structure was built in 1914 in the style of Europe's French gothic cathedrals, with intricate hand carvings and beautiful stained glass windows. Franklin A. Seiberling, the founder of The Goodyear Tire and Rubber Company, was a major influence in the construction of "The Little Cathedral on the Hill" at Park and Prospects Streets, while completing the construction of his new mansion, Stan Hywet Hall, Akron's Tudor masterpiece. The church's magnificent pipe organ was a gift from of the Seiberling Family, and today the expanded instrument is used by world-renowned guest artists during Trinity's Organ Recital Series. Holy Trinity Lutheran Church Information Retrieved August 4, 2013 from http://www.lutheran-church.org/index.php/about-us/history.html

I was not particularly religious, but somehow, I found myself engaged as an acolyte and member of the choir even though my vocal talents were lacking. My sister, Patricia, on the other hand was extremely gifted. Oh yes, catechism classes which after two years of classes led to confirmation and church membership were mandatory for Lutherans. Classes for twelve and thirteen year olds were held on Saturday mornings under the direction of the senior clergy. I remember being argumentative and challenging about the religious fluff regarding creation, the crucifixion, the afterlife, the trinity and virgin birth. The official response from the clergy was that we humans lacked the intellectual capacity to understand the teachings of the *Bible*. Apparently, I left an impression as an independent thinker, because after confirmation, the two senior ministers scheduled an appointment to visit our home to discuss the opportunity for me to consider a four-year scholarship to Wittenberg University and the seminary school. They told me that I had a calling from God. I thanked them and told them that I would need time to consider the opportunity. My parents and I discussed the matter and although, I was flattered, I told my parents that I had not heard the calling. The religious ideas simply were not consistent with my early understanding of science and technology and the fragile human condition. I wrote a polite letter to the senior clergy under my parent's guidance and thanked them for the invitation, but explained that I was not ready to make such a commitment.

Our father's mother and our paternal grandmother, Dorothy Getz Jestadt Sharp was agnostic. Grandmother Dorothy's mother's ethnic background was French and her father's family was German. She was quite talented in art, design, home decorating, fashion and music and played the organ, piano and accordion. She never remarried after the passing of my dad's father, William Walter Sharp, but had male friends throughout her entire life into her 80s. She was a bit guilty of favoritism toward her grand-children and favored me over my siblings and cousins, which she displayed by showering me with gifts for birthdays and Christmas. I loved her but the situation at times was very embarrassing. She had a beautiful home with magnificent gardens, a pond inhabited by magnificent water lilies and gold fish across from the Fairlawn Country Club Golf Course in the Fairlawn section of Akron. She eventually moved to Marietta, Ohio and purchased a home near her sister. Many years later after the passing of her sister and our mother, she returned to Akron and bought a condominium near my father and his new wife. My father's second wife was a widowed school secretary who was introduced to my father by my sister, Sally.

My early interests, as mentioned earlier, included model air-planes, soapbox derby race cars, reading magazines, working in my parent's store, swimming at the YMCA, and collecting stamps and coins. Gardening was always an important activity for me at home and at my grandparent's home in Columbus. This interest probably began at my great grandparent's farm, where I spent summers of my early childhood. My grandmother was an expert gardener and had a large backyard with an expansive array of huge gardens planted with exotic ornamental plants and vegeta-bles. She belonged to garden clubs and was knowledgeable about all the new varieties of vegetables and ornamental plants. I was fascinated by the diversity of color patterns, shapes, and sizes of gourds. My grandmother encouraged me to subscribe to garden-ing books and seed catalogues about gourds, pumpkins, squash-es and tomatoes. I learned that gourds were members of the *Cucurbitaceae* family and enthusiastically ordered an assortment of diverse cultivars and embarked upon breeding experiments. I learned how to pollinate using small hobby paint brushes and developed novel cultivars that, with my grandmother's encour-agement were entered into the County and The Ohio State Fair in Columbus and won a few prize ribbons. I had the good fortune to meet an Ohio State University graduate student contest judge from the Botany and Plant Pathology Department at The Ohio State University. We discussed the breeding experiments that produced the prize gourds and tomatoes. The graduate student encouraged me to pursue my breeding interests and to consider plant biology as a major in college and even to undertake graduate work in preparation for a career in research. I appreciated his encouragement and filed away the possibility of a research career in the back of my mind at that time.

The standout moments of elementary school years were a school trip to Washington D.C., and a four week cross-country automobile trip with my maternal grandparents. Our grandparents took each of the four of us on a one-month trip to the west coast and back, which expanded our horizons of the magnitude of the country and the diversity of the population. The route for my trip was extended for me to see as much as possible including Arizona, Indiana, Illinois, Missouri, Kansas, Colorado, Utah, Nevada, California, Oregon, Washington, Idaho, Montana, New Mexico, Wyoming, Nebraska, South Dakota, North Dakota, Minnesota, Iowa, and Wisconsin. The most memorable experiences were the Glacier National Park, Grand Canyon, Mount Rushmore National Park, Rocky Mountain National Park, Salt Lake City, The Badlands National Park, The Black Hills National Park, The Great Salt Lake, Yellow Stone National Park, and Yosemite National Park. We stayed at lodges in the National Parks and urban hotels, survived a dust storm in Cedar Rapids, took a day for car repairs, and pushed on to our final destination of Portland, Oregon. The trip was more than tourism and sightseeing. My grandmother provided insights on family history, etiquette, and with a particular emphasis on table manners. Both my grandfather and grandmother were history buffs and provided narratives that linked our travels with U.S. and World History. Other formative experiences included working at the Sharp Sporting Goods Company, The Hobby Shop, Saveway Supermarket and A&P Supermarket. These experiences developed important business skill sets that influenced my future career choices.

The stay in Portland, Oregon with my Aunt Mary Emmaline, and family was especially enjoyable. She and husband, Bill Culver met during service in the Officer Corps of the U.S. Navy during World War II. I was delighted in hearing the stories of their adventures and travels during their tour of duty with the U.S Navy. No doubt this influenced me about the decision later in life to enlist in the Navy. My uncle, Bill had served as a carrier pilot, and my aunt was in the administrative services. Bill at the time was an architect and member of an architectural firm in Portland. Mary Emmaline was at home caring for two children and somewhat crippled with Multiple Sclerosis. The M.S. continued to progress and confined her to a wheelchair after the birth of her third child. Bill and Mary Emmaline were always good to me and took an interest in my aspirations. Nine years later and after Bill's career change and their move to La Jolla, California, I spent time with them while serving with the U.S. Navy. After completing my military service, I visited them in La Jolla several times with my Grandmother Garrison and son, Jeff.

Sharp Sporting Goods Company closed following the 1951 holiday and the family's economic situation tanked. Sharp Sporting Goods Company enjoyed a busy November and December and their best sales record. Nonetheless, shortly after the holidays, our father announced that a going out of business sale would be launched and within three weeks all merchandise was cleared, including shelving, showcases and fixtures. I never understood what happened, but was saddened by the demise of the business that had been such an important part of my coming of age. Our Kenilworth Avenue home was placed on the market and sold within three weeks. We said farewell to our neighborhood friends and the search was began for locating a new home. Our family made a land contract purchase of a smaller frame home on Copley Road accompanied by a small one room retail store situated on land at the front of the property. We were enthusiastic about the move and quickly settled our six member family into the new home and began renovating the store. The store accommodated a retail hobby shop which was registered as the Sharp Model & Supply Company and my mother agreed to be the manager. The store inventory included model airplanes, u-control airplanes, gasoline airplane engines and HO gauge model trains and accessories.

My mother enthusiastically launched the new venture and my father assisted her on Saturdays. Some of the customers from the former Sharp Sporting Goods Company frequented the new hobby shop. This transition occurred during my second year of high school year. I decided to interview at the Saveway Supermarket one block from our new home and was hired as a stock boy/bag boy on a part time basis. After school, I typically relieved my mother at the hobby shop until closing time and then walked one block to the Saveway Supermarket to report to work at 5:30 P.M. I had a number of responsibilities including, stocking shelves, bagging groceries and serving as a part time cashier. The supermarket experience was important in developing an understanding of the retail food industry which proved invaluable to me later during my employment at the Campbell Soup Company.

Dad was offered a temporary sales position at a large hardware store that provided financial resources for our family while looking for alternative opportunities. Our father always had the uncanny ability to land on his feet. He was an impressive man, - athletic, good looking and smart - and soon landed a senior position in the human resources division of the Goodyear Aircraft Corporation, a descendent of the famous Goodyear Zeppelin Corporation and builder of the large rigid airships known as the

Akron and Macon. Today the Company is known as the Goodyear Aerospace Corporation.

The change of venue to Copley Road was not particularly disruptive and allowed me to continue enrollment at John R. Buchtel High School, which was about a twenty minute walk. My sister began high school the following year; and usually accompanied me on the morning walk to school. Schools for my brother and younger sister were located nearby. My academic performance was average where as my sister's was exceptionable. Patricia and I both enjoyed high school and often double dated for movies and school dances. I spoke to her on occasion about moving into together and sharing our lives because of our uncertain futures. I had one steady girlfriend throughout high school who was elected with me as - "Most Bashful" - in the senior year book. Much to my chagrin, she ran away from a premiere liberal arts college during the first week of classes and was committed to an asylum.

My younger brother excelled in sports. Our little sister was held back a year in second grade because of illness, and struggled in some subjects in school up until her junior year in college and then became an overachiever. Our mother began exploring management opportunities in the fashion industry. She interviewed for an executive buyer position with the Madison Company, a group of five upscale woman's apparel shops located in Ohio. She was immediately hired because of her knowledge of the fashion industry and background in retailing. The hobby shop in front of our home was shuttered and our mother excitedly assumed the new position. She excelled in the fashion industry and frequently boarded the New Yorker overnight train to Manhattan for buying trips to Manhattan and to attend the seasonal runway shows. She usually lodged at the Manhattan Hotel or the New Yorker Hotel, which were located in the heart of the garment district on Eighth Avenue. The Manhattan Hotel is currently undergoing extensive renovation and renamed "The Row." The Hotel is situated directly across the street from my current residence building.

My mother, Mary Louise, was a beautiful woman and a former beauty queen, who always dressed in the latest fashions, which prompted customers to emulate her. The position at the Madison Company was one of the golden moments of her life and career. The situation at home was somewhat complex, with our father crisscrossing the country interviewing senior management and technical personnel for the Goodyear Aircraft Corporation and our mother on frequent trips to New York attending fashion shows and buying merchandise for the coming season. My sister, brother, baby sister and I became proficient at managing household affairs during our parents' extended absences with the assistance of Bertha, a wonderful middle age woman who assumed responsibility for housekeeper tasks and dinner preparation. In the daytime, a younger woman, Vivian, served as a nanny care for our baby sister. Patricia, my older sister, was involved in a number school related activities and took dance and music lessons, and looked after our baby sister, Sally on the weekends. She and Sally were always close and Sally became Patricia's guardian angel in her senior years as her health gradually failed.

Our father's entrepreneurial spirit was soon in play again. He was recruited by a private sector HR firm with which he negotiated a partnership/equity deal and that subsequently appointed him to the position of executive vice president. The company, Nelson Employment Associates, focused on the recruitment of business executives and senior engineers and scientists for the plastic and rubber industry for which the hub at the time was centered in the Akron, Ohio area. Harold Nelson, his principal partner, unexpectedly died a few years later and my father was elected president and chief executive of the company. The company was successful and operated the corporate office in Akron and six branch offices in Northern Ohio.

Patricia and I were well aware that the success of our parents in the private sector had markedly improved our economic situation and for a better quality of life for all of us. We no doubt were guilty of excessive materialism. During this selfish period, we proceeded to plot ways to influence our parents to acquire a larger home and purchase a new automobile. My childhood obsession with reading newspapers and magazines continued through high school. I was an avid reader of the *Akron Beacon Journal*, The *Cleveland Plain Dealer, the Sunday, New York Times* along with my subscriptions to dozens of magazines. During my junior year in high school, I came across an advertisement in the real estate section of the Sunday, *Akron Beacon Journal*, for a ridiculously bargain priced lease, for a five bedroom corner property on Highland Square at Market Street and South Portage Path. The home turned out to be a mansion and carriage house adjacent to the home of John Knight, the publishing magnate, and owner of the Akron Beacon Journal and the Scripts Howard publishing empire.

We phoned the realtor representing the Highland Square property and made an appointment to see the property and quickly discovered the reason for the low rent price of the lease. The cost and physical labor required for heating the home with two coal fired furnaces made the property unattractive. The home was remarkably beautiful and a statue of a Portage Indian marking the Portage Path stood at the entrance. For decades the Indian statue had watched over this famous trail used by Native Americans to transport their canoes between the Cuyahoga and Tuscarawas Rivers. The Portage Path was part of the effective western boundary of the white and Native American lands from 1785 to 1805. When first erected by an area real estate developer, the Indian statue stood on the curb on West Market Street adjacent to our future driveway. The refurbished statue now stands on a landscaped site on the corner of Portage Path and West Market Street.

The entryway of the home led to a vestibule, off of which were a washroom and closet. The vestibule opened into a large foyer with a double stairway to a landing and a second double stairway to a second floor wrap around balcony overlooking the large foyer. On the second floor were five bedrooms, two bathrooms and a porch surrounded the foyer opening. The floors, woodwork and staircases were constructed out of beautiful solid cherry wood. A magnificent oil portrait of the former owner's daughter hung on the wall from the balcony to the first floor landing. The third floor was a full-sized ballroom. On the ground floor to the left of the entrance was a large family room with a door leading to a port-de-chez and to the right was a magnificent living room that connected to a study and dining room. Adjacent to the dining room were a pantry and a kitchen with a backdoor. The home had a double basement that housed the furnaces, coal storage room and laundry facilities. The driveway entry to the home from Market Street traversed the two-acre property between the home and carriage house with three garages on the ground floor to an exit onto South Portage Path.

We negotiated with our parents to lease the property on the condition that I assumed responsibility for the lawn care and shoveling coal during the winter months. My brother agreed to assist with the coal shoveling and our parents signed a two-year lease. My brother, sisters and I were ecstatic to have the privacy of our own bedrooms. The new home was convenient for the entire family because of its proximity to work and school. Our friends and teachers at school assumed that our family had inherited a fortune. My older sister's popularity quickly rose and she began dating the most popular guy at school. I continued my job at the Saveway Supermarket and the school years continued for the most part to be uneventful.

Although we had devoted and loving parents, living conditions at home were often disruptive. Our parents frequently exhibited outbursts of anger and attacked one another because of alcohol abuse at home that became extreme on the weekends. My parents' aberrant behavior sometimes occurred in public and during telephone conversations with coaches, friends, parents of friends and teachers. I was protected somewhat because of my jobs at the family hobby shop and supermarket. My parent's situation was understandable because of their early marriage, their lack of parental guidance, and the burden of rearing of four children. Basically the situation resulted from children raising children. My younger brother and sister provided therapy for each other during these difficult periods. However, our younger sister was alone when the problem worsened. My mother died at the age 57 of Cirrhosis of the liver and our father eventually stopped drinking and smoking and lived to the age of 70.

We always were grateful to our parents for raising us in neighborhoods with access to good schools and providing the guidance and financial support that enabled the four of us to succeed, attend college and achieve happiness. Our maternal grandparents kept an eye on us and intervened during the difficult times. The lesson to be learned for children raised in turbulent domestic environments is the need to remain focused on completion of K-12 and the university or, if necessary, consider living with a relative or enlisting in the military. The later provides opportunities for attending military schools and having educational expenses covered following completion of a tour of duty.

The University Dropout

High school was concluded uneventfully without any profound ideas about college programs or career choices. At the last minute, I registered at the University of Akron for the beginning Autumn Semester of 1954. I began my first university semester while continuing my job at the Saveway Supermarket. The required freshman courses were held in huge lecture halls and, although somewhat enjoyable, provided no clue as to my future career might be. During the semester, I visited my grandparents several times to discuss the possibility of enrollment at the Ohio State University beginning in January, 1955 to continue a liberal arts curriculum with a science major. I thought that the move away from home would be beneficial to me by providing greater independence. After some deliberation, I decided to move to Columbus and live with my grandparents, who offered me a furnished third floor apartment in their home which was a wonderful place to sleep and study. I was fortunate to gain employment at the local A & P Supermarket which provided income to supplement the assistance that my parents and grandparents always willingly gave me.

I landed a better paying job as an orderly though a friend at Doctor's Hospital, a rather large osteopathic medical hospital within walking distance of the university. I had the night shift from 11:00 P.M. to 7:00 A.M. and, as there was not much activity at that time, I was usually able to study at a large desk used by one of the administrators during the daytime. I enjoyed talking to the resident doctors and nurses about their careers and their interaction with the patients and issues pertaining to life and death. I was also entitled to cafeteria privileges which helped my budget. This experience further confirmed that a medical career was not of interest to me.

Ohio State was a huge place with a smorgasbord of academic programs and courses, enabling me to enroll in an assortment of required classes in the humanities, social sciences and sciences. I decided to join a fraternity to enhance my social life. The fraternity offered opportunities for me to assume creative leadership for developing and directing charitable events, sponsorship of dance bands, the annual rush program for membership recruitment, and designing and publishing marketing materials. Fraternity membership was no doubt a distraction from my studies, but instilled confidence in my abilities to be at the "tipping point" in development of important collaborative teams for the nurturing business and not-for-profit programmatic initiatives. I was fortunate to have a fraternity brother introduce me to Virginia Sue Riebel, a younger, smart and beautiful woman from Columbus enrolled in the College of Education at Ohio State University. We dated often and discussed our goals and aspirations. Our relationship evolved quickly but neither of us was quite ready for engagement and marriage. The relationship continued after I dropped out of college from Ohio State University and enlisted in the U.S. Navy. We wrote letters and talked by phone often when my ship was in port and saw one another during my weekend and annual leaves in Ohio and New York.

Coming to Grips

A self-appraisal during 1958 brought me to the realization that I was quite immature for my age and not yet ready to make decisions regarding an area of concentration at the university or a potential career. Therefore, I decided to enlist in the U.S. Navy and to attend the Navy Technical Schools and tour the world. The decision was precipitated by a recruitment billboard for the U.S. Navy. As you may recall, I was somewhat familiar with the U.S. Navy from the stories told by my aunt and uncle about their tours of service as officers. The morning, after making my decision, I went to the naval recruitment office at the main post office and enlisted. Virginia was quite disappointed, but we agreed to continue our relationship. Although, I felt remorse about disappointing Virginia, I wasn't yet ready for marriage. My grandparents gave their instant approval, as they always did to my decisions, and assured me that the experience would be worthwhile. I purchased a Greyhound ticket and rode the bus to Akron for a weekend visit with my parents to say goodbye and to inform them that I was heading to the Admiral Nimitz Naval Base in San Diego for basic training early during the next week. My father was in disbelief but the other members of the family were supportive of my decision. I believe that my father thought that entry into the military as an enlisted man was a dead end in regard to my education and career, and he could have been correct. Although as a young man during World War II, my dad had wanted to enlist in a pilot training program with the U.S. Air Force. However, my father was denied enlistment because of a health issue relating to a kidney infection which required surgery. My uncle was drafted into the U.S. Army toward the end of World War II for a short period of time.

The following Tuesday, I returned to Columbus, Ohio and my grandparents and Virginia drove me to Port Columbus and wished me well with hugs and kisses as I boarded a TWA flight for Chicago connecting to San Diego. Our treatment on arrival at the Admiral Nimitz Naval Base was a bit inhuman. We were herded much like prisoners of war, stripped for a medical examination, heads shaved, measuring and allocating uniforms. Groups of 60 recruits were assigned to individual squadrons and marched off to the barracks that was to become our home for the next eight to nine weeks.

A squadron commander and a recruit commander were assigned to each squadron. I served as recruit commander for three to four weeks until sickness with pneumonia from the damp weather and limited sleep put me in the hospital. After three weeks in the hospital, I was assigned to a new squadron and completed my basic training. The training was rigorous with emphasis on physical fitness, seamanship, firearms, firefighting, shipboard damage control, core values, teamwork, and discipline. Above all, emphasis was on cleanliness and organization. The rigorous training at times was life threatening because some squadron commanders were naive about the vulnerability of the human condition. Graduating recruits were required to take an examination covering all aspects of the boot camp training, including the U.S. Navy military history. I scored the highest grade among all recruits in the graduating class. My squadron commander asked if I was planted by the C.I.A. and I replied, "I wish that was true because my understanding is that they enjoy a higher pay grade." I took away lifetime lessons in leadership, organization and the importance of cleanliness which continue to be part of my character. During basic training, I had a few opportunities to visit my aunt, uncle and family. On one of the visits, they took me on an immensely enjoyable trip to Tijuana. On the Gold Coast of Baja California, Tijuana is the municipal seat, cultural, and commercial center of Tijuana Municipality. I have since visited there many additional times for business and vacation.

After, Navy Boot Camp, I was assigned to the Great Lakes Naval Based for enrollment in Fire Control School which was to provide background in electronics and mechanical computers. I flew to Chicago and boarded the train to the Great Lakes Naval Base. I enjoyed the classes and getting to know Chicago, Waukegan and Milwaukee. I especially liked Chicago, and many years later, my wife, son and I had an opportunity to visit Chicago often during my adjunct research appointment at Argonne National Laboratory.

My track record at Ohio State had been mediocre with the exception of the required classes in chemistry and biology. I elected to satisfy the biological sciences requirement by enrolling in botany classes. I had the good fortune to enroll in a general botany course taught by Professor Clara Weishaupt, a brilliant professor, with a superb understanding of plant anatomy, plant development and physiology. Professor Weishaupt was regarded as an expert in the *Gramineae* (grasses), served as curator and director of the Ohio State University Herbarium, developed a teaching manual and field book guide to Ohio plants, and co-authored an authoritative guide to Ohio plants.

Professor Weishaupt's classrooms of thirty students seated at laboratory benches were alive with scientific curiosity and excitement. She conducted science demonstrations and experiments in front of the classroom and/or had us make observations using microscopes at our laboratory benches. I was passionate about attending her courses and received top grades. She unselfishly opened up and shared her world of plant science with her many students. She voluntarily served as my academic mentor for many years and encouraged me to complete the B.S. Degree and attend graduate school. She was one of the important influencers of my future career. One could feel her passion for the plant sciences, the herbarium, and her office. My only wish was that somehow, I could have reciprocated and bestowed appropriate honors and recognition on her for her commitment and passion to mentoring and providing career guidance to me and so many other students. Maybe, I will still find an opportunity to honor her in some way at Ohio State University.

Navy service followed a positive track and accelerated my coming of age and development of the ability to make important decisions regarding my future. The Navy was both a refuge and training ground. I was delighted to have had the opportunity to attend the U.S. Navy technical schools in fire control which provided background in electronics and mechanical computers that were central to the operations of a military ship fire control systems for air and sea defense. I advanced rapidly though basic training at the San Diego Naval Base and the technical schools located at the Great Lakes Naval Base. Thereafter, I was assigned to the U.S.S. Compton (DD 705) at the Naval Station Newport, Rhode Island and again the importance of teamwork and collaboration was emphasized, Newport was the homeport for Cruiser Destroyer Force Atlantic (COMCRUDESLANT) until 1970.

My ship, the Compton (DD-705), a destroyer moored at the U.S. Navy Base in Newport, Rhode Island was launched 17

September 1944 by Federal Shipbuilding and Drydock Company, Kearny, New Jersey. It was named for, Lewis Compton, who served in active duty in the Navy during World War 1 and was Assistant Secretary of the Navy from 9 February 1940 to 13 February 1941. The Compton cleared Norfolk 17 February 1945 for training at Pearl Harbor between 16 March and 5 April, when she sailed to escort ships to Kwajalein and Eniwetok. Sailing on to Ulithi, she cleared for Okinawa 20 April, where, Compton provided gun-fire support to forces ashore and served in the anti-submarine and anti-aircraft screens protecting shipping off the island. On 12 May she covered the occupation of nearby Tori Shima, and while returning to her station off Okinawa was attacked by a lone Japanese plane which she attacked and shot down.

During her 1948-49 deployment to the Mediterranean, the Compton had duty with the United Nations Palestine Patrol. She returned to the Mediterranean in 1951, and in the late summer of 1952 cruised in European waters for the NATO Operation "Mainbrace." Assignment to duty with the 6th Fleet in the Mediterranean came once more in 1953 and 1955, and in the spring of 1956, Compton exercised off Bermuda with ships of the British Home Fleet in NATO Operation "New Broom V."

The Compton was serving in the Persian Gulf in the fall of 1956 when the Suez Crisis erupted, and stood by to evacuate American civilians in the Persian Gulf and Red Sea areas should that become necessary. With the Canal closed, Compton made her homeward passage by way of Mombasa, Durban, the Cape of Good Hope, Recife, and Trinidad, returning to Newport 8 January 1957. That fall, she again cruised off the British Isles in a series of NATO operations. From November 1957 to April 1958, she again served in the Mediterranean, the Persian Gulf and the Red Sea and that summer cruised to Rotterdam and Bergen with midshipmen on board for training. From that time to into 1960, her operations were coastwise and in the Caribbean, as she aided research and development projects, including meteorological research and gave service to the Fleet Sonar School at Key West. In August 1960 Compton again sailed to the Mediterranean for duty in the 6th Fleet.

Compton participated as a Secondary Recovery Ship for the Mercury-Redstone 3 (MR-3) on 5 May 1961, Mercury - Redstone 4 (MR-4) on 21 July 1961, and Mercury-Atlas 9 (MA-9) on 15-16 May 1963 spaceflight missions. Compton received one battle star for World War II service. Compton was decommissioned 17 September 1972. Was stricken and sold to Brazil 27 September 1972. She was finally stricken July 1990 and broken up for scrap. More about the USS Compton DD 705 Retrieved on August 5, 2013 at http://www.usscomptonassociation.com/history.html

My responsibilities on the USS Compton related to operations and maintenance of one of the advanced fire control systems. I reported to George Nicky, a senior non-commissioned second class petty officer from Ohio. He was bright and possessed superb engineering talents. He returned to Ohio following completion of his enlistment and enrolled at Ohio State University. He attended my wedding and served as an usher. I enjoyed the shipboard operations in the Caribbean, Europe and the Middle East. Captain Zimmerman, the commanding officer, mentored and encouraged me to apply for special programs and merit salary bonuses. I was granted leave during visits to the Italian cities of La Spezia, Naples, Rome and the Middle East for Amman and the Jerusalem Old City, Jordan (at the time), Damascus, Syria and Jerusalem. Our multiple port calls included Beirut, Cannes, Istanbul, La Spezia, Lebanon, Naples, Piraeus and Sicily. The time aboard ship, especially during evenings when alone in the fire control operations room or on the deck during good weather, gave me time to read, write letters and give thought to my return to civilian life

Dr. Clara Weishaupt, my former professor from Ohio State wrote to me often and I in turn shared my insights and sent her postcards from port calls. She always encouraged me to return to academia and complete the B.S. Degree following my tour with the Navy. The letters from my grandparents and parents were likewise encouraging. Our Division was known as the Fox Division and I was assigned to a compartment below the Mess Deck that served as our living quarters. Each of the sailors was assigned a bunk and storage space for his gear. Members of the Fox Division were Electronic Technicians, Fire Control Technicians, Torpedo Technicians and Sonar Technicians. My two closest friends were Don Brown and Bill Bradford, both of who were Sonar Technicians. We had many good times together during port visits seeing the tourist attractions, visiting the clubs and meeting the young local women. Unfortunately after leaving the Navy, we lost track of one another, something that would not have happened in today's world of e-mail, text messaging and social networking.

Captain Wertheim, succeeded Captain Zimmerman, of the USS Compton, as commander, after an unfortunate accident at sea in which the USS Compton rammed the fantail of a another ship in our squadron just off Block Island on its return to the Newport Naval Station from an exercise at sea. Captain Zimmerman was assigned to shore duty following a US Navy Board of Inquiry by a panel of admirals. Captain Wertheim encouraged me to apply for Class B Fire Control School at the Great Lakes Naval Base and then to the Submarine Service. The Class B Fire Control School provided advanced training for the next generation of electronic and computer based fire control systems.

My division officer and shipmates organized a ceremony on the deck for my sendoff to the Great Lakes Naval Command. It was indeed one of life's bittersweet moments, because duty on the USS Compton had been a rewarding period of my life. It had provided me with opportunities to visit ports in Europe and the Middle East and to attend the Navy Technical Schools where I establish new friendships.

The next step was a train from Providence to Chicago connecting to the Great Lakes Naval Base where I settled into barrack living quarters adjacent to the Class B Fire Control Training School. Shortly after classes began, I took the Submarine Service Interview and physical examination. All went well except for the physical examination which disqualified me for the Submarine Service. An explosion had occurred earlier in the year from a misfire while two shipmates and I were loading ammunition for the aft gun turret, puncturing my ear drums which resulted in stone deafness. Gradually, my hearing returned, but it was never quite the same because of nerve damage. The combined damage to my hearing resulting from the shipboard accident and a childhood illness caused challenges during my subsequent academic and career pursuits. The hearing loss abruptly terminated my naval career and I was subsequently transferred to the U.S. Naval Hospital in Philadelphia for discharge effective January 1.

On Track

After my discharge, I enrolled at the University of Akron. I soon became engaged to Virginia, and marriage plans were set for June of 1962. My family applauded the decision that I was going to return to the university and complete a degree program.

I chose to attend the University of Akron because at the time Ohio State University adhered to the quarter system which was already in progress at the time of my discharge. This decision turned out to be a blessing in disguise as to my choice of scientific research focus.

I would like to share a few words about the history of the University of Akron, Retrieved August 5, 2013 from http://www.ohiohistorycentral.org/w/University_of_Akron

The institution is now known as The University of Akron was founded as Buchtel College in 1870 by the Ohio Universalist Convention, which was strongly influenced by the efforts, energy and financial support of Akronites, particularly industrialist John R. Buchtel. From the outset, the college and the surrounding community were closely tied, with the college addressing the needs of the region as well as those of the Universalist Church and local entrepreneurs assisting the fledgling institution time and again. By 1907, Buchtel College's emphasis on local rather than denominational interests led it to become a private, non-denominational school. The combination of the college's strong community ties and its weak financial condition prompted Buchtel College trustees to transfer the institution and its assets to the city of Akron in 1913. For the next 50 years, the municipal University of Akron assisted by city tax funds, brought college education within the reach of many more young people. During those years, enrollment swelled from 198 to about 10,000.

The University's growth paralleled the remarkable expansion of Akron. People were drawn to the city, already a major manufacturing center, by the promise of jobs. Companies such as Goodyear, Firestone and Goodrich were headquartered in Akron, making it only natural that the world's first courses in rubber chemistry would be offered at the University, beginning in 1909. With the formation of the Rubber Technical Institute in 1942, University researchers and students were well-prepared to contribute to the development of synthetic rubber to aid the Allied war efforts.

A long era of expansion followed World War II. Overseeing much of this growth was the University's 10th president, Dr. Norman P. Auburn, the father of Richard Auburn, a classmate of mine at John R. Buchtel High School. It is interesting to note that my son had an opportunity to collaborate with David Auburn, son of Richard Auburn many years later in adoption of the book and stage play authored by David Auburn for the screenplay and movie version of *Proof* about mathematics professor from the University of Chicago and his brilliant mathematician daughter.

Under Auburn's leadership, the institution made the transition in 1967 from a municipal to a state university. In the years to follow, as tire production jobs left the Akron area, the University's pioneering research was instrumental in helping the once-undisputed Rubber Capital of the United States evolve into the polymer center of the world.

In 1988, the University established the world's first College of Polymer Science and Polymer Engineering - now the largest academic program of its kind in the world. Led today by its 15th president, Dr. Luis M. Proenza, The University of Akron recently completed the $300 million first phase of its New Landscape for Learning campus enhancement program. During the five-year project, nine new buildings were constructed, 14 major renovations were completed and 30 acres of green space were added to the 218-acre campus. The new facilities include two classroom buildings, an Honors Complex, Student Union, Student Recreation and Wellness Center, and Athletics Field House.

Following my return to Akron an appointment was scheduled with Dr. Roger Keller, Chair of the Department of Biology at the University of Akron and a geneticist with a doctorate from Indiana University. The meeting with Dr. Keller was encouraging. Also, having served in the military, he understood my situation. My credits from Ohio State University were accepted by the University of Akron and I was granted senior student status. The remaining courses were planned for completion of the B.S. and potential M.S. Degree in Arts and Sciences during spring, summer and autumn semesters with courses in advanced genetics, geology, literature, microbiology, micro-technique, plant anatomy, plant morphology, plant physiology, plant taxonomy, physics, independent studies and comprehensive German.

Dr. Keller and I discussed future career possibilities for me along with graduate school. I told him about Dr. Weishaupt from Ohio State University and her mentoring through my Navy years and the following years in Akron. Dr. Keller suggested it was time for me to get serious and focus on the academics and pursue graduate school, and a career in academia. I mentioned that I would probably seek a part time job and he offered me a position in the department that entailed managing the greenhouse facility, cleaning laboratory glassware and doing preparatory work for the introductory undergraduate classes. I was most grateful to him. After our meeting, he kindly drove me home and making the beginning of a lifelong friendship.

While in Akron, I enjoyed the lively discussions with my father on weekends over coffee and dinners with my family. My parents and Sally, my little sister, were overjoyed to have me home again. Sally and I, after helping our mother clear the table, usually retired to our bedroom desks to study and discuss our projects. We developed a lifetime friendship which still endures today. Weekends usually included travel to Columbus to visit my fiancée or her travel to Akron and making plans for our wedding in June following her graduation from the Ohio State University College of Education.

I quickly settled into a routine of going to downtown Akron with my father in the morning. His offices were located at the First National Tower Building, Akron's skyscraper which was in close proximity to the campus of the University of Akron. The campus was a 15 minute walk up the hill. My relationship with my parents was much different after my return from the Navy. They were no longer parents and mentors, but treasured friends whom I respected and loved. I was fortunate to have had the opportunity to live with them one more time prior to getting married and beginning my career. Life has not been the same since their passing. I deeply miss my parents.

I enjoyed my classes and the part-time job in the department. Most important was the opportunity to become acquainted with the faculty and their areas of interest. My sister Sally was attending high school at the time. Quickly, the first semester at the University of Akron came to a close and I received a congratulatory letter from the Office of the Dean stating that I was named to the College Honor Roll. This achievement instilled me with confidence in an area of specialization that somehow would lead to a career track. I was finally reaping the rewards of my maturation and development of self-discipline during my years in the Navy and being away from home.

Wedding Bells & Married Life

Our long planned wedding took place in Columbus at the King Avenue United Methodist Church, adjacent to the Ohio State University Campus. Virginia's and my family, Ohio State University fraternity/sorority friends, and a few of my Navy buddies were in attendance on a beautiful Saturday on June 1, 1962. My brother and Virginia's sister served respectively as best man and maid of honor. We all had a great time socializing and for me it was wonderful to again see some of my former fraternity brothers who had settled in the greater Columbus area. Afterwards, Virginia and I took a local honeymoon on the Lake Erie shore for a few days before heading to Akron to begin our married life. We had the good fortune of leasing an apartment on Buchtel Avenue across from the campus of the University of Akron which proved to be convenient for both of us. Virginia accepted an elementary teaching position at King School from which I, my sister Patricia, and my father graduated. She worked as a summer counselor at a nearby playground during the summer while I began my classes in second year comprehensive German and English literature. Virginia was quite helpful in the analysis of my literature assignments and in editing my essays. We socialized on occasion with a few friends in the building but were busy with my university projects and Virginia's lesson plans and lots of grading papers. Weekends were occupied by trips to Columbus for visits to Virginia's parents and sister's family or dinner at my family's home.

My brother and wife lived in Akron and we would occasionally get together. My older sister Patricia was now married to an older man with two adopted children and lived in Columbus. My sister and I were close in some respects but quite distant in others. She had a very difficult childhood because of her contentious relationship with my father and competition with her mother. She and I were in college together at Ohio State where she specialized in home economics, but her education was interrupted by a serious ski accident over the Christmas holidays. Her lonely recovery process at home was made even more unpleasant by my father's termination of her relationship with her boyfriend, who accompanied her on the ski trip. Patricia eventually returned to Ohio State after she recovered from the broken leg to drop out of college and marry an older man in Columbus. She should have pursued her passion the arts but no one in our family including myself had the insight to provide proper guidance. Such a tragedy!

My younger sister Sally was completing high school at this time and making plans for university studies. I showed her around the University of Akron and we visited Kent State University from which my father and uncle graduated from in the late 1930's. Virginia and I enjoyed our visits with her and her enthusiasm about enrollment at Kent State University. Sally was committed to acquiring a degree in education and she completed the degree program in just three years.

University of Akron

Dr. Roger Keller led me though the many hurdles toward the B.S. degree in the Biological Sciences followed by a MS degree. Dr. Keller and fellow faculty member, Dr. Paul Acquarone, were important in my scientific development. Dr. Acquarone and I had several discussions about my potential interests in plant biology and cell culture. He encouraged me to review the dissertation of Dr. Samuel Caplin, one of his former students. Dr. Caplin was a leading researcher in plant cell biology and cell culture. He earned a Ph.D. degree at the University of Chicago under Dr. F.C. Steward after completing his M.S. Degree at the University of Akron. I checked out the library both his dissertation and "The Handbook on Animal and Plant Cell Culture," by Philip White, a scientist at the Bar Harbor Institute. Drs. Acquarone and Keller encouraged me to undertake a special project in plant cell culture, helped me equip the laboratory appropriately, and persuaded me to remain at the University of Akron for an additional year to pursue the M.S. degree. Their support was critical to my eventual pursuit of the Ph.D. degree. Two other professors who were very important were Dr. Eugene Flammenhaft and Dr. Grace Kimble, both microbiologists and key to my involvement in plant cell culture research.

My dad and I bonded again and spent a weekend employing our skill sets from the soap box derby days to purchase lumber, craft and paint a sterile transfer chamber for the lab. This was before the availability of commercial laminar flow hoods with positive pressure for manipulating organisms under aseptic environmental conditions. I was especially interested in plant cancers, which mimicked animal cancers, and quickly developed skills in aseptic culture technology and succeeded in the culture of a number of plant-based cancer cell strains. I was surprised to discover that Dr. Acquarone was a classmate of Philip White at Johns Hopkins University and had been in a postdoctoral program with him at the United Fruit Company. Dr. Acquarone introduced me to Philip White and Samuel Caplin, two giants in the world of plant cell biology, who in turn recommended graduate program opportunities for me.

Regular library visits became the norm following the advice of my mentors. I soon became aware of the importance of library reading in staying abreast of the literature. I soon became engrossed in the literature and discovered landmark publications, which included:

Professor Haberlandt (1902) Growth and development of embryos and plants from cell explants grown in culture vessels was feasible;

Laibach (1925) succeeds in the maturation of immature embryos in cell culture; White (1934) succeeds in culturing tomato roots on defined culture medium showing for the first time, that an organ could be maintained in continuous culture in vitro;

White, Nobecourt and Gauthret (1939) succeed in establishing growing cultures of undifferentiated callus cells: tumor tissue cells for White, and carrot cambium on a medium containing auxin for the others.

Philip White (1943) publishes the Handbook of Plant Tissue Culture, with subsequent editions in 1954 and 1963.

R. Gautheret (1959) publishes a comprehensive volume entitled The Culture of Plant Tissues in 1959.

During the 1950s, plant cell culture acquired a momentum:

F.C. Steward, S.M. Caplin and F.K. Millar (1952) achieve success with continuous suspension cultures of carrot.

Muir, Hildebrandt and Riker (1958) introduce callus growth from single cells using the technique of nurse culture.

F. Skoog and C. Miller (1956) demonstrate the importance of cytokinin/auxin concentration ratios in the culture medium for the control of organogenesis.

C. Steward, M.O. Mapes and K. Mears, as well as J. Reinert (1958) define the conditions under which carrot cell cultures would initiate embryo formation which led to understanding of the importance of the culture medium formulation for the control of embryogenesis. The knowledge of cell culture and in vitro organogenesis (formation of organs originating from pluripotent stem cells) led to furthering the understanding of plant development.

I was particularly interested in the work of Armin Braun (1962) who became an important mentor during my doctorate studies and a lifelong friend. He was among the first to use cell culture as a tool for understanding an applied problem, the crown gall tumor problem in plants. G. Morel (1960) a colleague of my thesis advisor at Rutgers also ventured into the application of plant cell culture with the use of meristem culture for the micropropagation of orchids. The major breakthrough of the sixties was a paper by Murashige and Skoog (1962) describing a revised culture medium for rapid growth and bioassays of tobacco tissue cultures. This milestone was followed in 1965 by the Linsmaier and Skoog's paper on organic factor requirements of tobacco tissue cultures. These new formulations of culture medium provided the building blocks for the induction and analysis of the phenomenon of plant regeneration from tissue cultures that was to unfold during the next decade.

The First International Plant Tissue Culture Conference was held in 1963 at Penn State University while I was in undergraduate studies at the University of Akron. I was awarded a travel grant by the Akron Chapter of Sigma Xi to attend the Conference. Sigma Xi is a scientific research society, founded in 1886 to honor excellence in scientific investigation and encourage a sense of companionship and cooperation among researchers in all fields of science and engineering. The conference marked the 30th anniversary of the publication of Philip White's paper on the continuous aseptic cultivation of tomato root cultures (1933) and was organized by Philip White during his sabbatical leave at Pennsylvania State University. The gathering was attended by 40 to 50 dedicated scientists from around the world who spent the week discussing technical problems, successes and the future of this very young and promising field of plant biology. After the meeting, the scientists returned to their respective laboratories and the field began to blossom.

My professors supported my request to undertake an independent research project during the senior year using cell culture to develop a better understanding of crown gall bacterial-mediated plant tumors and genetic tumors occurring in certain interspecific genetic tobacco hybrids. It was my belief that plant cell culture held promise as a key research tool in unraveling the mysteries of plant cell differentiation and plant development, as well as having practical applications in genetics and plant breeding. This senior research project set the stage for my M.S. and Ph.D. dissertation research programs. I could not believe that I would soon be graduating with a B.S. degree. During the same week that I completed the form for graduation, a letter arrived from the Dean of the Graduate School that I had been admitted to the M.S. Biological Sciences Graduate Program at the University of Akron under the supervision of Drs. Paul Acquarone and Roger Keller.

I met with the members of the Graduate Committee to obtain approval of a graduate degree research proposal to study the influence of minor inorganic elements on the growth and differentiation of carrot cell cultures. Dr. Keller assisted me to procure the necessary scientific instrumentation, laboratory supplies and important analytical chemical assay protocols. My graduate committee included professors with expertise in minor nutrients from the Department of Chemistry. The experimental work progressed on schedule and provided me with good laboratory skill sets. The laboratory research experience was an important step in allowing me to become an independent researcher. I will always be appreciative of the mentorship of my professors who were so kind as to take an interest in my career building efforts.

The Next Step

Drs. Acquarone and Keller suggested that my next step should be a Ph.D. degree program. Dr. Acquarone discussed the matter with Drs. Philip White and Samuel Caplin and suggested that I apply to either Rutgers University for study with Dr. James Gunckel or Dr. Ian Sussex at the University of Pittsburg. I followed their advice and soon received a letter from Dr. Sussex explaining that he was taking a two year sabbatical and afterwards accepting a research professorship at Yale and would not be able to accommodate me until after he moved to Yale. Dr. Gunckel phoned me from Rutgers the following week. He suggested that Virginia and I visit Rutgers in New Brunswick, N.J. to discuss graduate school research opportunities. That same week a letter arrived from the office of Dr. Solon Gordon at Argonne National Laboratory informing me that I had been awarded a stipend for a summer research internship in the Plant Cell Biology Research Group. This grant was due to recommendation of Dr. Eugene Flaumenhaft, a former Argonne Research Scientist and microbiology professor with whom I studied with at the University of Akron.

Virginia requested two personal days of leave from her teaching assignment and we jumped in our aging English Ford and headed to New Brunswick, where we lodged at the Roger Smith Hotel. Early the next morning, I met with Dr. Jim Gunckel, who proved to be a truly amazing man. We had a wonderful discussion about the plant tumor problem, opportunities for cell biology and cell culture research and his research program at Rutgers and Brookhaven National Laboratory. He led me on a tour of the impressive laboratories and greenhouses at the Nelson Hall Biological Sciences Laboratory Building on the Piscataway campus. In addition, he introduced me to faculty members and graduate students in the department. I was awed by the opportunity to join the Rutgers research team. In the evening, Virginia, and I were invited to dinner at the Gunckel's magnificent home in Somerville, New Jersey. We met his wife, Jean Gunckel, who was an elementary school teacher, his daughter Nancy, his son Fred and their family dog, Gretchen. We enjoyed a gourmet dinner and wonderful conversation. Jean suggested teaching opportunities for Virginia in the New Brunswick area. Jean, Virginia and Nancy immediately bonded and became good friends. It turned out that both Jean and Jim Gunckel were graduates of Miami of Ohio University and afterward went to Cambridge for graduate school at Harvard University. What a small world!

After dinner, Jim Gunckel said that he would request the Graduate Committee and Graduate School to admit me with a fellowship that would be converted to a NASA Fellowship because of my forthcoming research at Argonne National Laboratory. He suggested that we meet the following morning to apply for student housing in the University Heights Graduate Housing Apartments adjacent to the Nelson Hall Biological Sciences Building and make arrangements for moving our furniture. Nancy Gunckel volunteered to be at the apartment on the day the movers arrived because of our summer internship at Argonne National Laboratory. The following year Nancy offered to baby sit after the birth of our son, Jeff. Our friendship continues today with Nancy, her husband Alan and family. Alan, a brilliant writer and editor, has collaborated with my colleagues at Rutgers on the development of program brochures and their two sons have collaborated with my son in the IT area.

The following morning, Virginia and I visited our future apartment in the graduate student housing complex on the Busch Campus, signed a rental agreement to take possession of the apartment in late August and accepted the graduate fellowship appointment. Virginia was offered a teaching position in the Piscataway School System. It is amazing how quickly our future fell into place. Then Virginia and I headed back to Akron for the completion of her teaching responsibilities and my completion of Spring Semester at the University of Akron. We were both looking forward to a pleasant summer at Argonne National laboratory and the opportunity to enjoy the Chicago environs. I completed the M.S. Degree requirements at the University of Akron in Biology but the dissertation defense was delayed until late summer because of the illness of one of the committee members.

Argonne National Laboratory

At the end of the semester, Virginia and I packed our possessions for storage and shipment to New Brunswick, N.J. and headed for Downers Grove, Illinois for the summer internship at Argonne National laboratory. Argonne was an amazing place. Here I had the opportunity to apply my knowledge in cell culture to understanding cell growth and differentiation in zero gravity environments. I had the opportunity to work with Drs. Richard Dedolph, Solon Gordon, and Jane Shen-Miller, all prominent researchers in plant physiology and plant biochemistry. Dr. Solon Gordon was director of the Division of Biological and Medical Research. Studies in plant growth and differentiation were of high priority for NASA at the time because of the need to understand the effects of gravitational-free environments on living organisms during prolonged space travel. I conducted experiments on carrot tissue explants grown in Erlenmeyer cell culture vessels clamped onto clinostats placed in gravitational free chambers. Interesting effects on organelles clumping and respiration were observed. My first research paper was published with Drs. Jane Shen-Miller and Solon Gordon. The Argonne internship lasted for three months but evolved into an academic visiting research fellowship during my years at Case Western Reserve University and Ohio State University. The opportunity to work with a world class team of scientists was invaluable to me in learning about experimental design, data crunching, the maintenance of laboratory notebooks, monthly/quarterly research reports and the importance of intellectual property.

Argonne was leading the quickly evolving field of cell biology and advances in microscopy, the electron microscope and cytology. The laboratory is situated on 1,500 beautiful, wooded acres and surrounded by the "Waterfall Glen Forest Preserve", Argonne is one of the nation's leading federally funded research and development centers and is the U.S. Department of Energy's oldest and largest national laboratory for science and engineering research. The facility employs about 3,200 employees including 1000 scientists and engineers, three quarters of whom hold doctoral degrees. The annual operating budget is about $630 million and supports 200 research projects. Since 1990, Argonne has developed collaborations with more than 600 companies and numerous federal agencies and other organizations (Argonne National laboratory History Retrieved August 5, 2013 from http://www.anl. gov/about-argonne/history

Rutgers University

The institution was chartered in 1766 as Queen's College, Rutgers is the nation's eighth oldest institution of higher learning and has a centuries-old tradition of rising to the challenges of each new generation. Soon after opening in New Brunswick in 1771 with one instructor and a handful of students, the college was caught up in the struggle for independence.

During the war, classes were suspended on several occasions as students, faculty, and alumni joined the fight for freedom. That revolutionary legacy is preserved today in the university's name in 1825, Queen's College became Rutgers College to honor trustee and Revolutionary War veteran Colonel Henry Rutgers. Rutgers is a leading national public research with more than 56,800 students from the 50 states and 125 countries including 42,300 undergraduates and 14,500 graduate students and more than 13,000 faculty and staff (Rutgers University History Retrieved from http://www.rutgers.edu/about-rutgers/rutgers-history

In the autumn of 1964, I entered graduate school at Rutgers University with a NASA pre-doctoral fellowship under the mentorship of Dr. James E. Gunckel. Jim was born and raised in Dayton, Ohio, graduated from Miami University, Oxford, Ohio, in 1938 and received his doctorate from Harvard in 1946, where he studied under Dr. Ralph Wetmore. He did his postdoctoral research with Dr. Kenneth V. Thimann.

At Rutgers, for many years, Dr. Gunckel chaired what was then called the Botany Department and did pioneering work in two important areas of study: tissue culture (plant cloning) and radiation biology, where he conducted benchmark studies on the effect of radiation on a variety of plant species. A prolific publisher of scientific articles, he presided at many national and international botanical meetings served as the translating editor of the seminal German botanical text, *General Botany* by Wilhelm Nultsch, and edited the textbook, *Current Topics in Plant Science.* A former president of the Torrey Botanical Society, the oldest botanical society in America, he also served many years as editor of *The Bulletin of the Torrey Botanical Society*, a refereed botanical journal. In 1959 having been awarded a Waksman Foundation Fellowship, he pursued meristem culture (plant tissue cloning) research at Station Centrale de Physiologic Vegetate, Versailles, France, under the tutelage of Dr. Georges Morel. His unequaled knowledge of radiation biology, much of it gained through his many summers of research in collaboration with Dr. Arnold Sparrow at the Brookhaven National Laboratory, led to his being called upon to provide expert testimony in a legal case pertaining to the Three Mile Island nuclear power plant accident.

Dr. Gunckel, my mentor, who became a lifelong friend, believed in the importance of cell culture technology as a tool for unraveling the mysteries of plant development. At Rutgers University, I continued the NASA-related work and investigations on the developmental physiology of plant tumors. During these graduate school years, I was privileged to know Armin Braun at Rockefeller University and meet with him often. He was an invaluable resource and the friendship with him continued during my professional career.

My classes at Rutgers provided an understanding of cellular and developmental biology. The study of organelle genetics was in its infancy. Dr. Charlotte Avers offered a graduate lecture and laboratory course in organelle genetics that significantly broadened my knowledge of cell biology. Another important course in advanced plant physiology that involved the isolation and study of organelles was taught by Dr. Carl Price. Dr. Gunckel and his colleagues nurtured a most unique environment for graduate students to interact with faculty and students in journal clubs, departmental seminars, and meetings with Dr. Bill Jacobs, a professor at Princeton University, and his research team. Dr. Jacobs had been a graduate school colleague of Dr. Gunckel at Harvard. In addition, Dr. Gunckel service as president of the Torrey Botanical Society, the oldest botanical society in the Americas. The Torrey Botanical Society was founded in the 1860s by John Torrey. The Society promotes the exploration and study of plant life, with particular emphasis on the flora of New York City and metro areas and until 1997 published the Torrey Botanical Bulletin. The Society was celebrating a 200 year anniversary and organized monthly meetings at important research universities, research centers and institutes including Boyce Thompson Institute, Brooklyn Botanic Garden, Columbia, Lehman College, New York Botanical Garden, New York University, Princeton, and Rockefeller University. This provided an opportunity for the graduate students to network with the leading scientists in the field which was important to their future careers.

Dr. Gunckel was an important mentor to newly hired tenure track assistant professors who received nine month appointments. He aggressively identified summer support to fund their summer research and travel which was key to contributing top quality research publication to the premiere refereed scientific journals. These resources and opportunities facilitated the professional development of young faculty members and resulted in their successful promotion and tenure. The result was the emergence of a world class academic department.

The first draft of my dissertation was completed during the summer of 1966. Dr. Gunckel gave me permission to accept a postdoctoral position at Western Reserve University (Now Case Western Reserve University) beginning in June 1966 with return trips to New Brunswick for revisions. Virginia, Jeff and I would usually lodge at the Gunckel's magnificent home in Somerset New Jersey during these sessions. I spent the time in the basement apartment with my typewriter and in the evenings, I would work with Dr. Gunckel on editing. These sessions were most important for developing my writing skills. I will never be able to thank him enough for his vitally important investment in my career. The dissertation was completed toward the end of summer but the defense was delayed until early autumn because one of the committee members was away on sabbatical leave in India. I graduated at the spring commencement of 1967 with U Thant, a Burmese diplomat, and the third Secretary General of the United Nations awarding my diploma. Virginia, Jeff and my grandmother from Columbus attended the ceremony. My parents were unable to attend because of my mother's illness.

Case Western Reserve University

Graduate school was followed by a post-doctoral fellowship at Case Western University from 1967-1969. Here I continued to study the regulation of plant tumor development and growth. Western Reserve College was founded in Hudson, Ohio, in 1826, about 30 miles southeast of where the campus stands today. In 1882, the college moved to "uptown" Cleveland and assumed the name Western Reserve University. The merger of the Case Institute of Technology during 1967 created a leading institution for academics and research, as well as one the nation's top-ranked universities. Information Retrieved from http://ech.case.edu/cgi/article.pl?id=CWRU

My time at the university was especially opportune because of the world-class faculty in the Department of Biology and the Developmental Biology Center. Dr. Howard Schneiderman, an eminent developmental biologist, academician, and university administrator, provided leadership for a multimillion dollar grant from the Ford Foundation to attract leading scientists in the biological Sciences to Western Reserve University. Dr. Schneiderman's proposal to the Ford Foundation provided the blueprint for the eventual merger of Western Reserve University and the Case Institute of Technology. The scientists recruited included Roger Bidwell, Douglas Davidson, Boris and Harriett Ephrusi, Michael Locke, Bodil Schmidt-Nielsen, Clifford Slayman and Carolyn Slayman. Howard Schneiderman invited me to attend his laboratory's weekly insect developmental research meetings broadened my background in cellular and developmental biology. The laboratory investigations focused on silkworm and *Drosophila* developmental genetics.

My approach to understanding the plant tumor problem was influenced by two faculty members with different approaches to development: Dr. R.G.S. Bidwell, a plant physiologist, and Dr. Douglas Davidson, a cytogeneticist and developmental biologist.

My interaction with them reinforced the importance of collaboration and continued learning. During this time, I developed expertise in thin layer chromatography and deeper understanding cell genetics and plant physiology. Among the highlights was the visit of Ruth Sager, a leading research scientist from Hunter College in the field of organelle genetics. She delivered a lecture series based on her course in organelle genetics at Hunter College. Cell biology lectures by Dr. Douglas Davison also were memorable. He was a superb lecturer and mentor. In addition to being eminent biologists, and prolific contributors to the scientific literature, Roger Bidwell and Douglas Davidson entertained many other diverse interests. Roger Bidwell was a master at Bridge, collected antique automobiles, played both the harpsichord and piano, and participated in several musical groups. Douglas Davidson appeared on the London stage with Maggie Smith during his graduate school days at Oxford, collected prints and organized a print group for collectors at Western Reserve. Lunch hours were spent at bridge tables throughout the laboratories. Graduate students, postdoctoral fellows and newly appointed faculty members were required to participate with the tenured faculty. Our lunches were ordered from a local delicatessen and consisted of bagels, cream cheese and lox. Roger Bidwell told us that playing bridge was an exercise in mental gymnastics important for improving our scientific thought process.

In 1969, the institutional grants and Ford Foundation grants were not renewed precipitating the departure of the best scientists. Howard Schneiderman's went to the University of California at Irvine to head the Department of Cellular and Developmental Biology to become the 3rd Dean of the School of Biological Sciences. Roger Bidwell moved to Queens University, Douglas Davidson to McMaster University and others assumed academic positions in Europe. I was offered a lectureship in the Department of Microbiology at the Ohio State University in Columbus.

The Ohio State University

Ohio State University's roots go back to 1870, when the Ohio General Assembly established the Ohio Agricultural and Mechanical College. The new college was made possible through the provisions of the Land-Grant Act, signed by President Lincoln on July 2, 1862. This legislation revolutionized the nation's approach to higher education, bringing a college degree within reach of all high school graduates. The Ohio State University's main Columbus campus is one of America's largest and most comprehensive. More than 56,000 students select from 170 undergraduate majors and more than 250 graduate and professional degree programs. As Ohio's best and one of the nation's top-20 public universities, Ohio State is further recognized by a top-rated academic medical center and a premier cancer hospital and research center (Ohio State University History, Retrieved August 5, 2013 from http://www.osu.edu/news/history.php

Following the mass exodus of senior faculty from the Department of Biology at Western Reserve University, an opportunity surfaced at The Ohio State University for appointment to an unconventional lectureship. The process began when at the request of the department chair, my wife and I represented Case Western Reserve University an open house at the Ohio State University College of Agriculture Research and Development Center in Wooster, Ohio. We were briefly introduced to the dean and associate dean of the newly formed College of Biological Sciences in Columbus. Dr. Don Dougall, the associate dean, had formerly been a research scientist at the Ohio Research and Development Center in Wooster Ohio, a division of the College of Agriculture at The Ohio State University, where he conducted plant biochemical research with plant cell cultures. We knew one another through our research papers and presentations at scientific conferences. The week after the open house, a secretary phoned me to schedule a visit to Ohio State to meet with the two senior administrators at the College of Biological Sciences.

The meeting and tour of facilities was impressive and, much to my surprise, I was offered a position as a lecturer beginning January of 1969 in the Faculty of Microbial and Cellular Biology of the College of Biological Sciences. The employment offer was a bit confusing because during the visit with the chair and a group of departmental faculty members, I was told that for budgetary reasons the department expected to make no additional faculty appointments in the current academic year. The verbal job description was to develop a plant based cell biology research program, participate in a course in general cell biology for upper division and graduate students and eventually develop an upper division advanced course in cell biology. Furthermore, I was told that the following autumn my appointment was to be converted into a tenure track assistant professorship with appointment to the graduate faculty.

The faculty at the time of my appointment was top heavy with tenured professors and had no tenure track assistant professors. In spite of the confusion in regard to the academic appointment, I was confident that I could win the support of the faculty because of the commonality of microbiology and plant cell biology research in employing *in vitro* approaches to the study of cellular genetics, physiology and development and the department's responsibility for teaching courses in general cellular biology. Two of the microbiology faculty members teaching general cellular biology had research expertise in protozoology and a third member with expertise in plant cell biology and biochemistry.

In January 1969, my wife, and three and one-half year old son settled with me into a huge two family brick home on Highland Street with a fenced backyard and three-car garage in close proximity to my laboratory and office at the Botany and Zoology

Building on Neil Avenue. The backyard was perfect for our young son. The downstairs consisted of two bedrooms, a solarium, study, bath and large kitchen. The second floor was occupied by a tenant who had been in residence there for 20 years. The home was purchased for $9,000 and the tenant's rent check covered the mortgage payment, taxes and insurance.

The laboratory space assigned to me was on loan from the Botany and Plant Pathology departments because of the lack of space in Edith Cockins Hall. The laboratory space was shared with Donald Dougall, the Associate Dean of the College of Biological Sciences, and a faculty member of the Faculty of Microbial and Cellular Biology. The laboratory consisted of two interconnected laboratory spaces, an office, aseptic transfer room, and carrels for the graduate students. Autoclave rooms, growth chambers and a greenhouse complex were located nearby. Don Dougall had three graduate students and a technician at the time, including two students from Brazil, a student from Maine and a talented technician with credentials from a laboratory technical school in the U.K. Dr. Dougall introduced the team and an instant repertoire was established. The timing was fortuitous in that publication of three of my research papers in the Journal of Plant Physiology during the next two months gave me credibility. My teaching responsibilities didn't commence until the following autumn, providing me time to prepare my lectures. I had the time to meet the departmental and neighboring faculty in the Botany and Zoology Building and to initiate a research program.

I enjoyed a spacious faculty office adjacent to the two connected laboratory spaces that had been occupied by Don Dougall prior to his appointment to the deanship. Case Western Reserve University allowed me to borrow some key pieces of equipment that enabled me to continue my ongoing research program. In addition to this, the departmental chair gave me permission to continue my collaboration relationship with the Argonne

National Laboratory. Quickly, I immediately set to writing research proposals and submitting grants applications. Meeting members of the Faculty of Microbial and Cellular Biology was somewhat challenging because of my unconventional appointment. Nevertheless, meetings were scheduled with each faculty member. I also arranged to monitor the winter quarter's course in general cellular biology for which I would assume responsibility in the autumn quarter.

The chaos of departmental reorganization worked to my advantage. The College of Biological Sciences was undergoing a revolutionary change in which the departmental administrative units were replaced by faculties under the direction of faculty administrators. All administrative fiscal responsibilities were thereafter assumed by the college administration. The former Department of Microbiology had evolved into the Faculty of Microbial and Cellular Biology. After my meetings with the faculty, the opportunities for collaboration with fellow faculty became evident.

Dr. Julius Kreier, a professor and expert in pathogenic protozoology with doctorates in veterinary medicine and immunology and originally appointed to the College of Veterinary Medicine, transferred his appointment and laboratory to the Faculty of Microbial and Cellular Biology. Dr. Kreier reached out to me and organized lunch meetings with fellow faculty members and invited me to participate in organizing the departmental weekly seminar program. He administered a large research operation with a global population of graduate students and postdoctoral fellows. He served as my mentor for recruitment of graduate students, grant writing, publications, committee work, organizing colloquium/symposium and development of a monograph series.

In the spring quarter, I was invited to present a research seminar to faculty and students. Knowing that a seminar was usually required of candidates for university academic appointments during the recruitment process, I was quite nervous. Apparently, the lecture was well received because shortly thereafter, the departmental chair informed me that I would be reappointed to a tenure track assistant professor position, which would allow me to be considered for a tenured appointment. Thus began a probationary period that could eventually lead to a vote for appointment to a tenured associate professor and thereafter professor. Tenure track assistant professors are expected to hit the ground running and demonstrate scholarship with publications in referred journals. Publication establishes the reputation of a scholar in his or her chosen field. The dissemination of knowledge through publications and teaching is the primary basis of national and international recognition and distinguishes the research university. Publication is a gauge that determines career mobility, national and international recognition. During the tenure approval process, an academician's record is evaluated by departmental senior faculty and panels of national and international experts in the field of specialization. The quality of the original research is judged based on originality, publication in leading journals and science index rankings. Other important factors include successful grantsmanship, graduate program participation, editorial review, and teaching. Congeniality and collaboration with senior academicians in the academic unit and the field are also important. Recently, intellectual property, patents and collaboration with the private sector have been included in the tenure evaluations.

The graduate students, Donald Dougall and I participated in weekly research meetings in which we reviewed our research programs and relevant journal papers. Don and I developed a few collaborative projects that proceeded well and we often traveled together to the annual research conferences and presented papers. The three graduate students were talented. Ruy Caldas was from a rural town in the state of Minas Gerias situated in the interior of Brazil and Henrique Amorim resided in the city of Piracicaba of the state of Sao Paulo. Ruy and Henrique graduated respectively from Versosa University and the University of Sao Paulo – ESALQ Campus both premiere research universities. Ruy's family had an agricultural background while Henrique's father was a medical doctor and his father and uncles were businessmen involved in a number of ventures. Harry Sommer, a graduate of the University of Maine delayed graduate school until fulfillment of his ROTC military obligation. He had been assigned to a biological warfare unit with the U.S. Army that he couldn't discuss because of secrecy issues. All three graduate students were passionate about their studies and research. Don had an invitation to conduct collaborative research in Australia in the summer months and departed at the end of spring quarter.

The summer was spent in the laboratory, writing papers for scientific meetings and preparing lectures for autumn semester. The autumn cell biology course included morning lectures and four afternoon teaching laboratory sections, each of which accommodated 25 students. The teaching laboratory was equipped with state-of-the-art lab benches, microscopes, balances, spectrophotometers, and accessory equipment. My wife and son accompanied me to conferences at which I presented research papers and that afforded opportunity for all of us to travel together.

Summer passed quickly. I soon was standing in a lecture hall, a bit overwhelmed, by the expectant faces of 100 upper division undergraduates and first year graduate students. The routine included daily lectures scheduled each morning at 8:00 A.M., afternoon teaching laboratory sessions, graduate student laboratory assistant meetings, student meetings and lecture preparation each evening. Tom Byers and Don Dougall were both helpful to me during the orientation period, as I had never been an instructor. My graduate school and postdoctoral fellowship activities had been limited to laboratory research funded by fellowships with no teaching requirement.

The Graduate Program

I was fortunate to have served on the graduate committees of Dr. Donald Dougall's three graduate students along with some of the students from the laboratory of Dr. Julius Kreier. This experience prepared me for acceptance of my first graduate students. I learned that not only should candidates for graduate school be academically qualified, but must possess a passion for conducting basic research and contributing to the body of scientific knowledge.

Rosa Raskin, a young attractive student with an exceptional academic background in the biological sciences, was my first graduate student. She had the advantages and disadvantages of working with a young assistant professor recently completing his Ph.D. degree and a postdoctoral fellowship. Shortly thereafter, John Peters, an honors student from Otterbein College in Westerville, Ohio joined the laboratory. John and Rosa were in the process of learning how to become effective independent research investigators while I was learning how to become an effective mentor. It is essential to teach graduate students with the importance of reading the scientific literature, observation, experimental design, data collection, analysis, and the drafting of scientific communications and manuscripts. Communication of discoveries through mass media was also vital. I always emphasized the importance of linking the laboratory bench to the pages of *The New York Times* and *Wall Street Journal*.

The important question is how laboratory research results might be translated into the development of products or services for the betterment of humankind. The goal is for graduate student to become independent and productive scholars and join the ranks of the professor's colleagues. I always followed the pattern set by my graduate school mentors, in allowing graduate students flexibility in determining their own research program within the guidelines of the funding agencies which dictate the parameters of a particular research program.

The years at Ohio State University were important to the development of a platform for launching my subsequent corporate career. I developed a friendship with Dr. Clyde Allison, a senior faculty member from the Department of Plant Pathology, who had an office and laboratory adjacent to mine. He was an expert in tropical plant pathology research and a veteran of global collaborations, particularly Brazil. He expressed interest in our laboratory's plant cellular research programs and suggested contacting a Brazilian scientist working in the field of coffee genetics. At the time, Coffee Rust, a fungal disease, was becoming a threat to coffee growing regions in the northern part of South America and quickly moving southward. He encouraged me to apply for a Fulbright-Hays Fellowship for the genetic improvement of coffee using plant tissue culture technology. Further encouragement came from Henrique Amorim and Ruy Caldas, the Brazilian graduate students.

I applied for a fellowship and, during the last week of May, I was surprised to receive a telegram from Washington D.C. stating that I was one of two Fulbright-Hays Fellowship Awardees selected for the Brazil program and that I should report to Washington D.C. for participation in a three day orientation program. The recently appointed chair of my faculty, Patrick Dugan, approved my acceptance of the fellowship. Dr. Dugan told me that although I was a recent hire, collaboration with the University of Sao Paulo and the Brazilian Nuclear Energy Commission would provide an important opportunity for the department. I was pleased by Dr. Dugan's forward thinking and for the opportunity for both the department and my career. About the same time, a new dean of the College of Biological Sciences was appointed who issued a mandate to regroup the various faculties into departments.

The Faculty of Microbial and Cellular Biology was renamed the Department of Microbiology with a chair responsible for undergraduate education, faculty teaching responsibilities and budgetary matters.

The Fulbright-Hays Program, University of Sao Paulo, ESALQ – Escola Superior de Agricultura "Luiz de Queiroz", CENA – Centro de Energia Nuclear na Agricultura.

A few words about the history of the Fulbright-Hays Program, Retrieved on August 5, 2013 from http://www.international.ucla.edu/pacrim/title6/Over2-Scarfo.pdf. During September 1945, the freshman senator from Arkansas, J. William Fulbright introduced a bill in the U.S. Congress that called for the use of proceeds from the sales of surplus war property to fund the "promotion of international good will through the exchange of students in the fields of education, culture and science." One year later President Harry S. Truman signed the Fulbright Act into law. This bill provided a life changing experience for so many of us in the fostering of global collaboration. From its inception, the Fulbright Program has supported bilateral relationships in which other countries and governments work with the U.S. to set joint priorities and shape the program to meet shared needs. The world has changed in the ensuing decades, but the fundamental principles of international partnership and mutual understanding remain at the core of the Fulbright Program's mission.

The 1960s: Education, the Democratic Spirit, Title VI and Fulbright-Hays.

As a new international system evolved, Title VI programs evolved and expanded accordingly. One important arena for U.S. security as well as foreign language and area expertise during the Cold War was Latin America. Countries throughout the region experimented with various forms of government spanning the political spectrum. In the early 1960s, the Cuban Missile crisis demonstrated the instabilities of bipolarity and its threats to international security as the world teetered on the brink of nuclear war. Reflecting the growing emphasis on and importance of our Latin American neighbors, in 1961 President Kennedy enacted the Alliance for Progress, which provided funds to combat illiteracy and promote education, and to support economic integration, the growth of the market economy, technical training and Peace Corps programs, as well as scientific and higher education collaborations in Latin America. In addition, the Kennedy Administration pledged defense of nations where independence was endangered. During this period, the NRC focus also expanded to include Latin America.

Also during this time in U.S. history, Senator J. William Fulbright succeeded in persuading Congress to pass the Mutual Educational and Cultural Exchange, or Fulbright-Hays, Act of 1961. An Executive Order ultimately assigned Section 102 (b) (6) of this act to the U.S. Department of Health, Education and Welfare because of the section's emphasis on creating an American international education infrastructure. Like Title VI, Fulbright-Hays was aimed at strengthening non-West European language and area expertise in the United States. These goals were achieved through focused opportunities for overseas study and research -- both essential for training language and area experts. Thus, Fulbright-Hays is viewed as the overseas counterpart to the domestic capacity-building Title VI programs.

The Fulbright-Hays Award Fellowship and the first research grant awards lent momentum to my university career. However, I was failing in my personal life. My marriage was rapidly collapsing because of the demands on my career building activities at the university. I was constantly preparing lectures, writing manuscripts and research grant proposals, and grading papers. Added to this arduous schedule was the stress of dealing with my wife's family's frequent visits during the week and on weekends.

Virginia and her family encouraged me to resign from the university and apply for a secondary teaching position that would entail less responsibility and permit more family time. That was all well and good but I had not attended the college of education or did I possess a teaching certificate. My participation in international scientific meetings and in the immediate Fulbright sponsored program in Brazil was not consistent with my wife's and her family's vision of family life in Ohio.

Virginia, Jeff and I will always be indebted to Jim and Jean Gunckel for their unexpected visit and intervention during this most challenging time on which I recently reflected on at a memorial service for Jim Gunckel in Jamesburg, NJ. The Gunckel's spent several days in Columbus with us and their council led to a separation agreement which Virginia and I would co-parent our son with frequent visitation and I would be obligated to provide continued support to Virginia. I was fortunate to have my maternal grandmother living in Columbus and her kind invitation for me to live in her beautiful home in the Clintonville section of Columbus allowed me to remain at the university and afford both alimony and child support. I felt bad about the fractured marriage and not being able to see my son on a daily basis. However, I was determined to continue to have him participate in all aspects of my life and to be his cheerleader during his maturation.

Brazil

I invited my Grandmother to accompany me on my first visit to Brazil in mid-June. We were greeted at the Viracorpos Airport, by Dr. Otto J. Crocomo, who at the time was the research coordinator of the Plant Biochemistry Sector of CENA (Center for Nuclear Energy in Agriculture), A National Atomic Energy Commission Laboratory situated on the University of

Sao Paulo ESALQ Campus in Piracicaba, S.P. We travelled to Piracicaba from the airport for an extraordinary welcome dinner hosted by Otto, Diva Crocomo and family. Our drive to Piracicaba provided a firsthand look at the Sao Paulo environs, the heart of the city and Piracicaba, which is a beautiful city located in the Brazilian state of Sao Paulo. The warm reception and spectacular setting immediately made us feel at home with the Crocomo family.

In 1970, 368,843 people resided in Piracicaba in an area of 1,369.511 km², at an elevation of 547 m above sea level. The name of the city originates from a word in the Tupi language that means "place where the fish stops" because of the waterfalls of the Piracicaba River that bisects the city where the "Piracema" (fish swimming upstream to reproduce) where the larger fish, such as the "Dourados," can be observed. "Mirante" a beautiful three level fish restaurant with delicious food resides on the bank of the Piracicaba River overlooking the waterfalls, which are illuminated by a light show in the evenings. The city is famous for sugarcane plantations, cane spirit production, and traditional music. The city houses the oldest agricultural school in Brazil, the Escola Superior de Agricultura Luiz de Queiroz – ESALQ of the University of Sao Paulo. This school is more than 100 years old and is located on a farm with a large collection of trees and plants. Piracicaba's economy is fueled in part by the rapid expansion of the market for cane sugar. The cane harvest yields many products, such as sugar, oil and alcohol / ethanol. Companies located in Piracicaba include Fermentech Ltda, which develops ethanol production technology and was founded by Dr. Henrique Amorim, colleague, friend and former professor of biochemistry at ESALQ, Copersucar, a large company focused on genetic improvement of sugarcane and several large industries including Aceolor Mittal, Caterpillar and the Didini Industries. .Piracicaba History Retrieved August 5, 2013 from http://en.m.wikipedia.org/wiki/Piracicaba, ESALQ

History Retrieved August 5, 2013 from http://www.en.esalq.usp.br/who-we-are

The University of Sao Paulo (USP) is the largest higher education and research institution in Brazil with over 57,000 undergraduate and 26,000 graduate students. USP is composed of seven campuses, 40 learning and research units, five hospitals, five museums, five specialized institutes, and multiple experimental laboratories and centers of scientific and cultural diffusion covering all areas of the human activity and offers approximately 700 regular including 230 undergraduate courses representing more than 3,400 disciplines, an average of 5,500 students graduate annually. Graduate studies at USP, with more than 500 fields of concentration areas (M.A.s and Ph.D.s), are an international point of reference in Science and Technology. USP History Retrieved on August 5, 2013 from http://en.m.wikipedia.org/wiki/University_of_S%C3%A3o_Paulo

My appointment was actually at Centro de Energia Nuclear na Agricultura (CENA) which is situated on the ESALQ Campus of the University of Sao Paulo in Piracicaba although in subsequent years my appointments at CENA were held jointly with ESALQ and the Agronomic Institute in Campinas, Sao Paulo. CENA was established by a group of ESALQ faculty members who envisioned the potential agricultural and environmental applications for nuclear techniques. Studies with radioisotopes and radiation were conducted by these faculty in collaboration with other research centers, e.g., IEA, currently IPEN) and USP Units after 1955. CENA History Retrieved August 5, 2013 from http://www.iaea.org/Publications/Magazines/Bulletin/Bull180su/18005493338su.pdf

Faculty members Admar Cervellini, Almiro Blumenschein, Akihiko Ando, André Martin Louis Neptune, Darcy Martins da Silva, Aeneas Salatti, Epaminondas Sansígolo de Barros Ferraz, Euripides Malavolta, Henrique Bergamin Filho, Klaus Reichardt,

Otto Jesu Crocomo and Renato Amilcare Catani submitted a proposal to the Technical Director of ESALQ for creation of the National Center for Nuclear Energy in Agriculture (CNEN). The establishment of the CNEN as an organ of ESALQ was endorsed by the congregation in September 1961. On 1 August 1962, the CNEN was formalized by the signing of an agreement between the ESALQ, University of São Paulo and the National Commission of Nuclear Energy (CNEN). In the ensuing two years, pioneering research was initiated, several international courses introduced and equipment acquired. However, the establishment of military rule in 1964 led to the suspension of all existing agreements, leading to the extinction of the CNEN.

But the idea and passion did not die. The project to create a unified installation for research on nuclear applications in agriculture was resurrected by the faculty of ESALQ, and culminated in the creation of the Center for Nuclear Energy in Agriculture by State Decree 46794, published on September 22, 1966 as an institute attached to ESALQ. The Scene (CENA) then went on to receive financial and material support from the National Commission of Nuclear Energy (CNEN) through an agreement signed in 1968, and which lasted five years. During this period, the first physical facilities of the CENA were constructed.

In 1972 an agreement was signed between CNEN and the *United Nations Development Programme* (UNDP), administered by the International Atomic Energy Agency (IAEA, Vienna, Austria), which provided for 1.3 million dollars to fund a five year scientific exchange. This program represented the only exclusive IAEA office operation in Latin America until the year 1991, This was instrumental in the development of the institution, not only by the expansion of the equipment facilities, but mainly for training and exchange of researchers, which led to a program of visiting experts and technicians from leading research centers for research collaboration and conducting training courses. CENA

/ USP currently has 37 faculty members, 31 fellows and 130 non-teaching staff. Since its inception in the 60s, CENA has emerged as an important research and teaching institute with both undergraduate and post-graduate programs.

The Fulbright-Hays Fellowship Grantee Award inaugurated a forty-three year research collaboration and friendship with Dr. Otto Crocomo and the University of Sao Paulo. I must say that many others were important to the melding of my relationship with Dr. Otto J. Crocomo including Dr. Henrique Amorim and Ruy Caldas, both graduate students in my laboratory at Ohio State University who returned to Brazil to assume professorship positions at the University of Sao Paulo ESALQ Campus in Piracicaba and the University of Brasilia in Brasilia. Dr. Amorim led successful programs in the improvement of consumer based coffee quality and subsequently pioneered the Brazilian ethanol program with the founding of Fermentech S.A. Today as CEO of Fermentech, he has developed a global technology company partnering with major fermentation companies in energy and consumer based products. Henrique Amorim possessed excellent business skill sets and advised me on numerous occasions on important business decisions.

Under Dr. Crocomo's direction, we equipped the plant cell culture laboratory and developed collaborative initiatives with colleagues at the University of Sao Paulo and other leading Brazilian research laboratories. Ruy Caldas, Otto Crocomo and I constructed cell culture transfer facilities because laminar flow hoods were not yet standard laboratory equipment. The carpentry skills developed from Soap Box Derby days and constructing tissue culture transfer chambers with my father came in handy. We immediately launched a program developing cell cultures for the important cultivars of tropical crops with special interest in citrus, cocoa, coffee, beans, palm, and sugarcane. These programs were initiated in collaboration with geneticists and plant breeders at the University of Sao Paulo/ESALQ Campus and the Institute of Agronomy in Campinas, Sao Paulo.

My grandmother and I resided in Piracicaba with Drs. Ruy and Linda Caldas at their recently acquired large home, which provided us with our own living quarters. Living with a Brazilian family facilitated our acclimation to the Brazilian society, social life, and incredible cuisine. My grandmother was so much fun and quickly made friends and learned to use the taxi service for beauty salon appointments and shopping. Diva Crocomo, the wife of Otto Crocomo, included my grandmother in a number of social groups which she enjoyed.

Toward the end of our visit in 1971, we had the occasion to book a cross country trip on a Pullman bus to Brasilia. We made frequent weekend trips to Sao Paulo and Rio de Janeiro. My grandmother returned to Columbus in September to attend to her social groups and family matters. She thoroughly enjoyed the visit and acquiring many new friends including Dr. Eva Wilson, a home economics professor from Ohio State University. Eva participated in an exchange between ESALQ and Ohio State University and elected to remain in Brazil as a faculty member in the Department of Home Economics.

The laboratory work proceeded extremely well. Otto and I developed a number of research projects which included undergraduate/graduate research scholars and submitted a two year training course proposal to the Organization of American States beginning in the summer of 1972. We cemented plans for continuing our collaboration after my return to Ohio State University. I confess, very happily, that my eagerness to apply plant tissue culture to agricultural problems was encouraged and sharpened during the years at CENA and the University of Sao Paulo ESALQ Campus. I was among the fortunate individuals to have the opportunity to travel abroad and forge lifelong collaborative relationships. I learned that successful research and teaching programs benefited from collaboration with colleagues at home and abroad. I have often discussed with Drs. Otto Crocomo and Henrique Amorim the value of the US/Brazilian collaboration in regard to globalization and development of new opportunities for our colleagues and students.

On my return to Ohio State in November, I became engaged in the team teaching approach to the general cellular biology course with daily lectures and afternoon student laboratory sessions. The course was quite popular with students from a number of colleges because of growing interest in the advances of cell biology and the importance to technology to multiple disciplines. My autumn lectures were delivered during November and December in the winter quarter. I assumed administrative responsibilities for the course and delivered about two-thirds of the lectures. I slowly improved my classroom lecture delivery performance.

In the spring quarter, I organized an advanced course on current topics in cell differentiation based on the mitotic cell cycle and differentiation, cancer, and animal and plant embryogenesis. I also had the good fortune to co-chair with Julius Kreier the weekly departmental seminar series, the largest such series on campus with an enrollment of over 150 students. The seminar series evolved into a one hour course and included a number of premiere scientists in various fields of cellular and microbiology. Julius and I were successful in attracting funds from major granting agencies and foundations for providing honoraria, lodging, and travel funds for visiting scientists. We invited students from the department and those in our classes to participate in evening fireside chats. Both faculty and students found these sessions to be most worthwhile. My cell differentiation course for upper division undergraduate and graduate students was taught at 8:00 A.M. (Monday thru Friday). Students were required to write research papers and attend the late afternoon and evening fireside chats for the visiting guest lecturers. The teaching assistant and

I ordered coffee, tea, water, donuts and bagels for the students each morning, a custom that improved the quality of life for all.

In 1974, a second advanced course was added entitled, "Advances in Plant Cell Culture and Differentiation," which included a laboratory session. The course drew a sizable audience of students from a number of colleges. The two advanced courses provided ideas for updating a number of the lectures contributed in the general cellular biology course and for the introduction of new laboratory exercises. These advances gave me more confidence regarding my professional abilities.

The plant cell culture field had gathered momentum and was now moving at a rapid pace. A number of important advances occurred in the late sixties and early seventies. First anther culture and isolation of haploid cells, Nitsch and Nitsch (1970), exploited the anther culture technique developed by Gupa and Maheshwari (1964) and laid the foundation for the subsequent introduction of methods for inducing haploid plant development from anthers. My dissertation mentor at Rutgers collaborated with the Nitsch team and I became a colleague and friend of Colette Nitsch. Other key developments were those of Takebe, Labib and Melchers (1971) who were successful in the isolation of protoplasts capable of regeneration. This was followed by breakthrough experiments by Carlson, Smith and Dearing (1972) achieving plant regeneration from the product of protoplast fusion, and thereby creating possibilities for the genetic improvement of plants. Philip Ammirato (1974) demonstrated that in vitro embryo maturation could be controlled experimentally.

In the summer of 1972, Dr. Crocomo, Dr. Linda Caldas, a recently minted Ph.D. from Ohio State University, and I, conducted a four week training course in plant cell culture sponsored by the Organization of American States for about 30 students throughout the Americas. We authored and published a handbook for the students. Morning, noon and night, we taught classes, held laboratory sessions, supervised student research projects, and led seminars and informal discussions on the potential of cell culture for basic and applied research. Two of the students from this training course were selected for participation in a two year research fellowship program. Maro Sondahl, the top performing student, accepted an offer to complete the Ph.D. degree in my laboratory at Ohio State University and continue to pursue the cellular genetics research program that he initiated at the Agronomic Institute, where he held the position of research associate.

About the same time Dr. Henrique Amorim and his wife Vera Amorim from the University of Sao Paulo and Dr. Vasantha Padmanaban, a postdoctoral fellow from India joined our laboratory as visiting faculty members. A number of superb graduate students joined our research team including Antonio Goncalves, Willie Loh, Sharon Maraffa, Marinez Molina, Bob Reisner, Wei Shen, and Karen Templeton. The students all launched successful careers in academia, medicine and industry. They all became leaders in their respective fields.

The visits to Brazil continued throughout 1972, 1993, 1974, 1975 and 1976. Grant support was provided by the Brazilian Nuclear Energy Foundation. Otto Crocomo and I generated a number of refereed journal papers and several books. The CENA laboratory at USP-ESALQ resulted in the education of a number of scientists from Brazil and the Americas who assumed senior positions in academia, government research laboratories and industry. Our collaboration expanded and included exchanges of a number of scientists from one another's institutions.

In 1974, the faculty and university promoted me to the level of associate professor with tenure. The process of promotion was becoming quite rigorous requiring extensive documentation that included prospectuses of courses taught, student evaluations, updated *curriculum vitae*, letters of recommendation from colleagues at leading U.S. research universities and institutes and recommendations from scholars at leading U.S. and foreign research universities and institutes and science citation index data on publications. Science citation index searches provided information on the significance of scientific papers by documenting the number of times the paper was cited in the scientific literature. Appointment to tenure provided a faculty member with more involvement in decision-making at the departmental and university levels, because service on a number of departmental and university wide committees was restricted to tenured faculty members.

In September 1977, an opportunity occurred for collaboration with three other faculty members at Ohio State University from the departments of botany, genetics and plant pathology to organize an international colloquium entitled "Plant Cell Culture – Principles and Applications." David Evans and Maro Sondahl, Ph.D. candidates at Ohio State at the time and my Brazilian colleagues contributed to the success of the colloquium. The colloquium was financed by corporate donors and major research foundations and attended by 350 scientists from around the world. It became clear that plant cell culture had come of age and would become one of the cornerstones of the age of biotechnology. Although, I was unable to visit Brazil in 1977 because of the colloquium, the Brazilian Nuclear Energy Commission nevertheless paid my summer stipend, which was crucial to my financial situation at the time. The outcomes of the colloquium were publication of the proceedings, academic promotion, publication of a monograph series and a career in the private sector.

1977 was a surprising year with the chair of the department requesting that I submit credentials for consideration for promotion to full professor. Shortly thereafter, our laboratory research team moved to the top floor of the biological sciences tower in one of the three laboratory suites designated for Nobel Prize Laureates in the design of the building. The Department of Microbiology had a distinguished faculty but lack of funding prevented recruitment of the Nobel Laureates. Our research team was overjoyed with the new laboratory facilities and quickly resumed our research activities.

The Corporate World

The phone began ringing with opportunities for consulting and in the late autumn an invitation was received to present a seminar at the Campbell Soup Company in Camden, New Jersey. This invitation was a follow-up from the recent visit of the Company's vice president for poultry research, who had participated in our weekly departmental seminar program and for which during autumn quarter the theme was corporate research. I remember at the time, his request to visit our laboratory and the graduate students and him mentioning that the Campbell Soup Company should be interested in our research programs. However, as much as I appreciated the seminar invitation, I was unable to accept because of the end of the quarter demands of examination grading, graduate student exams and research grant submissions.

Finally an agreement was in place for a brief trip to the Campbell Soup Company (CSC) for a seminar presentation. I quickly, assembled a slide show and packed my luggage for the trip. Rapid Rover Van Services shuttled me from the Philadelphia International Airport to the Cherry Hill Inn. Bill Ramer, the Research Vice President of the Campbell Institute for Agricultural Research, invited me and members of the research team from the nearby the Pioneer Research Laboratory in Cinnaminson, N.J., to his home for a reception. Bill and I had previously met at scientific meetings and at the recent research colloquium entitled *Plant Cell and Tissue Culture: Principles and* Applications held at Ohio State University on 6-9 September 1977. After the reception, we dined at a Hanover Trails restaurant, the CSC restaurant chain. Bill and his team provided me with a comprehensive understanding of the company and the research and development goals. Bill had been a senior scientist at the USDA Beltsville research facility in Maryland prior to joining the CSC and was quite interested in moving the company into biotechnology research. The evening had been quite enjoyable and I appreciated the hospitality. Little did I realize what the following day had in store! My feelings were that Bill was appreciative of the meetings at Ohio State and wanted to thank me and possibly engage me as an advisor in the future. The company provided my travel, lodging and a thousand dollar honorarium.

My seminar was scheduled for 9:00 A.M. the following morning at the Pioneer Research Laboratory in Cinnaminson, New Jersey. The Pioneer Research Laboratory research team and a few corporate vice presidents from Campbell Place Headquarters in Camden, NJ attended. The seminar session was followed by a stimulating question and answer session, and a tour of the facilities, and meetings with the senior research scientists. The Pioneer Research Laboratories conducted research related to basic research, development and commercialization of vegetable cultivars with improved processing characteristics. A nearby facility in Moorestown conducted research on advanced packaging and developed plastic soup containers used by CSC today. The Pioneer Research Laboratory Complex was situated on a beautiful campus in Cinnaminson, New Jersey consisting of three main research buildings, state-of-the-art laboratories, an electron microscopy facility, a tomato processing facility, a research greenhouse facility, two residential houses for the grounds keepers, miscellaneous storage sheds and a nearby 100 acre vegetable farm for breeding trials. The farm, which was about a fifteen minute drive from the Pioneer Research Laboratory, also had several research buildings. It was interesting to note that the Pioneer Research Laboratory campus had been the former summer home for the founding family of the Campbell Soup Company, the Dorrance family. John T. Dorrance was Chairman of the Board at the time. The family owned over fifty percent of the company even though CSC was a publically traded company.

Lunch was scheduled at Campbell Place Corporate Headquarters with additional corporate leaders in marketing, product development, agricultural research and strategic planning. We ordered lunch using individual order cards and of course, I ordered the Campbell Soup Special of the Day and a sandwich. Food was served and I observed that the soup was solidified. I proceeded to eat the sandwich and the product development executive asked if I enjoyed the soup. I answered that I did and was waiting for the soup to cool. As we finished lunch, I covered the small soup bowl with my napkin to avoid any embarrassment. Lunch was followed by a meeting with Harold Shaub, CEO, and the vice president of human resources in the president's office suite. Much to my surprise, after the introductory niceties, Mr. Shaub offered me the position of Director of Research for the Pioneer Research Laboratory, Cinnaminson, NJ, a division of the Campbell Institute of Agricultural Research, which included other laboratories in Arkansas, California, Maine, New Jersey and Ohio. My responsibility was to be the development of an agricultural biotechnology research program. The employment package included doubling my compensation and many perks including stock options. I was flabbergasted and requested two weeks to consider the offer. There was much to contemplate during the return trip to Columbus regarding my career, family, colleagues and obligations to my graduate students and colleagues.

The job offer was indeed a marvelous opportunity to leverage knowledge that had been gained over the past decade in my laboratory and around the world in plant biotechnology for enhancement of consumer products and food processing. Biotechnology, the new focus of the biological sciences was not a basic discipline, but rather represented a consolidation of disciplines, i.e., agronomy, cell culture, horticulture, plant biology, pathology, physiology, biochemistry, plant genetics and breeding, horticulture and molecular biology related to production of innovative goods and services.

On arriving home, I spoke with my son, Jeff, and Virginia, my former wife, about the opportunity. I invited Jeff to join me the following weekend in Philadelphia and before I knew it, we were off to Philadelphia. Snow and ice made the trip difficult... We lodged at the Holiday Inn on Market Street and visited the Campbell Place Corporate Headquarters, the Pioneer Laboratory in Cinnaminson, N.J., and enjoyed the environs of Philadelphia. We walked and talked about the future. Jeff was age 14, but quite mature for his age with many interests including acting, books, music, travel and track.

I remember, Jeff saying, "Dad, you know the opportunity would be good for both of us. Air tickets are inexpensive and I can visit you on weekends, during my school holidays and summer vacation." The salary increase made all things possible and sealed the deal between Jeff and me. Following the return trip home, Virginia and I negotiated an arrangement to allow Jeff to fly to Philadelphia for weekends and vacations.

The following week, I was busy with final examinations and grading, but managed to find time to talk with my graduate students about the opportunity. I promised to remain involved with their research programs and studies. Arrangements were made with the departmental chair, colleagues in the Microbiology Department and the College of Agriculture for a seamless transition of the student's graduate programs. One of the students moved to CSC with me to complete her graduate studies because her husband had accepted a position at Hahnemann Medical School in Philadelphia. The chair of my department and the dean of the college requested that I remain at Ohio State University and offered a significant increase in compensation in a display of generosity that I greatly appreciated. I will always be indebted to Patrick Dugan, the dean, and Robert Pfister, the departmental chair for their considerate efforts on my behalf. The career disruption was painful but deep down, I knew that I had to accept the new challenge and move on.

I had several conversations with my grandmother about that the CSC offer and she was in agreement that it was a once in a lifetime opportunity. She further agreed to accompany Jeff to Philadelphia on weekends and to continue to have Jeff come to her house on Wednesday evenings. The following week, I mailed my letter of acceptance to the Campbell Soup Company and the move to New Jersey was planned for January. Another visit to CSC in Cinnaminson, N.J. was followed by a second visit to the president. Mr. Shaub informed me that he wanted me to present a research plan two weeks after my arrival. Wow! No holiday vacation!

January 1978 was the start of a new life for me as Director of the Pioneer Research Laboratory at the Campbell Soup Company. I took up temporary residence at the Cherry Hill Inn while waiting to close on a home in Haddonfield, New Jersey. I signed a contract to purchase a duplex in Haddonfield, NJ adjacent to Cherry Hill, NJ, but was not able to close because the banks were waiting for approval of the new financing interest rates. The higher interest rates for financing were approved two months later and a closing was scheduled. The Campbell Soup Company director position was going to provide an important experience in linking the research laboratory to the marketplace. The experiences of working at CENA and the University of Sao Paulo-ESALQ in Piracicaba, Sao Paulo, Brazil had provided me with some background in applied research, but industrial research added the dimension of commercialization, which was the final link between the laboratory and the end user.

I landed on the ground in Cinnaminson New Jersey running and developed an excellent rapport with Bill Ramer, Vice

President of Research and the director and vice president of potato breeding and miscellaneous crops, Charlie Cunningham, a former university professor. Over the holidays, I worked on a biotechnology research plan that was tweaked by Dr. Bill Ramer, the group vice president at the Campbell Institute for Agricultural Research and Dr. Bud Denton, the Corporate Vice President for Research. We developed concise research and development goals and a budget for a five year program. The proposal allowed me to recruit five Ph.D. scientists with appropriate technical support to augment the research team at the Pioneer Research Laboratory. The CSC graphics unit was a superb creative team. The director of the CSC graphics unit provided invaluable assistance in preparation of the biotechnology research program slides for the upcoming research program presentation about providing a biotechnology focus for the Pioneer Research Laboratory. I had been advised to limit the presentation to fifteen minutes, which meant ten minutes for slides and five minutes for discussion. I rehearsed the presentation with Bill Ramer and Bud Denton several times prior to the meeting with Mr. Shaub.

The meeting was scheduled for the following Wednesday. I will never forget that the evening before, my older sister and family, were visiting Atlantic City and insisted that my younger sister, Sally, and Jeff have dinner with them. In those days, I attempted to fulfill all obligations. Patricia and family arrived late to Atlantic City, which meant that I didn't get home until the early hours of the morning.

The meeting in Mr. Shaub's office was interesting. All seemed to be going well until five minutes into my presentation, Mr. Shaub pulled out his car keys and began jiggling them.

I said, "Mr. Shaub, please allow me to show the final slide.

Afterwards, he responded, "Dr. Sharp, please don't ever do this again! I'm a busy man and time is of the essence to me. Please in the future limit presentations to five minutes! I hired you for your expertise. All you need to provide me is a brief plan and budget. Moreover, all letters and memos to this office should be limited to one paragraph or, better yet, one sentence. Do you understand?"

"Yes, Mr. Shaub, I will comply. Thank you."

This was my CSC baptism, but there was much more to come. Mr. Shaub approved the biotechnology research program and budget for the Pioneer Research Laboratory, which allowed me to begin implementation of the research plan and organization of the research team. One of the senior research scientists at the Pioneer Research Campus cut a side deal to be named director of food processing prior to my arrival. He viewed me as a threat and tried to sabotage my program at every turn by devious means ranging from lack of cooperation. On several occasions, he refused to process our biotechnology-derived vegetables, which was an essential step in evaluating our plant breeding programs. One of the senior vice presidents always defended him with excuses that the biotechnology vegetables were spoiled, or lost or time did not permit evaluation. The lesson learned is that the new guy on the block always has challenges from the skeptics and luddites who opposed advances in science and technology.

Meetings were scheduled with the plant breeders to better understand their programs and the important traits being bred into the CSC processing vegetables for the U.S. and overseas field production operations. Our strategy was to use somaclonal variation, which involved culturing tissues from commercial processing vegetables cultivars to generate genetic variation and subsequently selecting for improved processing traits in the field. The idea at the time was that this process, as opposed to conventional breeding programs that required seven to eight years for development of new processing cultivars, could shorten the time lines for releasing improved commercial vegetable cultivars. We immediately installed tissue culture growing racks in the laboratories and began to isolate and establish cell cultures from leaf explants from the CSC commercial tomato cultivars. Our goal was to have seed from greenhouse grown tissue culture-derived tomatoes for field trials in the summer of 1979.

I planned a weekend visit to Ohio State graduate David Evans in Binghamton, NY to discuss the opportunity for him to join the Pioneer Research Laboratory team and conduct research on cell culture-improved tomato cultivars. He had recently been appointed to a tenure track assistant professorship at the Center of Cell Genetics, SUNY Binghamton. In addition, we recruited a first rate molecular biology fungal geneticist to head a program on improvement of cultivated mushrooms.

One morning, a memorandum was circulated from corporate headquarters that a hiring freeze would immediately be implemented and only critical hires approved by the office of the president could be approved. At once, I knew that I must spring into action and requested permission to engage graduate students, postdoctoral fellows and visiting scientists to enable us to reach a full research team complement without using permanent positions. Approval was granted and within six months two world class plant scientists joined the company. One was from Venezuela, and the other one was a graduate student from Ohio State University. Also, two Rutgers graduate students joined our team.

A bomb was dropped within a few months into my assumption of the research directorship at CSC. The group vice president was departing the company for unknown reasons and would no longer be allowed on the premises. Shortly afterwards, I ran into Bill Ramer at the Philadelphia International Airport on his way to St. Louis to interview for an executive position with Ralston Purina to which he was later appointed. He requested that we ship to his home his personal books and files in storage at the Pioneer Research Laboratory. Understanding the value of personal research papers and books to the career of a research scientist, we immediately complied with his wishes.

After the announcement of the departure of Bill Ramer, my immediate report, I made an appointment with the CEO, Mr. Shaub, to ask if he would like me to submit my resignation. He requested that I remain with the company and report to him until the research operation reorganization was completed. I subsequently reported to Dr. Bud Denton and very soon afterwards to Dr. Charles Duncan, a former microbiology professor from the University of Wisconsin. We had an immediate rapport and the reporting arrangement was superb. We were always on the same wavelength.

Much excitement was created during the 1979 field trials demonstrating somaclonal variation for improving processing traits. The Vice President of Strategic Planning, Jim Imshoff and his team took a special interest in our work and recommended that strategic plans be devised for our research and development programs. This was an important experience for me in sharpening my business skill sets. The bean, carrot and mushroom programs were likewise taking shape with early successes. CSC acquired Lexington Gardens and soon we had improvement programs in place for herbs, spices and ornamental plants. The link to the strategic planning group kept us at the forefront of the CSC business aspirations.

Following Harold Shaub's acquisition of the Domsea Farms Inc., a sea farming, sea ranching and aquaculture operation in Seattle, Jon Lindbergh, the research vice president and son of Charles Lindbergh visited the Pioneer Research Laboratory to discuss possible collaborative research. He was clearly a world class underwater explorer and aquaculture pioneer who aspired to feed the world's people by harvesting the bounty of the oceans. He followed in the footsteps of his famous aviator father. Our research team enjoyed meeting Jon Lindbergh and developing a better understanding of the Domsea Farms Inc. research, development and production programs. The CSC Hanover Trails Restaurant chain began offering salmon from the Domsea Farms Inc. operations.

Jon Lindbergh was enthusiastic about our work and said,

"Rod, you are a pioneer much like my father in ushering in new technology for the betterment of humankind. Your new approaches to the development of improved plant cultivars will no doubt be important to the world food challenges."

Our scientific team initiated a number of important genetic improvement programs for tomatoes, carrots, mushrooms and ornamental crops focused on commercial product improvement. Linking the laboratory to product development was a very exciting opportunity for all of us.

In 1980, Gordon McGovern, former president of Pepperidge Farms, the Campbell Soup Company subsidiary, succeeded Harold Shaub as the new President and CEO of the Campbell Soup Company. He was quite interested in our work, as were several BOD (board of director) members, including Dr. Sterling Wortman, President of the Rockefeller Foundation, who met with our research team on a regular basis and fine-tuned our research programs. Sterling Wortman (1923-1981) was an eminent agricultural research scientist who made significant contributions to world food production. He earned a doctorate in plant breeding and genetics from the University of Minnesota in 1950 and joined the Rockefeller Foundation as a corn breeder assigned to its field office in Mexico (1950-1954). He served as the head of the plant breeding department (1955-1960) of the Pineapple Research Institute in Hawaii, and later directed that Institute (1964-1965). In 1960 he helped establish the International Rice Research Institute, for which he served as assistant director (1960-1961) and associate director (1963), and in 1966 helped create the Consultative Group on International Agricultural Research. He served the Rockefeller Foundation as director of the Agricultural Sciences Division (1966-1970), vice-president (1970-1979), and acting president (1979).

Gordon McGovern visited quite often and promoted fresh market opportunities that built on the Farm Fresh Mushroom Brand, such as farm fresh tomatoes, ready-to-eat pastas and salads. The pioneer Research Laboratory was green lighted to engage in research and development programs geared to the development of value-added fresh market products, including tomatoes and possibly Shitake mushrooms, other exotic mushroom cultivars, and strawberries. We scoped out a worldwide study of successful hydroponic operations in Holland, Japan, U.S. and the U.K. This was based on our contacts in Holland and the U.K. and collaborative relationships between the agricultural engineering department at Ohio State University and Professor Yasuyuki Yamada, Kyoto University. In the U.S., we relied on Merle Jensen and Paul Hodges, leading horticulturists and engineers at the University of Arizona and the Disney Epcot Land Pavilion. Our team worked with the engineering and strategic planning groups to development of an experimental hydroponic tomato production feasibility program.

At the same time as my move to the Campbell Soup Company, CEBTEC (Center for Biotechnology), the University of Sao Paulo – ESALQ Campus in Piracicaba was founded and placed under the direction of Dr. Otto J. Crocomo. CEBTEC ushered in a number of important scientific advances, including pineapple plantlet development, production of virus-free potatoes and strawberries, sugarcane tissue culture and somaclonal variation breeding programs (Crocomo, et. al.1984), Papaya disease eradication, embryo culture of palms, embryo rescue and interspecific genetic crosses in bean (Crocomo, 1976, 1978, 1979). Antonio Natal Goncalves et al. (1986) and Oliveira (1988) employed tissue culture in experiments tissue culture to elucidate the adult and juvenile states of *Eucalyptus* and *Pinus*.

More important breakthroughs occurred during in the late 1970's and early 1980's, the demonstrating how the accumulation and sharing of knowledge fosters new breakthroughs: for example:

Green and Philips (1975) Vasil and Vasil (1981) and Sondahl (1984); in cereals;

Collins and Philips (1982) in legumes;

Allavena (1984); in beans;

Bajaj (1984) in peanuts;

Sondahl and Sharp (1979) coffee;

Sondahl and Loh, (1988) oil palm;

Sharp and Evans (1981), tomato somaclonal variation' a tool for the introduction of single gene modifications, was advanced for use in plant breeding programs;

The recovery of new mixed organelle genetic types after protoplast fusion was a breakthrough. After Carlson's first success, all the world was attempting to create interspecific hybrid cultivars.

Belliard, Vedel and Pelletier's paper (1979) demonstrated recombinant DNA in cybrids. This discovery led to the transfer of cytoplasmic male sterility, which is important for the development of proprietary hybrid cultivars. This introduction of transformation and plant regeneration protocols provided the key link between cellular and molecular genetics. The realization that individual cells must be transformed and subsequently re-generated into plants brought an awareness of the dependence of genetic transformation on plant cell and tissue culture protocols.

The recovery of useful breeding lines by cell culture selection techniques in the laboratory could very well be characterized as a breakthrough, but it is limited to only those traits that can be se-lected in vitro, e.g. herbicide resistance, disease resistance, stress tolerance, etc. Roy Chaleff's work (1980) led the way.

Major advances in molecular biology of plants were being made:

Hoekema et al (1983) Development of t-DNA-based plant trans-formation vectors;

Bevan et al., 1983 and Van den Elzen et al., 1985; developed bac-terial genes for use as selectable markers of plant transformation;

Gridoni et al. (1985) Characterized controlling regions from high-ly expressed plant genes and demonstrated that these sequences me-diate regulated expression of foreign genes in transformed plants;

Dunsmuir et al. (1987) and Fraley et al. (1986) demonstrated genetic stability of introduced genes;

Federoff (1984) demonstrated use of transposable elements (Ac, Mu, Ds) for isolation of maize genes that were only phenotypically described (no biochemical understanding necessary);

Baker et al. (1986), another key advance was to demonstrate that maize transposable elements function normally in other plants and could be used for plant gene isolation and expression.

1981 – A Banner Year

1981 was my forty fifth year. I somehow thought that I had done it all and seen it all. Never did I believe while growing up in Akron that I would be appointed to a leading research position at a blue chip corporation. And little did I know that a most exciting and rewarding roller coaster ride was about to begin. These opportunities arose from past associations with Bill Jacobs, Princeton Professor and Bill Wardell, University of Maryland Professor, who happened to be sitting next to me and my graduate students on a bus from Helsinki to Leningrad en route to the International Botanical Congress in Leningrad. I had known Bill Jacobs from graduate school days at Rutgers when he and my mentor held joint research meetings for their graduate students. Bill Wardell had visited my laboratory at Ohio State University on a number of occasions as part of a collaborative recombinant DNA research program.

On or about Friday, April 3, 1980 in the late afternoon, my secretary announced that a gentleman by the name of Myer Blech from the Wharton Group in New York was on the phone. I assumed that he was bothering me about opening a brokerage account and asked Karen to inform him that I had no interest. She returned to report that he was calling about a new venture in agricultural biotechnology. I accepted the phone call and the roller coaster ride began.

Myer told me that his two sons, David and Isaac, had just founded Genetic Systems Corporation with scientist, Robert Nowinski, the former research director at the Hutchinson Institute in

Seattle, and they were currently exploring the possibility of forming an agricultural biotechnology company on the eastern seaboard. I inquired as to how he had come across my name and he told me that I was referred to his sons through Lloyd Schoen, a biochemist, at the Memorial Sloan-Kettering Cancer Center in New York. Apparently Lloyd was a consultant to the Blech brothers and had known Bill Wardell during postdoctoral days at the University of Wisconsin. Myer asked if I could act as a consultant to his two sons. I thanked him for the opportunity, but informed him that my engagement as a consultant would impose a conflict of interest with my employment at CSC. I would, however, be pleased to recommend potential candidates for a chief scientist for the new company.

The phone calls continued during the following months, but usually my schedule for Friday was filled with meetings at Campbell Place or in Cinnaminson. In late May, I agreed to meet Myer, Lloyd Schoen and Meyer's two sons in New York for a 6:00 P.M. Sunday dinner at Siegel's, a Kosher Restaurant, in midtown New York. Myer and I arranged to meet in front of the Park 'N Lock Garage across from the Lincoln Tunnel exit onto 42nd Street. I waited and waited but Myer and didn't appear. After thirty minutes, I phoned the restaurant and spoke with Lloyd Schoen. Lloyd said that Myer had mistakenly thought that I was parked at the nearby Port Authority Garage and he would contact Meyer shortly. A few minutes later, a twelve year old Cadillac pulled up to the curb and stopped. Myer stepped out of the car and shook hands. I immediately liked him. Because of the months of phone conversations, he was like an old friend. He reminded me of my father. On the way to Siegel's Restaurant, just a few blocks away, Myer mentioned that he had been driving the Cadillac and wearing some of the same clothing since he closed his first big deal. He was afraid that lady luck would abandon him if he departed with the precious automobile and clothing. Myer replaced the car engine a few years before and maintained the car in mint condition. But, I must tell you a bit more about Myer before moving on with the story. Myer at one time was rejected for admission to the Wharton School MBA Program at the University of Pennsylvania. He sought revenge by naming his New York brokerage firm the Wharton Group. The idea was to keep the Wharton School from opening offices in New York. He was passionate about the founding of Israel in 1948 and volunteered to serve as a rabbi after completion of his service in the Chaplain Corps of the U.S. Army during World War II. He was notified that Israel was interested in immigration of individuals with background in business, education, engineering, law, medicine, science and the trades, but unfortunately for Myer, Israel had more than enough rabbis. Myer immediately enrolled in auto mechanic school, but his wife Ester and family discouraged emigration to Israel. The young family stayed in Vineland for a while longer and then moved to Brooklyn for the launch of the Wharton Group.

We arrived at Siegel's Restaurant at about 7:00 P.M. The restaurant was quite noisy with a Sunday evening family crowd. Myer introduced me to Lloyd Schoen, David and Isaac, and after exchanging the usual niceties, Myer requested the Maitre 'D to seat us in a quiet corner. Lloyd and I hit it off immediately because of our research backgrounds. Lloyd told me that he and Bill Wardell were roommates during their postdoctoral fellowship days at the University of Wisconsin. He had recently spoken with Bill regarding possible candidates for a research director to provide leadership for an agricultural biotechnology company. Bill mentioned my name, our recent trip to Leningrad for the International Botanical Congress, and the collaboration between our laboratories at Ohio State University and the University of Maryland.

David and Isaac were a bit shy at first. David informed me that he had put his dad up to calling me at the Campbell Soup Company out of fear that I would not respond or take him seriously because of his age and lack of scientific background. He was 25 at the time and Isaac was 31. David and Isaac were Baruch College graduates with majors in music and film, respectively. After Baruch, David earned a MFA in music from Columbia University and assumed a position at the Wharton Group with his father. Isaac held a position with a marketing research firm in New York. The two brothers recorded over five record albums with Shanghai Surprise topping the charts in Hungary. David wrote the music and Isaac penned the lyrics. David told me about the founding of Genetic Systems Corporation in Seattle with Robert Nowinski, former scientific director at the Hutchinson Institute, and securing an investment from the Schroder Bank in New York. The way the deal was inked by David was quite brilliant. Bob Nowinski and colleagues had an offer to join a blue chip pharmaceutical company in New Jersey when David proposed forming Genetic Systems Corporation in Seattle and allowing the scientific team to remain in Seattle. David's offer cinched the deal and Genetic Systems Corporation was born.

David proposed that I resign from the Campbell Soup Company and join him and Isaac in founding an agricultural biotechnology company in New Jersey. I reminded David that I was a scientist not a businessman and my skill sets were inappropriate for the CEO task. He proposed a lunch meeting the following week to explore the possibility further with John Connor, Chairman of the Schroder Bank on Wall Street in New York. I agreed to bring David Evans, my research partner.

The next Monday, David Blech phoned to confirm a lunch meeting on or about Thursday May 28 at the Schroder Bank with John Connor and associates. David Evans and I were excited about the opportunity and discussed it at length. As mentioned earlier, CSC was in fact exploring the possibility of investing in one of the newly minted agricultural biotechnology companies.

We prepared a presentation for the Schroeder Bank with care not to disclose any CSC proprietary information.

Thursday morning was a bit hectic because of an early meeting at CSC Headquarters at Campbell Place. I returned to Cinnaminson at 9:30 A.M. and picked up Dave. Fortunately, the turnpike traffic was light and we were on time for the 12:30 P.M. lunch meeting. We were received by Mr. Connor, Stephen Petschek, the President of J. Henry Schroder Corporation, Jeffrey Collinson, Managing Director and David and Isaac Blech. The dining room was much like a fish bowl with an amazing panoramic view of the battery, Ellis Island, the Statue of Liberty and assorted maritime vessels.

John Connor was a seasoned senior business icon having served as President and CEO of Allied Chemical, Merck and U.S. Secretary of Commerce under LBJ. He had been very friendly with William Murphy, former CEO at the Campbell Soup Company during his tenure at Merck and Company in the 1970s. John Connor had been a pioneer in the implementation of strategic planning and during a golf match, advised William Murphy to consider the same, for the Campbell Soup Company. Mr. Murphy heeded the advice and encouraged basic research and vertical integration at CSC. The emphasis on basic research was the basis for the Campbell Institute for Agricultural Research and the hiring of me and Dave Evans by CSC. Dave and I enjoyed sharing our thoughts with Jack Connor. After lunch, we were invited to Jack Connor's enormous office suite. An endless display of awards including championship golf plaques and trophies, hung on the walls and filled glass cases. We were regaled with stories about his time in Washington, D.C. as Secretary of Commerce. One day, LBJ approached Mr. Connor about his passion for golf and requested that he retire from the sport. Mr. Connor offered to resign from the cabinet if his playing were going to create a problem for the White House, but the President didn't pursue the issue. One weekend, he was in the finals of a championship golf tournament located in a remote area of Bermuda when the President tried to telephone him. The White House aides had tracked down a local restaurant in proximity to the golf course.

The White House placed a phone call to the restaurant and the proprietor responded, "Ha, ha, the President."

After numerous phone calls, Mr. Connor was told of them and responded, "Please inform the president that I will return the phone call in about 20 minutes."

Mr. Connor phoned the White House and the President said, "Jack, it must be nice to be relaxing on the greens while my team is solving the nation's problems."

Mr. Connor then returned to the business at hand and said, "Rod, next time you visit, please bring a presentation on your science for us to discuss."

I replied, "Turns out that we brought a brief presentation to share with you today."

Dave and I shared a twenty minute presentation on the biotechnology opportunities for the agriculture, food and industrial sectors.

Mr. Connor asked thought provoking questions after summarizing the conversations of the afternoon, said, "Let's form a new agricultural biotechnology company and make it all happen."

I replied, "Wonderful opportunity, however, I do not have the business experience to make it happen! If we had a seasoned businessman like you to handle the business aspects, we could make it work!"

Mr. Connor immediately replied, "What if I serve in the Acting CEO capacity while we search for an appropriate CEO candidate?"

I agreed and we all shook hands on moving forward with the deal.

The Transition

The meeting at the Schroder Bank led to a surge of activity followed by a series of meetings at law offices in New York and drafting a prospectus. Then came meetings in New Jersey related to drawing up employment agreements for David Evans and me. I was fortunate at the time to have a tenant in my duplex residential dwelling in Haddonfield. The tenant, a vice-president of the Bank of New Jersey, was so kind as to recommend a law firm in Haddonfield to represent David Evans and me in negotiating the employment agreements. The law firm drafted five year employment agreements requiring that the company provide zero interest loans for the purchase of founders stock and establishment of an escrow account at the Bank of New Jersey for deposit of one year's compensation. The Company was incorporated on June 11, 1981 as the DNA Plant Technology Corporation (DNAP). On Monday June 15, David Evans and I resigned our positions at the Campbell Soup Company.

I had many regrets about leaving the Campbell Soup Company, but saw the new company as an extraordinary opportunity to pioneer applications of biotechnology for the agriculture, food, and plant based chemical industries. The severing of ties with the Campbell Soup Company was a stealth operation that involved leasing a Hertz moving van to remove personal effects from the Pioneer Research Laboratory under the cover of night during an early summer weekend. I joined forces with Dave Evans, his wife, Kitty, my son Jeff, and my sister Sally, to mastermind and execute the operation. We carefully, sorted through files and books on books to remove personal effects and leave behind corporate materials. We were nervous about being challenged by security forces and a possible arrest. These extreme measures were taken because only a few months earlier the personal effects of the former vice president of research were confiscated when he resigned from the company. In those days, research scientists accumulate massive paper files of scientific reprints and books. Today of course the files would be stored in electronic databases.

Dave Evans, and I, officially resigned from the Campbell Soup Company on the morning of June 15, 1981. That evening, I received a phone call from our CSC group vice president and Gordon McGovern, the CEO, to schedule a lunch meeting with our bankers and other investors in DNA Plant Technology Corporation. Arrangements were made for the bankers and other investors to meet at the Campbell Place Office of the President the following Thursday for a lunch. Dave Evans, David Blech, Issac Blech and I put our thoughts together and drafted a prospectus in preparation for the meeting. The named participants included John Connor, Jeffrey Collinson, David Blech, Isaac Blech, Dave Evans and Rod Sharp.

On Thursday June 25, the bankers and other investors arrived in two stretch limousines arrived at the Campbell Place Headquarters in Camden, while Dave Evans and I arrived by car and were escorted to the president's dining room, Gordon McGovern was most gracious to me. He and I were seated at opposite ends of the dining room table. The lunch meeting began with Gordon McGovern and Jack Conner recounting their experiences together at Harvard playing touch football and their early career acquaintances and milestones. The drafting of a final business plan for the DNA Plant Technology Corporation and the recruitment of a senior executive from the food industry for the CEO Position were the topics of discussion. Gordon McGovern informed John Conner that Campbell Soup Company was interested in taking a lead investment position in the new company and would invest a combination of cash and the Cinnaminson, N.J., laboratory facilities. Moreover, the Campbell employees would have an option to continue employment at the Campbell Soup Company or to join the new company. The next step was to move beyond a handshake to draft a definitive investment agreement, to conclude final negotiations and execute a formal investment agreement. Mr. Connor agreed to provide leadership to DNAP during the negotiation process.

It was apparent to all that the CSC investment of capital and the Cinnaminson New Jersey research complex, and the opportunity to retain the former Campbell Soup Company Pioneer Research Division personnel provided the new company with crucial momentum from day one. We started off as a full-fledged biotechnology company with highly skilled staff, farms, financial resources, greenhouses, and research laboratories.

There was much more work to be done prior to closing with CSC and other investors on a private placement. The important task of constructing a business plan lay ahead of us. This process began with regular meetings with Evelyn Berezin at her pied-a-terre in Greenwich Village. Dave Evans, Kitty Evans and my son Jeff would book hotels for the weekends in New York. Dave and I would meet at Evelyn's place to work on business plans while Jeff and Kitty hung out in Greenwich Village. Evelyn's husband Israel Wilenitz was always wonderful to us and often joined us for lunch.

Evelyn was a physics graduate from NYU who pioneered the word processor during graduate school at NYU by hooking up an IBM typewriter to an oscilloscope. She also developed appropriate software and demonstrated the word processing process. She was awarded a number of patents and founded Readactron, which designed, developed and manufactured word processing systems worldwide. She subsequently sold her company to the Boroughs Corporation (now Unisys) and was appointed to a vice president position. She served as president of Greenhouse Management Company and as General Partner of a group of venture capital funds and held a number of key positions at SUNY Stony Brook. Evelyn was named to Business Week's list of the top 100 Business Women in the United States.

John Connor assigned two of his senior vice presidents at the Schroder Bank to negotiate the investment deal with CSC. The stumbling block had to do with the biotechnology research program rights for CSC. Finally, an agreement was reached by which DNAP would provide quarterly options to CSC for investment in two major food crop biotechnology research and development programs with an appropriate review period. If CSC turned down an investment opportunity, DNAP was granted the technology and was free to negotiate a research development and or license agreement in the field with other major agricultural and food companies.

The negotiation process between the Campbell Soup Company (CSC) and DNA Plant Technology Corp (DNAP) to reach a final agreement for a private placement was time consuming. The joke at the time was that we would be closing in two weeks, but the two week closing date never seemed to happen. At one point, John Connor mentioned the possibility of renting incubator space at Rutgers University. On a handshake with Gordon McGovern, Dave Evans and I received permission from CSC to move back to our offices with continued access to our secretaries and support staff.

The final step was negotiation of a multiyear, multimillion dollar tomato research and development agreement that was essentially a continuation of the programs in progress prior to the formation of DNAP. On a late Thursday afternoon in March prior to closing time, a meeting was scheduled by the Chief Financial Officer for early the next morning to discuss financial aspects of the proposed research and development agreement. I showed up for the appointment Friday and the CFO explained to me that CSC would agree to a two year, two hundred thousand dollar contracts on a yearly basis on a take it or leave it basis. I hesitated for about thirty seconds, shook hands, and agreed to accept the CSC offer on behalf of DNAP. There were clearly forces within CSC that were in disagreement about the proposed DNAP investment. The expectation had been that CSC would enter into a multi-year 1-2 M million dollar research agreement.

Finally on March 23, 1982 all negotiations were complete. David Evans, David Blech and Isaac Blech and representation from the Schroder Bank Counsel met at the offices of Drinker Biddle & Reath at Broad and Chestnut Streets in Philadelphia and proceeded to sign hundreds of multiple original documents pertaining to the closing, a process that took about five hours. The transaction included title to the Cinnaminson New Jersey Research Complex which was the former Dorrance Family summer estate. John Dorrance II at the time was chairman of the board for CSC. Our former CSC 45 member research and technical support team was given an option to join the new company or to relocate to corporate headquarters at Campbell Place, Camden, New Jersey. All but two members of the team decided against joining the new company. Gordon McGovern, president and chief executive officer, offered all services of CSC's services to help the new company get up and running including access to the CSC health insurance plan. Our board of directors was stellar and included John Connor, Chairman, Schroder Bank and Acting DNAP CEO, Jeffrey Collinson, Senior Vice President, Schroder Bank, Henry Roberts, Chairman & CEO, Connecticut General, Evelyn Berezin, CEO, Greenhouse Ventures and pioneer in word processing, Gordon McGovern, President & CEO, Campbell Soup Company, the Blech brothers and myself.

Our team at DNAP began the process of reaching out to industry for additional investment and securing research and development agreements. In tandem, I worked closely with Gordon McGovern and John Connor in search of a CEO for DNAP. A number of candidates were interested in the position but were not quite suitable because of the lack of a science and engineering background. Finally, Henry Roberts, a DNAP board member and former Chairman and CEO of Connecticut General identified the perfect candidate. Henry was a member of the board of directors of General Foods Corporation (GFC) and a friend of Richard Laster, the executive vice president who was just retiring.

Henry Roberts tried unsuccessfully to reach Richard Laster by phone to discuss the DNAP CEO opportunity. He decided to drive to the Laster Home in Chappaqua and meet with Richard. Henry saw an airport transport vehicle in the driveway as he approached the house. Richard and his wife Lee were off on a trip to Europe and would not return until the later part of April. Richard agreed to phone Henry after his return to discuss the opportunity.

In the later portion of April, Richard Laster and Henry Roberts visited DNAP Headquarters in Cinnaminson, NJ. We arranged a meeting with the scientific team, brief research presentations and a tour of the facilities. Richard and I more or less knew of one another through the research team at GFC and hit it off immediately. Richard had begun his career at GFC at the Tarrytown Research Laboratories. Overtime, he received awards and honors for his innovative research and contributions to key GFC product lines. GFC had a portfolio of important consumer brand names including Crystal Lite, Entenmanns, Kool-Aid, Maxwell House, Minute Rice, Oscar Meyer, Post Cereals, Sanka Coffee, Shake N'Bake, and Tang.

Richard Laster agreed to join our management team as CEO and President. We were off and running. Dick was a marketing genius and brought invaluable creative energy in crafting research and development proposals for key companies. Under Richard Laster's leadership we began to explore aggressively joint ventures with blue chip agricultural and food companies.

Dick Laster and I met with Jim Imshoff, the Vice President of Strategic Planning and formerly a colleague of mine at the

Campbell Soup Company, about consultants for business plan development. Jim had been a professor at the Wharton School and recommended, a former colleague, Peter Davis. During the next few months, Dick, Dave and I met with Peter and a team of undergraduates to develop a definitive business and financial plan, which was necessary for development of a prospectus to be used for both the IPO and Secondary Public Offering.

In a short period, the company had inked important collaborative agreements with the American Home Products Corporation, Brown & Williamson Tobacco Company, CPC International Inc., DuPont Company, Farms of Texas, Fermenich Incorporated, General Foods Corporation, Hershey Food Corporation, Koppers Company Inc., Kraft Inc., Knorr and Maxwell House, a division of GFC. Under Richard Laster's leadership, the Company developed a creative approach to alliances with corporate partners for research, development and commercialization agreements with a joint venture. Upon successful development of product prototypes for commercialization, DNAP and the corporate partner would entered into a joint venture for market testing and subsequent product launch.

We expanded the board of directors to include Frank Carry, CEO, IBM Corporation, and Edward Hennessy, Jr., Allied Chemical Corporation, and completed the initial public offering (IPO) in January 1984 and a secondary offering in April, 1986. We named a scientific advisory board that included the most prominent scientists in plant biology including the preeminent Nobel Laureates Norman Borlaug and Melvin Calvin.

Shortly, thereafter, Dr. Maro Sondahl, a senior research scientist at the Institute of Genetics, Campinas, Sao Paulo, Brazil and a number of key scientists who had been at Ohio State University including David Grothaus, Willie Loh, Sally Miller and Donald Styer, were recruited to join the research team at DNAP. The Company forged collaborative research agreements with the

University of Sao Paulo-ESALQ Campus and the University of Campinas in Brazil.

My business skill sets were continuously being fine-tuned under the tutelage of David and Isaac Blech, John Connor, Jeffrey Collinson, and Richard Laster, but especially I am indebted to Richard Laster for his patience in sharing of his comprehensive knowledge of the food industry. Dick Laster made contact with Leslie Misrock, senior partner at the Pennie and Edmonds Law Firm about representing DNAP. Leslie Misrock was the biotechnology intellectual property guru and neighbor in Chappaqua who prosecuted the ground breaking patent for the patenting of life. Our team at DNAP enjoyed a productive relationship with Leslie Misrock and his team of intellectual property lawyers. Leslie opened the first biotechnology law practice and was known as the father of biotechnology law. Because of Leslie and team, the Company was successful in the prosecution of an impressive portfolio of plant biotechnology patents.

Arnold Jacobs, who, at the time, was a senior partner at Shea Gould, served as the DNAP legal counselor. He provided leadership to DNAP and collaborated with Lehman Brothers, our lead investment banker, management teams, during both the IPO and secondary offering. The preparation for an initial public offering is a very time consuming process involving drafting a prospectus and preparing an assortment of regulatory documents. However Arnold Jacobs and Shea Gould guided us through the process smoothly. Arnold Jacobs is a true business law genius with an understanding of both business law and technology. He holds degrees in engineering and law and has authored of a multivolume comprehensive treatise on business law. I have had the good fortune to enjoy a continued friendship with Arnie over the years. He is known as the "dean of securities law" by Crain's New York Business. He took Donna Karan International and Bear Sterns public, defended William Casey in a securities fraud case, when

Casey was head of the CIA, in a securities fraud matter. He represented the biotech company Celgene in a more than $1 billion equity public offering and set a world record by writing a law review article with 4,824 footnotes. Today, he advises my son, Jeffrey, in his business endeavors.

Lehman Brothers, our lead investment banker organized numerous road show meetings with potential investors in major U.S. cities and European capitals, including with Atlanta, Boston, Chicago, Los Angeles, Miami, New York, and San Francisco along with European Trips to Dublin, Edinburgh, Frankfurt, London, Munich, and Paris. We were much like an athletic team with events in multiple cities. Truly we were on a treadmill with a blur airports, air-train travel, hotels and limos. We participated in two or more road shows each day along with assorted breakfast, lunch, and dinner meetings. I marveled at the expertise of Lehman Brothers in processing the investor subscriptions and the launch of the IPO. When, the big day finally arrived, it seemed anticlimactic. All of the key people that participated in making the IPO happen joined the DNAP management at Club 21 in New York for the victory celebration.

DNAP was slowly maturing into the top echelon of agricultural biotechnology companies in the development of plant-based products for consumers and industry. The Company applied its pioneering work in plant cellular genetics, to create new plant varieties with characteristics desired by consumers, food processors, and industrial users and did so in less time than could be achieved by traditional breeding methods or by molecular genetics alone. The research and development focus was on products at the high end of the agricultural chain. For instance, the Company was involved in developing tomato cultivars with improved eating or processing characteristics, rather than tomato seed with improved agronomic properties. The Company identified such properties by conducting preliminary market research and analyzing its technological ability to develop new or improved products to satisfy the needs of the market place. The principal considerations in choosing product candidates for commercialization were market size, competition, and the length of time until revenues could be generated. Of course another requirement was the availability of the funds for successful commercialization. The lead products in development and near production at the time were branded vegetables such as carrots and celery to be marketed under the name VegiSnax, agricultural disease diagnostic kits, enhanced processing and fresh market tomatoes, improved consumer traits in coffee and popcorn, polyunsaturated cooking oils from canola, and enhanced plant derived raw materials for the flavor and fragrance industry.

A few unlikely challenges arose along the way. Production of the biotechnology-derived tomatoes were placed in a major field trial in Mexico with a contract farmer frequently used by the Campbell Soup Company and under the supervision of a Mexican Ph.D. plant breeder graduate from the University of California Davis. Our research team went to evaluate the tomatoes and discovered that the tomato fields had been plowed under by the farmer and that our supervising breeder consultant had been hired the week before by the Campbell Soup Company Research Station in Davis California. Neither the consultant nor the Campbell Soup Company had informed us of the hiring. For the record, the director of the Campbell Research Station in Davis had been a plant breeding professor at the University of California Davis. Our collaborators at the Campbell Research Station in Davis, CA and the consultant denied any knowledge of the matter. The farmer informed our team that he was ordered to plow under the tomatoes by an official who he assumed was affiliated with DNA Plant Technology Corporation.

The situation was sensitive because CSC had a significant equity position in the company at the time. Word of the incident would have been devastating to a public traded biotechnology company. It is amazing what actions are taken by individuals within the corporate world who view advances in science and technology as threat to their corporate existence and livelihoods.

But commercial success takes more than good science to complement the Company's scientific capabilities with manufacturing, marketing and distribution resources of major consumer and industrial companies. The Company entered into an array of joint venture and licensing agreements with such companies as American Home Products Corporation, Brown & Williamson Tobacco Company, Campbell Soup Company, CPC International Inc., E.I. du Pont Nemours and Company, Firmenich Incorporated, General Foods Corporation, Hershey Foods Corporation, Koppers Company, Inc., and Kraft, Inc.

In 1988, DNAP merged with Advanced Genetic Sciences, Inc. (AGS), an important agricultural molecular biology company to become the largest public plant biotechnology company. AGS was under the leadership of Joseph Bouckaert, the President and Chief Executive Officer, and John Bedbrook, Scientific Director. The new company was incorporated into DNAP Plant Technology Corporation under the leadership of Richard Laster and Joseph Bouckaert resigned and the boards of the two companies merged.

In the later DNAP years, a collaborative opportunity developed through Diyoung Wang, a DNAP scientist, Chinese Citizen, and Green Card holder. The opportunity was for the evaluation of Chinese traditional medicinal plants for pharmaceutical active compounds and the subsequent development of ethical drugs. We visited China and entered into collaborations with the Chinese Academy of Sciences (CAS) and two CAS Institutes: Shanghai Institute of Materia Medica and the Beijing Institute of Botany. Prior to joining DNAP, Diyoung had been a senior scientist at the Institute of Genetics a member organization of the Chinese Academy of Sciences. Diyoung and I decided to resign from

DNAP because the consolidation of the two merged companies in Cinnaminson and Oakland did not allow time for consideration of another investment opportunity. DNAP had capable leadership, which made me feel comfortable to explore new venture opportunities. I always made certain that a number two person was in place to assume my position. Diyoung, three former executives from the Ciba-Geigy Company, including the former president of the U.S. operations, joined forces with me in the founding DNA Pharmaceuticals Inc.

The DNAP operations were eventually consolidated and located in Oakland, California. The company was later acquired by the Pulsar International SA (Pulsar), a Mexican company. Pulsar made a decision to divest their tobacco business and invest in agricultural biotechnology leveraging their management expertise in agriculture. Subsequently, in addition to DNAP, Pulsar acquired a collection of vegetable seed companies under the umbrella of the Seminis Seed Company. Seminis is the world's largest development, production and marketing vegetable seed company. Their hybrids improve nutritional quality, increase yields, limit losses from pests and reduce the need for agrochemicals. In 2005, Seminis became a wholly owned subsidiary of Monsanto Company.

New Opportunities

Chinese Academy of Sciences, Beijing Institute of Botany; Institute of Materia Medica; Chengdu Institute of Biology, DNA Pharmaceuticals Inc. & Phytopharmaceuticals

One of the exciting corporate career moments was the opportunity to collaborate with the Chinese Academy of Sciences (CAS) and several of the CAS research institutes. A few words about CAS, Retrieved August 5, 2013 from The **Chinese Academy of Sciences (CAS)**, formerly known as **Academia Sinica**, is the national academy for the natural sciences of the People's Republic of

China. It is an institution of the State Council of China, functioning as the national scientific think-tank, providing advisory and appraisal services on issues stemming from the national economy, social development, and science and technology progress. It is headquartered in Beijing, with institutes all over the People's Republic of China. It has also created hundreds of commercial enterprises, Lenovo being one of them.

Diyoung Wang and I visited China on several occasions and met with Dr. Zhou Guangzhao, the President and Dr. Hu Hesheng, the Executive Vice President of the Chinese Academy of Sciences (CAS) and the Shanghai Institute of Materia Medica and Beijing Institute of Botany. The meeting at the headquarters of the Chinese Academy of Sciences occurred in a formal palace-like chamber furnished with large sitting areas in which the seating was arranged in a hierarchical pattern with Drs. Hu and Zhou seated in the center and the guests of honor on either side. We were served tea and biscuits.

Our conversation was relaxed, doubtless, because of our mutual backgrounds in academia and histories. We introduced one another with a brief account of our academic disciplines. Dr. Zhou was a world class physicist who escaped death during a visit to Moscow with two fellow leading nuclear physicists from China during the Nikita Khrushchev regime in the USSR. President Khrushchev told Chairman Mao that the USSR would assume leadership in nuclear energy research for both China and the USSR and invited three leading CAS physicists to Moscow to initiate collaboration. During the meetings, disagreement arose over the USSR's assumption of leadership of the Chinese nuclear energy research and development program. Dr. Zhou decided to return to China a day earlier by train to attend a meeting in a remote part of China while his companions returned the by plane. The plane exploded shortly after takeoff, killing the two Chinese nuclear scientists. Dr. Zhou, was safely in Chinese territory at the time of the explosion. Dr. Hu Hesheng was a renowned mathematician and former president of the Shanghai Mathematical Society who specialized in differential geometry. She led a group at Fudan University in the 1980s and 1990s. Her husband, Gu Chaohao, also a renowned mathematician, served as president of the University of Science and Technology of China.

Drs. Hu and Zhou agreed to consider participation of the two institutes of the Chinese Academy of Sciences (Chinese Institute of Materia Medica in Shanghai and the Chinese Institute of Botany in Beijing) in a collaborative research program with DNA Pharmaceuticals Corporation to evaluate botanical materials used in traditional medicines in pharmacological screens. The proposed collaborative relationship was to include an equity position for CAS in DNA Pharmaceuticals Inc. Dr. Hu agreed to assume leadership in negotiation of the proposed collaborative research and equity position. CAS after completion of negotiations acquired a significant equity stake and a Board of Directors seat.

Our company engaged consultants Professor Douglas Davidson from McMaster University and Dr. Bryce Douglas, former Vice President of Research for Smith Kline Corporation in Philadelphia. Dr. Douglas led a major research operation at Smith Kline Corporation for evaluating botanical materials for drug development. We next had the good fortune to hire a senior business development and research scientist from the neurobiology unit at Ciba-Geigy because of shuttering the U.S. neurobiology unit and consolidation of the unit with the corporate research headquarters in Switzerland. We now had in place, a team to develop a business plan and specific collaborative research programs with the Chinese Academy of Science and the two institutes.

DNA Pharmaceuticals Inc. was founded in 1988 in partnership with the Chinese Academy of Sciences, which held a significant equity position and board representation. Don MacKinnon, who had just retired as president of Ciba-Geigy U.S.A., was appointed to the CEO position. We held a celebration marking the launch of DNA Pharmaceuticals Inc. at the "Windows on the World" at the World Trade Center in New York attended by CAS President Zhou Guangzhao, CAS Executive Vice President Hu Hesheng, the DNA Pharmaceuticals Inc. officers and board of directors. Our offices were originally located in Cinnaminson, N.J. and later relocated to Stamford, Connecticut, in partnership with VimRx Corporation. Don Mackinnon was a member of the VimRx Corporation board. A possible merger of was being discussed when VimRx invited me to accept a senior position at the company to develop research relationships in China instead of merging the two companies. I was uneasy about possible legal implications regarding other employees at DNA Pharmaceuticals Inc., because of my being an employee and officer of DNA Pharmaceuticals Inc.

The Company's ongoing relationship in China allowed for collaborative sponsored research between DNA Pharmaceuticals, Inc., and the institutes of CAS. Genentech proposed forging a relationship with DNA Pharmaceuticals, Inc., that would allow investment and collaborative research, and relocate DNA Pharmaceuticals, Inc., to Genentech's headquarters in South San Francisco. The opportunity created excitement in the venture capital community. However, enthusiasm abruptly ended with the Tiananmen Square protests on April 15, 1989, and plans for collaboration and for financing by the venture capital community were cancelled. We scheduled a meeting in China following Tiananmen Square to request approval of merging the company with another company or selling the assets. The Chinese colleagues were in agreement. Xenova Ltd, Slough, United Kingdom entertained a possible merger with DNA Pharmaceuticals Inc. based on an earn-out proposition on achievement of successful clinical trials and commercialization of compounds in the Company's pipeline. After rejection of the offer, Xenova offered me a senior position as opposed to merger of the two companies. Don Mackinnon and I had a number of meetings with the executive team. Once again, I was worried about possible litigation in regard to closing on such a proposition because of my equity and officer role although I did like the company.

Ciba Geigy purchased the assets of DNA Pharmaceuticals, Inc. including the collaborative research agreements with the Chinese Academy of Sciences and the two institutes. The transaction allowed for payment of creditors and distribution of monies to the shareholders. The successive collaborative relationship with the two institutes of the Chinese Academy of Sciences evolved and continued through the merger of Ciba Geigy and Sandoz which resulted in the creation of Novartis, a 30 billion dollar pharmaceutical company. Subsequently, Novartis invested $100 million dollars in the CAS research collaborations during 2006 and more recently over $1 Billion Dollar investment in research facilities during 2009. DNA Pharmaceuticals, Inc. was dissolved the early 1990s, and I was uncertain in regard to the next opportunity. Or perhaps this had been the last dance for me.

Off to California
Chinese Academy of Sciences; CEBTEC
Biotechnology Center; PhytoPharmaceuticals Inc.

In August 1991, I received a phone call from a senior vice president of ESCAgenetics Corporation, San Carlos, California, about my potential interest in assuming leadership for a research and development project for the anticancer drug Taxol from *Taxus* cell cultures grown in bioreactors. This inquiry led to a trip to San Carlos in late 1991 during which I was offered an opportunity to launch PhytoPharmaceuticals, Inc., a spin-off company from ESCAgenetics Corporation to development plant based pharmaceuticals.

My son, Jeff, who at the time was located in Santa Monica, Los Angeles, was engaged in a San Francisco project mapping locations for a film project. He assisted me in finding a beautiful residence on California Street overlooking Chinatown and the magnificent Embarcadero. I moved to San Francisco during the morning hours of January of 1992 to assume responsibilities related to the launch of PhytoPharmaceuticals Inc., the drive to San Carlos from the San Francisco International Airport was about 25 – 30 minutes.

Life was hectic with the recruitment of a management and technical team to complement the scientific team already in place at ESCAgenetics. Dr. Tom Glenn was appointed to the CEO position because of his business and research leadership background. He had served as Vice President of Research for both Ciba-Geigy Corporation and Genentech and held CEO positions at several start-up pharmaceutical companies. Peter Hylands, a former biochemist at the London Kings College of Pharmacy was hired as the Chief Scientist. Tom Glenn was amazing. His work hours commenced at 4:30 A.M., requiring a change in my work schedule and the resetting of my biological clock, which remains unchanged to this day. I usually left my residence on California Street at 4:00 A.M., which allowed me to meet with Tom prior to the employee foot traffic, meetings and ringing telephones. I enjoyed the solitude of the early hours and the ability to organize and focus my thoughts on the projects of the day. The company negotiated collaborative agreements with the Shanghai Institute of Materia Medica, Beijing Institute of Botany and the Chengdu Institute of Biology (all members of the Chinese Academy of Sciences) to screen extracts from traditional medicines. A parallel agreement was negotiated with the Biotech Center, ESALQ-University of Sao Paulo, Piracicaba, Brazil, with Professor Otto J. Crocomo.

Our talented chemical engineering team successfully developed a bench-top Taxol bioreactor process for Taxol production and was awarded a substantial grant from the National Institutes of Health (NIH). The Company entered into a collaborative relationship with Genencor International for the scale-up and fine-tuning of the production process at that company's Rochester, New York, facility. Roy Stalhut, a senior scientist, was reassigned to Rochester and I assumed responsibility for managing the project, requiring me to attend management meetings in Rochester on Monday mornings and to return to San Francisco on Monday afternoons. My usual itinerary was a red eye flight on Friday to New York to visit family over the weekend and travel to Rochester on Monday mornings. Then Samyang Genex Corporation., Ltd., invested in a collaborative ethical drug development program with PhytoPharmaceuticals and the Taxol production process was transferred to Seoul, Korea.

Return to Academia
Rutgers University

In early 1992, a former colleague called to urge me to consider the position of Dean of Research at Cook College and Director of the New Jersey Experiment Station at Rutgers University. I politely responded that the opportunity was intriguing, but the timing unfavorable. A year and one phone call later, I agreed to meeting in New York with a member of the Rutgers search team. I was invited and agreed to visit Rutgers for two days for a seminar presentation and meetings with faculty groups and administrators. I later discovered that seven candidates were to be interviewed for the position. I researched the position and information about the college and prepared a seminar based on biodiversity and the prospect for leadership at Rutgers. My visit was hosted by Richard Lutz, a leading marine biologist known for pioneering research on the hydrothermal vents, and Chair of the Search Committee. The two of us had much to discuss because of our shared interest in biodiversity and secondary product research for discovery of lead compounds for ethical drug development. The visit was enjoyable, and I returned to San Carlos with no follow-up.

Daryl Lund, the Executive Dean, phoned me during late August to inform me that the search committee had decided to appoint me to the position of Dean of Research and Director of the Experiment Station. I requested a week to think over the opportunity, and to discuss it with my associates. ESCAgenetics at the time was in discussions with Pulsar SA, a Mexican company. I realized that my contribution to the launch of PhytoPharmaceuticals had been completed and a management team was now in place. My equity position in the company allowed me to enjoy the upside of the company's future success. I informed Ray Moshy, CEO, ESCAgenetics, Inc., of my decision to return to academia and willingness to continue a relationship with the company as a consultant. He endorsed the plan.

The following weeks were hectic. I was busy packing, arranging the move and transportation of my furniture, books, files and automobile to the east coast. I had found beautiful living quarters in Highland Park, a lovely suburb of New Brunswick, within walking distance to three of the New Brunswick campuses, including the Cook Campus College. My office suite was located on the first floor of Martin Hall, the administration building that had formerly housed the laboratories of Selman Waksman, the Nobel Laureate and his graduate student, Albert Schatz, the co-discovers of streptomycin.

With a faculty and staff of 350 and a student body of 3,500 undergraduates and about 800 graduate students, the opportunity at Rutgers Cook College was both challenging and exciting. The total student population of the university undergraduate population was about 40,000 and total university student population was over 52,000. The academic departments at Cook College during my time consisted of Agricultural, Food and Resource Economics, Ecology, Evolution and Natural Resources,

Animal Science, Biochemistry and Microbiology, Entomology, Environmental Sciences, Food Science, Landscape Architecture, Marine and Coastal Sciences, Meteorology, Natural Resources, Nutritional Sciences, and Plant Biology and Pathology. The departments were complemented by several research centers and institutes including the Center for Advanced Food Technology, the Agricultural Biotech Center, the Center for Environmental Prediction, Center for Marine Biology, Center for Remote Sensing, Center for Turfgrass Science, The EcoComplex, The Equine Science Center, The Institute for Marine and Coastal Sciences, and the IR-4 Project.

I committed the first few months to on-site visits with faculty members and developing an understanding of their research and teaching programs. It was crucial to devise a strategy for our office to understand how to catalyze the funding of faculty research and teaching efforts. I must say, that these two months were among the most memorable of my lifetime in providing an eagle eye view of academic research and teaching programs. It was an opportunity to envision interdepartmental, government, and industry collaborative opportunities on the Cook Campus and departments and research centers on other Rutgers campuses in New Brunswick, Newark and Camden and the sister academic institutions

Rutgers is unique in having more than 200 research centers on five campuses in New Brunswick and additional campuses in Camden and Newark. Our research leadership team in Martin Hall served as a catalyst for forming faculty teams for collaborative grant opportunities. The resultant grant awards led to creation of a number of new research centers, including, the EcoComplex, a collaborative research initiative between the Burlington County Resource Recovery Facility, the New Jersey Department of Environmental Protection, the Rutgers Cooperative Extension, with the Rutgers and Stevens Institute of Technology Faculty; NASA, and the New Jersey Department of

Commerce. Other successful outcomes included the Biodiversity Center, the Nutraceuticals Institute, the Center for Environmental Prediction, the Center for Remote Sensing, the Equine Research Center and the Food Policy Institute.

A number of global research programs were launched during this period, including a natural products collaborative research program with the world renowned scientist Monroe Wall. Our team organized the Monroe Wall Research Colloquium Series in New Brunswick and Venezuela. The Tripartite Biotechnology Collaborative Research and Teaching Program developed between, Rutgers, Ohio State University and the University of Sao Paulo was formed. The Program provided a new and innovative model for undergraduate, graduate, faculty research and teaching exchanges. The colleagues recently celebrated a fourteen year anniversary during June 2012. Other overseas programs were initiated in China and Mexico.

Looking back on my life, the five years at Rutgers were among the highlights of my career. Nonetheless, I decided to step down and turn the responsibilities over to a younger person and to pursue other interests in the biotechnology sector. The pace at Rutgers required appearances at meetings during the week and weekends in New Brunswick and numerous award dinners at statewide venues. A body can stand only so much rubber chicken. Three years into my tenure at Rutgers, Daryl Lund, the Executive Dean, accepted a parallel position at Cornell, which precipitated successive changes in the leadership at Rutgers. This was a playback consistent with other experiences at Case Western Reserve University, Ohio State University and the Campbell Soup Company. In all cases, my immediate supervisor departed early in my employment, creating new organizational and programmatic challenges. I was asked to serve as the interim executive dean after Daryl Lund's departure, but I turned down the offer. My hearing disability would have compromised my performance and made the position too demanding. The Executive Dean must make regular appearances at the State Board of Agriculture and the General Assembly to review the budget for the New Jersey Agricultural Experiment Station. Afterwards, Cook College was led by an acting dean, who was followed by an older non-tenured administrative.

I took stock of the situation and came to grips with the fact that the first five years of my tenure were coming to a close. The enhancements to the research establishment that I had planned had for the most part been accomplished under my leadership. Moreover, our team consisted of a group of capable young individuals who were destined to move on to positions with greater responsibility. I made the decision to resign and was offered a consultancy that afforded continued involvement with the W.K. Kellogg Mid-Atlantic University Collaboration, a successful collaborative research and teaching program among the mid-Atlantic universities and with several international program initiatives. The international programs included the Rutgers/Ohio State/University of Sao Paulo Tripartite Research and Teaching Collaboration and the Global Institute for Bio-exploration (GIBE). The Tripartite program was most innovative and supported annual undergraduate research scholars participating in one another's annual undergraduate research symposia that involved a one-week campus visit and presentation of research programs at poster sessions, lectures or round tables. The Tripartite program also encompassed collaborative faculty research, an undergraduate foreign exchange and a double diploma Ph.D. program, involving faculty on two campuses. GIBE program sponsored travel and participation in workshops in Botswana, Ghana, Nigeria, and South Africa and at Rutgers. The research and teaching efforts of the program were focused on the exploration of natural products from native flora for the development of dietary supplements, functional foods, nutraceuticals and ethical drugs. The program was an extension from a successful collaborative NIH grant held by Dr. Ilya Raskin involving a group of countries in Central Asia.

Now What?

Prior to stepping down from the deanship at Rutgers and vacating the Highland Park apartment, a young Wall Street trader offered to purchase my New York cooperative where I had intended to relocate. At first, I refused, but the trader continued to increase the offer until I finally agreed to sell. During times of economic downturn, the sale of cooperatives in New York is quite difficult. The ability to consummate a sales transaction is further complicated by the due diligence and interview process of the Cooperative Board which is quite thorough and involves review of the financial and personal matters of potential buyers. After purchase of a cooperative, the buyer is bound by a complex set of rules pertaining to limitations on renter lease agreements, restrictions on visitors, pets, deliveries, recreational and swimming pool regulations, limitations on redecorating without engineering blueprints and Cooperative Board approval. The buyer was interested in combining three units to create space for his family of four, which required engineering plans and approval by the Cooperative Board.

After the sale of the New York cooperative, I decided to purchase an ocean-front condominium in North Beach of Miami Beach on a side-street removed from the traffic congested Collins Avenue and located between Bal Habour and South Beach. The location seemed like a perfect setting for retirement and nurturing new hobbies and interests. I enthusiastically settled in the Miami Beach dwelling and enjoyed getting to know the neighborhood, greater Miami Beach and the city of Miami.

About the same time, I was contacted by a group of former colleagues about forming a LLC (Limited Liability Corporation) in New Jersey for the purpose of assisting start-up technology companies and the exploring opportunities for spin-off company opportunities from the research universities. The truth was that I found the thought of retirement quite boring and the pace of Miami Beach was too slow for me.

I relocated to New York with a move to a pied-a-terre in a building in the theater district that was a stone's throw away from the Port Authority Bus Station and Penn Station. I brought together a group of individuals with relevant experience to explore the LLC opportunity: a small business owner and former president of the New Jersey Farm Bureau, a lawyer and scientist with a background in biotechnology and senior licensing positions at research universities, and a CPA, who owned an accounting business in New Jersey. The CPA possessed extensive experience working with start-up technology companies. Gladstone New York Partners LLC (GNP) was successfully formed along with start-up ventures. New ventures included MedTower Inc., A Sickle Cell Anemia Company, the Norman Borlaug University, Jersey Flora, Inc., and Wellgen, Inc.

One of the interesting opportunities that arose developed by a young medical doctor and graduate of Cook College who had assembled a team of medical doctors and engineers to create MedTower Inc., a start-up company based on the idea of incorporating a medical school its teaching hospitals and medical practitioners into a virtual medical tower that would use a website to make medical services more accessible. I visited the company and met with the founders. They were housed in a two-room office on the second floor of a walk-up building in Summit, New Jersey. One could not but be impressed by the passion and commitment of the founders who, through the use of sleeping bags on their office floor, worked around the clock,

The team talked me into joining the board of directors to assist in forming collaborations with the pharmaceutical industry and academic medical institutions. Within, the next few months, we inked a collaborative agreement with UMDNJ, the University of Medicine and Dentistry of New Jersey through the office of the research vice president. The relationship evolved and we had the good fortune to engage the former president of the medical school as chair of the Company. The triumph was followed by expansion of the board to include a world distinguished scientist and former research executive of several leading pharmaceutical and medical biotechnology companies and a professor from the Tuck School at Dartmouth College. The Company had a momentum and soon developed an important collaborative research program with a Swiss pharmaceutical company and a contract with the New Jersey Department of Health to development and management of The New Jersey Homeland Security Website. Incorporation of medical practitioners and various medical services of UMDNJ into the Company's proprietary virtual medical tower were being negotiated. In tandem with Medtower Inc., another UMDNJ-based start-up was launched with support of a grant from the New Jersey Commission of Science and Technology to conduct trials on two experimental drugs for treatment of Sickle Cell Anemia. The experimental drugs were licensed from the College of Medicine of Southern Alabama University. The financial agreements required UMDNJ to manage the financial affairs of the two companies.

At the time the two companies were founded, the Governor was pushing consolidation of UMDNJ, the New Jersey Institute of Engineering (NJIT), a state-supported engineering university located in Newark, with the Rutgers University System. The plan called for installation of a chancellor and bureaucratic organization to manage the complex structure of educational research and teaching institutions. There was backlash from UMDNJ and Rutgers. Rutgers, which was originally a private university, is managed by a board of governors and a board of trustees. The later control matters pertaining to the grounds and buildings of the university and a number of the endowment funds, whereas the board of governors has responsibility for operating the institution. The Governor appoints the majority of the members of the board of governors, but has no control over the board of trustees. This arcane political organization provides a unique autonomy to Rutgers. The office of the Governor allegedly dismissed several dissenting UMDNJ executives and punished the chairmen of MedTower and the Sickle Cell Anemia start-up company by freezing their funds. It was a sad end to two promising medical technology companies.

I recently served as the principal investigator on two USDA SBIR grant awards involving the School of Biological and Environmental Sciences, formerly Cook College, the School of Engineering, and the School of Pharmacy. I continue to participate in the successful University of Sao Paulo/Rutgers/Ohio State University Tripartite Collaborative Research and Education Program(s). The program recently marked its fourteenth year anniversary which is quite unique among global research and teaching institutions.

Life now involves spending time together with grandson Jack, niece Emily an nephew Andrew. The special moments with the three of them are now an important part of my life. Correspondence continues to occur from former colleagues and students and a few requests for letters of reference. I guess my task at hand is to do my best to assist the next generation and possibly contribute in a small way to their success in this global world.

Otto Crocomo and Julius Kreier, colleagues respectively from the University of Sao Paulo, Ohio State University and Rutgers University are collaborating with me in editing an e-book series entitled: "Reflections & Connections". The book series includes about thirty contributors reflecting on their academic and/or corporate careers in the life sciences and the lessons learned. Our hope is that the lessons learned will be of value to students considering a career in the life sciences and those in the early stages of their careers. The casual reader may be interested in the making of scientists and their career navigational charts.

The most important lesson learned from my career was the value of collaboration at the departmental level, university level, and among universities for the development and delivery of programmatic initiatives. The same approach works between academia, government and industry. University research laboratories and research centers have the luxury of pursuing fundamental research supported by government granting agencies and foundations. Funding is available from multiple sources for the commercialization of academic research. Important sources of such funding include the state commissions on science and technology and the federal government small business research innovation awards which encourage partnerships between university research laboratories and the private sector.

The university research foundations and technology transfer offices play an important role in transfer of technology to the private sector and enjoy the upside potential of licensing fees and royalty streams. Universities, state government and industry have the potential to form key partnerships in providing students with valuable skill sets for becoming important participants in the private technology sector. These skill sets include knowledge of business law, regulatory affairs, research organization and marketing. A few academic institutions have set the standard for educating science entrepreneurs, including Columbia University, Dartmouth College, and Stanford. Dartmouth College offers an entrepreneurship classroom program for faculty and students that is filled to capacity every semester.

Looking Forward and Backwards

The journey has indeed been fulfilling. I feel fortunate to have had the good fortune of being raised in a household of loving parents, siblings and grandparents followed by a tour of duty with the U.S. Navy, guidance by university remarkable mentors and opportunities for employment in world class academic institutions in Brazil and the United States, a blue chip corporation, start-up technology companies and finally a return to the academy in an administrative post.

The diversity of experiences has prompted me to reflect on the lessons learned. These lessons have brought clarity to the importance of globalization and collaboration among scientists employed by the academy and private sector institutions. Alexander Von Humboldt 1769-1859 communicated the importance globalization and collaboration during his extensive travels to South America more than 200 years ago which his colleagues preserved in the founding of the Humboldt Foundation,

Important that the academy, the private sector and state and federal government understand that science and technology is an important driver of the global economy. Partnerships between these entities are important in providing financial resources for basic and applied research, scientists and students. This support is critical in allowing advances in science and technology along with the education of students with backgrounds in STEM (Science, Technology, Engineering and Mathematics). Federal immigration policy is another important factor ensuring that the human capital needs of academia, government and industry are satisfied. The best minds graduating from the research universities should be allowed by immigration services to adjust and seek employment (*Immigration and the Economy*, William R. Sharp, *Huffington Post*, The Blog, 04/08/2013)

Students with a STEM education are superbly prepared to contribute to the process of innovation and/or teaching future generations of students. STEM educated students are much like animal and plant stem cells which are pluripotent.

Photo 1. Family Portrait (Backrow - Left to Right) Rod, Mother, Sister Patricia, Great Grandmother Hale, Father, (Front Row – Left to Right) Brother Bill and Sister Sally.

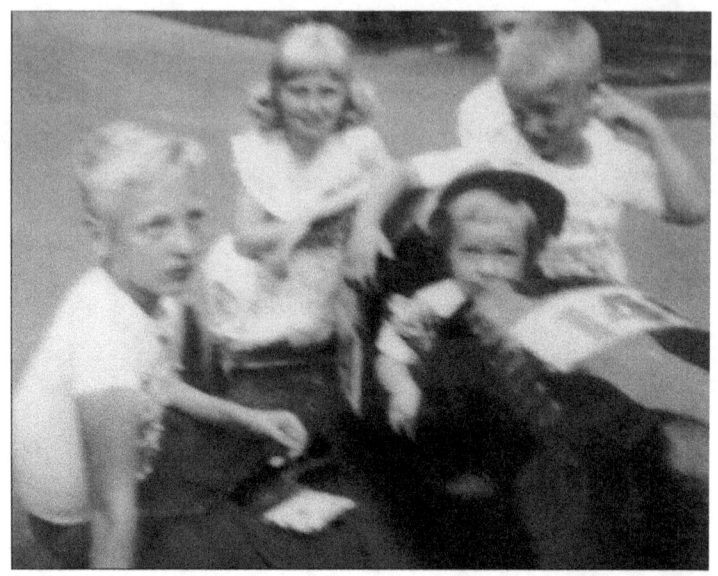

Photo 2. Another summer of Construction in the Basement with Dad Building a Soap Box Derby. (Left to Right) Rod, Sister Patricia, Brother Bill and (Seated in Race Car) Sister Sally.

Photo 3. USS Compton DD 705 Sailing in the Mediterranean Sea.

Photo 4. Syllabus prepared by Linda Styer Caldas, Otto J. Crocomo and William R. Sharp for Organization of American States Plant Cell and Tissue Culture Training Course

At Centro de Energia Nuclear na Agricultura, University of Sao Paulo, ESALQ Campus, Piracicaba, Sao Paulo during July 1973. Twenty Four Students Attended Orignating from Seven Countries.

Photo 5. Rod Sharp Recipient of the
1999 Medalha Luiz de Queiroz

Photo 6. DNA Plant Technology Corporation
Headquarters, Cinnaminson, New Jersey

Photo 7. DNA Plant Technology Corporation
Board of Directors Meeting.

(Seated Right to Left) John Connor, Chairman DNA
Plant Technology Corporation and Chairman Schroders
Incorporated, Richard Laster, President and CEO DNA
Plant Technology Corporation, Evelyn Berezin, President
Greenhouse Manangement Corporation, William Sharp,
Executive Vice President and Scientific Director DNA Plant
Technology Corporation, Henry Roberts, Former President
and Chairman Conneticut General Insurance Corporation,
David Blech, Chairman Technoven Corporation, (Standing
Right to Left) Isaac Blech, President Technoven Corporation,
Louis Basel, President and Director, Crawford and Russel
Corporation, Douglas Luke, Vice President, Rothchild Inc.,
Jeffrey Collinson, Chairman, J. Henry Schroder Corporation,
Bernard Meislin, Secretary, DNA Plant Technology Corporation,
Otto Wheeley, Vice Chairman, Koppers Company, Inc.,
Stephen Petschek, President, Amcon Group, Gordon
McGovern, President and CEO, Campbell Soup Company.

Photo 8. Kitty Evans (Wife of David Evans), Jeffrey Sharp and Rod Sharp Celebrating the DNA Plant Technology Corporation Initial Public Offering Onboard the New York DNAP Yacht Party.

Photo 9. (Left to Right) Rod Sharp, Norman Borlaug, Nobel Prize Laureate and Chairman of the DNA Plant Technology Scientific Advisory Board, and David Evans Discussing Field Trials.

Photo 10. DNA Plant Technology Corporation's Fresh World, a Pioneering Company, in the Introduction of Snacking Vegetable Products with "VegiSnax".

2 SEPTEMBER 1983 · VOL. 221 · NO. 4614 $2.50

SCIENCE

AMERICAN ASSOCIATION FOR THE ADVANCEMENT OF SCIENCE

Photo 11. Cover Article Science Magazine 2 September 1973 Entitled: Single Gene Mutations in Tomato Plants Regenerated from Tissue Culture, Authored by David A. Evans and William R. Sharp.

Photo 12. Celebrating Book Number Nine with the Completion of "Reflections & Connections ". And the Opportunity to Collaborate with Co-editors Otto J. Crocomo, Julius P. Kreier and Colleagues and Friends Contributing Manuscripts. Special Thanks to Julius P. Kreier for Encouraging the Publication of Books.

Photo 13. (Left to Right) Rod Sharp and David Evans Evaluating Somaclone Derived Tomato Plants at DNA Plant Technology Corporate Farm at Rancocoas, New Jersey.

Photo 14. Dr. James E. Gunckel, Rod Sharp's
Graduate School Mentor, Evaluating Somaclone
Coffee Trees at the DNA Plant Technology Corporation
Greenhouse Complex, Cinnaminson, New Jersey.

Figure 15. December 2007 Commencement Ceremony,
(Left to Right) William R. Sharp and Gordon Gee,
President, Ohio State University. Rod Sharp Recipient of
the Distinguished Board of Trustees Service Award.

Figure 16. 2007 Ohio State University Board
of Trustees Distinguished Award

Photo 17. Grandson Jack Exploring the World of Microbes.

Photo 18. Mary Garrison, My Maternal
Grandmother and My Angel.

Photo 19. (Left to Right) Niece Emily and
Nephew Andrew on Science Adventure at the
New York Hall of Science with Uncle Rod.

CHAPTER 15.

Growing up in Dad's Laboratory

Jeffrey Sharp

I've watched REFLECTIONS AND CONNECTIONS take shape with a sense of enormous excitement as my father, Rod Sharp, and his colleagues Otto Crocomo and Julius Kreier, have assembled more thirty contributors to create a work that will serve to define and honor a remarkable community of scientists from around the world and certainly inspire the next generation to come.

As the multi-year project was nearing completion, my father suggested that I contribute a few words to be included in the book. I was honored to take him up on the invitation as I felt that I might be able to share what it was like to grow up the son of a much-loved and respected scientist within this extraordinary world of scientists.

I have known most of the contributors of this book by their first names since I was a small child growing up in my father's laboratory at the Biological Sciences building at The Ohio State University and in later years, DNA Plant Technology, University of Sao Paulo Piracicaba, and Rutgers University. My father's colleagues were like surrogate parents, uncles and aunts; their children like brothers and sisters and his graduate students were my heroes

From the very beginning, Jim Gunckel and his wife Jean were kind of surrogate parents for our entire family. Jim had convinced Dad to make the move from Ohio to New Brunswick New Jersey for a NASA Research Fellowship at Rutgers University. And in bringing Dad, Mom and (eventually) Jeff to New Jersey, the Gunckel's took a real responsibility in looking after us. Their daughter Nancy and their son Fred were part of our family as was the Gunckel dog Gretchen. Jim remained my father's mentor for the rest of his life and the Gunckels - Nancy, husband Alan and sons Jeffrey and Matt, continue to be an important part of our extended family.

Julius Kreier was not just the brilliant doctor of veterinary medicine and medical microbiologist who specialized in parasitology but the kind and gentle man who welcomed me to his lab to visit his animals – pigs, sheep, mice, frogs and guinea pigs every Saturday morning. Through Julius, I had my own personal petting zoo. And throughout my childhood, Julius and his wonderful wife Ruth and their children Rachel and Jessie, were important members of our extended family with lunches and dinners at home and summer fishing trips at their favorite spot on the Jersey Shore. Rachel and Jessie were so kind to me, despite a sizable difference in age, and I imagined that they were both rock stars who led exciting and fabulous lives.

Otto Crocomo was not just the pioneering founder and Director of CEBTEC – The Center for Biotechnology but the warm-hearted uncle who encouraged me to learn Portuguese and get to know and love Brazil. Otto and his beautiful wife Diva and their children Adolfo Egidio, Maria Paula, Marco Augusto, Carla Maisa, Daniel and Marco Augusto, became family and best friends. We shared beautiful dinners together that lasted well into the night beneath the Flamboyant tree at Restaurante

Flamboyant, elegant lunches by the river at Restaurante Mirante with magical adventures traveling together around Sao Paulo state. Marco Augusto and I bonded over our mutual love of music and we often composed songs together late into the night. We often shared cassette tapes in the mail when we weren't together. I'll always remember those happy times playing and singing into the night making Diva's dogs bark and howl.

The fact that I spent so much time in my father's lab was largely circumstantial, my parents divorced when I was four years old and the vagaries of child custody often meant that Dad had no choice other than to leave me to my own devices while he caught up on his heavy work load over the weekends. I relished the independence that these Saturdays offered up as I conducted "science experiments" by mixing up potions in various beakers and test tubes – distilled water and masking tape was one of my favorite recipes. I made frequent trips to the building's vending machines for candy and soda and wrote newsletters about the various comings and goings starring my father's colleagues and graduate students. I remember Rosa Raskin, dad's first graduate student, followed by John Peters, Maro Sondahl, Natal Goncalves, David Evans, Marinez Alves de Lima, Willie Loh, Sharon Maraffa, Bob Reisner, Wei-Shen Shen, and Karen Sommers. The laboratory was shared with Donald Dougall and his students: Henrique Amorim, Ruy Caldas, Linda Styer Caldas and Harry Sommer.

At the core of this unique family was my Great Grandmother, Mary Garrison, or Momo as she was known to most of us. Momo would often host dinner parties at her elegantly set table in Columbus for Dad's colleagues and students. This is where I met Henrique Amorim and his beautiful wife Vera. She delighted in getting to know them and their families and her food was always delicious. And she travelled to Brazil where she was much admired for her loveliness and wisdom.

It certainly wasn't an ordinary childhood. Most kids my age would have been on a playground, taking swim classes at the YMCA, fishing, hiking, or going to the shopping mall and not sitting in a windowless laboratory on a college campus without other children in sight. However, once we left the laboratory at the end of a Saturday afternoon, the world opened up, and Dad and I would check out the latest movies (sometimes more than once), the latest exotic restaurants in Columbus (at that time, the Kahiki and the Jai Lai!), and the neighborhood bookstore called Little Professor – where we would both get lost in our respective sections for hours at a time. I still cherish the books that Dad and I collected during this time – a giant coffee table book on Airships remains my favorite and still sits proudly on my bookshelf at home. And on summer weekends, we would ride our bikes for miles and miles upon a newly built system of bike paths across the Columbus metropolitan region. While we caught up on everyday things and events, we mostly talked about the future and the great big world beyond Central Ohio. Once we ran out of bike trails sometime around my twelfth birthday, I think we both realized that our future lay beyond Central Ohio.

Dad started to invite me to join him at various conferences and business trips around the world…. Canada, Brazil, Japan, Malaysia, and Singapore to name a few. On these trips, I had the opportunity to reconnect with Dad's colleagues and former graduate students as well as meet Nobel Laureates and international dignitaries. Once I hit thirteen years old, I had grown to the point where I could go out on the town with Dad and his colleagues. I had my first glass of wine (didn't really like it), danced my first disco dance (not bad), ate my first sushi (delicious), was offered my first joint (didn't take it), and played my first slot machine (I won). Momo advised me to save some things for later years but to be honest, there didn't seem to be much time to waste in joining in these incredible experiences which ultimately shaped my love of travel and adventure forever.

It was on one of these scientific conferences that Dad got a tap on the shoulder and was asked to move to Cinnaminson, New Jersey to become director of the Pioneer Laboratory at the Campbell Soup Company. Before accepting the position, we had a real heart to heart, weighed the pros and cons, and ultimately decided together that this was a great opportunity for both of us. Nevertheless, it was a major transition in our lives. And while it involved a great deal of separation for us, it signaled the beginning of our journey to an exciting future. While Dad would often times fly back to Ohio to visit Momo and me, I started to fly regularly on my own to visit Dad at our home in Haddonfield, New Jersey. And while our weekends were much the same, Saturdays at the lab, the canvas was greatly expanded. We frequently attended the theatre, the wonderful museums and restaurants of Philadelphia and walked many, many miles from South Street to Broadway to the Benjamin Franklin Parkway. But just like Columbus, we started to run out of pavement and began to visit New York more and more. And the theatre became one of our favorite pastimes. I still have an enormous collection of Playbills from this time bearing witness to the many shows we saw together – Cats, Evita, Jesus Christ Superstar, Irene, My One and Only, Barnum, Big River, 42nd Street, Hello Dolly, West Side Story, Marilyn and many more.

Shortly into Dad's Campbell Soup Company run, came the founding of DNA Plant Technology Corporation which expanded out our family to include some of the titans of Wall Street such as David and Isaac Blech, Jeffrey and Sharon Collinson, Richard and Lee Laster and a team of world class biotechnology scientists. This was an incredibly exciting time in our lives. Dad's laboratory suddenly opened up into an entire company where he oversaw many laboratories in the pursuit of research and product development for world famous companies such as General Foods,

Campbell Soup and DuPont. Dad's success took on a new level of visibility when DNA Plant Technology stock started trading on the New York Stock Exchange. There were fancy parties in New York at places like Windows on the World at the World Trade Center and private yachts that sailed around Manhattan with the DNAP team and the Wall Street team celebrating the excitement of the biotech boom. And there were quiet get-togethers at the home of Evelyn and Israel Berezin in Greenwich Village.

Dad and I spent more and more time in New York with DNAP and it started to feel like home for us. We frequented hotels in the beginning, traversing the island of Manhattan up and down the avenues for miles at a time. And once I was in college, Dad finally took the leap and bought an apartment in the City.

While I always loved the world of the sciences – the multiethnic and multinational brew of graduate students and visitors, the excitement of traveling around the world, the thrill of discovery and with the biotech boom – the dreams of success – I did not gravitate towards the physical sciences in school. I loved the languages, the arts, the social sciences and travel. In fact, before I even graduated high school and into college, I was off on my own set of foreign adventures on summer study groups to Switzerland, Italy, France and the former USSR and had picked up Spanish, Italian and Portuguese.

I chose Colgate University for my undergraduate studies – a small liberal arts school – that provided the resources and opportunities to pursue what ultimately became my career in the arts. I spent most of my weekends at Colgate traveling with my acapella singing group – The Colgate Thirteen – across the country and back and down to the Caribbean every Spring Break. We recorded albums and performed on TV shows and sang the National Anthem before major sports events. Weekends were often spent visiting Dad in New York and Philadelphia, going to the theatre, watching movies and walking and talking.

After graduation, I moved to Los Angeles to work for film director Oliver Stone and worked on several of his major feature films including *Born on the Fourth of July, The Doors,* and *JFK.* Having little sense of how to pursue a career in the film industry, I would pester Oliver with questions all of the time about the benefits of going to graduate film school. Finally, he told me to apply and that he would support me. The only place that I wanted to attend was the Graduate Film Division at Columbia University School of the Arts. And as good fortune would have it, I was accepted in the fall of 1992.

During film school, I saw a short piece in the New York Times about a woman starting a new film festival in East Hampton, NY. And in time, we founded the Hamptons International Film Festival which is now celebrating its 21st year. The film festival provided a network of contacts through which I met my business partner John Hart – who was producing Broadway musicals at the time. I admired his success in raising money for theatrical productions and proposed to him that we might partner in raising equity for independent feature films such as the one my class mate Kimberly Peirce was developing called *Boys Don't Cry.* In fact, I ended up leaving graduate school (ultimately graduated in 2001) to begin producing movies with John for a company we formed called Hart Sharp Entertainment in 1996. Our combination of good taste, luck and proximity to extraordinary talent, allowed for an incredibly fruitful period of collaboration.

Our film productions included *Dark Harbor (1999),* Academy and Golden Globe Winner *Boys Don't Cry (2000),* Academy Award and Golden Globe nominated *You Can Count On Me (2001),* *Lift* – which introduced Kerry Washington to the world *(2001),* *Nicholas Nickleby (2002), P.S. (2005), Proof (2006), Evening (2007), Revolutionary Road (2008)* and many others. One of my favorite experiences during this time was the 2001 Academy Awards when Dad joined me as my date. As this was his first time at the Oscars, he wasn't quite aware of red carpet protocol and innocently joined actress Goldie Hawn's entourage. The following week, news papers and magazines had several photos of Dad with Goldie and Kurt and their family smiling away which provided lots of laughter for friends and family.

At Hart Sharp Entertainment, many of our best movies were based upon books. And when I left Hart Sharp to start my own operation – Sharp Independent at HarperCollins – I partnered with HarperCollins Publishers as a strategic partner to option and develop their books and authors for adaptation into feature films and TV shows. While at Sharp Independent, I developed my friendship with HarperCollins CEO Jane Friedman into a business partnership and when she departed HarperCollins in 2008, I approached her about starting a new company together. The company evolved to become "Open Road Integrated Media", which is now the largest independent digital publisher in the US. Our authors include William Styron, Pat Conroy, Michael Chabon, Michael Crichton, Alice Walker, Susan Minot and so many others. And we've developed a number of exciting film and TV projects from these author's works which are now in the process of moving into production.

Once Open Road Integrated Media was up and running, we formed a new company focused exclusively on film and TV production in Santa Monica, CA called Story Mining & Supply where I serve as President/CEO. With my business partner (with Jane and I at Open Road) and fellow filmmaker, Jim Kohlberg, we set up shop in 2013 to develop and produce material from our strategic partner Open Road as well as other publishing companies including Chinese language partners Shanda Cloudary and Jimmy SpA. And now as a result of my new company, I now split my time between New York and LA which has been a true adventure.

Books, reading and writing have remained my passion throughout the years. And that love of storytelling informed what was to become my journey – to produce feature films, TV shows and publish books. And curiously, a number of my films have been inspired by the sciences – especially *PROOF* – and we publish many authors from James Gleick to Albert Einstein as part of our digital e-book lineup for Open Road Integrated Media.

As the very proud father of my three year-old son Jack, I look back on my childhood years with a sense of wonder and amazement. I feel so fortunate to have grown up into a community of such extraordinary people. From my father's mentor Jim Gunckel and his wife Jean and their children and grandchildren to Raul Machado Neto, Roy Chaleff (who generously gave my son his first microscope for Christmas this year), Bob Pfister, Linda Styer Caldas, Geetha Ghai, Sally Miller, David Lee, John Peters, Harry Sommer and so many, many more, thank you. Thank you to the students who put me on their lap to read to me, thank you to the famous visiting scientists who took the time to ask me about my school interests, thank you to warm and generous wives who cooked for me and brought me presents, and thank you to the secretaries who let me use their typewriters and copy machines for my projects.

I hope that I too can instill my passion for life, career, colleagues and friends in such a way to make my son feel as much a part of my world as my father made me such a part of his world. Dad's successes, his happiness, his disappointments, and his legacy are part of mine. I was there every step of the way. I am so grateful to you Dad for sharing your incredible journey with me through the life sciences and throughout our lives together.

Congratulations to you and your colleagues on such an impressive achievement. I am so proud - Jeffrey Sharp, February 28th, 2014, Santa Monica, CA.

Photo 1. Jeff, Nancy Gunckel & Gretchen

Photo 2. Story Mining & Supply Company Ocean
View Location, Santa Monica, California

CHAPTER 16.

Fascinated by Plants

Judy Lyman Snow

Coming of Age

I remember having an interest in plants from a very early age, no doubt absorbed from my mother, who loved to garden. I was born in New Jersey in the early 1950's, and then moved with my parents and younger sister to a suburb near Rochester, New York, where I started school. There were large fields behind the house where my sister and I walked to and from school, and where I spent much time observing the variety of meadow plants. The first spring there, my mother gave me a couple of seed packets in my Easter basket and some space in the garden to plant radishes and lettuce. How proud I was of my first garden produce! I went on to try a variety of herbs and vegetables, help plant bulbs and annuals, and learn about propagation. It was the start of a life-long love of plants and gardening that has given me endless hours of pleasure and helped shape my career.

Developing an Interest in the Sciences

When my family moved to the Detroit area during my middle school years, I was thrilled to find that our retired next-door neighbor had a large greenhouse and an extensive garden. He was happy to share his knowledge with me, and by high school I was allowed to design and plant a section of his garden with annuals raised from seed. Transplanting the seedlings into flats was my particular job since his fingers were stiffened by arthritis. By this time I was reading horticulture magazines and was definitely bitten by the gardening bug. The idea that a small packet of seeds could result in a riot of colorful flowers and vegetables seemed to open limitless possibilities to me.

At some point I had also developed a parallel interest in fungi, probably because of their endless variety and colorful habits. I collected them in my rambles through the fields and woods, and brought them home to identify them and sketch them. My parents encouraged me in this interest as well, and even provided me with silica gel in which to dry and preserve my specimens. This unusual hobby attracted the attention of a friend of theirs, a botanist named Jim Wells who taught at the prestigious Cranbrook Institute of Science, not far from where we lived. He took me to the University of Michigan to meet Dr. Alexander Smith, a renowned mycologist, who kindly showed me around the herbarium and signed my copy of his book. That summer Dr. Wells arranged a part-time job for me at Cranbrook, cataloging the specimens of fungi in their herbarium. He also encouraged me to take a couple of summer courses in mycology at Douglas Lake, a biological station in northern Michigan run by the University of Michigan. No doubt he pulled strings to gain admission for a high school student to courses normally open only to undergraduate and graduate students. Because I already knew quite a bit about fungi, I did well in the lab courses, which helped my self-confidence tremendously when I went on to college. Very recently, I came across a list of my classmates from that summer, and among them to my amazement was the name of a long-time faculty colleague at Rutgers University. Neither he nor I remembered the other, but we both had vivid memories of the class.

As an aside, I should mention that I had a number of other interests that took me in different directions. I started violin lessons in grade school and continued to play throughout high school. When it was time to apply for college, I reluctantly decided that my talents were not enough to make a career of music, and that a career in science held more opportunities. However, music continues to be an important part of my life to the present day.

Meanwhile, I was also intrigued by other cultures and languages beyond the U.S. My mother, who had spent a year studying in Europe during college, belonged to an organization in Detroit that hosted international visitors. A procession of guests from all over the world came to dinner at our home during my high school years. So when I had the chance to go to Japan as a summer exchange student, I didn't hesitate. That was the first of many international adventures that I was fortunate to enjoy. Throughout my life, my parents were always supportive and very proud of my achievements. They encouraged me and followed my career with interest, so I always felt that I had a strong team behind me. As time went by and I learned about the struggles of others, I came to realize how lucky I was.

Choices of Undergraduate and Graduate School

When I began looking at college choices, my friend and mentor Dr. Wells did me another good turn by recommending that I consider Duke University, which had a strong Botany program at the time. I visited there and was warmly welcomed by the department chair, so I applied and was accepted through the early decision program. I enjoyed my years at Duke, with its mild climate and Piedmont ecology. I made a little extra money doing botanical illustration for one of the professors who studied mosses. This brought me into contact with the graduate students, who were friendly and encouraging to a new undergraduate. Since I had extra credits from advanced placement courses and summer classes,

I was able to graduate in three years. However, I knew that I was not finished with my education and ought to continue through graduate school. Recognizing that there were not a lot of job opportunities in botany, my father suggested agriculture as a practical application for my interest in plants. This wise advice put me on a new path that led to many new adventures and challenges.

I decided that plant breeding was the direction I wanted to pursue, and applied to several universities with strong programs. Cornell University was high on my list because it had not only a strong plant breeding program, but also a long history of international efforts. I was thrilled when Cornell accepted me with the offer of a graduate assistantship—but in horticulture, not plant breeding! They pointed out that I had had no undergraduate genetics courses. That was the down side of rushing through my undergraduate years so quickly. Nevertheless, I accepted the offer and headed to upstate New York in 1974.

Cornell's Horticulture Department was also a strong one, and Dr. Harold Tukey, Jr. became my advisor for a master's degree project that focused on plant physiology and propagation. Still thinking about plant breeding, I took genetics classes and other prerequisites for the plant breeding graduate program. In hindsight, I am surprised that my love of gardening did not seduce me into staying with horticulture and pursuing a career in landscape gardening or public garden management. When my master's degree was completed, I took a year off to consider my options.

Dr. Jerry Grant of the Rockefeller Foundation, who had recently retired as director of an international agricultural research center in Colombia (CIAT), arrived in Ithaca for a sabbatical year. He gave a seminar about the research underway at CIAT, with its focus on breeding beans, maize, rice, and pasture grasses for Latin America. When I talked with him afterward, he offered me a job as his assistant while he was in Ithaca. While this was an administrative position rather than a research position, it gave me the opportunity to learn about the world of international agricultural research. This convinced me that plant breeding in an international setting was still what I wanted to do. So I applied to the Plant Breeding Department to continue for a Ph.D. and was accepted. Dr. Henry Munger, a renowned vegetable breeder, agreed to take me on. After two years of coursework and participation in Dr. Munger's breeding projects on melons, cucumbers, and onions, I came up with a thesis project focused on breeding lima beans for tropical conditions, based at CIAT in Colombia.

CIAT, which I had visited with Dr. Grant, held the world collection of Phaseolus beans in its gene bank. The center's bean breeding program focused on *Phaseolus vulgaris*, but little attention had been given to the samples of *Phaseolus lunatus*, or lima beans. My thesis project evaluated the lima bean collection for disease and insect resistance in order to select promising lines for a breeding program. The presence of cyanide in the seeds was also checked, as some varieties had levels that might be toxic. I grew trial plots in several different locations and learned a great deal about field research during two busy years, from 1978-80. In hindsight, it was not the best arrangement to conduct my field research so far away from my major professor, as this was well before the advent of email and Skype. However, I got help from Dr. Mark Hutton, a retired professor and experienced breeder visiting from Australia, who very kindly spent time showing me some of the practical aspects of bean breeding.

One positive impact of my stay in Colombia was the exposure to team-based research programs with clearly stated goals. It was a very different strategy from the single-scientist approach that was then the model at most universities, and seemed to produce measurable results much more quickly—new varieties, better agronomic practices, and more effective economic policies. Now, of course, multi-disciplinary research teams are required for major grant-funded research projects.

I also met a diverse group of talented young scientists from around the world. They all spoke English, which was the common language at the research station, even among the staff. Nevertheless, I was determined to learn Spanish so I could communicate with the people outside the center walls. I had studied French for years all the way through college, but had only six weeks in which to take a few Spanish classes before heading to Colombia. CIAT employed a Spanish teacher who gave regular classes on site, and I became fairly fluent after a few months. I was proud that I gave a final seminar on my research in Spanish, rather than in English.

My project at CIAT was basically self-financed, since my assistantship at Cornell was state-funded and not available for international work. However, because of my experience working for Dr. Grant, I got a job as assistant to CIAT's director of research, Dr. Ken Rachie, also a Rockefeller Foundation employee. Although once again it was an administrative job rather than a research position, it gave me an inside view of the management of an international research station. I recorded the minutes for board meetings at which members from around the world debated the merits of research objectives, as well as budget constraints and priorities. It was a priceless experience that piqued my interest in the bigger picture of how international programs were organized, priorities determined, and funding obtained to carry out the vision.

Since the Rockefeller Foundation was one of CIAT's founding donors, the director of its agricultural sciences program, Dr. John Pino, was a member of CIAT's board. During one of the breaks in the board meetings, he asked me what I planned to do after completing my Ph.D. thesis. At that time the Foundation was concerned about plant genetic resource issues and the loss of diversity among crop plants. My experience with the bean germplasm collection at CIAT was very relevant, so Dr. Pino suggested that I come to the Foundation's New York headquarters as a junior scientist to work on these issues and help the Foundation determine where its funding should be focused. My decision to accept his offer was an important one, as it led me away from field research into the administration and management of research programs. I never regretted the move, which led onward to several long-term and rewarding positions.

People Influencing One's Career Pathway

I arrived at the Rockefeller Foundation in 1980, when the network of international agricultural research centers had expanded to build on the successful approach of the Green Revolution (and to address some of its shortcomings). The Foundation's Agricultural Sciences program, led by Dr. Pino, had more than 60 staff members stationed in centers around the world. The funding requirements for these centers, initially established by the Rockefeller and Ford Foundations, had grown dramatically. The Consultative Group on International Agricultural Research (CGIAR) was established in 1971 to manage the centers and provide coordination for the donors, and was a very complex organization. Dr. Pino was one of the visionaries who had played a key role in founding the CGIAR, and was naturally still very much involved as a member of its executive committee. How fortunate I was to learn about the critical issues facing the CGIAR from one of its architects, and to contribute to policy papers and proposals for deliberation at its meetings. Working with John was exciting. He would often have me write down ideas that flowed quickly as he talked. Then it was my job to fill in the details and edit the draft into a cohesive whole. It was good training for a young staffer and a steep learning curve.

My primary focus was on plant genetic resource conservation and management. The Rockefeller Foundation was providing some funding to a non-profit journal called *Diversity,* which reported on genetic resources issues. Bill Brown, President of Pioneer Hi-Bred, had helped establish the journal, and I became friendly with its editor, Deborah Strauss. Although there were a number of international agencies working on plant genetic resource issues, *Diversity*'s reporting focused attention on the subject in a sustained way and got the message out to a very broad audience. Along with the Keystone Conferences, which brought together the leaders in the field, *Diversity*'s efforts spurred a global interest in collection, preservation, use, exchange and other facets of this natural resource. Years later I served several terms on the board of *Diversity* as Secretary/Treasurer. This afforded me the privilege of meeting and working with Bill Brown's wife, Alice, and son Bill, who had come on the board after his father passed away. Both John Pino and Peter Day served many years on the board. In the meantime, I also contributed a couple of chapters to the book "*Managing Global Genetic Resources: Agricultural Crop Issues and Policies,*" which was published in 1990 by the National Research Council of the National Academy of Sciences, Washington, D.C. These efforts were the fruits born after I had left the Foundation in 1984, but they originated in friendships and connections made during my tenure at the Foundation.

The Rockefeller Foundation's Agricultural Sciences staff in New York consisted of a small group of senior scientists who had come up through the international research system, supported by a few younger staff members and several post-docs, of which I was the most junior. Dr. Gary Toenniessen, then a junior staff member, had an office right next to mine and was a great help in showing me around and explaining how things functioned. I appreciated his friendliness to me and was very pleased to see years later that he was eventually promoted to become Director of Agricultural Sciences—a well-deserved reward for his long experience and loyal service.

Another valuable experience I gained while in New York was learning how a foundation functions: how it sets its priorities and organizes its programs, how it crafts a call for proposals, and how it evaluates the proposals it receives. Reading a wide variety of proposals—from crude to highly-polished—was extremely valuable training for me which contributed directly to a major part of my responsibilities subsequently at Rutgers University.

Many people visited the Rockefeller Foundation officers in New York, hoping to persuade the Foundation to fund their efforts. Among them were two scientists I had met at CIAT—Dr. James Brewbaker and Dr. Jake Halliday. Both of them were then located in Hawaii, the former a distinguished plant breeder of maize and leguminous tree crops, and the latter a young microbiologist who headed the University of Hawaii's NifTAL program (Nitrogen Fixation by Tropical Agricultural Legumes). They successfully argued for support to organize an international conference on the genetic resources of nitrogen-fixing trees, at the Foundation's fabled conference center at the Villa Serbelloni in Italy. I was brought in to help coordinate the conference and produce the report. The event was one of the highlights of my tenure at the Foundation.

During the time that I worked at the Foundation, Dr. Pino and the senior agricultural scientists were already looking beyond the achievements of the Green Revolution for the next breakthrough. Biotechnology was in its very early stages, and they were debating how its potential could be applied to the problems of agriculture in developing countries. Along with Dr. Pino, I attended a meeting led by Pioneer Hi-Bred's President Bill Brown. The meeting brought together many leading scientists to discuss which directions biotechnology research should take, which crops could serve as model systems, and whether the results would have a practical impact on the needs of the developing world. Once again, I was privileged to listen and learn from some of the key people involved in the early days of the technology.

Back in New York, I did a lot of literature research on the potential of biotechnology as background for position papers that the staff used in its internal deliberations. I no longer recall how it came about, but I received an invitation to participate in a small workshop in Rome at the Vatican. The purpose was to present recommendations to the Pope on the potential of the biotechnology and how it might contribute to society. Accordingly I prepared a paper and presented myself in Rome for a dream-like week of meetings in a small room inside the Vatican walls. It was hard to focus on the papers due to the beautiful frescoes decorating the ceiling and walls, but we managed to produce our report as charged. One of the other participants was Dr. Peter Day, then head of the Plant Breeding Institute (PBI) at Cambridge, England, which was in the forefront of plant molecular biology research at the time. Later I went to a research conference hosted by Dr. Day at the PBI, since the Rockefeller Foundation was funding some of their research. Dr. Day and his wife kindly invited me to stay at their home, which I later learned was just one example of their generous hospitality to visitors. Little did I know that as a result of that brief stay, Dr. Day would recruit me to a position as his associate that would last more than 20 years!

However, there were many other changes underway in the meantime. There was a new president at the Rockefeller Foundation, and a feeling that the Agricultural Sciences division with its large and costly field staff should downsize to support more lab-based research in biotechnology. Dr. Pino left for a position in Washington, which was a big blow to me. Losing one's mentor is always unsettling. Fortunately, my personal life took a lucky turn at that time when Jim Snow, a fellow Cornell graduate student, asked me to marry him. By coincidence, he had taken a job not far from my parents' house in New Jersey at the United States Golf Association. We married in 1984, and I embarked on a new path familiar to so many women who have to learn how to juggle family lives and careers. I am lucky that my husband has been so supportive through the years, and we will soon celebrate our 30th anniversary.

Between the downsizing underway at the Rockefeller Foundation and a new home with my husband in New Jersey, it was clearly time to look for another job closer to home. However, the options in the metropolitan area for someone with international agriculture experience were limited. I had a couple of interviews with big agricultural chemical firms, but they were not a good fit. Fortunately, I found Dr. Reed Hertford, an economist and former Ford Foundation staff member who headed the International Agriculture and Food Program (IAFP) at Rutgers University. He offered me a staff position split between the IAFP and a new biotechnology initiative that was underway. At that point I recognized that I was heading for an administrative career, rather than a faculty position of research and teaching. It was not a decision to take lightly, but the exposure I'd had to research policy-making at CIAT and then at the Rockefeller Foundation had been intriguing, so I accepted the offer and started a new chapter at Rutgers in late 1984.

Like most university-based international agriculture programs, Rutgers' IAFP operations were funded through contracts from the U.S. Agency for International Development (USAID), other government agencies, and some non-profit foundations. Dr. Hertford had won a contract for a field project in Panama, helping the ministry of agriculture evaluate its policies and improve the effectiveness of its national agricultural research stations. I spent several weeks in Panama on a couple of occasions with a multi-disciplinary team, visiting research stations and talking with the Panamanian scientists. Back at Rutgers we sifted through large amounts of economic data and wrote a lengthy report. It was a challenging assignment and good experience. That contract and some smaller projects kept me busy for about two years, but it was becoming more difficult to land contracts as USAID funding dried up.

In the meantime, Rutgers had convened a committee to develop a biotechnology center at Cook College, where the New Jersey agricultural experiment station was located (as well as the IAFP). The groundwork had been laid for support from the state legislature through a series of bond issues. As the plans matured, Peter Day was recruited from the Plant Breeding Institute in Cambridge, England to lead the new center. When he visited Rutgers in late 1986 to finalize his appointment, he stopped by my office at the IAFP and offered me a position to work with him in getting the new center off the ground. Naturally I was delighted to see him and both flattered and excited by the offer. However, there were two issues to grapple with. The first was that I felt I owed Reed Hertford some loyalty for his help in creating a position for me, and the second was that I was then pregnant with my first child. After much deliberation, I realized that the IAFP was on a downward path, while the biotech initiative was about to take off. Another deciding factor was that Dr. Day was willing to let me work part-time after my baby was born. While it was difficult to tell Dr. Hertford of my decision, subsequent events proved that I had made the right choice. After another year or so, he moved on as well.

Accepting the position with Peter Day in the Biotech Center in 1987 was the start of a long and exciting trajectory that would continue more than 20 years. I had the good fortune to be in the right place at the right time, and to have such a great boss. Peter was already a senior statesman in his field and a real gentleman--it would be hard to imagine a more considerate person. His style was to lead by expecting great things of his staff and faculty, and no one wanted to let him down. He was genuinely interested in the personal lives of his staff, and always asked about family members. He was also willing to help when the chips were down. Several times when I was working late to get out a grant proposal, he offered to stay and help copy and package the documents. I just naturally wanted to do my best for him.

In the beginning, there were three of us sitting in a small office in the college administration building—Peter, myself, and Phyllis Telleri, the executive secretary. My title was something like Program Associate at the time. We prepared endless documents and budgets for the university and the legislature, held many meetings within the university, visited New Jersey companies interested in biotechnology, and finally began recruiting faculty. The first few hires were placed temporarily in labs in other units of the university until we could build our own space. In 1989 an extra floor was added to a new wing of the Environmental and Natural Sciences building, so our growing center could be housed together. The dream was becoming reality.

After some debate, the name chosen for the new initiative was the AgBiotech Center. Our mission was to "pioneer in agricultural and environmental biotechnology research; educate and train students to build a high-tech workforce; and to share knowledge and technologies that improve the quality of life." As the center expanded, some of the areas the faculty worked in were agricultural and environmental applications of genomics, mechanisms of gene regulation, bioprospecting for useful microbes and plants, discovering novel plant pharmaceuticals and nutraceuticals, disease resistance in plants, bioremediation of contaminated environments, endocrine regulation of animal development, animal models for gene therapies, and ethical issues associated with biotechnology.

Another important team member came on board at that time—Roger Grillo, who was our financial and administrative guru. He was also our resident computer and equipment expert. In fact, he did just about anything except wash lab glassware during those early days when our staff was so small. One Thanksgiving weekend he was on call when the building water system developed a leak, and he and Peter had to come in to stop the leak and mop the floors! Roger was a terrific colleague and a great person to work with—again, a lucky break for me. With Peter at the helm and Roger and Phyllis to work with, it was a "dream team" for me. They helped me keep all the balls in the air when my second child was born in late 1991. I was still working part-time, but tried to be flexible and come in for meetings even if they took place on my "off" hours. My husband traveled extensively, so it was a challenge to keep all the bases covered. My parents had moved to Massachusetts some years before and were not available to help with daycare, but I found a wonderful woman (a former teacher) who took care of kids in her home and remains a close friend to this day. Many times I realized how critical it is for women to have a supportive husband, a flexible boss, and good day care for their kids in order to pursue their career. I was very fortunate.

Back at the office, Peter, Roger and I were working on a major proposal to the US Department of Agriculture (USDA) for funds to help construct the center's eventual home. Most of the funds came from New Jersey state bond issues, but an additional $10 million was awarded by USDA. With that in place, planning for the building could go forward. In 1992 we held a ground-breaking ceremony, inviting all the leaders from the legislature, the university, industry, and government agencies who had helped move the project forward. I was in charge of coordinating the event and had to write speeches for Rutgers' president, the Cook College dean, and for Peter Day. Naturally the timing was carefully worked out. Imagine my dismay when the president began ad-libbing and strayed far off-script! However, all ended well.

Although located at Cook College (the ag school), the AgBiotech Center reported directly to the university's Vice President, Joe Seneca, who was a key backer and gave us much of his time. A number of college deans held office during the life of the center. Initially they were supportive and worked with the center on college-wide efforts. Dr. Rod Sharp, with a background in molecular biology and corporate experience with DNA Plant Technologies, was hired as the Dean for Research in 1993 and was especially enthusiastic about the Center's work. He was very encouraging and took the time to get to know the AgBiotech center staff. It always made a big impression on me when senior administrators had time for staffers.

At Rod's request I coordinated a weekend event at Cook College in 1994 with Johns Hopkins for 500 talented middle school students and their parents, introducing them to biotechnology research through lectures and hands-on lab exercises in labs around the campus. I solicited volunteers from the faculty and students to help with the labs and logistics. There was a student biotechnology club called Designer Genes, which offered to help. Just before one of the club's meetings in the AgBiotech Center's conference room, someone threw butyric acid (which smells like vomit) in the halls of the building. It reeked for days. Around that time there were protests nationally against biotechnology and "Frankenfoods," and I was worried that whoever had vandalized our building might try to do the same at the Johns Hopkins event. I could imagine the reports on the evening news! So in addition to all the other planning to take care of, I had meetings with the campus police, who did some investigation and learned that the perpetrator was the disgruntled boyfriend of a female student who worked in our labs. They warned him off and sent a plainclothes officer to the event. Thankfully it went off without incident, but it was a nerve-wracking experience that I will never forget!

Meanwhile, Peter and Roger spent untold hours from 1992-1995 with the architects, poring over building plans, and then with the contractors as the new building went up. In 1995 it was finally finished, and we moved all of the faculty and staff into 34,000 square feet of labs and offices in our three-story wing. The building housed not only the AgBiotech Center, but also the Plant Science and Plant Pathology Departments, and a new science library named for donors Stephen and Lucy Chang. Ultimately the Biotech Center had around 13 core faculty, and a staff of about 100 that included postdocs and graduate students.

When the building was nearing completion, I asked if there would be a plaque in the lobby honoring Walter Foran. I was told that yes, there would be a plaque with a date and the names of the President and Board of Governors at the time. But what about Walter Foran? I was told I would have to take care of that. So I called the Dean's office, who contacted Mrs. Foran. She graciously agreed to fund the plaque and provided some photos of Senator Foran. I worked with a local business to design the plaque, which included a bas-relief bust of Mr. Foran and a description of his important role in siting the biotechnology initiative at Cook College. I took pleasure every day in seeing the plaque, just to the left inside the main entrance, as I came and went from the building.

The new building was dedicated in October 1995 and named in honor of the late Walter Foran, a New Jersey Senator who had been a key supporter. Guests included Senators Frank Lautenberg, John Ewing, Mrs. Foran, and other dignitaries. There were 400 guests and many speakers to include. Naturally the planning for this event was another major project! It was a festive day and a major milestone reached in the growth of the Center.

Up to this point, I have related some of the highlights and big events that took place in the life of the AgBiotech Center, but I haven't described what occupied most of my time on a regular basis. My role and responsibilities changed as the center grew, but there were always reports and proposals to write. In the beginning these were focused on funding to get the center up and running. Then we worked on recruiting faculty, which required the drafting of position descriptions and policy statements. As the faculty came on board and began their research, I devoted more time to helping them with grant proposals and preparing budgets. Grant development became my specialty, and I worked closely with the university's grants office to streamline the Center's submissions. In addition, I wrote annual work plans to accompany budget requests to the New Jersey Commission on Science and Technology (NJCST)—our state sponsor--as well as annual reports documenting results. I found that my science training was essential to understand the faculty's research initiatives, and to summarize them in terms intelligible to a broader audience. Eventually, over a period of more than 20 years, those reports filled a whole shelf in my office.

Working with the faculty and the graduate students was very satisfying. It was intriguing to hear their ideas for new projects as I helped them with grant proposals, and then follow their progress when the grants were awarded. There were a number of international projects that were challenging to administer, particularly one led by Dr. Ilya Raskin that had partners in four Central Asian countries. There was also a tripartite exchange program between the Universities of Sao Paulo/Brazil, Ohio State, and Rutgers with annual meetings that rotated between the participating universities. Rod Sharp was the key person who brought Rutgers into the exchange. Graduate and undergraduate students participated, along with the faculty. It was exciting to see the sparking of new ideas during the discussion following research presentations, and to get to know the Brazilian researchers. Later Dr. Raul Machado, Dean of Escola Superior de Agricultura "Luiz de Queiroz" (ESALQ), the agricultural college of the University of Sao Paulo,

Brazil, spent a sabbatical year at Rutgers in an office across from mine, and I was privileged to work with him daily.

Tours and outreach efforts were another of my responsibilities. The Center hosted a number of annual workshops and events for high school students in cooperation with other units at Rutgers. I coordinated with the faculty to run hands-on lab experiences for the kids, as well as tours through the center. As I gave each group a welcome and brief summary of the center's activities, I always encouraged them to ask the faculty and graduate students about their careers, and what got them interested in science. I also urged them to think about other applications of biotechnology beyond medicine (their usual interest), and gave examples that they would see during their visit.

Sometime after the AgBiotech Center had moved into Foran Hall and its programs were expanding, funding from NJCST--the key state sponsor--began to dwindle, and eventually to dry up. The rationale was that the state funding was intended to get the Center up and running, and then the agency would move on to other new initiatives. Fortunately, the faculty had developed successful grant programs that funded more than 50% of the center's budget. The university agreed to pick up some of the building expenses which had been paid by the NJCST. Around the same time, the Center's name was changed to the Biotech Center for Agriculture and the Environment, to better reflect the focus on environmental microbiology that was a major part of its programs.

Meanwhile, family life continued to present its rewards and challenges. My kids moved on to elementary school, which meant finding afterschool programs and summer camps. In many ways, that required more juggling than when they were in year-round preschool care. More serious was the bad news that the cause of my mother's increasing difficulties was diagnosed as Alzheimer's disease. I encouraged my parents to move back to New Jersey so that I could help. Amazingly, there was a house for sale on my street that fit their needs, so they were close by. With Peter Day's blessing, I shuffled my schedule to spend more time with my mother. It was very difficult to see someone so smart and talented struggle with such a debilitating disease, but despite that, I have some good memories of the time we spent together. Later, when more care was needed, we found some wonderful caregivers from Kenya, and I helped my Dad coordinate with them for my mother's care.

Life was moving on at Cook College and the Biotech Center as well. Rod Sharp had left around 1998 when the college administration changed once again. Peter served as Acting Dean of Research and Director of the Experiment Station from 1999-2000 after Rod's departure, while continuing as Director of the Biotech Center. Finally the day came that all of us had been dreading: Peter announced his retirement. He must have been in his mid-70's at the time, and still working at a pace that would have exhausted someone half his age. However, he told us that his wife, Sue, was getting tired of waiting! Naturally the faculty and staff wanted to honor his contributions, so planning got underway for a retirement event in October 2001.

In the late summer and fall of that year, my mother—who had developed cancer in addition to Alzheimers—slipped into a coma. She wanted to stay at home, so we had the support of a hospice program. On top of it all, the 9/11 event shocked everyone, particularly in my town, which lost 19 people in the disaster. It was a very emotional time. I was working from home and going in to the office for essential meetings leading up to the retirement event. I was hoping that my mother would hang on a bit longer, but she passed away the day before Peter's retirement program. Naturally Peter was very sympathetic and said that he did not expect me to come, but I wanted to see him honored and his contributions celebrated. So I went and told no one else that my mother had died, because I didn't want to distract attention from Peter's well deserved send-off. It was a double blow, losing my mother and my revered boss and mentor.

As years went by, resources for Cook College (and the university) were strained, and later deans wanted more control over the Center's resources. With Peter Day's retirement in 2001, and Joe Seneca's departure from the central administration in 2003, it became more difficult to maintain the Center's autonomy. Eventually the Center's reporting line was shifted from the VP's office to the Cook dean, who began moving personnel and funding away from the Center to start a new initiative. It was a shame to see the Biotech Center's programs dismantled after all that had been accomplished. My position was eliminated, and after a short and frustrating stint in the college's central administrative office, I took an early retirement. It was not the way I would have chosen to end my career, but the outpouring of appreciation from the faculty was uplifting and proved that my efforts through the years had been valued.

Proudest Achievements and Lessons Learned

I was very fortunate to have found such able and supportive mentors throughout my career. They were all men, as there were few senior women at the time in the settings where I worked. I attribute my good fortune to showing an interest in the careers of others, and a willingness to ask questions and talk to people more senior than I. They were usually pleased to talk about their careers and their current responsibilities, so I could learn from their experiences or at least judge whether or not they were relevant to my interests. I was lucky to find positions where I could learn and contribute in exciting and challenging programs and institutions. It was important to me to be in a stimulating academic setting where learning goes on every day, and every day brought some new challenge.

As the years passed, I wanted to do what I could to help younger people get started in their careers, particularly the young women. There was a young Chinese woman who came in as a postdoc, and then took a technician's position where she seemed to get stuck, even though she was basically running the lab. She was an extremely productive worker and full of ideas—it was interesting to listen to her talk about her research plans. I encouraged her to publish more and look for a tenure-track assistant professorship. She took on more and more initiatives and made herself indispensable in a new lab where she is making better progress. I look forward to following her career. Another opportunity came along when one of my daughter's high school friends was looking for a summer lab internship. I went back to my Chinese friend, who agreed to take the high school girl on and gave her a couple of interesting projects to work on. I was gratified that she, in her turn, was helping a younger woman get a start.

After my early experience in Colombia, where I was able to travel through the rural countryside and observe the poverty and limited opportunities, it was natural to feel sympathetic toward the plight of the immigrants in the United States. Through the years I have become good friends with several of my Hispanic cleaning women, and helped them out with references for other jobs, tuition for classes, and legal issues. The same rapport developed with the Kenyan ladies who helped care for my mother, and later for my grandmother as well. I was grateful for their loving care, and they appreciated having a job where they were respected and well treated. I have always believed in the old adage, "Treat others as you would have them treat you."

There were many other experiences in the work environment that helped me grow along the way. One was learning to find some common ground with people who were from very different backgrounds than mine, so that we could work comfortably together. The diversity of cultures added richness and also complexity to the work environment, and was a good reminder that there are many ways to look at problems. One behavior that I had no tolerance for was the backstabbing that went on at all levels, but was worse in some units than in others. I have never understood the attitude that you have to step on others in order to get ahead. I often observed that the leadership of the unit set the standard for the rest of the group, and was grateful once again to work in an office where civility and consideration were the norm.

Of course, affairs did not always go smoothly, and there were pressures both internal and external that led to considerable stress. Learning to handle stress was probably the most difficult of all the challenges I faced. When it was possible to take action to resolve the problem, even if it meant extra work or disagreeable tasks, that seemed better than doing nothing at all and letting the problem fester. At other times, having the patience to let things work themselves out without interfering was the best approach, but that was also difficult. And there were many times when there was nothing at all that I could do to remedy the situation—they were the most frustrating. That's where relaxation techniques come in: deep breathing, a long walk, or a soothing cup of tea.

This brings me back to gardening--always an absorbing and soothing activity for me. I have had plants around me ever since childhood, and now that I am no longer working I have more time for serious gardening. My husband and I have developed extensive gardens around our current home, where there was almost nothing aside from foundation shrubbery when we arrived 15 years ago. Now there are hundreds of different varieties of trees, shrubs, perennials and annuals. We support a nearby arboretum, visit gardens in the U.S. and abroad, and go to conferences for serious plant geeks. I have also designed and planted gardens for a couple of friends. It's the ideal hobby because it gives pleasure to yourself, your family and friends; it beautifies your home; and it provides lots of physical activity. There is always something to do-- and a new plant to add to your collection. What's more, I have met many interesting friends through gardening, so it's a nice social connection as well. Last but not least, a friend of my daughter's has been helping me recently, so I am able to teach her some skills and hopefully pass on a love of gardening to her.

I will add a few thoughts on retirement to these reflections, as it seemed like uncharted territory since I had not expected to get there so soon. Advertisements that show pictures of laughing seniors enjoying a carefree life make you think that retirement will be like embarking on a very long vacation. Obviously financial planning is essential, particularly in the current economic circumstances. I was surprised by the number of decisions to make and the mountains of paperwork required to set our affairs on the new track. Not to mention health insurance options, health issues, related paperwork and online screen time. It's a good thing I was a professional administrator!

However, there is certainly a much greater degree of freedom to choose how you want to spend your time. Some people elect to downsize their homes and move somewhere completely different, or to begin another career. Volunteer options are numerous and can be very rewarding, as well as connecting you to a new network of friends. I volunteer in the book department of a huge semi-annual rummage sale, which affords me an unlimited supply of reading material. I am also playing in a chamber music group during the winter and taking piano lessons. Travel is another mind-expanding option, and an opportunity to reconnect with family and friends. In short, time is flying by and the days are never long enough for all that I want to do. With a little luck, I hope that the next chapter in my life will be as full and rewarding as the ones already completed.

Judy Snow and John Pino at a *Diversity* event, 2001.

In the garden, 2012.

Peter Day, Judy Snow, Roger Grillo, and Phyllis Telleri, 2003.

CHAPTER 17.

Harry Edward Sommer 1941-2002

Margaret F. Sommer

When Dr. Sharp asked me to write about Harry, he mentioned that he and the other members of the lab at Ohio State, where Harry was a graduate student, used to say that that Harry was their own branch of the Smithsonian. Ruy Caldas used to call him Mr. Library. I do not know when Harry developed the ability to remember references and to pull the information that people were looking for out of his memory banks. It was a skill he already had when I met him in high school.

As I was researching files for this article, I found a short biography that Harry wrote for some reason and I thought it was worth sharing. *"I was born on a poultry and truck farm."* Perhaps that had something to do with his memory. Joshua Foer in his book <u>Moonwalking with Einstein: The Art and Science of Remembering Everything</u> mentions the profession of chicken sexers as one that requires a good memory and eye for detail. Helping his dad sort baby boy chicks from the girls was something Harry mastered early.

"In 1946 my father was hired for two weeks to substitute for the (high school) English teacher. Twenty-five years later he retired as Guidance Director. My high school education was college preparatory." One of my favorite memories of Harry in High School was the friendly competition he and a best friend had to see who could get the highest grades on the regents exams New York required. Harry got the highest score on the history test and his friend, who went on to major in history at college, got the highest score in the science one.

Although we grew up 8 miles from each other, we attended different school systems. We met through 4-H activities and started dating in high school. On a questionnaire he filled out once he said that his favorite holiday was Labor Day because that marked the end of summer and the beginning of fall and the County Fair. He used to show chickens and work in the poultry exhibit. I used to work at the 4-H concession stand. We would spend spare time wandering the fairgrounds with his sister and her boyfriend. We were married in 1964 and spent the first part of our married life exploring bogs and working on Masters Degrees at the University of Maine.

"After obtaining a BS in Agriculture with a major in Botany at the University of Vermont (1963), a Masters in Agricultural Biochemistry was earned at the University of Maine (1966). There I served two years as a teaching assistant and 1 year as an assistant instructor. In both cases my primary responsibility was teaching the organic and biochemistry laboratories for third year nurses and 4th year nurses and animal science majors." Harry graduated from the University of Vermont with honors. His honor's topic was "Quinoe Occurrence in Hemlock" He maintained a life-long interest in the university's Proctor Maple Research Center. Harry acquired his interest in tissue culture working as a graduate trainee under Dr. Philip R. White at the Jackson Laboratory in Bar Harbor Maine. His project there was "Glucose Metabolism of Spruce Tissue Culture" Harry's master thesis at the University of Maine was "Purification and Characterization of Fructose 1, 6-diphosphatase."

"In 1966 I entered the U.S. Army Chemical Corps, leaving in 1968 as a Captain. After the army I earned a PhD at Ohio State working on a problem in plant tissue culture." We spent our army years at Dugway Proving Ground located in a rural area of western Utah. We both enjoyed exploring a new environment. Harry's PhD thesis was entitled <u>Influence of 2,4 Dichlorophenoxyacetic Acid on Nitrate Reductase and Protein in Wild Carrot (Dauchus Carota L.) Tissue Culture</u>. One purpose of Dr. Sharp's project, as I understand it, is to encourage young people to keep working when the going gets tough. Therefore, I will tell you more about our time at Ohio State. I took advantage of the fact that Harry was in school to work on my PhD in history. It meant that we were both poor students with similar schedules, pressures and needs to study. I remember it as an interesting time of learning, loving and making new friends. It ended on a sour note however. I am proud of the way Harry did not let it keep him from finishing up and going on with his dreams.

He picked Ohio State because he wanted to work with a certain professor that he had met previously and thought his work was interesting. He worked in this professor's lab for 4 years on a problem involving tissue culture of Queen Anne's lace. The other students in this lab and Dr. Sharp became permanent friends as well as professional colleagues. Harry's major was biochemistry, but he also took several botany courses because he has always been interested in the subject. Many of our dates in high school and college involved taking hikes with his camera so he could photograph plants. He also spent his free time during his time in the army at Dugway Proving Grounds wandering the surrounding area photographing and collecting a herbarium of what he found. Before he finished his thesis work, his major professor moved on to another job. The two of them had made arrangements for another professor in the biochemistry department to handle the necessary details involved in presenting his thesis. We both thought that everything would continue as before, but after Harry wrote his thesis and presented it, he was told that his project 'was not biochemistry' and that the professor would not recommend him to earn his degree with it. Since the original professor was no longer on campus, there was no one in the biochemistry department to turn to for help. Harry already had a summer job in Brazil and arrangements for a post doc position at this point. I knew he was angry, disappointed and frustrated, but he did not let it get him down. It was always important to him to be able to roll with the punches that life handed out and to figure out a way to overcome and keep going. So, instead of giving up or starting over with a new project, he looked around and tried to figure out another way to graduate.

We were lucky because Harry's work had impressed Dr. Sharp and other professors in the Botany department. Dr. Sharp approached Dr. Carroll Swanson, Chair of the Department of Botany who agreed to become Harry's advisor. Dr. Richard Popham also became a member of his committee. Dr. Popham authored <u>The Laboratory Manuel for Plant Anatomy</u> - the Manuel was used in the classroom at all major research universities. Carroll Swanson authored <u>Laboratory Plant Physiology</u>. Although it took several months to make all the necessary arrangements and we did not get to go to Brazil for the summer, I graduated in May and Harry earned his degree in Botany in August of 1972.

"The next step was 18 months at UGA as a Research Associate." Our time at the University of Georgia was productive. Harry worked with Dr. Claud Brown on studies on the growth and metabolism of gymnosperms in tissue culture. He figured out the appropriate culture to get long leaf pine tissues to differentiate and produce plantlets. This was a major accomplishment, but Harry was not one to blow his own horn. He reported the results at a meeting with a test tube containing one of the plantlets in his pocket but had to be reminded by Dr. Paul Kormanik, his colleague at

the presentation to pull it out of his pocket to show people. His was the first report ever published demonstrating that pine trees could be clonally propagated using tissue culture. While we lived in Athens our first child, Henry David Sommer, was born on the 4th of July.

"...and 18 month with Weyerhaeuser." Weyerhaeuser hired Harry in hopes that he would be able to accomplish the same thing with Douglas fir. This resulted in a move to Centralia, Washington, where our daughter, Katrina Helen Sommer, was born not long after our arrival. Harry found working for industry frustrating. Since his work involved basic research, it was difficult to produce practical results as quickly as the company wanted him to. Although he never said so, I also suspect that he missed the contact with students and teaching that occurs in an academic setting. So, when Dr. Brown, who missed his presence in his lab, called to say he had found funding for a research position and would like Harry to return to UGA, Harry accepted.

"In 1975, I joined UGA doing research and teaching in tree growth and development and plant tissue culture (without having any forestry courses) in the School of Forest Resources. I retired in December 2000." We moved to nearby Oconee County, which was considered country at the time while Harry continued his research and guided numerous graduate students and I became involved with the Girl Scouts and the children. Although teaching positions in history were hard to find locally, I eventually used my history degree when I was hired by Curtis Publishing Company to oversee the collection of stories that became <u>The History of Oconee County, Georgia</u>

One project at UGA that Harry enjoyed was helping Dr. Brown, the Founder of the Thompson Mills Forest, develop the arboretum there. He helped locate trees, plan where to plant them and watched them grow. Today the arboretum is the state's official arboretum. It contains about 90% of Georgia's native trees, and a major pinetum with more than 100 conifer species. The Eva Thompson Thornton Garden (7 acres) contains more than 100 ornamental trees from around the world. Several of the trees at the entrance of this section of the forest are longleaf pines that Harry started in tissue culture. The photo that accompanies this story shows Harry with one of these trees in 2001. One of the extra trees, a Big Leaf Magnolia ended up in our front yard. It became something that people stopped to ask what it was since it is not common this far north in Georgia. It was my granddaughter's favorite climbing tree and one neighbor's child was delighted when she took a leaf to school as part of a tree identification project and discovered she was the only one able to identify it. Even the teacher did not know.

When Harry passed away May 23, 2002, the following summary of his work was written by his daughter Katrina for his memorial service. "Dr. Harry Sommer was a pioneer in the field now known as plant biotechnology. In particular, Dr. Sommer was recognized world-wide as an expert in the area of woody plant cell and tissue culture. Although he worked with a number of forest trees, he was most famous for his ground-breaking work with the tissue culture of pine trees. His research formed the basis for in-vitro propagation of conifer species in laboratories throughout the world. Dr. Sommer trained and mentored a number of students from the United States and other countries who went on to become very successful researchers at universities and forest product companies. His depth of knowledge and willingness to share his knowledge with others, usually on a one-on-one basis, made a huge difference in the education of many students."

Harry collaborated with Dr. Hazel Wetzstein of the UGA Horticulture Department to investigate the propagation of sweet gum trees in the same manner. They also enjoyed great success. Harry also worked closely with many of his graduate students.

When Harry retired in December of 2000, his coworkers at the lab presented him with a plaque which reads: "Harry is internationally recognized for his pioneering and innovative research in plant tissue culture, organogenesis, and cloning of woody plants. His knowledge of the world's literature in plant physiology, tissue culture and morphogenesis is truly phenomenal. He has shared much knowledge, untold time, and assistance with students, faculty, and colleagues in solving many problems in the broad field of forest biology and biotechnology."

I was recently asked by a young college student who is interested in scientific research but dating my granddaughter who is an art major, what it was like being a non-scientist historian married to a scientist. For us it was a good match. Harry did not want to bring home his problems from work. Discussing the issues he was working on was something he would rather do with colleagues and students. We both shared an interest in plants, (though I often got frustrated when a 'what plant is that?' question got answered by the scientific name rather than the common name I was looking for), nature, the children and history. I would not have been able to do all I have done with the Girl Scouts without his encouragement and advice. He was especially serious about safety issues and I still hear his voice in my head cautioning me to be careful as I plan activities. I still miss his quiet suggestions and his company.

Harry Edward Sommer Photograph

TRIBUTES

Volumes I & II

Our Influencers & Mentors
Dr. Norman Borlaug Tribute
Dr. Murray F. Buell Tribute
Dr. Linda Styer Caldas Tribute
Dr. Otto J. Crocomo - For Those Who Made A Difference In My Scientific Life
Dr. Otto J. Crocomo - Para Aqueles Que Influenciaram Minha Carreira Cientifica
Dr. David A. Evans Tribute
Dr. David E. Fairbrothers
Dr. Percy Cyril Garnham Tribute
Dr. Leonard George Goodman Tribute
Dr. James E. Gunckel & Jean Longworth Gunckel Tribute
Dr. Roger F. Keller Tribute
Dr. Miodrag Ristic Tribute
Dr. Maro R. Sondahl Tribute
Dr. T.S. Subramanian and Meenakshi Subramanian Tribute
Dr. Clara Gertrude Weishaupt

Norman Borlaug Tribute
By MATT CURRY and BETSY BLANEY –
Associated Press Writers
Submitted by: William R. Sharp

Agricultural scientist Norman Borlaug, the father of the "green revolution" who won the Nobel Peace Prize for his role in combating world hunger and saving hundreds of millions of lives, died Saturday in Texas, a Texas A&M University spokeswoman said. He was 95.

Borlaug died just before 11 p.m. Saturday at his home in Dallas from cancer complications said school spokeswoman Kathleen Phillips. Phillips said Borlaug's granddaughter told her about his death. Borlaug was a distinguished professor at the university in College Station.

The Nobel committee honored Borlaug in 1970 for his contributions to high-yield crop varieties and bringing other agricultural innovations to the developing world. Many experts credit the green revolution with averting global famine during the second half of the 20th century and saving perhaps 1 billion lives.

Thanks to the green revolution, world food production more than doubled between 1960 and 1990. In Pakistan and India, two of the nations that benefited most from the new crop varieties, grain yields more than quadrupled over the period.

"We would like his life to be a model for making a difference in the lives of others and to bring about efforts to end human misery for all mankind," his children said in a statement. "One of his favorite quotes was, 'Reach for the stars. Although you will never touch them, if you reach hard enough, you will find that you get a little 'star dust' on you in the process.'"

Equal parts scientist and humanitarian, the Iowa-born Borlaug realized improved crop varieties were just part of the answer, and pressed governments for farmer-friendly economic policies and improved infrastructure to make markets accessible. A 2006 book about Borlaug is titled "The Man Who Fed the World."

"He has probably done more and is known by fewer people than anybody that has done that much," said Dr. Ed Runge, retired head of Texas A&M University's Department of Soil and Crop Sciences and a close friend who persuaded Borlaug teach at the school. "He made the world a better place _ a much better place. He had people helping him, but he was the driving force."

Borlaug began the work that led to his Nobel in Mexico at the end of World War II. There he used innovative breeding techniques to produce disease-resistant varieties of wheat that produced much more grain than traditional strains.

He and others later took those varieties and similarly improved strains of rice and corn to Asia, the Middle East, South America and Africa.

"More than any other single person of his age, he has helped to provide bread for a hungry world," Nobel Peace Prize committee chairman Aase Lionaes said in presenting the award to Borlaug. "We have made this choice in the hope that providing bread will also give the world peace."

During the 1950s and 1960s, public health improvements fueled a population boom in underdeveloped nations, leading to concerns that agricultural systems could not keep up with growing food demand. Borlaug's work often is credited with expanding agriculture at just the moment such an increase in production was most needed.

"We got this thing going quite rapidly," Borlaug told The Associated Press in a 2000 interview. "It came as a surprise that something from a Third World country like Mexico could have such an impact."

His successes in the 1960s came just as books like "The Population Bomb" were warning readers that mass starvation was inevitable.

"Three or four decades ago, when we were trying to move technology into India, Pakistan and China, they said nothing could be done to save these people, that the population had to die off," he said in 2004.

Borlaug often said wheat was only a vehicle for his real interest, which was to improve people's lives.

"We must recognize the fact that adequate food is only the first requisite for life," he said in his Nobel acceptance speech. "For a decent and humane life we must also provide an opportunity for good education, remunerative employment, comfortable housing, good clothing and effective and compassionate medical care."

In Mexico, Borlaug was known both for his skill in breeding plants and for his eagerness to labor in the fields himself, rather than to let assistants do all the hard work.

He remained active well into his 90s, campaigning for the use of biotechnology to fight hunger and working on a project to fight poverty and starvation in Africa by teaching new drought-resistant farming methods.

"We still have a large number of miserable, hungry people and this contributes to world instability," Borlaug said in May 2006 at an Asian Development Bank forum in the Philippines. "Human misery is explosive, and you better not forget that."

Norman Ernest Borlaug was born March 25, 1914, on a farm near Cresco, Iowa, and educated through the eighth grade in a one-room schoolhouse.

"I was born out of the soil of Howard County," he said. "It was that black soil of the Great Depression that led me to a career in agriculture."

Murray F. Buell Tribute
Submitted by: William R. Sharp

Murray F. Buell died July 2, 1975 while on a field trip in the New Jersey Pine Barrens. He had recently assumed the leadership of a natural resource study and at the time of his death was actively engaged in activities that delighted him throughout his lifetime: working with students, studying vegetation, and advancing the cause of conservation.

Murray Buell's influence on ecology was deep, constant and long sustained. This quiet, patient scholar was reared in a liberal New England family, and studied at the Loomis School, Cornell University, and the University of Minnesota. After studying under W. S. Cooper, he started his professional career at North Carolina State University in 1935. There he began his notable work on the paleoecology of bogs, plant succession, and tension zones between vegetation types. In 1947, he moved to Rutgers University where he eventually became Professor of Botany and Director of the William L. Hutcheson Forest. He devoted great effort in setting aside this forest and in making it into a major ecological study area and one of the best studied woods in North America. Well before it was fashionable, he initiated important studies linking ecology to land-use management. Two decades ago he and his students studied the impact of people on park ecosystems, investigated the ecology of power line right of ways and the use of fire on forest and hydrologic management. He also made intensive studies of the structure and dynamics of vegetation in and around New Jersey, and now the State is among the best known ecological regions in North America. Among his last works is the book *Vegetation of New Jersey* coauthored with Beryl Robichaud.

After his retirement from Rutgers in 1971 he served as a visiting professor of ecology at Yale, the University of Minnesota, Georgia, Arizona, California Davis, California Santa Barbara, Montana and Colorado State.

Perhaps Murray Buell's greatest impact on ecology was achieved as a teacher. A gentle and thoughtful man, he was considerate of his students, yet demanding of excellence. His influence on undergraduates resulted in a steady stream of students flowing to graduate schools, while the ecology program he initiated at Rutgers attracted scores of students from throughout North America. In Murray Buell they found a stimulating teacher concerned not only about the study of ecology, but about them as individuals. His life touched many hundreds of North American ecologists through the Rutgers Ecology Seminar that he initiated and sponsored. In the many summers he taught at the University of Minnesota's Lake Itasca Biological Station, he recognized exceptionally promising young students. Often the fortunate person was hired as an assistant, transported across the country in his car, fed chicken dinners and given a thorough introduction to life as a field ecologist. A summer at Itasca was the beginning for at least a half dozen current full professors of Ecology. Murray Buell's relationship with his students did not end with the award of a diploma. He actively followed their careers, acted as a sounding board for ideas and decisions, and provided wise counsel when asked.

No recounting of Murray Buell's career could be complete without mention of his wife, Dr. Helen Foot Buell. Murray and Helen Buell worked as a team and between them maintained a lively and inquisitive interest in all things around them. Many of us were fortunate to pass through their sphere of interest.

Murray Buell labored long and hard for the Ecological Society of America. He served as Associate Editor of Ecology and Ecological Monographs, Secretary, Vice President and as President in 1961-62. At the time of his death he was Chairman of the ESA Awards Committee. Less obvious, but nonetheless important, was his contribution to the drafting of the new constitution and bylaws of the ESA and his contribution to the early development of the Institute of Ecology. For his multifaceted contributions to ecology, Murray F. Buell was named Eminent Ecologist by the Society in 1971.

The loss of this scholar is great, but there is some satisfaction in knowing that Murray Buell died while fully active and in pursuit of the things he loved.

F. Herbert Bormann
Paul G. Pearson

List of Paleoecology Graduate Students
Murray F. Buell
provided by Allen M. Solomon
June 25, 2003

- John Cantlon
- Peter Comanor
- Ralph Good
- Kathy Harmon
- <u>Calvin Heusser</u>
- William A. Niering
- Bill Reiners
- Allen M. Solomon

Partial Bibliography of Palynology and Paleoecology Papers of Murray F. Buell
provided by Allen M. Solomon June 25, 2003

- Buell, M.F. 1939. Peat formation in the Carolina Bays. Bull. Torrey Bot. Club 66:483-487.
- Buell, M.F. 1945. Late Pleistocene forest of southeastern North Carolina. Torreya 45:117-118.
- Buell, M.F. 1946. Jerome Bog, a peat-filled "Carolina Bay." Bull. Torrey Bot. Club 73:24-33.
- Buell, M.F. 1946. Size-frequency study of fossil pine pollen compared with herbarium-preserved pollen. Am. J. Botany 33:510-516.

- Buell, M.F. 1946. A size-frequency study of *Pinus banksiana* pollen. J. Elisha Mitchell Scientific Soc. 62:221-228.
- Buell, M.F. 1947. Mass dissemination of pine pollen. J. Elisha Mitchell Scientific Soc. 63:163-167.
- Buell, M.F. 1970. Time of origin of New Jersey Pine Barrens bogs. Bull. Torrey Bot. Club 97:105-108.
- William S. Cooper and Helen Foot. 1932. Reconstruction of a late-Pleistocene biotic community in Minneapolis, Minnesota. Ecology 13:63-72. ESA Bull. 56(4): 26, 1975. Information retrieved on October 19, 2013 from http://www.palynology.org/murray-f-buell

A Tribute to Dr. Linda Styer Caldas
Authored By Sally Miller, Chip Styer and Sandy Styer

Linda Styer Caldas was born December 10, 1945 in Appleton, Minnesota, USA, the eldest of five children of Dr. Donald James and Carol Opal Hancock Styer. Linda's father was a career army dentist, and as a typical military family, the Styers moved often. Linda lived in several different states and in Germany prior to beginning undergraduate studies at the University of Colorado at Boulder. When her family returned from Germany and settled in the Washington, D.C. area, Linda transferred to George Washington University, where she studied biological sciences. Linda was a brilliant student and highly enthusiastic about science. She graduated from GWU in 1967 and entered the Department of Botany at The Ohio State University as a graduate student the same year in the laboratory of Dr. Carroll Swanson. She studied elm tree physiology, and completed her M. S. research in 1969 with her thesis entitled "Diurnal changes in radius of trunk and water potential of leaves of *Ulmus americana* L.".

Linda then entered the burgeoning field of plant tissue culture, specifically wild carrot tissue culture, for her PhD research. Dr. Rod Sharp, then a lecturer in the Department of Microbiology at OSU, met Linda in 1969 when they shared basement labs in the Botany and Zoology building. Rod remembers Linda as a star graduate student, extremely bright, energetic and passionate about her graduate program. "Linda was more like a faculty member than a graduate student. I remember Linda's participation in the weekly departmental seminar program when she delivered a remarkable presentation about plant embryology followed by a huge round of applause and a robust question and answer session. She was a gifted science writer and was always editing manuscripts for fellow students." While a graduate student, Linda met Ruy de Araujo Caldas, another bright and capable OSU graduate student in the same lab. Linda and Ruy married in 1971, and then moved to Brazil to begin their academic careers. Linda defended her dissertation entitled "Effects of Various Hormones on the Production of Embryoids of Wild Carrot (*Daucus carota*)" before a committee composed of both OSU and University of Sao Paulo - ESALQ faculty following her move to Piracicaba.

With Rod Sharp, Dr. Otto Crocomo (who at the time was the research coordinator of the Plant Biochemistry Sector of the Center for Nuclear Energy in Agriculture), and Ruy, Linda pioneered plant tissue culture research in Brazil. They launched collaborative research programs developing cell cultures for important cultivars of tropical crops with special interest in citrus, cocoa, coffee, beans, palm, and sugarcane. These programs were initiated in collaboration with geneticists and plant breeders at the University of Sao Paulo ESALQ Campus and the Institute of Agronomy in Campinas, Sao Paulo. Rod marveled that "Linda developed fluency in Portuguese within six weeks of her arrival and was participating in research meetings at CENA and USP-ESALQ – an amazing feat!"

In 1972, Linda and Ruy moved to the University of Brasilia (UnB), where Linda served on the faculty of the Botany Department until her retirement in 1998. She became a naturalized Brazilian citizen, and with Ruy had three children: Pedro, Cristina and Juliana.

Linda was a highly productive researcher in plant physiology and continued her pioneering work in plant tissue culture at UnB. She published more than 40 peer-reviewed journal papers, seven book chapters and five books, and served on several editorial boards. She chaired the UnB Botany Department from 1991 – 1993, and served as vice-director of the Biosciences Institute of UnB from 1994-1998. She was also a member of the Brazilian Agency for Education Development and Research (CNPq). She remained very active in academia after her retirement, including organizing the VII Brazilian Congress of Plant Physiology in 1999. She was awarded the title of Professor Emeritas of UnB in 2006. She spent two years as a professor at Catholic University of Brasilia after her retirement helping the university develop its biological sciences curriculum. Linda also collaborated with Bioplanta Technology Ltd., one of the first biotechnology companies in Brazil.

Linda was a dedicated teacher. Remarkably, after only a few years in Brazil, she wrote a biology textbook in Portuguese *"Principios Biológicos- uma Introdução"* to fill a much-needed niche in undergraduate education. This was the first textbook of its kind in Portuguese in Brazil. She also advised many M.Sc. and Ph.D. graduate students in plant science, particularly in botany and ecology. Linda was an early adopter of student-centered learning and committed her considerable energy and enthusiasm to her students.

Within a few years after their arrival in Brasilia, Linda and Ruy purchased a farm in Cristalina, where she fell in love with and became dedicated to the preservation of the Cerrado, the beautiful plains with tropical woodlands and scrub vegetation, an ecoregion found only in Brazil. Linda also recognized the risk of loss of Cerrado plant species by aggressive agricultural development. She created, and served as President of, the Cerrado Seeds Network, a preservation group funded by the National Fund for the Environment/Ministry of Environment in 2001[3]. On August 30, 2007, the Botanical Garden of Brasilia dedicated a garden in her memory to recognize her many contributions to the preservation of the Cerrado. Accounts of the dedication of the space are excerpted as follows:

"The space dedicated to Dra. Caldas is located in the Cerrado Medicinal Garden, next to the Visitor Center, an area of 500 m^2 of Cerrado maintained with native trees and species catalogued for medical use. Also, another seed she planted, the education efforts about the Cerrado, will be immortalized in the area, which will be visited over time from students and external audiences for environmental education programs. 'She was and always will be a model in the Federal District for research on the Cerrado. This is a way to make her energy and dedication to the perpetuation of the Cerrado known to future generations' said Jeanito Gentilini, director of Botanical Garden.

Dr. Kumiko Mizuta, 67, arrived in the federal capital in 1969 to join the faculty of UnB. In 1972 she met Linda, and together they built a relationship of friendship and dedication to biology and the Cerrado. 'What struck me the most was her dedication to work, her willingness to share knowledge and the captivating way she treated everyone around her' recalled Kumiko. Kumiko was beside Linda until the time that breast cancer, detected at an advanced stage, took the life of the researcher. 'She fought until the last moment and did not stop working. I remember her walking through the Cerrado and teaching even with the pains of the disease' said her friend.

Her children, Pedro, 35, Cristina, 33, and Juliana, 22, chose to follow in Linda's footsteps. 'My mom passed on her dedication and love for education and the environment as examples, always present in our experience', said Juliana Caldas, recently graduated in biological sciences from UnB. Pedro, who takes care of the family farm, opted for a major in agronomy and Cristina also followed the career of biologist and science writer. 'We are flattered by the honor of the garden dedication. For us it is very important to see her remembered the way she would like: in the Cerrado and sharing her knowledge with others', said Juliana.

Cerrado Seeds Network

With young researchers from Acesita Energetica, Linda developed mini-cutting (a cultural practice) for clonal multiplication of eucalyptus, used today industrially in Brazil, which leads the world in this technology. 'With a sliver of the plant, she could produce thousands of clones', Kumiko explained. Linda also published important books such as *Tissue Culture and Genetic Transformation of Plants*, a reference in biotechnological applications throughout Brazil. One of her final achievements was the creation of the Non-Government Organization (NGO) Cerrado Seeds Network.

For Gustavo Souto Maior, UnB professor and president of the Environmental Institute Brasilia, the inauguration of the Cerrado garden space represents the rapprochement between academia and conservation areas in DF. 'There are places to do research that are very rich, but have been forgotten. We want to revive the interaction between academia and the environment, following the examples that Linda has left us. Hundreds of people who visit the garden can read and reflect on a phrase written by Cristina, the daughter of the researcher, which summarizes the history of her mother: Cerrado: respect, know and love'." [1]

"About ten seedlings were planted at the inauguration of the space. Linda's friend Celina de Oliveira Martin believes the Botanical Garden is indeed the ideal place to celebrate Linda's achievements and contributions to biology. 'The wind, the water, the trees and the calm of this place sing her name. Whoever comes here meditating will be able to feel her energy and love of nature's beauty. Hopefully it touches other people and lets them continue the work of preserving the Cerrado she began' said Celina." [2]

In 2005, Linda returned to the US to spend several weeks with her father in Huntsville, Alabama, and with her siblings and their families across the US. With her usual energy, she took on the task of organizing hundreds of Styer family photos and documents, and wrote detailed histories of the family. These histories are a treasure to Linda's children, grandchildren, siblings, nieces and nephews, and to future generations. After Linda's death, her children found that she had written down the things for which she would like to be remembered: "the ability to be original and advocate innovative ideas, unrestricted love dedicated to friends and family, and contribution to biology, education and conservation of the Cerrado".[2]

We will always remember the kindness, creativity, energy, charisma and beautiful soul of our sister and friend, Linda Styer Caldas.

We thank Alba Clivati McIntyre and Claudio Vrisman for assistance with translation.

Notes

[1]http://pib.socioambiental.org/pt/noticias?id=48967

[2]http://www.secom.unb.br/unbagencia/ag0807-76.htm

[3]http://www.iesambi.org.br/parcerias_arquivos/triste_perda_lindacaldas.htm

Linda Styer Caldas Photograph 1

Linda Styer Caldas Photograph 2

For Those Who Made a Difference in My Scientific Life
Authored By Otto J. Crocomo

Excerpt from the Address delivered in the ceremony of granting the Fernando Costa Medal by the Association of Agricultural Engineers of the State of São Paulo on 15 June 2012.

The streetcar went down XV de Novembro Street, snaking through José Pinto de Almeida Street, reached the São Joao Street and entered the sacred territory of the agricultural sciences and parked next to a building bearing the word "Chimica" surmounting the facade of the Hall of Chemistry at the Escola Superior de Agricultura "Luiz de Queiroz"–ESALQ, University of São Paulo, in Piracicaba.

I got out of the streetcar, walked up the staircase, and entered its spacious lounge. As I did many times, my gaze followed once again the top of the wall bearing several bronze medallions showing the faces of illustrious ancestors whose decisions were crucial to further substantiate the already magnificent Agriculture School. The Fernando Costa Medallion had been there since the Pavilion was opened in 1930.

I was, at that time, attending the first year of ESALQ and I looked for Professor Euripedes Malavolta. I started working in his laboratories as an intern under his orientation. It was in that same laboratories that five years later, in June 1958, already an Agronomy Engineer, I was invited by Prof. Malavolta to submit to a contest for the "Livre Docente Degree" although not yet employed by the University of Sao Paulo.

That moment was a very important fact of my entire life. Had it not been for the confidence placed on me by Professor Malavolta I would have never been a recipient of the Fernando Costa Medal. So now, I pay a tribute to his memory (photo 1).

We all need one another. We are not juxtaposed individuals, each one living their own lives oblivious to what happens to their surroundings. As in all areas of the universe of human knowledge, in the world of biological sciences in which I live, an idea can arise in a single mind, but the realization of it requires many other minds to reason and observe with critical and keen eyes the results of the experiments: there needs to be many other hands to manipulate the test tubes and culture flasks, and many other feet to support the bodies that stand by the laboratory counters or walk between the lines, as if they were backstreets, and to separate and identify creeping or slender plants in the agricultural fields with the desirable agronomic characteristics.

As I use that metaphor I remember Demosthenes Santos Correa, my first Professor of Chemistry. With him I entered the realm of chemical reactions and laboratory techniques, which involved long hours, and I recall how he enthusiastically encouraged my participation in debates of Chemistry during the three years of my high school course. I am also remembering now Jose Dall Pozzo Arzolla, my Professor of Organic Chemistry at ESALQ. With him I got in touch with the dynamism of theoretical and practical classes in the laboratory, developing refined procedures for conducting experiments and the statistical interpretations of the results. I am remembering also Admar Cervellini, who greatly supported me in my research using radioisotopes, as Diretor of the Center Nuclear Energy in Agriculture – CENA. Much of that research such as mineral nutrition of plants was conducted at CENA in collaboration with my very good friend Andre Martin Louis Neptune. Also to the memories of these four great men I pay my tribute.

In the 50s and 60s of the last century the biochemistry of plants was emergent in Brazil and there I was entrenched in the biochemical intricacies responsible for the life of the plant cells. I remember one of my brightest student, Luiz Carlos Basso: long hours, day and night, spent making and repeating experiments, including the one that led to the discovery of a new enzymatic activity. My tribute to him.

I would like to pay tribute to William Rod Sharp who, in June 1971, arriving from the United States of America, joined me to introduce in Brazil the techniques of cell and plant tissue culture in agriculture, building since then, a fruitful relationship and scientific exchange between us (photo 2).

At that time, in June 1981, it was created at ESALQ, the Center for Agricultural Biotechnology –CEBTEC, thanks to the unconditional support I received from Aristeu Mendes Peixoto, Joaquim Jose de Camargo Engler and Paulo Fernando Cidade de Araujo. CEBTEC has always been supported by various ESALQ and FEALQ directors, among them Joao Lucio de Azevedo, Antonio Roque Dechen and Justo Moretti Filho who is no longer among us. I pay also tribute to them.

With the collaboration of my teammates and technicians at CEBTEC, mainly Helaine Carrer, Enio Tiago de Oliveira and Luiz Antonio Gallo, it was possible to give life to the real meaning of plant biotechnology: from cell to a viable plant, using the traditional improvement technology if necessary, and finally to commercialization. This was done with various plant species at CEBTEC in a very happy union among biochemistry, molecular biology and cell biology in our R & D projects followed by the transfer of technology to the private sector.

I need to emphasize the *sine qua non* collaboration I received from one of my greatest friends that began when we were undergraduate students at ESALQ, in the 50s, Ary A. Salibe, expert in vírus citrus. This collaboration made it possible for CEBTEC at ESALQ and the Faculty of Agronomy of UNESP, in Botucatu, State of São Paulo, in the 90s, the tripartite project with a private company, using the technique of micrografting, to produce "orange pear" plants resistant to the "tristeza" vírus. These plants are now being cultivated in an agricultural field – a striking example of a genuine biotechnological product. To Ary A. Salibe, who passed away in 2013, my tribute (photo 3).

There is not enough space to mention each one of my undergraduate and graduate students, my collaborators in Brazil and abroad; nevertheless, each one is present in my memory and in my heart.

I'd like to quote Albert Einstein: "A hundred times every day I remind myself that my inner and outer life depend on the labors of other men, living and dead, and that I must exert myself in order to give in the same measure as I have received."

PHOTO 1
OTTO (LEFT) AND MALAVOLTA (RIGHT), NOVEMBER 1990

PHOTO 2
ROD AND OTTO WORKING ON THE BOOK, AND DIVA.
OTTO'S HOME. PIRACICABA, SEPTEMBER 2013

PHOTO 3
ARY (LEFT) AND OTTO (RIGHT). SHOWING SAMPLES
OF THE ORANGE "PEAR" FREE OF THE "TRISTEZA"
VIRUS. IN THE FIELD. VOTORANTIM CITRUS
FARM. ITAPETININGA, SP, BRAZIL, 2003.

Para Aqueles Que Influenciaram Minha Carreira Carreira Cientificia
Authored By Otto J. Crocomo

Baseado no Discurso de agradecimento ao ser agraciado com a "Medalha Fernando Costa" pela Associação dos Engenheiros Agrônomos do Estado de São Paulo em 15 de junho de 2012.

O bonde desceu a Rua XV de Novembro, serpenteou pela Rua José Pinto de Almeida, alcançou a Rua São João, adentrou o território sagrado das ciências agronômicas e estacionou próximo a um prédio que ostenta a palavra "Chimica" encimando a fachada do Pavilhão de Química da Escola Superior de Agricultura "Luiz de Queiroz" da Universidade de São Paulo, em Piracicaba.

Desci do bonde, subi a escadaria, entrei em seu amplo saguão. Meu olhar uma vez mais dirigiu-se para o alto da parede que ostenta vários medalhões de bronze mostrando as faces de ilustres antepassados cujas decisões foram cruciais para alicerçar ainda mais a já, naquela época, magnífica Escola Agrícola. O medalhão de Fernando Costa lá está, desde quando o Pavilhão foi inaugurado em 1930.

Estava eu, naquele momento, cursando o primeuiro ano da ESALQ e naquela manhã de novembro de 1953 procurei pelo Professor Eurípedes Malavolta. Passei a trabalhar nos seus laboratórios como estagiário, sob sua orientação. Foi nesse mesmo Pavilhão que, 5 anos mais tarde, em junho de 1958, já sendo eu engenheiro agrônomo, recebi o convite do Professor Malavolta para prestar concurso para "Livre-Docente", apesar de ainda não pertencer ao quadro de contratados pela Universidade de São Paulo. Não fora a confiança em mim depositada pelo Professor Malavolta a mim não teria sido outorgada a Medalha "Fernando Costa". A ele, portanto, a minha sincera homenagem (foto 1).

Todos nós precisamos uns dos outros. Não somo indivíduos juxtapostos, cada um vivendo sua própria vida alheio ao que se passa em seu entorno. Como em todas as áreas do universo do conhecimento humano, também no mundo das ciências biológicas em que vivi e ainda vivo, uma ideia pode surgir em uma única mente, mas a concretização e as provas de sua veracidade exigem muitas outras mentes que raciocinem e observem com olhos críticos e clínicos os resultados dos experimentos, muitas outras mãos que manipulem os tubos de ensaio e os frascos de cultura e muitos outros pés que sustentem os corpos que ficam diante das bancadas dos laboratórios ou caminhem pelas entrelinhas, como se ruelas fossem, a separar e identificar plantas esguias ou rasteiras, nos campos agrícolas, com características agronômicas desejáveis.

Ao usar essa metáfora, lembro-me de Demósthenes Santos Correa, meu primeiro Professor de Química. Com ele penetrei no reino das intrincadas reações químicas, com horas e horas de práticas de laboratório, incentivando-me a participar de debates de Química durante os 3 anos de meu Curso Colegial. Lembro-me também de José Dall Pozzo Arzolla, meu Professor de Química Orgânica no Curso de Graduação na ESALQ. Com ele familiarizei-me com o dinamismo das aulas teóricas e práticas, desenvolvendo refinados processos experimentais e aplicando métodos estatísticos na interpretação dos resultados. Estou também lembrando-me de Admar Cervellini que, como Diretor do Centro de Energia na Agricultura – CENA, me proporcionou todas as facilidades para a realização de experimentos utilizando radioisótopos. Muitos desse experimentos, como aqueles sobre nutrição mineral de plantas, foram conduzidos coma colaboração de meu grande amigo André Martin Louis Neptune. A todos esse 4 homens as minhas homenagens.

Nas décadas de 50 e 60 do século passado a bioquímica de plantas era incipiente no Brasil e lá estava eu entranhado nos meandros bioquímicos responsáveis pela vida das plantas. Recordo-me de um dos meus mais brilhantes orientados Luiz Carlos Basso: longas horas, de dia e de noite, passamos fazendo e repetindo experimentos dentre os quais aquele que nos levou à descoberta de uma nova atividade enzimática em células vegetais, nos finais dos anos 60 e início dos anos 70. A ele minhas homenagens.

Alguém a quem não poderia deixar de prestar minha homenagem é William Rod Sharp que, em junho de 1971, vindo dos Estados Unidos da América do Norte, colaborou comigo na introdução no Brasil das técnicas de cultura de células e tecidos de plantas em agricultura. Essa nossa atividade foi importante para que a ESALQ e o CENA contribuíssem sobremaneira para o desenvolvimento da biotecnologia de plantas no Brasil (foto 2).

Nessa época, em julho de 1981, foi criado na ESALQ o Centro de Biotecnologia Agrícola –CEBTEC, graças ao apoio de Aristeu Mendes Peixoto, Joaquim José de Camargo Engler e Paulo. Fernando Cidade de Araujo. Esse Centrp sempre foi apoiado pelos vários Diretores da ESALQ e da Fundação de Estudos Agrários Luiz de Queiroz – FEALQ, entre eles João Lúcio de Azevedo, Antonio Roque Dechen e Justoi Moretti Filho, o qual já não mais se encontra entre nós. Minhas homenagens a cada um deles.

Com a colaboração de Helaine Carrer, Enio Tiago de Oliveira e Luiz Antonio Gallo, meus atuais continuadores no CEBTEC, é que foi possível dar vida ao real significado da biotecnologia de plantas: da célula à planta viável, seguido, se necessário, da metodologia tradicional de melhoramento, e finalmente à comercialização. Assim foi feito com várias espécies de plantas, vivenciando-se no CEBTEC uma feliz união entre a bioquímica, a biologia molecular e a biologia celular nos nossos Projetos de Pesquisa e Desenvolvimento e de transferência tecnologia à iniciativa privada. Aqui eu lhes presto minhas homenagens.

Tenho de enfatizar a colaboração *sine qua non* de Ary Aparecido Salibe na realização de um projeto tripartite entre a USP, a UNESP e empresa privada, para obtenção de plantas de laranja "pera" livre do "vírus da tristeza", utilizando a técnica de cultivo de microenxertos. As plantas estão sendo cultivadas em campo agrícola, e seus produtos comercializados. Exemplo marcante de um produto genuinamente biotecnológico. Salibe faleceu em novembro de 2013. Para ele, a minha homenagem (foto 3).

Não há espaço suficiente para mencionar e homenagear todos os meus alunos de graduação e de pós-graduação, os meus colaboradores do Brasil e do Exterior. Cada um deles está presente em minha memória e em meu coração.

Quero citar Albert Einstein:

Centenas de vezes todos os dias eu me conscientizo de que a minha vida interior e exterior depende do trabalho de outros homens, vivos ou que já não mais estão entre nós, e que eu devo devolver à humanidade na mesma medida o que eu dela recebi".

PHOTO 1
OTTO (À ESQUERDA) E MALAVOLTA (À
DIREITA). NOVEMBRO, 1990

PHOTO 2
ROD E OTTO EDITANDO O LIVRO, COM DIVA.
RESIDÊNCIA DE OTTO. PIRACICABA, SETEMBRO 2013

PHOTO 3

ARY (À ESQUERDA) E OTTO (À DIREITA). LARANJA "PERA"
LIVRE DO VIRUS DA "TRISTEZA". NO CAMPO. FAZENDA DE
CITRUS DA VOTORANTIM. ITAPETININGA, SP, BRASIL, 2003.

David A. Evans Tribute
Submitted By William R. Sharp

NEW BRUNSWICK, N.J., June 5 /PRNewswire/ -- David A. Evans Ph.D., Chief Executive Officer of WellGen, Inc., passed away on June 1, 2006, following a brief illness. Dr. Evans, 54, was a well-regarded scientist and business leader in the food, biotechnology and nutrigenomics industries. WellGen, Inc. is a privately owned biotechnology company using nutrigenomics to develop ingredients that reduce the risk and severity of disease for the food, therapeutics, and dietary supplement markets. Richard Laster, Chairman of WellGen, said, "We are all shocked and deeply saddened by this terrible, unexpected occurrence. Dave Evans was a wonderful colleague and friend. He was highly intelligent and filled with energy and new ideas. Over the years, Dave had earned tremendous respect throughout our industry as a serious scientist and businessman, and had been doing an outstanding job guiding WellGen towards a very promising future. We will greatly miss him." Mr. Laster stated that the Board of Directors will meet soon to decide on the longer-term leadership of the Company. In the interim Mr. Arthur Finnel, WellGen's Chief Financial Officer, will assume day-to-day operating responsibilities for the Company. Dr. Evans joined WellGen in 1999 as CEO and President. From 1981-1999, Dr. Evans was with DNA Plant Technology Corporation, where he was Co-Founder and Executive Vice President of Business and Product Development. While at DNAP, he was on the Board of Directors of several joint ventures with DuPont, Union Carbide, and others. He introduced several new products into DNAP's $300 million produce operation, from research through production to market introduction.

Dr. Evans earned his B.S., M.S. and Ph.D. from Ohio State University in Genetics, and completed the PMD at Harvard Business School. He was formerly Assistant Professor, State University of New York, and Research Manager, Campbell Soup

Co. Dr. Evans was an inventor on 12 US patents and published over 100 scientific papers. He was an adjunct faculty member of the Department of Biology at Rutgers University from 1983-1994. In addition to his business activities, Dave pursued many interests. He was an avid birder, completed the 2005 New York City Marathon and played racquetball daily. WellGen, Inc., is based in New Brunswick, NJ, and developing products for food, therapeutics, and dietary supplement markets. WellGen's technical platform is a method of screening the effect of food and related substances on the expression of genes associated with human health conditions. The company has developed proprietary substances that help reduce risk and severity for a variety of diseases.

Source:

Information retrieved October 19, 2013 from http://m. prnewswire.com/news-releases/dr-david-a-evans-chief-executive-of-wellgen-inc-passes-away-after-sudden-illness-55878842.html

David E. Fairbrothers, 1925-2012 Tribute
Authored By David Lee and Dennis Stevenson

David E. Fairbrothers, a long-time Rutgers professor and eminent botanist and systematist, passed away on October 29th, 2012,
after a lengthy illness. He had a distinguished academic career
in the field of plant molecular systematics, and was a leader in
the conservation of plants and natural areas, particularly in his
home state of New Jersey. David was born and raised in Absecon
during the depression years, part of a family of commercial fishermen and duck hunters, and he grew up close to nature. Soon
after graduating from high school in 1943, he joined the Army
when he turned eighteen. An excellent marksman as a young
man, he became a Sergeant Rifleman and Squad Leader of the L
Company, 376th Regiment of the 94th Infantry Division. Landing
at Utah Beach on the second day of the Allied invasion, his company fought its way across France and was in the middle of the
infamous Battle of the Bulge during the bitterly cold winter of
1944-45. In a frozen pothole during that battle, he suffered severe
frostbite of his lower legs and barely escaped having both of them
amputated, injuries that affected him the rest of his life. After
hospital recuperation, David was stationed in Prague; after the
German surrender on May 8th of 1945, he helped supervise train
convoys of starving and ill refugees returning to their homes.

David was discharged from the Army in February of 1946 at the
age of 20, and he took advantage of the G.I. Bill to enter Syracuse
University. He met his future wife Marge while in school, and they
married in 1949. He graduated in 1950 and immediately enrolled
as a graduate student in Botany at Cornell University, under
the direction of Robert T. Clausen. David worked in the area of
grass systematics, completing a Master's degree (*A Cytotaxonomic
Investigation within the Genus Echinochloa*) in 1952, and his Ph.D.
(*Relationships in the Capillaria Group of Panicum*)[1] in 1954. David
was recruited for a faculty position at Rutgers University by the

eminent plant morphologist and head of the Graduate School, Marion L. Johnson. David began his 34 year career at Rutgers, the autumn semester of that year. He was the department's taxonomist and Director of the Chrysler Herbarium; during his tenure, its collections increased from 37,000 to over 140,000 specimens. His successful career at Rutgers was marked by two traits: (1) an intimate knowledge of plant and habitat diversity in the small but biologically rich state of New Jersey; and (2) an interest in employing new and multiple techniques (perhaps influenced by his earlier use of cytogenetics) in plant systematics. He was in the right place at the right time. Rutgers was the home of the Serological Museum, which had contributed significantly to advances in zoological systematics, particularly among birds, from the leadership of Alan Boyden and his junior colleague Ralph De Falco.[2] With the help of Marion Johnson, David learned the immunological techniques Boyden and De Falco had employed in studying animals, and they applied them to problems in plant systematics.[3] Boyden died in 1962, and Johnson in 1964; Fairbrothers then developed an independent research program in plant molecular systematics that utilized a growing arsenal of techniques, starting with immunology, adding polyacrylamide gel electrophoresis (PAGE) and isoelectric focusing, and secondary compounds (terpenoids and flavonoids) to a range of taxonomic problems, from population variation within species, to hybridization and introgression between closely related species, and to phylogenetic relationships among different families. This laboratory operated until his retirement in 1988, and it was the setting for the training of 29 graduate students. David was an excellent mentor, supportive of students and always available for discussions. His students will treasure the memories of field trips in his station wagon, cruising up the New York Thruway as David ticked off the names of roadside grasses and composites. The lab was also the temporary home of nine faculty members visiting on sabbaticals, and six post-doctoral fellows. The majority of his 122 peer-reviewed articles were in the field of plant molecular systematics. This body of research, along with the laboratory at the University of Texas, formed the backdrop for the revolution in systematics that came with the application of techniques for the analysis of DNA, starting with DNA hybridization (actually quite reminiscent of the serological research) and then sequencing.[4] Although David is considered by most to be a flowering plant taxonomist, his interests were actually more eclectic as evidenced by his co-authorship of the *Ferns of New Jersey* and, at the time of his illness, his interest in and study of the marine algae of the state parks in New Jersey.

Two of David's close friends were Arthur Cronquist (1919-1992) of the New York Botanical Garden and Armen Takhtajan (1910-2009) of the Komarov Institute in St. Petersburg. Whenever Armen was in New York, David visited Armen there and/or he travelled to Rutgers. David frequently attended the Torrey Botanical Club meetings at the New York Botanical Garden accompanied by Rutgers students. He and the students often went for the day, and thus the students could use the herbarium and interact with other botanists. David met Art Cronquist for discussions about Art's system of classification and David's deep knowledge of the New Jersey Flora when Art was revising Gleason and Cronquist's *Manual of the Vascular plants of North-eastern United States and Adjacent Canada*. As part of his commitment to north-eastern botany, from 1990-1998, David Fairbrothers served as the Torrey Botanical Club representative on the Botanical Science Committee of the Board of Managers at the New York Botanical Garden. He was also a long-time mentor for the Flora of New Jersey Project (www.njflora.org) which has close links with both NYB and the Torrey Botanical Society.

David's deep knowledge of New Jersey botany was a mother lode for projects that his students pursued, and some of them involved work on endangered species and habitats. In time, he became more focused on practical issues of endangered species and habitat management, and this coincided with the environmental movement of the 1970s. For decades, the Chrysler Herbarium had grown in the range of collections, particularly of endangered species and habitats. It eventually became the resource that allowed Fairbrothers and the herbarium manager, Mary Hough, to complete (to our knowledge) the first state description of threatened and endangered plant species.[5] This publication was influential in the modification of the Endangered Species Act, first passed by Congress at the end of 1973 and modified in 1975, to include plants, and to stimulate other states to conduct similar surveys. As a south Jersey native, David had great affection for the Pinelands and deep knowledge of its plants and natural history; he worked with others to protect this special area. He helped prepare "A Plan for a Pinelands National Preserve," and presented it to the U.S. Senate (through its Parks and Recreation Sub-Committee) and assisted substantially in the passage of the act authorizing it in 1978. His study of endangered and threated plants in the pinelands led to two publications that were instrumental in the establishment of the comprehensive management plan for the reserve,[6,7] which explicitly mentioned the initial 54 species to be protected. This act established the first Federal Reserve, similar in intent to the Catskills and Adirondack Parks in New York, but partly under the umbrella of the National Park Service and managed by the state of New Jersey. Later, the pinelands were added to the Federal Natural Preserve system, and then to the UNESCO Global Biosphere Reserve system in 1988.

Later in his tenure at Rutgers, he performed more administrative service, inaugurating the establishment of the Department of Biology (with 89 faculty members) as its first chairperson. Although he continued to be involved at Rutgers, through advising and consulting, he retired as a Distinguished Professor in 1988. His activity in endangered plant and pinelands issues at the state and federal levels continued well into his retire-ment. He and Marge moved south to Toms River, and David lent his support to the protection of natural communities in nearby Island Beach State Park. There, he helped with the development of the Emily de Camp Herbarium at the Forked River Interpretive Center, and helped to document plants and communities at the park. David and Marge frequently visited their son and daughter, spouses, and five grandchildren. They pursued new interests. Familiar with the history of New Jersey, David became interested in the antique glass produced in the state, then in silver overlay antique glass, and they both contin-ued studying and collecting other antiques further afield. He became a sought-after lecturer on these subjects. Because of his declining health, he and Marge moved to Lebanon, NH, in 2010 to be near their son. David died two years later, at the age of 87 and after 63 years of marriage.

In recognition of his accomplishments, David received sev-eral awards. In addition to a variety of teaching, research and administrative awards at Rutgers, he was awarded the Rutgers Medallion in 1988. The Chrysler Herbarium and other collec-tions were re-organized as part of the university biodiversity col-lections, and a symposium and banquet were held in his honor in 2005, to launch the fundraising effort to establish the David E. Fairbrothers Plant Resources Center. The Botanical Society of America presented him with its Merit Award in 1989, in com-memoration of his research discoveries and service to the society. For his contributions to conservation in New Jersey, The Garden Club of New Jersey awarded him its Gold Medal in 2008, and the Pinelands Preservation Alliance placed him in its Pine Barrens Hall of Fame, also in 2008.

He had a long, productive and happy life, and he will be deep-ly missed by his family, many former students, professional col-leagues and personal friends.

David Lee
Department of Biological Sciences
Florida International University
Miami, FL 33155

Dennis Stevenson
New York Botanical Garden
Bronx, NY 10458
Republished from Plant Science Bulletin

References

Fairbrothers, David E. 1953. Relationships in the Capillaria Group of *Panicum* in Arizona and New Mexico. *American Journal of Botany* 40:708-714.

Strasser, Bruno J. 2010. Laboratories, museums and the comparative perspective: Alan A Boyden's quest for objectivity in serological taxonomy, 1924-1962. 2010. *Historical Studies in the Natural Sciences* 40:149-182.

Fairbrothers, David E. and Marion A. Johnson. 1961. The precipitin reaction as an indicator of relationships in some grasses. *Recent Advances in Botany* (University of Toronto Press, Toronto), pp. 116-120.

1. Jensen, Ü. and David E. Fairbrothers, eds. 1983. *Proteins and Nucleic Acids in Plant Systematics*. Springer Verlag, Heidelberg, 408 P.
2. Fairbrothers, D. E. and M. Y. Hough. 1973. Rare or endangered vascular plants of New Jersey. *New Jersey State Museum of Science Notes* 14:1-53.
3. Fairbrothers, D. E. 1979. Endangered, threatened, and rare vascular plants of the Pine Barrens and their biogeography, pp. 395-405. In R. T. T. Forman, ed.: *Pine Barrens: Ecosystem and Landscapes*. Academic Press, New York. (Forman was at Rutgers 1966-1988, and was an

important ally in the campaign to save the Everglades; he then moved to the Harvard School of Design as its landscape ecologist).

4. Caiazza, N. and D. E. Fairbrothers. 1980. Threatened and endangered vascular plant species of the New Jersey Pinelands and their habitats. Prepared for the New Jersey Pinelands Commission, New Lisbon, NJ.

Figure 1. David with long-time friend Armend
Takhtajan at Rutgers University in 1968.

Figure 2. David and Marge Fairbrothers relaxing at Frazer's Hill, Malaysia, in 1975, after he gave a keynote address at the symposium inaugurating the Rimba Ilma, still the only scientific botanical garden in Malaysia. David was fifty years old at the time.

Figure 3. David Fairbrothers leading a tour of the Webb's Mill Bog in the Pinelands Preserve, the day following the symposium in his honor, in June of 2005. David was 80 years old at that time.

Percy Cyril Claude Garnham Tribute
Authored By Francis Edmund Gabriel Cox

Percy Cyril Claude Garnham, known by his intimates as Claude, was born in London on January 15th 1901. He was educated at the Paradise School in London and St Bartholomew's Hospital where he graduated in medicine in 1923. In 1928 he was awarded the degree of MD for which he received the Universality of London Gold Medal. Between 1923 and 1925 he studied at the London School of Hygiene and Tropical Medicine and travelled to Paris, Amsterdam and Rome where he worked with some of the most eminent parasitologists of the time. In 1925 he joined the Colonial Medical Service at the Medical Research Laboratories in Nairobi, Kenya, and in 1928 became Director of the Division of Insect-Borne Diseases where he worked on malaria, plague, yellow fever, leishmaniasis and river blindness and their vectors. In 1947 he joined the staff of the London School of Hygiene and Tropical Medicine first as Reader in Medical Parasitology and later as Professor of Medical Protozoology and Head of the Department of Parasitology. It was during his time at the London School that he, together with Professor Henry Shortt, began his search for the enigmatic stages in the life cycle of the malaria parasite between the injection of sporozoites by a mosquito and the appearance of parasites in the blood. Garnham's interests in malaria focussed on the parasites themselves rather than the disease and he had an encyclopaedic knowledge of the malaria parasites of primates. His understanding of the life cycle of *Hepatocystis kochi* led him to the discovery of the liver stages of this parasite and subsequently those of the primate malaria parasite, *Plasmodium cynomolgi*, and the human parasites, *P. ovale*, *P. vivax* and *P. falciparum*. These discoveries revolutionised the treatment of malaria and opened up the possibility of a vaccine against malaria, an aim still unfulfilled. Garnham retired in 1968 but continued to work as a Senior Research Fellow at Imperial College, London, until 1980.

Cyril Garnham's achievements have been recognised by his election to a Fellowship of the Royal Society in 1964 and his appointment as a Companion of the Order of St Michael and St George (CMG) in the same year. Garnham never sought honours but many were showered on him including fellowship or membership of more than 12 overseas learned societies. He also served parasitology and tropical medicine as President of the Royal Society of Tropical Medicine and Hygiene from 1967-1969, President of the British Society for Parasitology from 1970-1972 and President of the European Federation of Parasitologists in 1971.

Garnham's publication list is impressive, over 400 publications, including his scholarly and classic book, *Malaria Parasites and other Haemosporidia* (1966). He was always in great demand as a speaker at international congresses and was an inspired and generous teacher and many of his students have themselves reached the higher echelons of tropical medicine and parasitology and carried on his tradition.

I first met Cyril Garnham while working as a temporary lab boy at the London School of Hygiene and Tropical Medicine during my university vacations and I must have impressed him because he invited me to work with him after I had graduated. When I was studying for the Diploma in Parasitology and Applied Entomology at the London School he supervised my dissertation and later acted as an informal advisor for my PhD thesis on Host-Parasite Relationships in the Haemosporidia. Afterwards he guided my career with a real paternal interest and I was fortunate to be able to serve with him on the Council of the Royal Society of Tropical Medicine and Hygiene an experience from which I learned a great deal about committee work. He was one of the most formal people I have ever met and it wasn't until I was a professor that he deigned to call me 'Cox' instead of what he regarded to be the more formal Dr Cox. He only once called me Frank!

Cyril Garnham died on December 25th 1994 at and his memorial service at the church of St Bartholomew, appropriately close to St Bartholomew's Hospital Medical School, was crowded with many of the most eminent scholars of tropical medicine and parasitology from the UK and overseas.

Leonard George Goodwin Tribute
Authored By Francis Edmund Gabriel Cox

Leonard George Goodwin, always known as Len or LG, was born in Wood Green in North London on July 11 1915 and went to school at the William Ellis School in London. He read Botany and Zoology at University College London and, after graduation, switched to Pharmacy in which subject he qualified in 1935 and later qualified in medicine. In 1939, shortly before the beginning of the Second World War, he went to work at the Wellcome Bureau of Scientific Research where he investigated the chemotherapy of leishmaniasis, then a serious problem among British troops particularly in Sicily. It was while he was working on the use of pentostam for the treatment of leishmaniasis that he realised that the criteria used for estimating drug dosages were inadequate and began to devise a more rational approach to drug usage. He also worked on the chemotherapy of malaria, sleeping sickness and bilharzia (schistosomiasis) and developed an index to be used for testing drugs. After the war, he remained at the Wellcome Bureau and in 1958 became Director of the Wellcome Laboratories' of Tropical Medicine. In 1964 he was appointed Director of the Nuffield Laboratories of Comparative Medicine attached to the Zoological Society of London and then Director of Science at the Society. He retired in 1980.

Len Goodwin's contribution to science was acknowledged by his election to the Royal Society in 1976 and the award of the Companion of the Order of St Michael and St George (CMG) in 1977. He was President of the Royal Society of Tropical Medicine and Hygiene from 1979-1981 and President of the British Society for Parasitology from 1964-1966.

One little known fact about Len Goodwin is that, while he was working at the Wellcome Laboratories, he pioneered the use of the Syrian, or golden, hamster, *Mesocricetus auratus*, for medical research in the UK. The importance of these animals is that they are all derived from a single litter and are therefore genetically identical. All the pet hamsters in the UK are derived from this first Wellcome colony. He also kept wallabies and was a very good sketch artist.

Len Goodwin was a very private and gentle person who never actively sought any honours or public acknowledgement of his work which was all carried out with the minimum of fuss. He was very much liked by his staff and everyone else who took the trouble to get to know him.

I first met him while serving on the Council of the British Society for Parasitology and later when I was editing the *Wellcome Trust History of Tropical Diseases* where he became a frequent and welcome visitor to my office at the Wellcome Trust Building in Regent's Park and it is with his help and guidance that this book, that he had instigated, came to fruition in 1996.

Len Goodwin died on November 25th 2008.

James E. Gunckel 1914-2011 & Roberta Jean (Longworth) Gunckel 1918-2012
Authored By Alan Knight

Dr. James E. Gunckel, a retired professor of botany at Rutgers University, died September 19, 2011, at Monroe Village, Jamesburg, N.J. He was 97. He was survived by his 93-year companion and wife, R. Jean (Longworth) Gunckel, a son, Fred James Gunckel of Albuquerque, N.M., and daughter, Nancy Gunckel Knight, of Duanesburg, N.Y., as well as two grandsons, Jeffrey A. Knight, Brooklyn, N.Y. and Matthew James Knight, Ithaca, N.Y., and four great-grandchildren. Born in Dayton, Ohio, he graduated from Miami University, Oxford, Ohio, in 1938 and received his doctorate at Harvard in 1946, where he studied and began his research career under Dr. Ralph Wetmore and Dr. Kenneth Thimann. At Rutgers, he chaired what was then called the Botany Department for many years and did pioneering work in two important areas of study: tissue culture (plant cloning) and radiation biology, where he produced benchmark studies on the effect of radiation on a variety of plant species. A prolific publisher of scientific articles, he presided at many national and international botanical meetings, served as the translating editor of the seminal German botanical text General Botany by Wilhelm Nultsch, and edited the textbook Current Topics in Plant Science. A former president of the Torrey Botanical Society, the oldest botanical society in America, he also served many years as editor of The Bulletin of the Torrey Botanical Society, a refereed botanical journal. In 1959/60, having been awarded a Waksman Foundation Fellowship, he did meristem (plant tissue cloning) research at Station Centrale de Physiologic Vegetate, at Versailles, France, under the tutelage of Dr. Georges Morel. His unequaled knowledge of radiation biology, much of it gained through his many summers of research at the Brookhaven National Laboratory, led to his being called upon to provide expert testimony in a legal case pertaining to the

Three Mile Island nuclear power plant accident. Dr. Gunckel was particularly proud of the career achievements of his many graduate students at Rutgers. They always wanted to thank him for his commitment to preparing them to assume leading research positions in academia. Professor Gunckel would say, "Don't thank me. Just pass it on to the next generation." His quiet devotion to his home gardens in Somerville and to the Second Reformed Church of New Brunswick, where he was an ordained elder and served in volunteer leadership roles, was well known to his close friends.

Roberta Jean (Longworth) Gunckel

Roberta Jean (Longworth) Gunckel died Friday, April 13, 2012 at Monroe Village, Jamesburg, N.J. She was 94. The daughter of a mining engineer, she was born in British Columbia and grew up in Copper Hill, Tennessee. She attended Duke University and graduated from Miami University (Ohio) with a degree in elementary education. Her career, like her retirement years, was devoted to children. She taught first grade for 23 years in the Highland Park and Bridgewater-Raritan school districts, a career she said she enjoyed every day.

A long-time resident of Bridgewater, N.J, she is survived by daughter, Nancy Gunckel Knight, of Duanesburg, N.Y.; son, Fred James Gunckel, of Albuquerque, N.M.; two grandchildren, Jeffrey Knight, of Brooklyn, N.Y. and Matthew Knight of Ithaca, N.Y. and four great grandchildren. She was predeceased by her husband, James Eugene Gunckel, who passed away in September.

The Team

Jim and Jean Gunckel were the backbone of the Rutgers' Department of Botany during Jim Gunckel's leadership years as professor and chair. The two of them and their children Nancy and Fred hosted multiple events at their magnificent Somerville home for faculty, new faculty and graduate students which often included lodging. These social events included gourmet dinners, dinner parties, bridge tournaments, departmental picnics and receptions. These events promoted strong bonds among the faculty and graduate students which led to the department's premiere national and global reputation. Jean Gunckel enthusiastically shared these leadership responsibilities at Rutgers in addition to her many responsibilities as an educator.

Information Resources:

Information retrieved October 17, 2013 from http://www.legacy.com/obituaries/app/obituary.aspx?pid=157540580Information retrieved October 16, 2013 from http://www.legacy.com/obituaries/mycentraljersey/obituary.aspx?pid=153779072

Roger F. Keller Tribute
Submitted By William R. Sharp

DR. ROGER F. KELLER JR., born in Manchester, NH, passed away at age 89 on December 28, 2011 in Akron, OH. He was preceded in death by his wife, Arline; and his son, John Roger. He is survived by his children, Nancy (Todd) Kislak of Agoura Hills, CA., and Brian (Connie) Keller of Hudson, OH.; grandchildren Heather (Sonny) McClinsey, Michelle and Sarah Kislak, Kendra and Kyle Keller; and great-grandsons, Nicholas Baker and David McClinsey. Roger enlisted in the U.S. Army Oct. 1942, was in the ROTC 3 years at UNH, and was enrolled in Office Candidate School in Ft. Knox, KY., April 1944. He served active duty as a Lieutenant from April 1944 until 1946. He was in the 11th Armored Division in General Patton's 3rd Army during WWII. He received a Purple Heart due to injuries sustained in the Battle of the Bulge. During his time at Walter Reed Army Medical Center in Washington, D.C., where he met the love of his life, his wife Arline. Originally from Milwaukee, WI. Arline served as an occupational therapist at the Walter Reed. Roger received his PhD in Zoology from Michigan State College in 1953. He served as professor and chair at the University of Akron in the Biology Department and provided research guidance to undergraduate and graduate students in the classroom and laboratory until his retirement in 1985. He was a Professor Emeritus of Biology and Professor Emeritus at the Community and Technical College. Roger served on the Board of Trustees at the Akron Zoo for decades. He was a member of the Sons of the American Revolution in his retirement. He was a charismatic and well-respected professor and dearly loved by his family and friends. He unselfishly mentored significant numbers of students seeking medical and graduate degrees.

Dr. Keller provided important leadership in building the University of Akron into a world class research institution. He possessed the uncanny ability to develop cross-campus collaborations in the recruitment of a top ranked research faculty, provide the essential resources for their advancement and develop significant undergraduate and graduate research initiatives.

Information sourced from the *Cleveland Plain Dealer*, Retrieved October 16, 2013 from \http://obits.cleveland.com/obituaries/cleveland/obituary.aspx?pid=15525260

Miodrag Ristic Tribute
Authored By Julius Kreier

Miodrag Ristic was a professor of veterinary pathology and hygiene in the veterinary college of the University of Illinois at Urbana Champaign Illinois. I became his first graduate student shortly after he joined the veterinary college of the University of Illinois. Our relationship persisted for many years after I finished my graduate studies with him and became a professor of microbiology at the Ohio State University.

Dr. Ristic was born in Serbia. He became a prisoner of war when the Germans occupied Serbia before they invaded Russia. He was liberated from the Camp by Canadian troops participating in the Allied invasion of German occupied Europe.

He was quite a linguist. He spoke English, German, French and Russian in addition to his native Serbo-Croatian. An officer of the Canadian troops who he contacted after the camp collapsed recognized his language skills and he remained with the Canadian officer has translator until the war ended. After the war ended he remained in Germany, married a German woman, and with Canadian support he enrolled in a veterinary college. Sometime after he completed the veterinary program he immigrated to the United States where he obtained a position in the veterinary college of the University of Florida at Gainesville. There he developed a successful well-funded research program. At about the same time the veterinary college at the University of Illinois hired a new Dean who wished to bring new faculty active in research and bringing in research grants to enhance research at the school. One of the people he wished to hire was Dr. Ristic.

There arose a problem in his plan when he found that the University of Illinois, unlike the University of Florida, required all faculty to have a PhD, a veterinary degree alone was not sufficient. To solve this problem the new Dean enrolled Dr. Ristic in a PhD program at Illinois. As Dr. Ristic was a well-established research professor he already had enough material for a thesis it just needed to be written up. It still was necessary for him to pass the various examinations required for a PhD degree. He passed the language requirement simply by taking an examination. The various other examinations he also passed when they came up with little trouble. One requirement however required him to spend time on the Illinois campus. He had to take certain courses required of all PhD students. It was because of this requirement that I met him. At the time I was also taking those courses and we ended up being laboratory partners in several of then, I of course at this time knew him only as a fellow graduate student.

After he completed the required courses and passed the various examinations required for his PhD degree he returned to the University of Florida. I did not expect to see him again.

At this time the man who was my advisor left to go to the University of Colorado. When you are a graduate student your advisor is a major factor in your life. I did not know what to do. We had started some research on the development of a vaccine for a viral disease of cattle called shipping fever. All we graduate students did however was to inject a vaccine which our advisor had made but we were not told what the vaccine was nor given much information about the whole project. As a result of the position I was in I became upset and a bit depressed. Fortunately however the chairman of the Department, J.O. Alberts a fine man who had invited me to join his department when I had completed my Master's degree and planned to move to another University assured me that things would work out and that I should continue to work on my degree requirements.

Some months later I was quite surprised when Dr. Ristic walked into my office, told me he was a professor authorized to train graduate students at Illinois and asked me to join him as his student. It will tell you something about him when I tell you that he came into my office in high good humor, said that as we had gotten along well as fellow graduate students he was sure we would get along 1 well in our new relationship. He then said he had funds for research on anaplasmosis and *Vibrio fetus* and then asked me on which I would prefer to work if I join him. I chose to join him without hesitation and chose to work on the anaplasmosis as it was a disease of the blood and I had been working on hematology in my Master's program. It will tell you something more about him to know that our first joint project was to clean up the mess in the laboratory left by my previous advisor.

In the years that followed. His constant optimism and joy in his work was a pleasure to behold. He informed us all about every aspect of the research going on in the laboratory. He gave help to every student without stint. If a student having problems joined his group they would soon be on the track again to obtaining their degree. It is a tribute to him that in my career I attempted to treat my students as he had treated me.

Maro Ran-Ir Sondahl Tribute
Authored By Antonio Figueira
Submitted By William R. Sharp

I regret to inform the INGENIC Newsletter readership that Dr. Maro Söndahl died early this year in a tragic car accident in Brazil. Maro was not a frequent attendant of cocoa meetings, and probably most of our cocoa research community did not know him well, except for the biotechnologists. Maro was more popular with the coffee community since he dedicated 35 years of his life to this crop, working in many aspects of physiology, breeding and biotechnology.

Maro is widely recognized for his great contributions and pioneering work on tissue culture of various tropical crops. He developed the first protocol for somatic embryogenesis of coffee in the late 1970's, followed by other great achievements in maize, oil palm and roses during the 1980's. He was also a pioneer in cocoa tissue culture. In the 1980's, his team at the DNA Plant Technology Corp. (Cinnaminson, NJ, USA), with support from a chocolate manufacturer, developed the first protocols to obtain cocoa somatic embryos from sporophytic tissues (nucellus and floral parts). Before that, somatic embryos had only been obtained from immature zygotic embryos, with obvious limitations for propagation and genetic transformation. He was granted a US patent for somatic embryogenesis and plant regeneration of cocoa in 1994. His protocol opened the possibility for further developments. In fact, improved somatic embryogenesis protocols were published in 1993 by Nestlé and CIRAD, culminating with the advances developed in the Penn State group in 1998, all derived from Maro's pioneering work.

I first met Dr. Maro Söndahl in 1982 in Rio de Janeiro, Brazil during my last year in college, when he gave a talk about the use of tissue culture in plant breeding. His seminar definitively helped to direct my career to biotechnology. Maro had attended the same school (Brazilian Federal Rural University of Rio de Janeiro), graduating 15 years earlier (1968). He got his Masters degree in 1972 at the Center for Nuclear Energy in Agriculture of the University of São Paulo, where I currently work. His PhD. degree in Cell Biology was from the Developmental Biology Program of Ohio State University (1978).

He started his successful scientific career in 1970 as a researcher at the Agronomic Institute of Campinas, a state owned research center of São Paulo, where he worked mainly with coffee physiology. After concluding his PhD. in the US, he returned to the same Institute, where he became the chairman of the Department of Plant Genetics, and later he was indicated to be the Director of the Biology Division. In 1983, he moved to the US, joining DNA Plant Technology Corporation in Cinnaminson, NJ, as research manager, supervising work with somaclonal variation (coffee, popcorn), protocol development for somatic embryogenesis (cocoa), anther culture (rice, sweet corn) and breeding (sweet corn, popcorn). He became Senior Research Director in 1987 with technical and business responsibilities in cell genetics and breeding on the following crops: coffee, cocoa, oil palm, pineapple, banana, corn, sweet corn, popcorn, rice, oats, watermelon and rose. He later became Director of New Business and Product Development of DNA Plant Technology. In 1993, he started his own company, Fitolink Corp. More recently (1997), he started a new company in Brazil, Bionova (www.bionova- mudas.com.br), working with commercial micropropagation of sugarcane, banana, and pineapple. The tragic accident occurred during a business trip to establish new contracts to provide micropropagated plants to growers in Mossoró. Maro will be remembered for his great contribution to biotechnology of tropical crops and to plant sciences in general.

During his career, Maro was very successful in combining science, publishing important breakthrough articles, with a business oriented entrepreneur perspective. Maro had a great sense of humor and was an entertaining person to have around meetings. He was born in Brazil, but his family was originally from Iceland. He served as an Honorary Consul for Iceland in Curitiba, Brazil since 2000. He was survived by his wife Dr. Clemencia Noriega, who continues to run their business in Brazil, and three children.

Source:

Information retrieved on October 19, 2013 fhttp://ingenic. cas.psu.edu/documents/publications/News/10.pdf

Dr. T. S. Subramanian and Meenakshi Subramanian Tribute
Authored By Geetha Ghai

My parents Dr. T.S. Subramanian (Toppur Seethapathy Subramaian) and Meenakshi Subramanian were my earliest mentors and influencers.

Dr. T.S. Subramanian obtained his high school diploma and Undergraduate degree in Madras (today known as Chennai) India from PS High School and Presidency College, respectively. He obtained his PhD in organic chemistry from Liverpool, England in the 1930s and stayed in Liverpool during the 2nd world war. He was involved in the discovery of DDT scientific and intellectual achievement that went wrong in application. He came back to India in 1945 and was the first Indian director after India obtained freedom from British colonization of the ordinance laboratory in Kanpur India. He brought science to the rural area teaching farmers good agricultural practices along with proper use of herbicides. He then led the Textile research and Jute research. His motto for life was hard work, ethics, helping humanity. He played a major role representing India in FAO, UNESCO, and in establishing science education policies for the country. He played an international role by participating in various Common wealth Conferences, Natick, Canada, Australia and Russia representing Indian science and productivity interests. He led through example and died on September 19, 1985.

Meenakshi Subramanian obtained her undergraduate degree in chemistry from Queen Mary's College Madras, India. Her life was mingled with spirituality and science. She was a magnanimous person filled with compassion for the under privileged. Always providing a helping hand by cooking nutritious meals for the needy children and encouraging them to continue school. Even today when walking on the streets in Chennai (formerly

Madras) people walk up to me and state what a wonderful lady she was. She died on July 5, 2005.

My parents Dr. T.S. Subramanian and Mrs. Meenakshi Subramanian 1945 in front of their first house in Kanpur, India

Clara Gertrude Weishaupt
Submitted by William R. Sharp

Clara Gertrude Weishaupt, age 93, died at Greene Memorial Hospital in Xenia, OH 12 August 1991. She was for 22 years an outstanding professor of general botany and local flora in the Department of Botany, The Ohio State University. Simultaneously for 18 years, Dr. Weishaupt provide dedicated leadership as curator of the University Herbarium and conducted research on Ohio flora, culminating in her authoritative book, Vascular Plants of Ohio, 1960, 1968, 1971, and two publications on the grasses of Ohio (1967, 1985).

Born 20 July 1898 to Peter and Elizabeth Barbara (Weisflock) Weishaupt, who lived on a farm west of Lynchburg in Dodson Township, Highland County, OH, Miss Weishaupt was educated there in a one-room elementary school and graduated from the Lynchburg High School (1916). She received a diploma in bookkeeping, shorthand, and typing from Bliss Business College, Columbus, Ohio. At The Ohio State University she completed three degrees, B.S. in Home Economics (1924), M.S. in Botany (1932), and the Ph.D. in Botany (1935).

Her professional career began as a stenographer with the Department of Agricultural Education at The Ohio State University and with the Goodyear Tire and Rubber Company in Akron followed by eight years of teaching biology, mathematics and related subjects in the Lynchburg High School. While at Ohio State University, she was a graduate assistant in the Department of Botany and Plant Pathology (1932-35). Her college teaching career initially was at the State Teachers College, Jacksonville, AL (1935-46), where, while holding the rank of assistant professor and later associate professor of biology, she taught courses in biology, nutrition, field botany, human physiology, industrial arts, and physical science for elementary teachers. At the time she was the only woman on the faculty with a Ph.D. degree. In the Department of Botany and Plant Pathology at The Ohio State University, Dr. Weishaupt served as instructor (1946-51), assistant professor (1951-1960), associate professor (1960-1968), curator of the herbarium (1949-1967), and emeriti associate professor.

Dr. Weishaupt's early interest in the plant sciences was initially fostered in high school while taking an excellent course in botany, but as an undergraduate she developed her education in the areas of home economics and biological chemistry. As a graduate student in the OSU Department of Botany and Plant Pathology, she specialized in plant physiology and completed her master degree thesis on the effects of ultra-violet light on plants, and her Ph.D. dissertation on diffusion of water vapor through multiperforate septa, both completed under the direction of Professor Bernard S. Meyer. While teaching local flora at The Ohio State University, Professor Weishaupt early saw a need for a new field and laboratory manual of Ohio plants that would be useful to the students. Her first effort was a *Guide to Ohio Plants,* co-authored with three other members of the Department. Later she developed her own book, *Vascular Plants of Ohio* 0960), with a revised edition (1968), and a third edition 0970), followed by several subsequent reprinting's. The book is still quite popular and is used by students in local flora classes at various colleges and universities in Ohio and adjacent states.

Not trained as a plant taxonomist and with no experience in herbarium curatorial procedures, Dr. Weishaupt, upon being appointed curator of the OSU Herbarium in 1049, learned quickly the methods necessary to rejuvenate the herbarium. The facility had suffered neglect in the early 1940s during World War II. She brought order to the collection, including the identification of numerous specimens, updating the county distribution maps for the Ohio flora, and conducting extensive field work throughout Ohio to obtain specimens of species from those counties not well represented in the herbarium

She focused on the State Herbarium, adding to its collection through her own field work and through the contributions of others. Renewed interest in the flora of the state was stimulated by the initiation in 1951 of the Ohio Flora Project, sponsored by the Ohio Academy of Science. The OSU Herbarium was to be the primary resource for this project. During Weishaupt's tenure, two volumes of the Ohio Flora were published by E. Lucy Braun -- *The Woody plants of Ohio* (1961) and *The Monocotyledoneae of Ohio* (1967). Weishaupt's own research on the state flora resulted in the publication of her *Vascular Plants of Ohio* (1960), written for beginning students. Long popular in local flora courses in Ohio and neighboring areas, the book is still in use. Subsequent focus on Poaceae (also called Gramineae or true grasses) a family of obvious agricultural importance, led to her contribution of a treatment for this family to Braun's Monocotyledoneae volume, and to the publication of her *Descriptive Key to the Grasses of Ohio Based upon Vegetative Characters* (1985).

Professor Weishaupt held memberships and offices in many scientific and honorary societies. As a devoted and conscientious professor of research and teaching, Professor Weishaupt will be remembered by many of whom she touched in this capacity. He exciting lecture and demonstration research experiments in the classroom encouraged countless numbers of students to explore careers in the life sciences. She was the recipient of many honors: one of five awarded the Annual Ohio State University Distinguished Alumni Distinguished Teacher Award and the Highland County American Association of Women Distinguished Service Award from the Centennial Honoree Award of the Ohio Academy of Science.

She once said, "I've really has a very ordinary life" Her contributions to the botany of Ohio and the service she gave to so many individuals in the state and the nation are achievements from more than an "ordinary" life.

Information retrieved October 16, 2013 from https://kb.osu.edu/dspace/bitstream/handle/1811/23480/V091N5_221.pdf?sequence=1

Clara G. Weishaupt

AUTHOR CONTRIBUTORS

Volumes I & II

(Author Abbreviated Curriculum vitae listed in
alphabetical order)

Jeff Alder

Jeff Alder, alder.11@osu.edu, Home town: Mount Olive, New
Jersey, Business address:

Bayer HealthCare, 100 Bayer Blvd., PO Box 915, Whippany,
NJ 07981-0915

Academic Institutions: The Ohio State University; B.S., M.S.,
Ph.D., University of Wisconsin (post-doctoral), Journal papers:
approximately 80, Book chapters: 2 Patents: 2, Organizations and
Awards: Contributed to successful development of four antimicro-
bial agents used to treat people with serious bacterial infections;
Chairperson, American Society for Microbiology, Antimicrobial
Chemotherapy, 2012-2014 term; NIH/NIAID and grant Reviewer
for Biodefense, session Chair for Biodefense contract review, NIH/
NIAID 2002 – present; Clinical Laboratory Standards Institute;
Antimicrobial Susceptibility Testing Subcommittee Executive
member; Reviewer, Antimicrobial Agents and Chemotherapy,
Infectious Disease Society of America.

Societies: American Society for Microbiology (ASM);
Infectious disease Society of America (IDSA); European Society
for Clinical Microbiology and Infectious Disease (ESCMID);

Various awards from Abbott Labs, Cubist Pharmaceuticals, and Bayer HealthCare, including the President's Award (Abbott), Scientific mentor of the Year (Abbott), Chairman's Award as employee of the Year (Cubist), and Special Recognition Award (Bayer); Named one of the top 20 notable people in Research and Development; *Research and Development Directions*; February, 2006; Jeff Alder lives with his wife Lisa in Mount Olive, New Jersey. They hope to settle in their "Mountain House" in the Catskills one day. Jeff is the Senior Director, Global Clinical Development for Bayer HealthCare, based in Whippany, NJ.

Henrique V. Amorim

Henrique Vianna de Amorim. Hometown, Piracicaba, SP. Brazil. Graduated in Agricultural Science 1966, ESALQ, the University of São Paulo, Brazil. Master of Science, Ohio State University, Columbus, Ohio – USA, Ph. D. at Univ. São Paulo. From January 1970 to 2001, he was an associate professor in the Biological Sciences Department at University of São Paulo in Biochemistry, Piracicaba, S.P,. Brazil. He launched Fermentec in 1977. Amorim has published over 85 refereed journal articles, book chapters and abstracts, and the authoritative book entitled Alcohol Fermentation Technology: Science and Technology, 2005. **Awards:** Ambassador Medal (2004), Ohio State University, for his achievement in biological Science an entrepreneurship. Entrepreneur of the year 2010, Piracicaba, SP, Brazil. **Professional Membership**: STAB – Brazilian Society sugar and alcohol technicians. IBD – Institute Brewing & Distilling – London UK., **Number of Patents**: 3, **e-mail:** amorim@fermentec.com.br, **Homepage**: www.fermentec.com.br. **Facebook**: Henrique Amorim

Carolyn Brooks

Dr. Carolyn Branch Brooks is a native of Richmond, VA. She received her B.S. degree and the M. S. Degree in Biology from

Tuskegee University and a Ph.D. in Microbiology from The Ohio State University. Dr. Brooks joined the University of Maryland Eastern Shore (UMES) in 1981 and rose through academic and administrative ranks through the years to become a full professor and to serve in the positions of Director, Coordinator, Department Chair, Executive Assistant to the President and Chief of Staff, Research Director of 1890 Land Grant Programs and Dean of the School of Agricultural and Natural Sciences. Since July, 2007 she has served as the Executive Director of the Association of 1890 Research Directors (ARD) which is composed of the research administrators in the food and agricultural sciences at eighteen historically black land grant universities. Among the professional awards she has received are the, **George Washington Carver Public Service Hall of Fame Award**, the **William A. Hinton Award from the American Society for Microbiology,** recognized as **one of Maryland's Top 100 Women**, featured as one of the **100 Distinguished African American Scientists** in *"Distinguished African American Scientists of the 20th Century"*, UMES **National Alumni Association's Faculty Award for Excellence and Achievement, Outstanding Educator Award** from the **Maryland Association for Higher Education, White House Initiative for Historically Black Colleges and Universities - Faculty Award for "Excellence in Science and Technology," the "Woman of the Year Award"** from the Maryland Eastern Shore Branch of the National Association of University Women, **"Chancellor's Research Scholar Award"**, the School of Agricultural Sciences' **Outstanding Faculty Award for Research and the 2005 Spirit of Excellence Award for Community Leadership.** She has published more than 50 journal papers and continues to serve on numerous panels, councils, boards, task forces etc. which has allowed her to serve as a consultant or research and academic program evaluator for universities in California, Michigan, Oregon, Idaho, South Dakota, Florida, New Jersey, New York, Puerto Rico, Washington State, South Africa,

Costa Rica, Honduras, the U.S. Virgin Islands, the Dominican Republic, Tanzania, and Malawi. Her funded research projects allowed her to conduct research in Egypt, Cameroon, Togo, Nigeria and Senegal. Carolyn is extremely active in the Links, Inc., a national service organization of professional African American women. Cbbrooks78@comcast.net

Helanie Carrer

Dr. Helaine Carrer is Associate Professor of Plant Biochemistry and Molecular Biology at the University of Sao Paulo, Agriculture College at Piracicaba (ESALQ). Has a degree in Agronomy Engineering graduated by ESALQ, University of São Paulo in 1983. Obtained her MSc in Agriculture Sciences at The Institute of Nuclear Energy in Agriculture (CENA), University of São Paulo advised by Prof. Otto Jesu Crocomo in 1988. Received her PhD at Rutgers, The State University of New Jersey, USA working with Prof. Pal Maliga at the Waksman Institute on Plastid Transformation Technology in 1994. Actually, she teaches biochemistry and plant molecular biology and conducts research in plant genetic transformation and functional genomics of photosynthetic genes with the goal to develop new sugarcane varieties with drought resistance and higher sugar content in leading a project in the biomass division of BIOEN, Brazil's public consortium for sugarcane to bioenergy R&D. During her career she has published 52 scientific articles, 6 book chapters, participates in 4 patents, advised 17 MSc and 9 PhD students. She received a Medal of Scientific Merit as a leading researcher for the contribution to the DNA Sequencing of *Xylella fastidiosa* plant pathogen bacteria by the governor of the State of Sao Paulo. Current address: Department of Biological Sciences, ESALQ-University of São Paulo. Av. Padua Dias, 11. Piracicaba-SP. 13418-900.

Email: hecarrer@usp.br

Roy Chaleff

Education: Amherst College, B.A., 1968, Yale University, Ph.D., 1972, Brookhaven National Laboratory, post-doctoral fellow, 1972-1974, *Employment:* John Innes Institute, England, Senior Scientific Officer, 1974-1976, Cornell University, Assistant Professor, Depts. of Genetics and Plant Breeding, 1976-1980, E. I. DuPont & Co., Central Research & Development Dept., 1980-1987, American Cyanamid, Director of Plant Biotechnology, 1987-1995, Rutgers University, Professor of Plant Biology, 1995-1998, University of Medicine and Dentistry of New Jersey, Office of the Vice President for Research, Central Administration, Director of Patents and Licensing, and Research Dean for NJ Dental School, 1998-2004, Ben Franklin Technology Partners, 2004-2005, *Currently:* Retired and residing in Pennington, New Jersey.

Frank Cox

Professor Cox is primarily a parasitologist. He graduated in Zoology with Parasitology as his Special Subject at the University of Exeter, UK, and, after postgraduate training at the London School of Hygiene and Tropical Medicine, joined King's College London as Lecturer and Reader in Parasitology in the Department of Zoology and subsequently Professor of Parasite Immunology in the Department of Cell and Molecular Biology. From 1986-1990 he served as Dean of Science in the University of London. He then joined the staff of the London School of Hygiene and Tropical Medicine as a Senior Research Fellow in the Department of Infectious Diseases until his retirement in 2013. He has published over 120 original research papers, reviews and congress proceedings and has authored or co-authored seven books including The *Wellcome Trust Illustrated History of Tropical Diseases* and also an interactive CD, *Six Thousand Years of Tropical Medicine.* Professor Cox has worked on various WHO and other international and national expert committees and has held visiting professorships at UK and overseas universities. He has been Editor of *Parasitology, Trends in Parasitology* and the *Transactions of the Royal Society of Tropical Medicine* and has also served on a number of editorial boards. He holds degrees of PhD and DSc of the University of London. His current interests are in the history of tropical medicine and parasitology. E-mail address: francis.cox1@btinternet.com

Otto J. Crocomo

Full Professor of Biochemistry at University of São Paulo, Rockefeller Fellow with C.C. Delwiche at UC Davis campus, and British Research with L. Fowden, University College, London, visitor professor with D. Boulter, University of Durham, England. Founder of the Center for Agricultural Biotechnology (CEBTEC) at the University of São Paulo, Piracicaba. Lives in Piracicaba, SP, Brasil with his wife Diva and has 5 children and one grandson. **Adolfo Egidio Lovadino Crocomo, Carla Maisa Lovadino Crocomo, Daniel Lovadino Crocomo, Maria Paula Lovadino Crocomo, & Pedro Augusto de Toledo Almeida Crocomo**

Adolfo Egídio graduated in Medicine at the Federal University of the State of Santa Catarina in 1992. He is an ear, nose, and throat specialist. He lives and works in Piracicaba with his wife, Kátia. **Maria Paula** received her B.A. in English Translation at University Ibero-Americana in 1992, in São Paulo. In 1994, she founded in Piracicaba "Interaction-School of Languages" which she directed until 2000. In 2000, she moved to London, UK, where she received her M.A. in Tourism Management from Westminster University. She is currently the Pedagogical Coordinator at "Self School of Languages," in Piracicaba. **Carla Maísa** currently teaches ESL at City College of San Francisco in San Francisco, California. She received her B.A. from Unimep in Piracicaba, where she taught for more than 10 years. In 2000, she decided to move to the San Francisco Bay Area. There she received her M.A. degree in TESOL from San Francisco State University. In addition to teaching, Carla has love for yoga and the arts. **Daniel** lives in Piracicaba. In 1998, he graduated in Tourism at the University Anhembi Morumbi, in São Paulo. In 2004, he received his M.B.A. in Marketing from Unicef in Piracicaba. He is currently a trade representative of high added value agricultural products. He also works with photography, advertisement, and he is a musician.

Pedro Augusto lives in Piracicaba and is Daniel's only son. He is a high-school student, and in his free time, he enjoys playing the guitar and practicing sports.

Joaquim José de Camargo Engler

Joaquim José de Camargo Engler was born in Campinas, State of São Paulo, Brazil. In 1960 he passed the entrance examination for admission in Agricultural Engineering at Escola Superior de Agricultura "Luiz de Queiroz", Universidade de São Paulo (ESALQ/USP) in Piracicaba, São Paulo. Completed college in 1964 and obtained the title of Agricultural Engineer, with specialization in food technology. In 1965 he began his academic career at ESALQ/ USP as a teacher and researcher in the Chair of Economics, current Department of Economics, Administration and Sociology. During his academic career he obtained the titles of "Doctor of Agronomy", "Associate Professor", and "Professor at the University of São Paulo (USP) and completed the degrees "Master of Science" and "Doctor of Philosophy" at The Ohio State University. He was approved in all Teaching Career competitions at USP, hitting the post of full professor. His international experience involves, the Coordination of the Agreement between USP and The Ohio State University for the development of the Research Program on Capital Formation and Technological Innovations in Agriculture; the Coordination of the Agreement between USP, the Ministry of Education and Michigan State University to develop the "Program of Higher Education in Agriculture" (PEAS); the Coordination of the Agreement between USP and the Ford Foundation to develop the Program for Research and Graduate Program in Rural Social Sciences, at ESALQ/USP; the Coordination of Planning and Finance of the Agreement between USP and the Inter-American Development Bank; the Coordination of the International Agreement between USP and European and Latin American Universities (Project UNIBEUR-INFO) member of the International Affairs Committee of the Department of Agricultural Economics and Rural Sociology, The Ohio State University, USA. During his professional activity he received numerous awards and honors, including the Medal "Fernando Costa", the OSU-International Alumni Award-1994, the Medal of Merit in Science and Technology in São Paulo State - 2001; Biography transcribed in the International Directory of Business and Management Scholars and Research of Harvard Business School; the title of "Piracicaba Citizen" in 2007, the Trophy "O Semeador", awarded by ESALQ/USP in 2010 in recognition of significant contribution to education development in the area of Agriculture. He is currently Administrative and Finance Director of the São Paulo State Research Foundation (FAPESP) having been elected and reelected by the Board of Trustees and appointed by the Governor of São Paulo State since 1993, for seven consecutive three-year periods. In addition to the position: And, he serves as Visiting Professor of Economics at Gulbenkian Institute of Science.

Geetha Ghai

Geetha Ghai lives in New Providence, New Jersey, USA. She retired after 13 years as the Associate Director at the Center for Advanced Food Technology Rutgers University, New Brunswick, New Jersey. Prior to this she worked in industry Ciba-Geigy Pharmaceuticals, Summit New Jersey and before that was a faculty member at the Pharmacology Department at the University of Southern Alabama Medical School, Mobile Alabama. She obtained her PhD from the Maharaja Sayajiroa University Vadodra, India in biochemistry and a MBA through the executive program at Rutgers University Business School Newark, New Jersey. She did a postdoctoral fellowship at the State University of New York, Buffalo, New York. She has authored over 57 papers some as primary and others as co-author, a few book chapters and 7 patents to her credit. She was on the grant review board of American Heart Association New Jersey Chapter. She was a member of the American Society of Pharmacology and Experimental Therapeutics and the Federation of American Society of Experimental Biology. For three years she served on the Advisory Board of the State Minority and Multicultural Office an appointment by the Governor of New Jersey. She is a founding member of the South Asian Total Health Initiative based at Rutgers University Medical School, and SKN Foundation both located in New Jersey. Recently, she has been elected to the board of SAGE Eldercare situated at Summit New Jersey. She has been invited to speak at many scientific and non-scientific conferences. Some salient speaking assignments include the motivational keynote speech for Johns Hopkins University Center for Talented Youth, Institute of Food Technology Scientific Lecture series, National Agri-Marketing Association and others. Email: geethaghai@gmail.com

Joseph Hamburger

Joseph Hamburger, Born: Haifa Palestine/Land of Israel, Residence: Jerusalem, Israel, Prof. (research), Member of the Kuvin Center, Hebrew University Hadassah Medical School, P.O.B 12272, Jerusalem 91120, Academic Institutions attended: Hebrew University of Jerusalem (HU), B. Sc., 1960-63, Microbiology and Parasitology, Hebrew University of Jerusalem (HU), M. Sc., 1964-67, Parasitology, Hebrew University of Jerusalem (HU), Ph.D., 1968-73, Parasitology, PROFESSIONAL EXPERIENCE: 1973-1975 Post-Doctoral Fellow at Ohio State University Department of Microbiology - Research subjects: Immunity in rodent malaria, 1975-1976 Research Fellow at Case Western Research University, Department of Geographic Medicine. Research subjects: Identification and characterization of schistosome egg antigens inducing granulomatous hypersensitivity, 1976-1978 Research associate at HU, Division of Helminthology. Research subjects: Characterization of schistosome egg antigens inducing immunopathology. Acquired resistance in schistosomiasis (bilharzia), 1978-1983 Lecturer at the HU, Department of Parasitology. Research subjects: Characterization of schistosome egg antigens inducing immunopathology. Acquired resistance in schistosomiasis (bilharzia), 1984 visiting scientist at Case Western Research University, Department of Geographic Medicine. Research subjects: The use of monoclonal antibodies and recombinant DNA technologies in schistosomiasis research, 1984-1991 Senior Investigator at the HU Department of Parasitology. Research subjects: Characterization of schistosome egg antigens. Identification of infected snails by detecting specific schistosomal antigens. Repetitive sequences in the genome of schistosomes, 1991-1991 Visiting Scientist at Smith College Department of Biological Sciences. Research subjects: Repetitive sequences in the genome of filariae, 1992-to date Prof. (Research) at the HU Department of Parasitology. Research subjects: Molecular markers of infection in schistosomes and filariae- Structure and significance for molecular monitoring of infection in human and intermediate hosts, MEMBERSHIPS: Israel Society of Parasitology (Elected President 1989-90), American Society of Tropical Medicine and Hygiene, The Royal Society for Tropical Medicine and Hygiene, PUBLICATIONS number: 51 and PATENTS number: 1.

Julius P. Kreier

Julius P. Kreier, Professor Emeritus, Department of Microbiology, The Ohio State University, Columbus Ohio; 2047

Iuka Avenue Columbus Ohio, 43201. Telephone Contact: 614-294-6832, Birthplace: Philadelphia, PA 1926; Education. Philadelphia Public Schools, 1932- 1945. Temple University 1945-1948, University of Pennsylvania, 1949-1953, V.M.D. University of Illinois, 1956-1962, MSc. Ph.D.; Employment. Veterinarian, Cooperative Mexican American Commission for the Eradication of Foot and Mouth Disease, Mexico 1954-1955. Veterinarian, Tuberculosis and Brucellosis Eradication Campaign, 1956 Maryland, USA. University of Illinois, Urbana Illinois, Research Associate 1956-1962. Ohio State University, Assistant Professor, Associate Professor and Full Professor of Microbiology 1962-1989; Professor Emeritus of Microbiology, 1989 to present. Publications 150 in reviewed journals. Books 22; Teaching. Parasitic Protozoology, Rodent Surgery, Introductory Immunology, Infectious Diseases; Mentoring of graduate students: PhD students 24, MSc students 38, Postdoctoral fellows and visiting professors 4; Fulbright Award University of the Republic of Uruguay, invited lecturer, Campinas University Brazil, Veterinary College, Ankara Turkey, Institute of Parasitology Shanghai China, Haryana Agricultural University, Madras, India; My research was primarily on the pathogenesis of infectious diseases of the blood. These were primarily caused by protozoa although some were caused by Rickettsial type organisms. From my work the greatest pleasures I had were from teaching. In fact I also believe that my greatest contribution to science was the students trained. My students are now scattered around the world and most are continuing to carry out scientific investigations.

Jesse Kreier

Jesse Kreier is a lawyer with the Word Trade Organization in Geneva, Switzerland. Since joining the then-GATT Secretariat in 1992, Mr. Kreier has supported multilateral negotiations on international trade issues, has served as legal advisor to numerous dispute settlement panels and has provided technical assistance to assist WTO Members to implement their obligations under the WTO Agreement. From 1987 to 1992, Mr. Kreier practiced law in Washington, D.C., where he specialized in international trade regulation. Previously, he clerked for the Idaho Court of Appeals. Mr. Kreier holds a J.D. degree, *magna cum laude*, from Georgetown University Law Center, a Master of Science in Foreign Service from Georgetown University, and a Bachelor of Arts from Johns Hopkins University. He is admitted to the Bar in California and the District of Columbia USA. Mr. Kreier and his wife Susan Schorr live in the lakeside town of Nyon, Switzerland, a short distance outside Geneva. Their two children, Jacob and Freda, were born in Geneva, and are now in attending college in the United States.

Rachel Kreier

Health economist Rachel Kreier (Rachel.Kreier@gmail.com) lives in Port Jefferson, New York. She received her doctoral degree from Stony Brook University in 2004 at the ripe old age of 48. She has worked as a health rights activist, union staff member, editor, journalist specializing in the health care industry, and professor at Hofstra University in Hempstead, NY. She speaks and teaches frequently about health care reform, and was co-director of the 2010 conference, "New Directions in American Healthcare: Innovations from Home and Abroad." Her published work includes hundreds of newspaper articles, and three peer-reviewed journal articles. Her article, "A dynamic model of health plan choice from a real options perspective," (with B. Sengupta) received the 2011 Best Article Award from the *Atlantic Economic Journal.*

Jerry Ladman

Prior to his retirement in 2007 Jerry served seven years as associate provost for international affairs at The Ohio State University, where he had the broad responsibilities for the University's international programs, area studies, study abroad and international students. As a professor in the Department of Agricultural, Environmental and Development Economics he previously served as director of the Latin American Studies Program and as the director of the Ohio Leadership, Educational and Development Program. In addition, he coordinated the OSU College of Food, Agricultural and Environmental Sciences' program in Mexico and was the founding resident director of the OSU study abroad program held at the Mexican Postgraduate College of Agricultural Sciences. Prior to coming to Ohio State in 1990 he was professor of economics and director of the Center for Latin American Studies at Arizona State University. He has lived twice in Mexico where he was assistant program officer with the Ford Foundation and was a visiting professor at the Postgraduate College. He also spent five years in the Dominican Republic as chief of party of the OSU Agribusiness Partnership Project at the Instituto Superior de Agricultura. He led major projects in Bolivia and El Salvador; was a Fulbright Scholar in Ecuador; and a visiting scholar at the Food Research Institute of Stanford University. During his academic career his research has focused on economic and agricultural development with a special emphasis on rural finance in Latin America, especially in Mexico, Bolivia and Central America. He is author of more than 60 articles and monographs on rural finance, the political economy of Bolivia, the Mexican economy, Mexican migration, the U.S.-Mexican border region and is the editor of two books on Mexican economic and political topics, and one book on Bolivia's political economy. He has received numerous grants totaling more than seven million dollars. The most recent was, during the first five years after his retirement, to work with the Postgraduate College in bringing microfinance to rural areas in Mexico. In the past he served as a member of the Board of Directors of the Research Program on Mexico (PROFMEX),

President of the Pacific Coast Council for Latin American Studies, and President of the Borderlands Scholars Association. He has testified before congress, served as a consultant to a number of organizations, including the World Bank and the United States Agency for International Development. He served as a member of the Executive Committee of the NASULGC Commission on International Programs and was named an Honorary Professor the Catholic University of Bolivia. Since his retirement he has been professor emeritus in the Department of Agricultural, Environmental, and Development Economics and is currently serving as a docent at the Columbus Museum of Art and a member of the Board of Directors of the Columbus Council on World Affairs. He spent most of his youth in Clarion, Iowa. Upon graduation from Clarion high school he studied at Iowa State University, where he received the B.S. degree in Farm Operations and the Ph.D. in economics.

David Lee

David Lee is Emeritus Professor of Biology at Florida International University and resides in Miami, Florida. He was born in Wenatchee in 1942 and raised in Ephrata, on the Columbia Plateau of eastern Washington State. He attended Pacific Lutheran University (B.S. in Biology, 1966) and Rutgers University in New Brunswick for graduate work in Botany (M.S. in 1968, Ph.D. in 1970, in biochemical plant systematic under the direction of David E. Fairbrothers). Following postdoctoral research in plant cell biology and tissue culture with Rod Sharp and Donald Dougall at The Ohio State University (1970-1972), he and his wife Carol moved to Malaysia, where he worked as a Lecturer at the University of Malaya, in Kuala Lumpur, 1973-76. There he developed life-long research interests in the functional ecology of tropical plants. He then worked with Francis Hallé at the University of Montpellier, 1977-78. Following a couple of years of work as a carpenter and landscaper in upstate New York, he moved to Miami to work as Assistant (and then Associate and Full) Professor at the young Florida International University. There he developed a research program in tropical botany, and he helped the institution develop strength in tropical biology, partly through collaboration with local institutions, particularly Fairchild Tropical Botanic Garden. He conducted field work in tropical Asia, Latin America and West Africa. He is best known for discoveries concerning the basis and function of color in vegetative organs (as autumn coloration in temperate trees and structural colors in tropical plants), and the plastic developmental responses of plants to understory shade. This research has resulted in the publication of 87 peer-reviewed articles and book chapters, and 10 books (three edited). His 2007 book, *Nature's Palette* (University of Chicago Press) won the AAP award for scholarly publication in the life and biomedical sciences for that year. He is presently finishing a companion book, *Nature's Fabric*, about leaves, for the same publisher. He received the Bessey Award for Excellence in Teaching by the Botanical Society of America in 2005, and the Alumni Association Outstanding Faculty Award at FIU in 2007. Just prior to his retirement in 2009, he served as Director of The Kampong of the National Tropical Botanical Garden, in Coconut Grove, and recently published a book about its founder, *The World as Garden. The Life and Writings of David Fairchild* (CreateSpace, 2013). Email: leed@fiu.edu.

Raul Machado Neto

Raul Machado Neto, lives and works in Piracicaba, his hometown, and the current institutional address is – Universidade de São Paulo, Escola Superior de Agricultura Luiz de Queiroz (USP/ESALQ), Av. Pádua Dias 11, Piracicaba, São Paulo, CEP 13418260, Brazil. Higher educational training includes BS in Agricultural Science (1973), MS in Animal Science (1977), both at Escola Superior de Agricultura Luiz de Queiroz, the College of Agrculture of Universidade de São Paulo, USP/ESALQ, Piracicaba, São Paulo, Brasil, PhD in Animal Physiology, University of Illinois at Urbana Campaign, USA, in 1980, and Postdoctoral Fellow at Agricultural and Food Research Council-AFRC, Institute for Animal Health, England, (1989-1990). About the professional career, always at USP/ESALQ, started in 1974 in the position Assistant Professor of Universidade de São Paulo/ Escola Superior de Agricultura Luiz de Queiroz, USP/ESALQ USP/ESALQ (1974-1980), Doctor Assistant Professor (1980-1985), Associate Professor (1985), and Full Professor (1997). He received in 2001 the Scientific Merit Medal of State of São Paulo Governor, is currently Research Fellow of CNPq (National Council for Scientific and Technological Development), has delivered numerous lectures and published sixty two papers in scientific journals. His current email is raul.machado@usp.br

Sally Miller

Sally A. Miller, Professor, Department of Plant Pathology, The Ohio State University, Ohio Agricultural Research and Development Center, Wooster, Ohio USA, Email: miller.769@ osu.edu, Twitter: @OhioVeggieDoc, Website: www.oardc.osu. edu/sallymiller, Dr. Sally Miller is a Professor of Plant Pathology in The Ohio State University College of Food, Agriculture and Environmental Sciences. She was born in Canton, Ohio and currently lives in Wooster, Ohio. She received a B. Sc. in Biology from The Ohio State University and M.S. and Ph.D. degrees in Plant Pathology from the University of Wisconsin-Madison. Her career efforts have been centered on plant disease diagnostics and sustainable disease management. Dr. Miller teaches graduate level, laboratory-intensive courses in diagnostic field plant pathology and vegetable disease management, and a short course "Pest and Disease Diagnostics for International Trade and Food Security".

She also serves as State Extension Specialist for vegetable disease management, focusing on integrated disease management in conventional and organic systems. Dr. Miller has published 95 peer-reviewed journal articles, 250 technical reports, 25 book chapters and one co-edited book, and is a co-inventor on seven patents. She is a Fellow of the American Phytopathological Society (APS), served as Director of its Office of International Programs Board from 2007-2013, and was elected APS Vice President in 2013. She was awarded the APS International Service Award in 2002, the OSU Gamma Sigma Delta International Award of Merit in 2007, and the OSU-OARDC Distinguished Multidisciplinary Team Award in 2013.

Mark Mueller

Dr. Mark Muller joined the faculty at The Ohio State University in 1980, initially in the Department of Microbiology. In 1988, along with several colleagues, he helped create the first Department of Molecular Genetics in the country. Dr. Muller was Professor of Molecular Genetics until 2004 when he was recruited to join the Biomolecular Science Program at the University of Central Florida in Orlando. Dr. Muller was involved in forming a new College of Medicine at UCF and joined the Medical School faculty in 2007. He is currently Professor of Medicine at UCF and runs a cancer research group. The laboratory is an established group of researchers working on the molecular biology of cancer and gene regulation. Specific research focus areas include studies on epigenetics, gene silencing, DNA repair and telomerase regulation. Dr. Muller has nearly 100 publications in the fields of cancer and virology and multiple patents (H-index >30). In addition, he has over 400 national and international presentations and abstracts. Dr. Muller currently has national collaborations with researchers at NIH, Mayo Clinic, MD Anderson, as well as international collaborations with the University of Napoli (Naples, Italy) and the

University of Kwazulu-Natal (Pietermaritzburg, South Africa). Dr. Muller has started several for-profit biotechnology companies, including TopoGEN, Inc., Visual Genomics, Inc., Methylation, Ltd., DNA Protein, Ltd., and is active on Scientific Boards in the US and EU. He has worked with Nobel Laureates including Howard Temin (University of Wisconsin) and Michael Smith (University of British Columbia). Dr. Muller is an active member of the American Association for Cancer Research, American Society of Molecular Biology and Biochemistry, and American Association for the Advancement of Science. He reviews manuscripts for multiple journals and grant applications and is a member of the editorial board for The Journal of Plant Pathology and Microbiology. Email contact: Mark.Muller@ucf.edu, Facebook:https://www.facebook.com/mark.t.muller, LaboratoryWebsites:www.biomed.ucf.edu/mtmuller (general public), http://med.ucf.edu/biomed/directory/profile/dr-mark-t-muller/ (scientific), Current Address: Dr. Mark T. Mueller, Ph.D., UCF, College of Medicine, 6900 Lake Nona Blvd,, Orlando, FL 32127.

John E. Peters

John E. Peters grew up in Dover, Ohio and received his Bachelor's degree from Otterbein College (Westerville, Ohio) and his Masters and Ph.D. degrees from The Ohio State University (Columbus, Ohio). His Master's degree research was in plant tissue culture and his Ph.D. dissertation included the discovery and characterization of a novel proteolytic enzyme from Pseudomonas aeruginosa. His research in these two areas resulted in 15 papers. He served 20 years as an officer in the United States Air Force as a clinical laboratory supervisor and as a faculty member at the United States Air Force Academy Biology Department. Following retirement from the Air Force he began a second career as a faculty member and Biology Department Chair at McHenry County College (Crystal Lake, Illinois). His primary area of expertise has been clinical microbiology education. He is a member of the American Society for Microbiology and the American Society of Clinical Pathologists (ASCP-MT),

Email-johnpampeters@hotmail.com, Facebook – john.e.peters.9

Robert Pfister

I joined the faculty at the Ohio State University at the end of summer in 1966. I was hired as an assistant professor of microbiology and after seven years rose to the rank of Professor. I became chairman of the department of microbiology in 1973 and resigned from the chairmanship in 1985. I retired from the Ohio State University in 1993. The courses that I taught during my tenure as a professor were introductory microbiology, microbial cytology, electron microscopy, general biology, and microbial seminar. During the time that I was in the department, I was also chairman of the graduate committee.

I published 70 scientific articles in various professional journals, and presented numerous papers at society meetings. I developed a "numerical taxonomy" program using" Fortran" for the IBM 1620 computer at Lamont Geological Observatory while working in the laboratory of Dr. Paul Burkholder. We studied and identified various microorganisms In the Antarctic using this numerical taxonomy program. With Dr. Burkholder's help I started a sponge collecting program in Puerto Rico to look for drugs from the sea. I was one of the early scientists to recognize microbial forms in ancient sediments, and to study the movement of toxic heavy metals in native soils using the interaction of microbes, visualizing them in both transmission and scanning electron microscopy. I also had numerous cooperative programs with fellow scientists.

I mentored 65 doctoral students and about 80 Masters degree students during my tenure at the Ohio State University. Most importantly, I had a wonderful wife and raised four wonderful children during my scientific career. My life has been blessed, yours can also be if you try.

James A. Quinn

Jim Quinn graduated from high school (Valedictorian) in Guymon, Oklahoma, was active in 4-H (State President, Oklahoma Hall of Fame), and received a B.S. in Crops and Soils at Oklahoma Panhandle State University (Valedictorian, Student Body President). He received a M.S. (Range Science) and a Ph.D. (Botanical Science) at Colorado State University. Immediately after the receipt of his Ph.D., he joined the faculty at Rutgers University (Assistant Professor, 1966-71; Associate Professor, 1971-77; Professor, 1977-2000; Professor Emeritus, 2001-Present). He taught undergraduate and graduate courses in biology, botany, and ecology, serving on 176 graduate student committees and as Chairperson of the graduate committees for 12 M.S. and 13 Ph.D. degree recipients. He was an author of 167 publications and published abstracts of papers at meetings (62 articles in referred journals), and was an invited symposium speaker at national (4) and international (4) meetings. Other recognitions and awards include External Examiner for Ph.D. dissertations in Canada (2) and Australia (4) and for three university graduate programs, Alumni Hall of Fame at Oklahoma Panhandle State University, and a Torrey Botanical Society Life Member Award for "long and dedicated service." A. QuinnProfessional memberships are the Botanical Society of America (Vice-Chairman, 1977 and Chairman, 1978, of the Ecological Section; Editorial Board, Amer. J. Bot., 1980-82), Ecological Society of America (Certified "Senior Ecologist", 1990-2011), Nature Conservancy, NJ Conservation Foundation, NJ Academy of Science (Council, 1972-76; Treasurer and Executive Committee, 1976-80), Pinelands Preservation Alliance, Sigma Xi (Chapter Vice-President, 1985-86; President-Elect, 1991-92; President, 1992-93), Society for

Range Management, and Torrey Botanical Society (Council, 1970-79; Director, 1980-81; President, 1982-83; Associate Editor, Bull. Torrey Bot. Club, 1983-85; Director, 1998-2000). Email: (quinn@aesop.rutgers.edu)

Rosa Shine Raskin

Rosa Shine Raskin has a background in technical and business information analysis, and clinical medicine. She has worked in academic medical centers throughout Ohio, the federal government, and industries in the Fortune 100 for over 30 years. She has held positions as a Research Assistant, Environmental Biologist and Technical Information Specialist in Materials Science, Tires, Engineered Products, Pneumatic Tools, Airplane Safety, Food Science, and Instructor in Clinical Medicine. She co-founded a start-up company and is the principal consultant at Rosa S. Raskin & Associates LLC, empowering companies with the information they need to succeed. She participates in the publication of articles and books in medicine and psychology. She serves as Contributing Author for the leading international trade journals on coatings in the Asia Pacific Region, Middle East, Africa, and Europe. Her articles often appear on the cover of the journals for which she writes. She published a book on Amazon.com about her experience as a Displaced Person and the continued search for her lost sister entitled, *Walk Forward,* a different Schindler's List, book's trailer is at www.youtube.com/watch?v=Zp7uQap6p2M

Her company website is at www.raskinfo.com and she writes three blogs published both on Amazon.com and on the web, including "Information Specialist Secrets" at preciousinformationspecialist.blogspot.com, "Precious Cooking" at preciouscooking.blogspot.com, and "Most Precious Memories" at mostpreciousmemories.blogspot.com. She has an M.L.S. in Library Science from Kent State University, an M.S. in Microbiology and a B.S. in Biology from The Ohio State University.

Contact information, rosaraskin@gmail, rosa@raskinfo.com,
Skype: rosa.raskin1,
Twitter handle: *RosaSRaskin,*
Linkedin Profile: www.linkedin.com/profile/view?id=15895388,
Pinterest: www.pinterest.com/rosaraskin/
Personal Facebook page: www.facebook.com/rosa.s.raskin,
Facebook page for her book, *Walk Forward:* www.facebook.com/
walkforwardbook

Thomas Seed

Paterson, NJ. Current residence: Bethesda, MD, USA.
Communication address: tmseed@verizon.net. Education: *1964-
1968* – University of Connecticut, Storrs, CT, BSc (Bacteriology);
1968-1972 – Ohio State University, Columbus, OH, MSc (1969 -
Microbiology), PhD (1972 – Microbiology; *1972-1973* – Case
Western Reserve University, Institute of Pathology, Cleveland, OH,
(Postdoctoral Fellowship in Cellular Ultrastructure). Positions &
Institutions: *1998-present* – Consultant, Tech Micro Services Co.,
Bethesda, MD; *1995-1997* - Associate Chief of Research, Radiation
Effects Research Institute, Hiroshima, Japan; *1993-1995* – Research
Professor, Vitreous State Laboratory, Department of Physics,
Catholic University of America, Washington, DC; *1996- 2003* –
Senior Scientist/Group Leader, Radiation Casualty Management,
Armed Forces Radiobiology Research Institute, Bethesda, MD;
1975-1995 – Research Biologist & Group Leaders for Radiation
Hematology (1982-1995), Radiation Leukemogenesis (1979-1981),
Cellular Indicators (1977-1978), Electron Microscopy (1975-1995),
Divisional of Mechanistic Biology and Biotechnology, Argonne
National Laboratory, Argonne, IL; *1973-1975* – Assistant Biologist
& Head, Ultrastructure Group, Blood Research Laboratory,
American National Red Cross, Bethesda, MD. Professional works:
Journal articles (107 as per 'Pub Med' listing); Total cited pub-
lications, including books, book chapters, guidance documents,

etc. (~186 publications cited ~2800 as per Google Scholar listing); Patents: 3 patents- 'Radiation Countermeasures' (US Patent US 7,919,525 B2; US Patent 7,665,694 B2; EP Patent EP 1,767,215); Awards/honors (select listing): Appointed, The Ohio State University Research Fellowship, 1971-1972; Appointed Head, US Delegation, NATO Research Study Group-23, 1996-1998; Elected, Chairman, NATO Research Task Group TG-006 (1999-2001); Awarded, Distinguished Seminar Speaker (2006); Elected, Council member (2005-2010) & Consociate member (2010-present), National Council on Radiation Protection and Measurements; Professional affiliations: American Association for the Advancement of Science (AAAS- emeritus); Microscopy Society of America (MSA- emeritus); American Society for Microbiology (ASM- emeritus); International Society of Experimental Hematology (ISEH- emeritus); Radiation Research Society (RRS- retired member). General research interests: Nature and mechanisms of radiation injuries; Medical countermeasures; Hematopoiesis (structure/function/pathologic mechanisms); Leukemogenesis (nature/mechanisms); Low level radiation/chemical toxicity.

William R. Sharp

Dr. Sharp, birthplace, Akron, Ohio, has a background in biotechnology, translation of science into business ideas, spawning start-up companies and extensive technology transfer experience in the Americas and Asia. He has authored over seventy refereed research papers, abstracts and eight books in the field of plant cell biology including the five volume series entitled the Handbook of Plant Cell Culture. He previously held the positions: Dean of Research, Cook College (Now SEBS – The School of Environmental & Biological Sciences) and Director of Research, New Jersey Agricultural Experiment Station, Rutgers University; Executive Vice-President, DNA Pharmaceuticals, Inc.;

Executive Vice-President for Research, DNA Plant Technology, Corp; Research Director, Pioneer Research, Campbell Institute for Research & Technology, the Campbell Soup Company; Full Professor, Ohio State University; and Fellow, Argonne National Laboratory. He was a Fulbright Grantee during 1971 and 1973. Dr. Sharp holds a Ph.D. in Plant Cell Biology from Rutgers University. He is the recipient of the title Eminent Professor and the Luiz Queiroz Distinguished Service Medal award from the University of Sao Paulo and more recently, The Ohio State University Board of Trustees Honorary Services Award.

Jeffrey William Sharp

Jeffrey Sharp, an award-winning producer and publishing entrepreneur, is President/CEO of Story Mining & Supply Co., a Los Angeles-based production company committed to acquiring, developing, and financing and producing multi-platform premium content through unique access to quality material. SMS combines deep financial resources, strong creative relationships and an executive team that has produced award winning and commercially successful films. Sharp has produced a series of Academy Award winning and Golden Globe nominated films over the past ten years like Boys Don't Cry, You Can Count on Me, Nicholas Nickleby and Proof. With those movies as well as other renowned adaptations, including A Home at the End of the World, The Night Listener, and Evening, Sharp has worked to develop new models for integration of the publishing and film industries. Prior to Story Mining, Sharp co-founded the digital publisher Open Road Integrated Media with former HarperCollins CEO Jane Friedman. Story Mining and Open Road have a strategic partnership to develop and co-produce feature films and television shows based upon Open Road titles including: Lie Down in Darkness, based on the novel by William Styron, written and to be directed by Scott Cooper (Crazy Heart) and Cocoa Beach, a television series adapted by Andre and Maria

Jacquemetton (Mad Men) based on the life of NBC news reporter Jay Barbree's recollections of the early days of the U.S. space program in Cape Canaveral, Florida. Sharp holds an MFA from Columbia University and a BA from Colgate University. In 2005, the Columbia University School of the Arts honored him with the Andrew Sarris award for his contribution to independent film. Sharp currently serves as Chairman of the Hamptons International Film Festival Advisory Board, member of the Executive Board of Literacy Partners, Special Advisor for the Book Meets Film Forum at the Taipei International Book Expo as well as a member of the Advisory Board of BookExpo of America. He is a member of BAFTA (British Academy of Film and Television Arts).

Harry E. Sommer

Harry E. Sommer, Associate Professor, D.B. School of Forest Resources, University of Georgia; Education: B.S. in Agriculture with honors, University of Vermont, Honors Topic: Quinone Occurrence in Hemlock, 1963; M.S. in Biochemistry, University of Maine, 1966, Thesis Title: "Purification and Characterization of Fructose 1, 6-diphosphatase"; Ph.D. Ohio State University 1972, Dissertation: "Influence of 2, 4 Dichlorophenoxyacetic Acid on Nitrate Reductase and Protein in Wild Carrot (Dacucus carota L.) Tissue Culture; Selected Publications: Sommer, H.E., C.L. Brown and P.P. Kormanik, 1975, Differentiation of Planets in Longleaf Pine (Pinus palustris Mill) Tissue culture in vitro, Bot. Gaz, 136: 196-200; Brown, C.L and H.E. Sommer, 1975, An Atlas of Gymnosperms Cultured in vitro, 1925-1974, Georgia Research Council, Macon, Georgia, p 271.; Sommer, H.E. (1975) Differentiation of Adventitious Buds on Douglas Fir Embryos in vitro, Proceedings of Imperial Plant Prop. Soc. 25; 125-127.; Sommer, H.E. and C.E. Brown, 1979, Application of Tissue Culture to Forest Tree Improvement in W.R. Sharp, P.O. Larsen, E.F. Paddock and V. Raghavan, eds. Plant Cell and Tissue

Culture: Principles and Applications, pp 461-491.; Sommer, H.E. Organogenesis in Angiosperm Trees, Bull. Soc. Bot. Fr. Actual Bot. 130: 79-81; Brown, C.L. and H.E, Sommer, 1983, Shoot Growth and Histogenesis of Trees Possessing Diverse Patterns of Shoot Development, Amer. Journal Bot. 79: 335-346.

Judy Lyman Snow

Judith Lyman Snow, Basking Ridge, NJ, Judysnow99@gmail. com, *Employment* Rutgers University, Cook College--1985-2011, Biotechnology Center for Agriculture and the Environment (AgBiotech); Rockefeller Foundation, Agricultural Sciences-- 1980-1984, *Education,* Cornell University, MS in Horticulture--1976, PhD in Plant Breeding—1980 Duke University, BA in Botany, Magna cum Laude—1974, *Publications:* Author, annual reports and work plans for the AgBiotech Center at Rutgers for over 20 years, Author/co-author/editor of 2 books, 4 book chapters and 7 journal articles, Editor, Cook College *Grants Alert* weekly email newsletter 1993-96, *Awards:* Junior Science & Humanities Symposium Service Award, Rutgers University—2002 Cook College/NJAES Individual Impact Award, Rutgers University—1996 Rutgers University Merit Award—1994, *Associations/Affiliations*: Missouri Botanical Garden, William L. Brown Center for Plant Genetic Resources: Secretary of the Advisory Board—2002-2008, *Diversity* journal for plant genetic resources: Secretary/Treasurer—1996-2002.

Dan Tomas Spira

Born: 1932. Nitra Czechoslovakia; Emigrated to Israel 1949; Address: 35 Ben Zvi Blv. Jerusalem 96260; Tel: +972 (0)544 599 576. E-mail dant@mail.huji.ac.il; Institution: Hebrew University Hadassah Medical School P.O.B. 12272 Jerusalem 91120 ISRAEL; Study: The Hebrew University in Jerusalem; 1954 Begin. 1959 M.Sc.: Thesis: Blood Loss and Replacement in Malarious Rats. 1966. Ph.D.: Thesis Antigenic Analysis of

Plasmodia; Awards: 1975 The Royal Society, London, Bruno Mendel Fellow at National Institutes for Medical Research Mill Hill, England; 1983 Minerva Fellowship, DFG. Germany at: Dept. of Physiological Chemistry, Eberhard Karls University, Tubingen Germany; Publications: about 100, the first one: Spira D. and Zuckerman A. 1962. Science 137: 356-357; two books edited; Professional activities: 1997 elected President of International Congress of Protozoology and appointed Chairman of The XI congress of ICOP. Retired: October 1, 2001; Prof. Emeritus Dan T. Spira; The Kuvin Centre for the Study of Infectious and Tropical Disease; Hebrew University School of Medicine, Jerusalem, ISRAEL; Tel. +972-54-4599576.

Donald Styer

Donald J. (Chip) Styer II, Systems Developer, The Ohio State University, Ohio Agricultural Research and Development Center, Wooster, Ohio USA, Email styer.21@osu.edu, Dr. Chip Styer is a Systems Developer at the Ohio Agricultural Research and Development Center, The Ohio State University College of Food, Agriculture and Environmental Sciences. He was born in El Paso, Texas, and currently lives in Wooster, Ohio. He received a B. Sc. in Microbiology from The Ohio State University and Ph.D. in Plant Pathology from the University of Wisconsin-Madison. Dr. Styer's doctoral research was in phytobacteriology, and as a research scientist at DNA Plant Technology Corporation, he developed plant tissue culture bioreactor systems. His principle interest at DNAP was in the development and implementation of data management systems for plant biotechnology applications. After moving to Ohio, he worked in computer hardware and software marketing and later joined OSU as a Systems Specialist. He was responsible for implementation of a College-wide online faculty reporting system. His current efforts include database management, web development, and data compilation and analysis.

ABOUT THE EDITORS

(Editors Abbreviated *Curriculum vitae* Listed in
Alphabetical Order)

Otto J. Crocomo

Full Professor of Biochemistry at University of São Paulo, Rockefeller Fellow with C. C. Delwiche at UC Davis campus, and British Research with L. Fowden, University College, London, visitor professor with D. Boulter, University of Durham, England. Founder of the Center for Agricultural Biotechnology (CEBTEC) at the University of São Paulo, Piracicaba. Lives in Piracicaba, SP, Brasil with his wife Diva and has 5 children and one grandson.

Julius P. Kreier

A Short Not Completely Academic Biography of Julius Kreier was born in Philadelphia in 1926. I attended the public school system there. In high school, I followed the academic program but with a supplemental program in agriculture. At 18, in 1944, I was called up for the draft but rejected for service because of severe arthritis of my left hip joint caused by a bacterial infection as a young child. After high school, I enrolled in Temple University, with a major in biology. Actually the program, I followed was the premedical program.

During the third year in Temple University, I applied for the College of Veterinary Medicine at the University of Pennsylvania. I was accepted and entered the veterinary college in the fall of next year. I therefore did not get a Master's degree from Temple University.

In 1953, I graduated from veterinary school and got a job with the US department of Agriculture to work on foot and mouth disease eradication project in Mexico. When the program ended successfully in 1955, I transferred to the eastern shore of Maryland and worked there on tuberculosis and brucellosis control program. Just before reporting to the job in Maryland, I married Ruth Casten, a woman, I met during my second year in veterinary college. We had maintained a relationship during the five years in which I completed my veterinary degree and worked in Mexico. Despite our frequent geographic separation during the time we continued contact. After our marriage, we remained together until her death in 2010. In 1957, I enrolled in a graduate program at the University of Illinois. There I received a Master of Science degree and a PHD. Perhaps of more significance during these years, my wife presented me with a little girl and then a little boy. I was hired by the department of microbiology at the Ohio State University in 1963. I retired from the position on December 31, 1988. I am now a professor emeritus of microbiology. I occupy myself now with making wooden bowls on a lath, sculpturing animal and human figures in clay and other materials and tending my garden greenhouse. I visit my children as often as possible and go to movies and ballet with friends. I am now also involved in producing this book with my good friend Rod Sharp. I do all of these things to help fill the void created by the absence of my wife for 55 years. It has been said that the paths of glory lead but to the grave. This is true but it is also true that all the other paths lead to the same place. What is important is the journey not the end. While we are making the journey, we should do our best to make it pleasant not only for ourselves but also for those with whom we travel it. This last consideration is of importance particularly to those of us who are to be teachers because our behavior may be imitated by our students.

William R. Sharp

Dr. Sharp has a background in biotechnology, translation of science into business ideas, spawning start-up companies and extensive technology transfer experience in the Americas and Asia. He has authored over seventy original research papers, abstracts and books in the field of plant cell biology including the five volume series entitled the Handbook of Plant Cell Culture. He previously held the positions: Dean of Research, Cook College and Director of Research, New Jersey Agricultural Experiment Station, Rutgers University; Executive Vice-President, DNA Pharmaceuticals, Inc.; Executive Vice-President for Research, DNA Plant Technology, Corp; Research Director, Pioneer Research, Campbell Institute for Research & Technology, the Campbell Soup Company; Full Professor, Ohio State University; and Fellow, Argonne National Laboratory. He was a Fulbright Grantee during 1971 and 1973. Dr. Sharp holds a Ph.D. in Plant Cell Biology from Rutgers University. He is the recipient of the title Eminent Professor and the Luiz Queiroz Distinguished Service Medal award from the University of Sao Paulo.

APPENDIX

Academic and Scientific Cooperation between the ESALQ/USP and OSU
Joaquim José de Camargo Engler

The academic and scientific collaboration between the College of Agriculture "Luiz de Queiroz", Universidade de São Paulo (ESALQ/USP) and the College of Food, Agricultural and Environmental Sciences at The Ohio State University (FAES/OSU) began formally in 1964 with the signing of an Institutional Agreement between the two entities, with financial support from the United State Agency for International Development.

The relationship between these two academic institutions already existed, but individually and informally. The said agreement was aimed at institutional improvement of higher education and research in agriculture in Brazil. This general objective should be achieved through integration of teaching, research and extension and development of a postgraduate program of high level, as well as research that could support government policies aimed at economic and social development of Brazil.

A number of long-term and mid-term projects were developed to achieve the goals of the Program of Cooperation between the research and teaching faculty who participated as consultant collaborators with the faculty at ESALQ/USP. The projects included development and offering of new graduate courses, improvement of existing facilities with emphasis on expanding and strengthening relationships in education and research exchanges with institutions in Brazil and abroad.

OSU Research and Teaching Faculty assigned to Piracicaba, as participants in the Cooperation Agreement were the following, with their respective areas:

- Allen Steinhauer, Entomology
- Alvin Moxon, Animal Science
- Claire Young, Agricultural Education
- Clyde Allison, Plant Pathology
- David O. Hansen, Rural Sociology
- Donald Larson, Agricultural Economics
- Eva Wilson, Home Economics
- Fred Deatherage, Food Technology
- John Parsons, Agronomy
- John Sitterley, Agricultural Economics
- Kelso Wessel, Agricultural Economics
- Olen Leonard, Rural Sociology
- Paul Clayton, Poultry Science
- Richard L. Meyer, Agricultural Economics
- Robert Welsh, Agricultural Economics
- Roger Williams, Entomology
- Trevor Arscott, Agronomy
- Walter Slatter, Food Technology

In order to expand and strengthen the competence of its faculty, ESALQ/USP, with the active participation of OSU, emphasized their training activities in human resources with emphasis on Graduate Research and Teaching. In this sense 29 ESALQ faculty members were sent to American universities including OSU and other institutions for the implementation of programs of Master of Sciences (MS) and 24 programs for Doctor of Philosophy Degree (Ph.D.). Other professors participated in collaborative research as post-doctoral fellows and visiting professors.

For obtaining evidence concerning the MS or Ph.D. program at the American universities during the Agreement the following

ESALQ faculty served as counterparts in their respective areas of expertise:

- Adilson Paschoal, Zoology
- Antonio Galvao, Forestry
- Arare Pedroso, Entomology
- Avany Santos Correa, Home Economics
- Gaius Octavius Nogueira Cardoso, Plant Pathology
- Roberto Cássio Melo Godoy, Statistics
- Cesario Pires, Forestry
- Cyro Paulino da Costa, Horticulture
- Delmar Antonio Marchetti, Agronomy
- Diva Resende, Home Economics
- Elke Jurandy B. Nogueira Cardoso, Plant Pathology
- Fernando Perez Curi, Agricultural Economics
- Geraldo Tosello, Agronomy
- Gilberto Casadei Baptist Entomology
- Helena Teixeira Martins, Home Economics
- Henrique Vianna de Amorim, Plant Physiology
- Humberto de Campos, Statistics
- Ignacio Dal Fabbro, Agronomy
- Iracema de Sa, Home Economics
- Irenaeus Umberto Packer, Animal Science
- John Nunes Nogueira, Food Technology
- Joaquim José de Camargo Engler, Agricultural Economics
- Joaquim Oliveira, Food Science & Nutrition
- Jose Molina Son, Rural Sociology
- Keigo Minami, Horticulture
- Luiz Antonio Balastreire, Agricultural Engineering
- Dulce Maria Bergamin Bandeira, Psychology
- Max Lázaro Vieira Bose, Animal Science
- Mitsue Hironaka, Home Economics
- Moacyr Corsi, Agronomy
- Murilo Graner, Food Technology

- Oriovaldo Fall, Rural Sociology
- Paul F. C. Araujo, Agricultural Economics
- Paulo Roberto Cantarelli, Agricultural Biochemistry
- Paul Martin Soder, Ecology
- Randolph Custodio, Poultry Science
- Raul Dantas d'Arce, Animal Science
- Ricardo Shirota, Agricultural Economics
- Roberto Cobbe, Agricultural Journalism
- Roberto Dias de Moraes e Silva, Poultry Science
- Rose Higaki, Home Economics
- Rubens Valentini, Agricultural Economics
- Ruy Caldas, Biochemistry
- Sergio Brandt, Agricultural Economics
- Sérgio Paranhos, Agronomy
- Toshiaki Kinjo, Agronomy
- Valdomiro Bittencourt, Agronomy
- Vidal Pedroso de Faria, Animal Science
- Violet Coast, Home Economics
- Walter de Paula Lima, Agronomy
- Wilson Roberto Soares Mattos, Animal Science
- Zilda Matos, Agricultural Economics
- Zilmar Mark Ziller, Agronomy

Although the coordination of activities for improvement of the ESALQ faculty in the USA were made by OSU, they were held in various Universities and Institutions aiming to offer the best working conditions in their respective areas of expertise.

In addition to creation of formal programs of graduate training, several short-term activities were developed, including seminars and exchanges of members of ESALQ and OSU with each other, in order to increase collaboration and competence in teaching, research, the extension service and university administration. Among these activities, the "Agricultural Marketing and Food

Technology Project," in which 14 students and five teachers of the newly created area of specialization in Food Technology from ESALQ attended conferences and visits to Ohio State University, California, Louisiana and Florida, and USDA research laboratories and industrial production equipment and food in the United States of America as a way to complement the offerings at ESALQ.

Another collaborative research activity between ESALQ and OSU was entitled the "Capital Formation and Technological Innovations in Agriculture in Developing Countries". This was held in the period 1971/73. Upon completion of the Agreement ESALQ OSU-USAID-funded academic-scientific collaboration program between these institutions, This program continued, including the realization of joint programs, such as the PICA (Program of Assistance to the Inter-University for Agricultural Sciences), which aimed at Inter-University cooperation between U.S. and Brazilian institutions in partnerships, in which ESALQ and OSU worked with the College of Agriculture of Para for improving their teaching and research programs.

Another ESALQ-OSU project occurred as part of "Higher Agricultural Education Program (PEAS), developed by the Ministry of Education, in which important joint teaching and research efforts were conducted. The partnership between OSU and ESALQ including the realization of research projects together with funding agencies to American and Brazilian, as the National Council for Scientific and Technological (CNPq) and Foundation for Research Support of São Paulo (FAPESP), especially in the areas of Agricultural Engineering, Soil Science, Seed Biology, Agricultural Economics.

ESALQ today annually receives students under the auspicines of the Tripartite Research and Teaching Program from OSU as well as at Rutgers University for cooperative programs.

The collaboration between the academic and scientific faculty of ESALQ / USP and FAES / OSU had a great impact on both institutions. The ESALQ expanded and improved its program of Graduate Studies and Research at OSU and the internationalization of its teaching and research. Thus both institutions continued to maintain this fruitful partnership.

www.ingramcontent.com/pod-product-compliance
Lightning Source LLC
Chambersburg PA
CBHW051435170526
45166CB00001B/1